Nanobiostimulants in Innovative Agriculture

Riyaz Sayyed • Noshin Ilyas
Dilantha Fernando
Editors

Nanobiostimulants in Innovative Agriculture

Springer

Editors
Riyaz Sayyed
Department of Biological Science and Chemistry
College of Arts and Science, University of Nizwa
Nizwa, Oman

Noshin Ilyas
Department of Botany
Pir Mehr Ali Shah Arid Agriculture University
Rawalpindi, Pakistan

Dilantha Fernando
Department of Plant Science
University of Manitoba
Winnipeg, MB, Canada

ISBN 978-981-95-2476-1 ISBN 978-981-95-2477-8 (eBook)
https://doi.org/10.1007/978-981-95-2477-8

This Springer imprint is published by the registered company Springer Nature Singapore Pte Ltd.
The registered company address is: 152 Beach Road, #21-01/04 Gateway East, Singapore 189721, Singapore

The book ***Nanobiostimulants in Innovative Agriculture*** *is dedicated to our mentors, teachers, colleagues, and students whose contribution enhanced innovative agricultural practices through transforming the knowledge of agriculturally beneficial nanobiostimulants.*

Riyaz Sayyed
Noshin Ilyas
Dilantha Fernando

Foreword

Achieving global food security to reach Zero Hunger Goals is the key challenge of this millennium. Soil nutrient loss, drought conditions, decreased soil fertility, and various stresses pose significant challenges in achieving this target. Mitigation of these challenges requires sustainable approaches to improve global food production with minimal environmental impact while maintaining the resilience of agroecosystems against a rapidly changing climate.

Nanotechnology has been recognized as a successful technology that can sustainably improve global food security. Nanotechnology holds significant potential for achieving agricultural sustainability by enhancing the efficacy of farming inputs, thereby increasing crop yields and food production. Nanomaterial encapsulates beneficial crop microbiomes and delivers pesticidal or nutritive active ingredients in a controlled, targeted, and synchronized manner. These features offer new opportunities to enhance the efficacy and efficiency of nanobiofertilizers, biostimulants, and pesticides compared to their chemical counterparts. The development of nanobiofertilizers, nanobiopesticides, and nanoformulations has revolutionized traditional agricultural practices, making them more sustainable, innovative, and environmentally friendly. Further, the NMs application impacts the development and physiology of crops by modulating the phytohormone signaling networks. Additionally, controlled delivery systems and the regulated use of NMs can help mitigate the pollution associated with them.

The book, *Nanobiostimulants in Innovative Agriculture*, covers the role of NPs and the crop microbiome as nanobiofertilizers, nanobiopesticides, and biostimulants to address the goal of climate-smart agriculture comprehensively. This book offers in-depth insights into the functional role of the rhizomicrobiome in enhancing plant growth, development, pest control, and bioremediation of contaminated agricultural soil. The book also highlights the benefits of using crop microbiome-based nanobiofertilizers, nanobiopesticides, and nanoformulations. The book features an excellent compilation of chapters from leading experts, making it a valuable resource for students, faculty, researchers, and entrepreneurs involved in nanobiostimulants.

President
Bashan Institute of Science
Dadeville, AL, USA

Luz Bashan

Foreword

Agriculture faces several challenges, one of which is striking a balance between feeding a growing population and protecting the environment. Although this is very challenging, this can be overcome through the use of nanobiostimulants, which are nano-enabled materials and microbial systems that can enhance plant performance, improve nutrient delivery, and increase tolerance to stresses with precision that is difficult to achieve using conventional farming methods.

The book, *Nanobiostimulants in Innovative Agriculture* to be published by Springer-Nature, brings together experts from different parts of the world to offer their expertise on how nanobiostimulants can help shape the future of agriculture.

The chapters cover various topics, including nanobiofertilizers, nanopesticides, the interaction between nanoparticles, soil, and microbes, seed biopriming, and the application of transcriptomics, proteomics, and metabolomics to study plant responses. They also explore plant hormone signaling, stress tolerance, safety regulations, consumer acceptance, and nanotoxicology. This combination makes the book both a scientific reference and a practical guide. In addition to this, the authors highlight how nanobiostimulants can improve yields and reduce chemical use, while also considering safety, environmental impacts, and public trust.

The book provides students with an introduction to an emerging and vital field, presenting researchers' state-of-the-art knowledge and practitioners' clear explanations of how nanobiostimulants function, as well as their applications and limitations. It is challenging to find a single technology that can address all the challenges facing agriculture. However, the work presented in this book shows how nanobiostimulants can become an essential part of the solution.

I congratulate the editors and contributors on this wonderful book that will inspire and guide all those committed to a more resilient and sustainable agriculture.

Abdullah M. Al-Sadi

Professor
University of Al-Dhaid
Sharjah, UAE

Professor and Former Dean
Sultan Qaboos University
Muscat, Oman

Preface

Achieving sustainable agricultural productivity and global food security improvement to reach Zero Hunger Goals, particularly under unpredicted climate changes, is the biggest challenge of this millennium. These circumstances demand efficient and sustainable agricultural expansion. However, reduction in crop yield, soil nutrient loss, drought conditions, decreased soil fertility, and various stresses are significant challenges in agriculture. Ensuring a continued supply of ample and safe food is a challenging task. Handling such difficulties requires sustainable approaches to improve global food production with minimal environmental damage while maintaining agroecosystems' resilience against a rapidly changing climate.

In the present era, nanotechnology has been recognized as a successful technology that can sustainably improve food security. Nanomaterial encapsulates beneficial crop microbiomes and delivers pesticidal or nutritive active ingredients in a controlled, targeted, and synchronized manner. These features offer new opportunities to increase nanobiofertilizer, biostimulant, and pesticidal efficacy and efficiency compared to their chemical counterparts. The encapsulation of nanoscale biostimulants has resulted in the development of various novel products such as nanobiofertilizers, nanobiopesticides, and nanoformulation, which have revolutionized the traditional agro-practices making them more sustainable, ingenious, and environmentally efficient.

Nanotechnology offers a broader potential in achieving agricultural sustainability by improving the efficacy of farming inputs, thereby increasing crop yield and food production. Nanotechnology has emerged as a promising approach to boost agricultural productivity by utilizing efficient nanobiopesticides and nanofertilizers and regulating soil nutrient content. Nanotechnology has great potential to solve issues related to global food security by boosting food availability and providing quality agri-products.

Improving crop yield with conventional farming practices has been a continuous challenge in agricultural production. Recently, attention has been given to crop microbiomes to enhance plant nutrient content and disease control. Crops and their microbiomes have established their roles in combating challenges to agricultural sustainability, such as increasing crop yields and reducing hunger globally. Increased awareness of the importance of microorganisms for plant and soil health has fueled a boom in research of agricultural microbiomes. This has resulted in the reinvigoration of microbial biotechnologies that utilize microbial benefits to improve agroecosystem functioning through enhanced soil health, crop vigor, and pest protection.

At the optimal concentrations, microbiome-based nanomaterials (NMs) enhance the seed germination, growth, and yield parameters and suppress phytopathogens. Further, the NM application impacts the growth and physiology of crops by modulating the phytohormone signaling networks. Various advanced molecular studies like transcriptomic, proteomic, and metabolomic approaches have been conducted recently to examine the influence of NMs and the related responsive genes, which help develop nutrient-rich, high-yielding, and stress-resilient plants. In addition, controlled delivery systems and the regulated use of NMs can reduce the pollution associated with them.

This book covers the role of NPs and crop microbiome as nanobiofertilizer, nanobiopesticide, and biostimulants to address the goal of climate-smart agriculture comprehensively. This book provides a comprehensive guide to discovering, assessing, and utilizing beneficial microbiomes and their metabolites for the sustainable promotion of crop nutrition, crop growth, and bioremediation of contaminated soil in the face of climate change. The present book will provide deep insights into the functional role of rhizomicrobiome and microbiome or their products of different plant parts as nanobiofertilizers, nanobiopesticides, and bioremediators in augmenting plant growth, development, biocontrol of pests, and bioremediation of contaminated agriculture soil. The book also highlights the benefits of using crop microbiome-based nanobiofertilizers, nanobiopesticides, and nanoformulations.

The book will be helpful for students, faculty, researchers, and entrepreneurs involved in nanobiostimulants. With up-to-date research from leading experts, the book will help develop an eco-friendly, sustainable agricultural system.

Nizwa, Oman — Riyaz Sayyed
Rawalpindi, Pakistan — Noshin Ilyas
Winnipeg, MB, Canada — Dilantha Fernando

Acknowledgments

The book *Nanobiostimulants in Innovative Agriculture* has been an incredible journey for its editors. The whole process of turning an idea into a book has been equally challenging and rewarding for us. Foremostly, we are thankful to our mentors who introduced us to this fascinating world of nano-microbiome biostimulants and made us aware of the challenges in innovative agriculture practices. We are also thankful to our students, colleagues, and institutions whose continuous support inspires us to reach our goals.

Gratitude goes to our authors and contributors who put their best into their writing. We acknowledge your expertise, dedication, and hard work. We thoroughly enjoyed your ideas and believe our readers will be equally impressed. May you all keep adding more to science and play your role in innovative agriculture.

We are pleased to extend our gratitude to the senior editor and publisher, the production team, and the entire Springer family. We are thankful for your continuous trust and cooperation, which made this task smooth and pleasant.

Riyaz Sayyed
Noshin Ilyas
Dilantha Fernando

Contents

Editors and Contributors

About the Editors

Riyaz Sayyed is a Professor in the Department of Biological Sciences and Chemistry at the University of Nizwa, Nizwa, Oman. He has over 21 years of teaching, research, and administrative experience in Microbiology. He is an Associate Editor of various journals, has authored over 360 research papers in high IF international journals, and edited 34 books. He is a fellow of the Royal Society of Biology, Indian Phytopathological Society, and Association of Agricultural Technology in Southeast Asia. Elsevier has consecutively listed him among the Top 2% of highly cited researchers for three years, and Clarivate has listed him in the Top 1% of Highly Cited Researchers (2024) among 6,886 researchers worldwide. He is a visiting scholar at INTII International University, Malaysia, and an Expert consultant at Qassim University, Saudi Arabia. He has been an invited speaker at many international conferences in India, Southeast Asia, and other European countries.

Noshin Ilyas currently serves as Chairperson of the Department of Botany at PMAS-Arid Agriculture University, Rawalpindi, Pakistan. She is recognized among the World's Top 2% of Scientists by Stanford University (2024). She has built an exceptional career marked by pioneering research, academic leadership, and global outreach. She is a Visiting Scholar and Research Fellow at INTI International University, Malaysia, a National Expert on Women Empowerment in Agriculture for the Asian Productivity Organization, Country Reporter for IFOAM-Organics International, and President of the Pakistan Chapter of the Asian PGPR Society. She has proudly represented Pakistan at premier scientific platforms in over ten countries. She has been a recipient of many prestigious national and international awards. With over 400 impact factor points, 5,000+ citations, 3 patents, and a strong record of funded national and international projects, she has supervised 11 PhD and 40 MPhil students and has served as an editor of four international books and an editorial board member for top-tier journals.

Dilantha Fernando is a Professor and Dean of Studies at the University of Manitoba, Canada. He is an internationally renowned host-pathogen interaction researcher. He has trained 15 PhD students and 17 master's students, who have graduated and found excellent employment in academia, industry, and government research institutions. He is a frequent keynote speaker at conferences and grower meetings and sits on several national and international boards. He has received various national and international prestigious awards such as Fellow of the Asian PGPR Society, Merit Award for Excellence in Research and Service from the University of Manitoba, the highest award for research from the Canadian Phytopathological Society, CPS Outstanding Research Award, and Fellow of the American Phytopathological Society, USA. He has been successful in receiving research funds from various groups. He has published over 200 research articles and 17 book chapters. He has co-developed 37 canola/HEAR cultivars, bringing prestige and royalty funds to the University of Manitoba, adding millions to the Canadian economy and growing profits.

Contributors

Nosheen Akhtar Department of Botany, PMAS Arid Agriculture University, Rawalpindi, Pakistan

Athikho Kayia Alice College of Agriculture, Central Agricultural University (I), Pasighat, India

Muhammad Ahsan Altaf Key Laboratory for Quality Regulation of Tropical Horticultural Crops of Hainan Province, Sanya Nanfan Research Institute, Hainan University, Sanya, China

Abdel Rahman Mohammad Al-Tawaha Department of Biological Sciences, Al-Hussein Bin Talal University, Maan, Jordan

Abdel Razzaq Al-Tawaha National Agricultural Research Center (NARC), Baq'a, Jordan

Ayman Azad Division of Agronomy, FOA-Wadura, SKUAST, Kashmir, India

Malik A. Aziz Division of Basic Sciences and Humanities, FOA-Wadura, SKUAST, Kashmir, India

S. Ramesh Babu Rajasthan College of Agriculture, Maharana Pratap University of Agriculture and Technology, Udaipur, India

Hujjat -ul Baligah Division of Soil Science and Agricultural Chemistry, FOA, Wadura, SKUAST, Kashmir, India

Veluru Bhargav MTTC &VTC, Central Agricultural University (I), West Kameng, India

Tanmaya Kumar Bhoi Forest Protection Division, ICFRE-Arid Forest Research Institute (ICFRE-AFRI), Jodhpur, India

Fatima Bibi Department of Botany, PMAS Arid Agriculture University, Rawalpindi, Pakistan

Gökhan Boyno Department of Plant Protection, Faculty of Agriculture, Van Yuzuncu Yil University, Van, Turkey

Gloria Sarahí Castañeda-Ramírez Laboratorio de Estudios Ecogenómicos, Centro de Investigación en Biotecnología, Universidad Autónoma del Estado de Morelos, Cuernavaca, Morelos, México

Saoli Chanda Earth and Environment Department, Florida International University, Miami, FL, USA

Fernando Cuevas-Remigio Centro de Investigación en Biotecnología, Universidad Autónoma del Estado de Morelos, Cuernavaca, Morelos, México

Younes Rezaee Danesh Department of Plant Protection, Faculty of Agriculture, Van Yuzuncu Yil University, Van, Turkey

Edgar Dantán-González Laboratorio de Estudios Ecogenómicos, Centro de Investigación en Biotecnología, Universidad Autónoma del Estado de Morelos, Cuernavaca, Morelos, México

Zaffar Mahdi Dar Division of Basic Sciences and Humanities, FOH-Shalimar, SKUAST, Kashmir, India

Khalid Derraz Laboratory of Functional Ecology and Environment, Sciences and Technology Faculty, Sidi Mohamed Ben Abdellah University, Fez, Morocco

Laishram Olivia Devi College of Horticulture & Forestry, Central Agricultural University (I), Pasighat, India

Rihab Djebaili Department of Life, Health and Environmental Sciences, University of L'Aquila, L'Aquila, Italy

Ramgopal Dudwal Rajasthan Agricultural Research Institute, Durgapura, Jaipur, India

Nehal M. El-Deeb Pharmaceutical Bioproducts Research Department, Genetic Engineering and Biotechnology Research Institute (GEBRI), City of Scientific Research and Technological Applications SRTA-City, Alexandria, Egypt

Hesham Ali El Enshasy Innovation Centre in Agritechnology for Advanced Bioprocessing (ICA), Universiti Teknologi Malaysia (UTM), Pagoh, Johor, Malaysia

Faculty of Chemical and Energy Engineering, Universiti Teknologi Malaysia (UTM), Johor, Malaysia

City of Scientific Research and Technological Applications (SRTA-City), New Borg El-Arab, Egypt

Naïma El Ghachtouli Microbial Biotechnology and Bioactive Molecules Laboratory, Sciences, and Technology Faculty, Sidi Mohamed Ben Abdellah University, Fez, Morocco

Beatrice Farda Department of Life, Health and Environmental Sciences, University of L'Aquila, L'Aquila, Italy

Ramila Fares Department of Life, Health and Environmental Sciences, University of L'Aquila, L'Aquila, Italy

Laboratory Biotechnology, Water, Environment, and Health, Abbes Laghrour University of Khenchela, Khenchela, Algeria

Laboratory of Mycology, Biotechnology, and Microbial Activity, Brothers Mentouri University of Constantine, Constantine, Algeria

Mohamed Ferioun Microbial Biotechnology and Bioactive Molecules Laboratory, Sciences, and Technology Faculty, Sidi Mohamed Ben Abdellah University, Fez, Morocco

Natural Resources and Environmental Laboratory, Taza Polydisciplinary Faculty, Sidi Mohamed Ben Abdellah University, Fez, Morocco

K. Gayathri PG and Research Department of Microbiology, Sacred Heart College (Autonomous), Tirupattur, India

Kavisa Ghosh Department of Biotechnology, The Oxford College of Science, Bengaluru, Karnataka, India

Armando Hernández-Mendoza Laboratorio de Bioinformática y Genómica Evolutiva, Centro de Investigación en Dinámica Celular, Universidad Autónoma del Estado de Morelos, Cuernavaca, Morelos, México

Ting Ho Back2Nature Organic Farm, Kuala Pilah, Malaysia

Noshin Ilyas Department of Botany, Pir Mehr Ali Shah Arid Agriculture University, Rawalpindi, Pakistan

Shayesta Islam Division of Environmental Sciences, FOH, Shalimar, SKUAST, Kashmir, India

Aqsa Jabeen Department of Botany, Pir Mehr Ali Shah Arid Agriculture University, Rawalpindi, Pakistan

L. Jahnavi Reddy Credora Life Sciences, Bangalore, Karnataka, India

Muhammad Azhar Jameel Department of Zoology, Wildlife & Fisheries, PMAS Arid Agriculture University, Rawalpindi, Pakistan

Rozaliana Abdul Karim Innovation Centre in Agritechnology for Advanced Bioprocessing (ICA), Universiti Teknologi; Malaysia (UTM), Pagoh, Johor, Malaysia

Arun Karnwal Department of Microbiology, Graphic Era (Deemed to be University), Dehradun, India

Ahmed M. A. Kenawy Nucleic Acids Research Department, Genetic Engineering and Biotechnology Research Institute (GEBRI), City of Scientific Research and Technological Applications (SRTA-City), New Borg El-Arab, Egypt

J. Komal Basic Tasar Silkworm Seed Organization (BTSSO), Central Silk Board, Bilaspur, Chhattisgarh, India

L. Lee San Department of Agriculture Sarawak KM20, Kuching, Sarawak, Malaysia

Lisnawita Program Study of Agrotechnology, Faculty of Agriculture, Universitas Sumatera Utara, Medan, Indonesia

B. Lokeshwari PG and Research Department of Microbiology, Sacred Heart College (Autonomous), Tirupattur, India

Said Louahlia Natural Resources and Environmental Laboratory, Taza Polydisciplinary Faculty, Sidi Mohamed Ben Abdellah University, Fez, Morocco

Deepak Kumar Mahanta Forest Entomology Discipline, Forest Protection Division, Forest Research Institute (ICFRE-FRI), Dehradun, Uttarakhand, India

Johnny Meddy Department of Agriculture Sarawak KM20, Kuching, Sarawak, Malaysia

Mailin Misson Biotechnology Research Institute, University Malaysia Sabah, Sabah, Malaysia

Intan Baizura Mohamed Innovation Centre in Agritechnology for Advanced Bioprocessing (ICA), Universiti Teknologi; Malaysia (UTM), Pagoh, Johor, Malaysia

Hina Mustafa Department of Botany, PMAS Arid Agriculture University, Rawalpindi, Pakistan

Neelum Naheed Department of Botany, Pir Mehr Ali Shah Arid Agriculture University, Rawalpindi, Pakistan

Hadjadj Naima Agri-food Department, Faculty of Natural and Life Sciences, University of Blida1, Ouled Yaïch, Algeria

Natalia Nesterova National University of Life and Environmental Sciences of Ukraine, Kyiv, Ukraine

Muktiningsih Nurjayadi Department of Chemistry, Faculty of Mathematics and Natural Science, Universitas Negeri Jakarta, Jakarta, Indonesia

Amesha Panwar Best Agrolife Ltd, New Delhi, Delhi, India

Raqib Rasool Parray Division of Basic Sciences and Humanities, FOA-Wadura, SKUAST, Kashmir, India

Marika Pellegrini Department of Life, Health and Environmental Sciences, University of L'Aquila, L'Aquila, Italy

Vishnu D. Rajput Academy of Biology and Biotechnology, Southern Federal University, Rostov-on-Don, Russia

Mohamed Rashad City of Scientific Research and Technology Applications, New Burg Al Arab, Alexandria, Egypt

Maimona Saeed Department of Botany, Rawalpindi Women University, Rawalpindi, Pakistan

Rosalba Salgado-Morales Laboratorio de Estudios Ecogenómicos, Centro de Investigación en Biotecnología, Universidad Autónoma del Estado de Morelos, Cuernavaca, Morelos, México

Ipsita Samal ICAR-National Research Centre for Litchi, Muzaffarpur, Bihar, India

P. Saranraj PG and Research Department of Microbiology, Sacred Heart College (Autonomous), Tirupattur, India

Saman Sarfraz Government College Women University, Sialkot, Pakistan

Riyaz Sayyed Department of Biological Sciences and Chemistry, College of Arts and Science, University of Nizwa, Nizwa, Sultanate of Oman

Sibte Sayyeda Department of Plant Protection, Faculty of Agricultural Sciences, Aligarh Muslim University, Aligarh, India

Saima Shafi Division of Basic Sciences and Humanities, FOA-Wadura, SKUAST, Kashmir, India

Asim Shahzad The College of Geography and Environmental Sciences, Henan University, Kaifeng, China

Kuldeep Sharma Forest Protection Division, ICFRE-Arid Forest Research Institute (ICFRE-AFRI), Jodhpur, India

Aditya Pratap Singh Faculty of Agriculture, Bidhan Chandra Krishi Viswavidyalaya, Mohanpur, Nadia, India

Sangeeta Singh Forest Protection Division, ICFRE-Arid Forest Research Institute (ICFRE-AFRI), Jodhpur, India

Siddhartha Singh College of Horticulture & Forestry, Central Agricultural University (I), Pasighat, India

Vaibhav Pratap Singh Department of Plant Protection, Faculty of Agricultural Sciences, Aligarh Muslim University, Aligarh, India

P. Sivasakthivelan Tamil Nadu Agricultural University, Agricultural Research Station, Bhavanisagar, India

Dalia Sukmawati Department of Chemistry, Faculty of Mathematics and Natural Science, Universitas Negeri Jakarta, Jakarta, Timur, Indonesia

Virgilio Gavicho Uarrota Instituto de Ciencias Agroalimentarias, Animales y Ambientales (ICA3) Campus Colchagua, Universidad de O'Higgins, San Fernando, Chile

Muhammad Umer Department of Botany, Pir Mehr Ali Shah Arid Agriculture University, Rawalpindi, Pakistan

Gilberto Villa-Rojas Laboratorio de Estudios Ecogenómicos, Centro de Investigación en Biotecnología, Universidad Autónoma del Estado de Morelos, Cuernavaca, Morelos, México

Vipula Vyas Forest Protection Division, ICFRE-Arid Forest Research Institute (ICFRE-AFRI), Jodhpur, India

Tasso Yatung MTTC & VTC, Central Agricultural University (I), West Kameng, India

Nouhaila Zouine Microbial Biotechnology and Bioactive Molecules Laboratory, Sciences, and Technology Faculty, Sidi Mohamed Ben Abdellah University, Fez, Morocco

Ilham Zouitane Microbial Biotechnology and Bioactive Molecules Laboratory, Sciences, and Technology Faculty, Sidi Mohamed Ben Abdellah University, Fez, Morocco

Nano-Biofertilizers: An Eco-Friendly Practice for Sustainable Agro-Ecosystem Enhancing Plant Responses

1

Vaibhav Pratap Singh and Sibte Sayyeda

Abstract

Fertilizers are commonly used by farmers to fulfill the nutritional requirement to enhance plant growth and play an important role in increasing agricultural production globally. Over a period of time, prolonged use of chemical fertilizers causes an adverse impact on the environment by releasing the chemical residue and agricultural runoff. It affects soil fertility, reduces crop production, causes the emergence of new races of pests and pathogens, and causes low nutrient efficiency, high leaching of minerals, and depletion of soil organic content. Due to the drastic effect on the environment and human health, various researchers, scientists, and environmentalists have looked forward to searching for an eco-friendly approach to achieve desired goals and increase agricultural food production. Nanotechnology is the most promising tool in agriculture and other related fields. Biofertilizers, nano-biofertilizers, nanofertilizers, and nanopesticides are nanotechnology-based fertilizers. Nano-biofertilizers are safe, low-cost, non-hazardous, and eco-friendly in nature. Globally, the biofertilizers market was estimated at around $1.49 billion in 2019, developing at a compound annual growth rate of around 12.8% from 2020 to 2027. Nano-biofertilizers have the ability to increase crop productivity and minimize environmental residues. They boost the water-holding capacity of soil and provide vital plant nutrients due to the coating of a nanomaterial as well as microbial revitalization due to organic content having plant growth promoters either directly or indirectly (bio-fertilization, rhizoremediation, disease resistance, etc.). Contrary to traditional chemical fertilizers, nano-biofertilizer gives a better result. It upgrades the texture and function of soil and improves various attributes of plants such as physi-

V. P. Singh (✉) · S. Sayyeda (✉)
Department of Plant Protection, Faculty of Agricultural Sciences, Aligarh Muslim University, Aligarh, India

R. Sayyed et al. (eds.), *Nanobiostimulants in Innovative Agriculture*,
https://doi.org/10.1007/978-981-95-2477-8_1

ological, biochemical, morphological, and, most importantly, yield attributes. There is a scarcity of literature on the formulation and application of nano-biofertilizer at the ground level. This smart fertilizer necessitates consideration, as it diminishes the utilization of chemical fertilizer and makes the environment, soil, and crop healthier. This chapter highlighted the niftiness of nano-biofertilizers as an eco-friendly approach to sustainable agriculture. It covers the status, formulation, application, and constraints of nano-biofertilizers and their sustainable future in agriculture. The chapter also focused on the role of nano-biofertilizer in augmenting the growth and development of crops and their protection.

Keywords

Nano-biofertilizers · Plant growth. · Sustainable agroecosystem. · Yield attributes.

1 Introduction

The use of conventional fertilizers plays an essential role in intensifying agricultural production globally and also in gratifying the uprising food demand of the continuously growing world population (FAO 2015). As a result, over the past few years, especially after the green revolution, there has been a consistent increase in the consumption of three major nutrients, namely, nitrogen (N), phosphorus (P), and potash (K). Over the past 10 years, it was reported that the consumption of ammonia for industrial and agricultural sectors consistently increased worldwide at a compound annual growth rate of 2%. During 2014–2018 the demand for phosphorus (P_2O_5) increases by about 39×10^5 tons (FAO 2015). Although P is present in the soil in a relatively higher amount, its accessibility to the plants is frequently low (Mahanta et al. 2014). In spite of being a key factor in increasing crop yield and productivity, the use of inorganic fertilizer is less advocated for from a future agriculture perspective. Its poor assimilation of nutrient potency and higher loss in percentage of nutrients through the soil and water runoff lead to environmental pollution, pest and disease emergence, reduction in crop yield, loss of biodiversity, and its biomagnification in humans (Savci 2012). In addition to this, chemical fertilizer contributes a lot to the eutrophication of the water system and its deposition in soil, which results in reduced plant growth, increased salinity, exhaustion of soil nutrients, depletion of soil fertility, and reduced water-holding capacity of plants (Wilson et al. 2008; Mahanty et al. 2017). Many agricultural and scientific articles cited that excessive reliance on conventional fertilizers, i.e., urea, phosphate, ammonia, or nitrate compounds in agricultural practices, causes inevitable and irreplaceable damage to human health and environmental ecology (Mahanty et al. 2017). Over the past decades, to overcome these effects and substitute chemicals with better eco-friendly options, great efforts have been made by researchers.

Nanotechnology has appeared as the sixth most revolutionary technology after the green revolution in the 1960s and the revolution of biotechnology of the 1990s and has become the fast-emerging field consisting of multiple disciplines of current science (Battard 2012; Rahayu et al. 2020; Yorseng et al. 2020). Lately, numerous developments have brought through nanotechnology in the agricultural sector via nanofertilizers (Fatima et al. 2020b); green nanomaterials (Khan et al. 2017); nano-herbicides (Kumar et al. 2020); nanoagrochemicals (Gomez et al. 2021); nanopesticides (Grillo et al. 2021); nano-biosensors (Usman et al. 2020); and nano-based treatment of agricultural waste to escalate the food productivity by foliar and soil appliance along with the establishment of sustainable agriculture management. One of the best alternatives that has earned impetus and become progressively acceptable in the agricultural sector is nano- and biofertilizer (Mikhak et al. 2017; El-Ghamry et al. 2018; Rasouli et al. 2019). Recently, nano-biofertilizers have been preferred over chemical fertilizers, as they increase the potency of resource utilization, minimize consumption of fertilizer, sustain the release of agrochemicals in a slow-release manner, and increase crop production and yield with the least intrusion of soil structure (Jatav and De 2013). It is considered a step ahead in the advanced agricultural field toward the combined formulation of nanoparticles and biofertilizers into nano-biofertilizer, a more successful substitute for eco-sustainable agriculture (Duhan et al. 2017). The role of nano-biofertilizer in agriculture is depicted in Fig. 1.1.

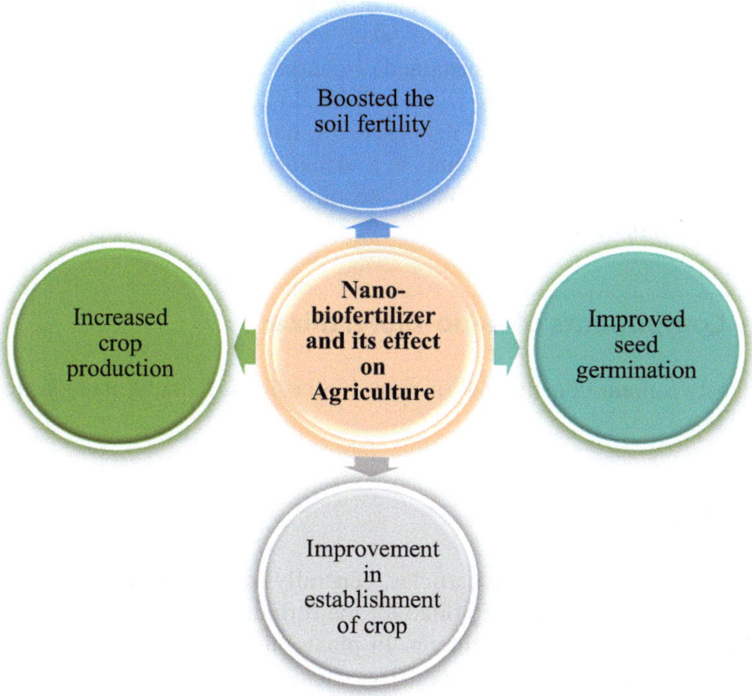

Fig. 1.1 Role of nano-biofertilizer in agriculture

"Nano-biofertilizer" is a hybrid amalgamation of nanoparticle and biofertilizer obtained by the formulation of organic fertilizer or biofertilizer (*Azolla*, *Azotobacter*, *Azospirillum*, *Azotobacetera*, *Bacillus*, *Cyanobacteria*, *Pseudomonas*, *Beijerinckia*, *Rhizobium*, etc.) to nanosize (1–100 nm) with the aid of certain nanomaterial coatings (Thirugnanasambandan 2019). Although in past literature, nano-biofertilizer terminology is not well-defined, few scientists explained and quoted its definition in scientific publications (Mardalipour et al. 2014; Thirugnanasambandan 2019). According to the literature, nano-biofertilizer could be defined as "cocktails" of nanomaterial and biofertilizer consisting of higher potency of both fertilizer benefits, facilitating the slow and controlled release of nutrients for longer periods of crop growth stages along with improved efficiency of nutrient use and eventually assisting in elevated productivity and yield of the crop (Shukla et al. 2013). Some salient features of nano-biofertilizer are large-scale utilization, low quantity of fertilizer required, enhanced nutrient use potency, cost-effectiveness, shorter time, mass-scale production, renewable fertilization, and eco-safe prospects (Duhan et al. 2017; Thirugnanasambandan 2019). The bioorganic components of nano-biofertilizer, i.e., urea and growth-promoting microorganisms, provide diversified advantages to soil and plant systems because of the stimulation of nitrogen-fixing ability, the solubility of phosphate, the escalated concentration of plant growth hormone generation, and the improvement of the condition of microorganisms in the soil (Dikshit et al. 2013; Shukla et al. 2015). Coating of biofertilizer with nanomaterial may aid in refining the solubility and diffusion of insoluble nutrients in the soil, lessen soil absorption and fixation, escalate nutrient bioavailability to plants and soil, and permit their slow and sustained discharge with their direct internalization and uptake by crops (Naderi and Danesh-Shahraki 2013). Because of the collective benefits of both biofertilizer and nanoparticles, nano-biofertilizer is best fitted for effective release of nutrients to plants in a sustained and slow-release way, imparting the accessible nutrients to plants steadily over the longer growing period (Thirugnanasambandan 2019).

2 Components of Nano-Biofertilizer

There are two main constituents of nano-biofertilizer, namely, nanoparticles and biofertilizers.

2.1 Nanoparticles (NPs)

Nanoparticles are microscopic particles, generally ranging from 1 to 100 nm. The most common nanoparticles exploited for the formulation of nano-biofertilizer are zinc, copper, iron, silver, and silicon. In plants, the accumulation and uptake of nanoparticles depend on their shape, size, chemical composition, and agglomeration state. The relation between the translocation of nanoparticles and their size is inversely proportional. Due to the variation in receptors from plant to plant, it is also linked with the plant species. A plant may accumulate one type of nanoparticle and eliminate another kind of nanoparticle. To improve plant growth, different nanoparticles used different mechanisms (Fatima et al. 2020a, b; Baazaoui et al. 2021). After the application of nanoparticles in the rhizosphere, they penetrate the roots of the plant by endocytosis, through carrier proteins, or by plasmodesmata. While if applied in the form of a foliar spray, they simply diffuse via stomata and make their way to vascular bundles. Inside the plants, they travel through apoplastic and symplastic pathways. The movement of nanoparticles is based on the type of nanoparticles and the type of plant (Pérez-de-Luque 2017; Anjum et al. 2019). The details of nanoparticles used in the synthesis of nano-biofertilizer are given below.

2.1.1 Nanoparticles of Silicon (SiNPs)

Silicon comes in the category of beneficial substances for plants and is present in ample amounts in the soil. It has a formula of $Si(OH)_4$ and is present in the form of monosilicic acid. The amount of silicon in soil decreased due to the extensive use of agricultural lands, which leads to its deficiency in plants (Caubet et al. 2020). According to an estimate, every year, approximately 210–224 million tons of silicon are removed. In plants, the application of silicon results in enhancement in metabolite production, improvement in growth, and development of resistance or tolerance under stress conditions. To transport the silicon against the concentration gradient, Si transporters are present in plants. The movement of Si to the cortical cell was allowed by SIT1, while from the cortical cell to the xylem by SIT2. The type of plant species (accumulator and non-accumulator of Si) and various other factors decide the silicon's uptake and translocation in plants. The uptake of Si is less than 1% of the dry biomass of the plants in silicon accumulators, while greater than 1% in non-accumulator plants (Meunier et al. 2018; McGinnity 2015).

Silicon (Si) is categorized as a quasi-essential element in developing stress-tolerant plants by controlling the genes positively (Greger et al. 2018). The nutrient uptake and translocation from root to aerial plant parts were enhanced by reducing the harmful effects in plants due to oxidative stress. It was reported that the plant with silicon application shows an enhanced photosynthetic rate and reduced transpiration rate. It also reduces the damaging effects of insects and pathogens (Ali et al. 2020). In a study in *Oryza sativa* L., it has been documented that the silicon application results in a reduction in the damage caused by stem borer and leaf folder (Han et al. 2018). Plants under drought stress show positive results after their application. A significant escalation in antioxidants and growth hormone production resulted in plants that have received Si application exogenously (Helaly et al. 2017).

The performance of Si nanoparticles is better than silicon salt. The nanoparticle synthesis can be green and better if done using plants or their metabolites. Biowastes and microbes can be used as bioactive agents. The stem of *Bambusa bambos* (L.) Voss. was cut into pieces subjected to pyrolysis. Once pyrolyzed, it was treated with HCl, H_2SO_4, and HNO_3 and then calcinated in an electric furnace for green synthesis of SiNPs (Snehal and Lohani 2018). Another example of green extraction of SiNPs was using the leaves of *Azadirachta indica* A. Juss. and rice husk. The nanoparticle's characteristics can be evaluated using different characterization techniques (Ghorbani et al. 2015; Jabeen et al. 2017).

The better performance of SiNPs after their exogenous application compared to Si salt has been documented in the literature. In *Zea mays* L. under heavy metal stress, SiNPs augment the activities of ascorbate peroxidase, dehydro-ascorbate reductase, superoxide dismutase, ascorbic acid, and glutathione reductase (Tripathi et al. 2016). In *Triticum aestivum,* SiNPs efficiently alleviate more UV stress than Si salt. The antioxidant enzyme activities (ascorbate peroxidase, guaiacol peroxidase, superoxide dismutase, and catalase) improve with the Si application. It also augmented the content of non-enzyme antioxidants such as phenolics, ascorbate, proline, and flavonoids. In plants, improvement in the synthesis of antioxidants aids in scavenging reactive oxygen species and maintaining the photosynthetic rate (Tripathi et al. 2017). Under drought stress conditions, improved relative water content was reported in *Fragaria vesca* L. with the foliar spray of silicon and selenium (Se) nanoparticles. These treatments also improved plants' ascorbic acid, anthocyanin, and total phenolic compounds (Zahedi et al. 2020). In food crops, heavy metal uptake and accumulation were reduced by utilizing SiNPs. In the plants, they compete with the absorption and translocation of heavy metals by immobilizing the metals in the soil. They also translocate the nutrients such as S, Mn, P, and K from the soil to aerial plant parts, thereby enhancing plant growth (Han et al. 2018). Moreover, in *Ocimum basilicum* L., by osmotic alteration facing salinity stress, they increase dry weight, fresh weight, and concentration of chlorophyll a and b (Kalteh et al. 2014; Khan et al. 2020).

2.1.2 Nanoparticles of Zinc (ZnNPs)

In plants, Zn acts as an integral part of many different enzymes. It is a micronutrient that triggers various enzymes concerned with the metabolism of carbohydrates. It acts as a cofactor of enzymes and plays a significant role during DNA transcription. It is essential in biochemical pathways related to photosynthesis, protein, auxin carbohydrates, and starch metabolism. It also preserves biological membrane structure and functions. In addition, it regulates the stomatal opening and closing and adjusts the level of K in the cells (Snehal and Lohani 2018).

The crops show signs of chlorosis and stunted growth due to deficiency of Zn, as it inhibits the tillering. It also inhibits the production of pollen and affects the growth of the pollen tube, thereby reducing the fertilization rate and seed set. In the extracellular matrix of pollen grains, it causes structural and biochemical alterations and downregulates esterase (Ghorbani et al. 2015; Jabeen et al. 2017). Zn application activates the antioxidant enzymes that scavenge reactive oxygen species and hence reduces the damage due to photo-oxidation. In a study in *T. aestivum*, it was observed that out of four different levels of Zn fertilizer, 11.4 kg/ha concentration of Zn significantly improves the surface area of roots, length of root, and dry weight of root and shoot (Liu et al. 2019; Danish Toor et al. 2020). In another report in *Zea mays*, Zn application at 15 kg/ha escalated the plant height, tasseling, and silking of the plant and improved the protein content and grain yield (Tripathi et al. 2017). Under unfavorable environmental conditions, Zn utilization increases plant growth

(Rafiq et al. 2010). Though the quality and quantity of crops improve with foliar application of Zn fertilizer, the excessive use of chemical fertilizer reduces the growth and yield of crops. Also, it leads to an imbalance in the environment (Ma et al. 2017).

The plant that receives the application of Zn nanofertilizer releases nutrients slowly in a controlled way with enhanced translocation. The conversion of salts to NPs alters their physical and chemical characteristics, and after their application, the plant gives better outcomes (Siddiqi et al. 2018). In a study, the leaf extract of *Camellia sinensis* (L.) O. Kuntze was used for the green synthesis of ZnNPs. In the fresh leaf extract, zinc acetate was dissolved and positioned on a magnetic stirrer, followed by centrifugation. At 80°C, the ZnNPs were washed and dried (Shah et al. 2015). The preparation of ZnNPs having the size of 18 nm was done by treating zinc acetate with NaOH and thioglycerol (Lawre et al. 2014).

The adverse impacts of biotic and abiotic stresses were reduced with the application of ZnNPs. It aids in the detoxification of heavy metals, increasing the efficiency of water use, maintaining stability, and balancing the nutrient uptake. In an experiment, the yield of *Sorghum bicolor* (L.) increases by 22–183% facing drought stress compared to untreated due to the ZnNPs application at different concentrations (Dimkpa et al. 2019). In the root and shoot of *Camelina sativa* L. cultivated in saline soil, its application enhances the DPPH (2, 2-diphenyl-1-picryl-hydrazyl-hydrate) radical scavenging activity (Akhavan Hezaveh et al. 2020). With the foliar application of Zn and Fe nanofertilizers, improvement in the morphological attributes and also a significant increase in the yield and yield component were reported. In *Cicer arietinum* L., they enhanced the yield to 34% compared to the control (Talebi et al. 2016). The exogenous application of ZnNPs enhances the photosynthetic rate and also improves the anatomical structures of the stem and leaf. It is reported that in *Solanum melongena* L. cultivated in the presence of a limited water supply in saline soil, ZnNPs application increased the yield from 50% to 66% (Semida et al. 2021). *Mangifera indica* L., facing salinity stress if treated with Zn and SiNPs, observed a significant increase in the number, size, and weight of its fruits. The development of floral parts can be associated with the role of Zn. However, the phytohormonal content synthesis and balance in plants also play an essential role (Shah et al. 2015).

2.1.3 Nanoparticles of Copper (CuNPs)

For the normal growth of the plant, both structurally and functionally, copper is an essential micronutrient. In plants, its concentration is based on mobility and bioavailability in the soil (Lawre et al. 2014; Dimkpa et al. 2019). The optimum level of Cu is maintained by plants in various tissues that are involved in numerous processes inside mitochondria and chloroplasts. Its role comprises protein trafficking, iron mobilization, mitochondrial respiration, cell metabolism, and hormone signaling (Elsheery et al. 2020; Lopez-Lima et al. 2021). The structural component of plastocyanin is Cu. A plant develops stress tolerance or resistance as it enhances the production of signaling molecules, i.e., H_2O_2. It also provides strength to the cell wall, thereby reducing the wilting of the plant. In plants, exogenous Cu application escalated biomass production and improved the quality and yield of the plant (Lafmejani et al. 2018; Priyanka et al. 2019).

Cu deficiency in plants results in curling of leaves, chlorosis, stunted growth, and interruption of the process of photosynthesis. Its shortage escalated the reactive oxygen species production and reduced nitrogen and carbohydrate metabolism. It disturbs the photosystem, and electrons flow in the electron transport chain (Shabbir et al. 2020).

Cu deficiency can be surmounted by applying Cu exogenously to plants. CuNPs application improves the growth and yield of crops. In a study, the leaf extract of *Azadirachta indica* is used in the synthesis of CuNPs. In the $CuCl_2$ solution, the leaf extract was added with continuous stirring and observed until the color started changing from yellow to dark brown. The change in color specifies CuNPs formation. The solution was centrifuged, and after washing, CuNPs were dried. Various techniques of characterization, namely, Transmission Electron Microscope (TEM), Scanning Electron Microscope (SEM), X-ray Diffusion (XRD), Fourier Transform Infra-Red (FTIR), and Zetasizers, were used to affirm the synthesis and purity level of copper nanoparticles (Nagar and Devra 2018). To minimize the Cu salt size to the nanoscale, fruit extract of *Citrus limon* (L.) Burm. f. was used (Akl and Amer 2021). In the literature, it was also documented that the leaf extracts of *Eclipta prostrata* (L.) L. (Chung et al. 2017), *Plantago asiatica* L. (Nasrollahzadeh et al. 2017), and *Tinospora cordifolia* (Willd.) Miers. (Sharma et al. 2019) were used in the synthesis of CuNPs. Two important characteristics of CuNPs, i.e., rapid absorption and slow release, make them an exceptional candidate to be used. According to a report, the morphological and physiological parameters in *Ocimum basilicum* L. were improved with the application of Cu and ZnNPs (Abbasifar et al. 2020). Creating drought tolerance improves the status of water, maintains the photosynthetic rate, and enhances the antioxidant activity in crops. It was documented that the fruit quality of *Solanum lycopersicum* improves with the application of CuNPs. It also enhances the activity of catalase and superoxide dismutase and escalates the synthesis of bioactive compounds (López-Vargas et al. 2018).

2.1.4 Nanoparticles of Iron (FeNPs)

Iron is another essential micronutrient required for plant growth. As it is a cofactor, therefore, in various biochemical reactions, it acts as a catalyst. It is necessary for the synthesis of various enzymes and regulation of photosynthesis and respiration. Iron deficiency causes a reduction in chlorophyll, resulting in chlorosis and necrosis in the leaves. The appliacation of Fe as fertilizer reduces the detrimental consequences of environmental stress and improves the growth of the plant (Vaghar et al. 2020; Pourjamshid 2021).

The adverse impact of chemical fertilizer reduces with the nanofertilizer application if applied in combination with the soil and foliar method, which is found to be beneficial for plant growth enhancement (Mohamadipoor and Mahboub Khomami 2013). It was reported that the plant extracts of *A. indica*, *C. sinensis,* and *Eucalyptus tereticornis* (Sm.) were used in the green synthesis of FeNPs (Herlekar et al. 2014). The size of ferrous salt was reduced, and the reaction stabilized due to the phytochemicals of plant extracts. The frequently used iron nanoparticles are ferric oxide, ferrous ferric oxide, iron oxide hydroxide, and iron mineral complex (Mohamadipoor and Mahboub Khomami 2013). A positive response was observed with the application of FeNPS in plants facing stress (Ebrahiminezhad et al. 2017). Due to its application, the toxic effects of ions were reduced as it gets readily absorbed by the plants. A significant increase in the content of glutathione, phenol, and vitamin C in *S. lycopersicum* was due to fertilization using FeNPs (El-Desouky et al. 2021). The selective uptake of substance from the root membrane was induced with the FeNPs application. It makes the plant salinity-resistant by enhancing the K content in shoots and reducing the Na absorption. It was observed in *F. vesca* that the combined application of FeNPs induces the characteristics of drought tolerance (Mozafari et al. 2018). In *T. aestivum,* the use of FeNPs stimulates germination and increases the shoot and root length compared to untreated plants (Yasmeen et al. 2015). Sharma et al. (2025) have reported the potent antifungal activity of α-Fe_2O_3NP and *Trichoderma* sp. against Fusarium wilt. It has been documented that cadmium (Cd) stress results in oxidative stress and reduces the photosynthetic rate and yield of *T. aestivum,* resulting in grains with a high concentration of cadmium. An enhancement in the Fe content and reduction in the Cd uptake was observed in the grains and tissues of the plant with the application of FeNPs. It also improved plant growth by suppressing the harmful impacts of cadmium stress (Adrees et al. 2020).

2.1.5 Nanoparticles of Silver (AgNPs)

The role of Ag ions is significant in somatic embryogenesis, genetic transformation, and root and shoot formation. Its application in plants decreases the chances of chlorosis and chlorophyll destruction and reduces ethylene synthesis. It also augments the secondary metabolites production, downregulates the aging process, and improves crop growth and grain yield (Yasmeen et al. 2015; Adrees et al. 2020).

In plants, the role of nanoparticles is determined based on their properties, such as surface reactivity, composition, and rate of application (Kumari et al. 2017a; Öktem and Keleş 2018). Nanoparticles' high concentration causes detrimental effects. However, a positive response was developed in plants with the appropriate dose. The concentration of dose is associated with the plant species type. As documented in the literature, nanoparticles are more efficient (Mahendran et al. 2019). Green synthesis of AgNPs has been done using various plants because of their phytoconstituents, which decrease the Ag salt size. These metabolites include flavonoids, aldehydes, tannins, ketones, amides, terpenoids, and carboxylic acid (Ahmed and Mustafa 2020). The leaves of *Syzygium cumini* (L.), *Ocimum tenuiflorum* L., *Citrus sinensis* (L.) Osbeck (peel), and *Centella asiatica* (L.) Urban were boiled, and the product was gradually poured into silver nitrate solution (1 mM). The color change from yellow to dark brown confirms the nanoparticle synthesis. Its characterization was achieved by employing techniques such as atomic force microscopy, UV-visible spectroscopy, X-rays diffraction, and scanning electron microscopy (Logeswari et al. 2015). However, Jakinala et al. (2021) also reported the synthesis of AgNPs from *Mang mao* wings extract and evaluated its antioxidant and antimicrobial activity.

The antioxidant activity and germination were improved in *Z. mays*, *Allium cepa* L., and *Trigonella foenum-graceum* L. with the AgNPs application, synthesized using *Eucalyptus globules* Labill. (Soliman et al. 2020). In a study, it was observed that the hydrogen peroxide production and hyper-hydricity reduced to 50% and 13%, respectively, in the culture of *Dianthus chinensis* L. with the utilization of AgNPs (Sreelekshmi et al. 2021). In *Glycine max,* they improved the root lengths and diminished the production of reactive oxygen species (Mustafa et al. 2020). Its application in *S. lycopersicum* infected with *Alternaria alternata* increased the polyphenol oxidase, chlorophyll, and carotenoid content to 33%, 45%, and 46%, respectively. In addition, disease severity was reduced by 58% compared to untreated plants (Mahawar et al. 2020). In plants, the uptake of water is not consistent; however, the nutrients are actively absorbed by the leaves if applied as a foliar spray. Compared to soil application, it is a more feasible and inexpensive method (Ahmadi-Majd et al. 2021). It was reported that fresh weight and chlorophyll content increased to 32% and 23%, respectively, with the foliar application of AgNPs in *S. lycopersicum* infected with *A. solani.* While the fungal spore count was reduced to 48%, it aids the plants in developing stress tolerance by enhancing the activity of stress enzymes (Kumari et al. 2017a). In *Satureja hortensis* L., under salinity stress, its application improves the shoot length, root length, rate of seed germination, and biomass (Bastami et al. 2021; Nejatzadeh 2021). In another study, Nayeri et al. (2021) also proved that green synthesis of AgNPs using medicinal plants can augment the quality of prepared particles for exhibiting trustworthy functionalities. Here is the list of some fungal species that are used in the biosynthesis of nanomaterials (Table 1.1).

2.2 Biofertilizer

Biofertilizers consist of beneficial microorganisms such as plant-growth-promoting rhizobacteria (PGPR), fungal mycorrhizae, blue-green algae (BGA), or processed plant material (Vafa et al. 2021). Ancient Roman writings and classical Greeks used the notion of biofertilizer to delineate the various agricultural methods of improving crop yield. The marketing and use of microbes as biofertilizers commenced in the late nineteenth century (Malusà et al. 2016). One of the significant causes of agro-ecosystem pollution is the extensive use of chemical fertilizers. It is one of the main components of air, water, and soil pollution. The most pressing need is the development of cheap, non-hazardous, and eco-friendly alternatives for achieving high agricultural productivity without any consequences (Sagar et al. 2022). The biofertilizers are cost-effective, eco-friendly, sustainable, and have long-lasting impacts. They enhance the availability of nutrients by colonizing the rhizosphere and improving plant growth (Enrico et al. 2020; Nasab et al. 2021).

They release substances that promote plant growth, execute solubilization of phosphate (Sharma et al. 2013; Ahmed et al. 2021b) and potassium (Sethi et al. 2025; Baba et al. 2021), and atmospheric nitrogen fixation in the soil (Kusale et al. 2021a). They inhibit the pathogen growth by sequestering the Fe through a

Table 1.1 List of some fungi used in biosynthesis of nanoparticles (NPs)

Fungus	Nanoparticles (NPs)	Mode of biosynthesis	Country	References
Cochliobolus lunatus	Silver (Ag)	Extracellularly	India	Salunkhe et al. (2011)
Aspergillus flavus	Silver (Ag)	Extracellularly	Iran	Moharrer et al. (2012)
A. clavatus	Gold (Au)	Intracellularly	India	Verma et al. (2011)
A. niger	Silver (Ag)	Extracellularly	India	Gade et al. (2008)
Neurospora crassa	Silver (Ag)	Intra-/extracellularly	Mexico	Castro-Longoria et al. (2011)
Alternaria alternata	Gold (Au)	Extracellularly	India	Sarkar et al. (2011)
A. alternata	Silver (Ag)	Extracellularly	India	Gajbhiye et al. (2009)
Trichoderma reesei	Silver (Ag)	Extracellularly	USA	Vahabi et al. (2011)
T. harzianum	Silver (Ag)	Extracellularly	India	Singh and Balaji (2011)
T. viride	Silver (Ag)	Extracellularly	India	Fayaz et al. (2010)
Pleurotus sajar-caju	Silver (Ag)	Extracellularly	India	Nithya and Ragunathan (2009)
Penicillium spp.	Silver (Ag)	Extracellularly	Poland	Sadowski et al. (2008)
Verticillium spp.	Magnetite	Extracellularly	India	Bharde et al. (2006)

siderophore released by some biofertilizers, making it unavailable for pathogens (Kusale et al. 2021b; Nithyapriya et al. 2021). They secrete various phytohormones such as gibberellins, cytokines, indole-3-acetic acid, abscisic acid, etc. (Hamid et al. 2021), which aid in the development and growth of the plant (Sumbul et al. 2020; Fasusi et al. 2021). At the same time, disputes with biofertilizers comprise low quality, the requirement of a high dose to cover a large area, substandard shelf life, and low performance in a fluctuating environment (Kalam et al. 2020; Basu et al. 2021). Several issues have been resolved with the use of nano-biofertilizers. Agriculture depends on nanotechnology for maintaining the natural ecosystem and improving the growth and yield of crops. They keep the quality and augment the slow release of fertilizer. Biofertilizers have long-lasting effects in the field and are a renewable approach that lessens the cost of chemical fertilizer (Du et al. 2018).

3 Formulation of Nano-Biofertilizer

The amalgamation of nanoparticles and biofertilizer is nano-biofertilizer. In plants, it augments the nutrient use efficiency via the slow release of nutrients. It imparts a long-lasting impact on soil's biological, physical, and chemical characteristics and hence improves the properties of soil. The nano-biofertilizer formulation involves nanoparticle encapsulation with biofertilizer. Its efficiency is enhanced by the slow release of nutrients, and it further aids in protecting bacterial strains from mechanical stress. The process by which biofertilizer cells are accommodated in the nanomaterial capsule is called encapsulation. It uses materials that are biodegradable and non-toxic, like starch and calcium alginate. Starch escalates the bacterial stain growth (Du et al. 2018; Vafa et al. 2021).

Nano-biofertilizers synthesis comprises three crucial stages: (a) biofertilizer culture growth; (b) encapsulation of biofertilizer with the nanoparticles; and (c) the appraisal of its quality, purity, effectiveness, and shelf life (Panichikkal et al. 2021b). Nano-biofertilizer can also be formulated as a microcapsule. It was synthesized by amalgamating the PGPR suspension with the concoction of 4% bentonite, 3% starch, and 1.5% sodium alginate in the ratio of 2:1. The crosslinked $CaCl_2$ solution was used for coating the concoction. Sterile distilled water was used to wash the microcapsules (Saberi-Rise and Moradi-Pour 2020). Nano-biofertilizer has also been formulated using the amalgamation of nanoparticles and salicylic acid. In this process, the biofertilizer is combined with ZnOPs (1 μg/ml), salicylic acid (1.5 mM), and sodium alginate (2%). The mixture is coated with the 3% solution of $CaCl_{2,}$ and beads of size 1 mm are created, which are air-dried and incubated at 4 °C (Panichikkal et al. 2021a). It has been documented that the combination of nanoparticles and organic waste like cow dung, flowers, and kitchen waste was used to develop an efficient nano-biofertilizer that escalates the fertility of the soil. The organic waste was washed, sliced into tiny pieces, and exposed to pyrolysis or decomposition to remove the impurities. Nanoparticles were then integrated with this partially decomposed or pyrolyzed waste to develop a nano-biofertilizer (Singh et al. 2019). Nowadays, nano-biofertilizer are easily available in the market. Some of the commercially available nano-biofertilzer is given in Table 1.2.

4 Responses of Plants to Nano-Biofertilizer

The growth of the plant significantly improves with the application of nano-biofertilizer. The encapsulation of nanoparticles with biofertilizers leads to the sustained and slow release of nanoparticles in the rhizosphere of plants and enhances the efficiency of biofertilizers. They sustain the fertilizer's dissolution and diminish the chances of leaching (Ali et al. 2021; Akhtar et al. 2022). They augment the production of secondary metabolites such as enzymatic (superoxide dismutase, peroxidase, and catalase) and non-enzymatic antioxidants (flavonoids and phenols), thereby enhancing the crop's quality. These metabolites have many health benefits and escalate the shelf life of fruits and vegetables (Paschalidis et al. 2021). Plant responses get intensified with its application due to the mechanism of synergistic action of nanoparticles and biofertilizers. In plants, it triggers various mechanisms accountable for the improved development and yield of plants. Nano-biofertilizers replenish the essential soil nutrients by assisting in bioremediation. They maintain the function and structure of the cell, diminish the detrimental impacts of ROS on plants, and upregulate the genes responsible for producing osmolytes, antioxidants, and proteins related to stress. They also uphold the transporters of the membrane and enhance the production and activities of hormones (Shcherbakova et al. 2017;

Table 1.2 List of some commercially available nano-biofertilizers in the market (Prasad et al. 2017a)

Commercial name	Remark	Company	Country
Nano-Gro™	Regulate the plant growth and improve immunity	Agro Nanotechnology Corp.	FL, United States
Nano Green	Having extracts of corn, grain, soybeans, potatoes, coconut, and palm	Nano Green Sciences, Inc.	India
Biozar Nano-Fertilizer	Mixture of organic materials, macromolecules, and micronutrients	Fanavar Nano-Pazhoohesh, Marakazi Company	Iran
Nano Max NPK Fertilizer	Having organic acids chelated with major nutrients, amino acids, organic carbon, micronutrients, and vitamins	JU Agri Sciences Pvt. Ltd.	New Delhi, India
Nano-Ag Answer®	Having sea kelp and mineral electrolyte	Urth Agriculture	CA, United States
Master Nano Chitosan Organic Fertilizer	Comprises organic acid, salicylic acids, water-soluble liquid chitosan, and phenolic compounds	Pannaraj Intertrade	Thailand
TAG NANO fertilizers (NPK, Zinc, Cal, etc.)	Contains chelated micronutrients, vitamins, probiotics, humic acid, and seaweed extracts	Tropical Agro-system India (P) Ltd.	India

Vedamurthy et al. 2021). The antioxidant systems of treated plants become active, protecting the cell membrane and cell organelles from the adverse impacts of stress. They also increase the growth hormone concentrations in plants, including cytokinin and indole acetic acid; however, they reduce abscisic acid (stress hormone) production. These alterations develop stress tolerance in plants and enhance the probability of crop establishment under adverse environmental conditions (Bibi et al. 2022). An increased amount of proline, carbohydrate, and anthocyanin in *Hibiscus sabdariffa* and *Foeniculum vulgare* was observed after the application of the nano-biofertilizer in Iran (Bibi et al. 2022; Faridvand et al. 2021).

Nano-biofertilizers application in *Zea mays* enhances the number of rows per comb, the number of grains per comb, and biomass yield. It improves the harvest index in a limited water supply (Farnia et al. 2015). An accelerated performance was observed in *Pseudomonas elgii*, *P. fluorescens* (57%), and *Bacillus subtilis* (33%) with the application of gold nanoparticles. If used as a nano-biofertilizer formulation, its application also improves the growth of plants (Shukla et al. 2015). It has been documented that the accessibility of nutrients and nanoparticles in the soil is controlled by biofertilizer (PGPR). A study was carried out to appraise the function of nanoparticles on *S. lycopersicum* with or without the inoculation of *Azotobacter salinestris*. *A. salinestris* synthesizes exopolysaccharides (EPS), which entrap the nanoparticles. In the rhizosphere, they sustained their concentration and reduced the adverse effects of the high concentration of some nanoparticles (Ahmed et al.

2021a). Enhanced plant biomass, photosynthetic pigments, phenolics, flavonoid content, stomatal conductance, and transpiration were observed in *Cucumis melo* L. facing Cd stress after inoculating the combination of ZnNPs and *Bacillus fortis*. They also minimize Cd metal uptake, accumulation, and translocation in plants (Shah et al. 2021). In *Z. mays* facing drought stress, the combined application of PGPR and SiNPs has improved the activity of the antioxidant enzyme, photosynthetic rate, and relative water content. They also augmented the uptake and translocation of nutrients in plants and boosted the yield attributes (Hafez et al. 2021). The adverse effects of arsenic toxicity were reduced with the use of a *Bacillus subtilis* and FeNPs combination in *Cucurbita moschata* Duchesne. It escalates the production of catalase and superoxide dismutase and improves plants' photosynthesis and gaseous exchange (Mushtaq et al. 2020).

In *Carthamus tinctorius* under salinity, reductions in the concentration of Na^+ by 74% and 60% in leaves and roots, respectively, were observed with the co-application of biofertilizer and ZnO NPs. In addition, they also amplify the number of leaves, the number of pods per plant, and the weight of the capitulum. These variations can be associated with zinc, as it is needed for the production of chlorophyll and biosynthesis of auxin. It plays a vital role in germination, pollen performance, and soil fertility. They collaborate perfectly, producing significant results (Yasmin et al. 2021). An increment was observed in the soil's available P, K, and N levels by 283%, 126%, and 38%, respectively, due to the interaction of nitrogen nanoparticles and biofertilizer. They augmented the availability and uptake of nutrients by interacting with the plant roots and soil. Biofertilizers can fix atmospheric nitrogen and improve the fertility of the soil. They release some nutrient-dissolved organic acids such as propionic acid, citric acid, ketogluconic acid, lactic acid, gluconic acid, and succinic acid. They enhance the plant nutrient levels, including P, N, and K, and improve the rate of photosynthesis (Rahman et al. 2020).

5 Applications of Nano-Biofertilizer

5.1 To Improve the Nutrient Status and Crop Growth

The application of nano-biofertilizer imparts varied effects on soil and plant systems, which promotes nutritional quality and plant growth. Nanoencapsulation of organic nutrients utilizing nanomaterials, i.e., zeolite, polymers, and chitosan, aids in the slow and sustained release of nutrients to plants (Qureshi et al. 2018). The stability of nutrients is enhanced because of the surface coating of the biofertilizer with nanoparticles. The nanoparticle-coated fertilizer has an enlarged surface area, nanosize, and higher reactivity, which permits enhanced interaction and active nutrient uptake for crop fertilization. Moreover, at different growth stages, they also sustainably offer bioavailable nutrient supply at their need to the plant (El-Ghamry et al. 2018). Fertilizer's dissolution rate is reduced due to this stability, which permits the slow and sustained nutrient release by nano-biofertilizer (Manjunatha et al. 2016). In nano-biofertilizer, its bioorganic components (such as fungal inoculants or beneficial bacterial PGPR) benefited synergistically by elevating the nutrient status of soil by various means. For instance, atmospheric nitrogen fixation via plant roots by rhizobacteria, siderophore production that chelates the metal elements and makes them accessible for plant roots, solubilization of phosphate because of the existence of P-solubilizing fungal and bacterial strains, and synthesis of phytohormones (Ahemad and Kibret 2014; Mala et al. 2017).

Several studies showed that the synergistic benefits of both the components of nano-biofertilizer result in an exaggerated response to improving crop growth, production, yield, and quality attributes of crops.

5.2 To Promote Plants' Development Morphologically and Physiologically

The crop growth intensifies with the nano-biofertilizers application, as it optimizes nutrient absorption efficiency and photosynthesis of plants. Moreover, it also elevates photosynthate accumulation and translocation of nutrients to the economic parts of the plant, resulting in an increment in the quality and productivity of the plant. In a study, it has been documented that the application of nano-biofertilizer formulated by encapsulating biofertilizer (including PGPR, *B. subtilis*, *P.* fluorescens, and *Paenibacillus elgii*) with Ag and Au NPs was found to be significantly efficient in promoting the growth of various crops (Dikshit et al. 2013). The co-application of nano-encapsulated neem cake fertilizer and PGPR potentially fosters the productivity of some leguminous crops by stimulating the potency of seedling germination and transporting nobbled nutrients to plants in an efficient way (Ferioun et al. 2025; Rahman and Zhang 2018).

5.3 To Enhance the Plants' or Soils' Nutritional Security

The constant use of chemical fertilizers exhausts the essential nutrients from the soil. The primary cause of deterioration of soil fertility is excess soil acidification (Rahman and Zhang 2018). The main challenges that the farmers face are nutrient imbalance and low soil fertility. These deficiencies lead to a decline in crop productivity and inadequate food nutrition (Monreal et al. 2016). However, nano-biofertilizers offer sustainable, low-cost, and integrated disease management. It optimizes the absorption and assimilation of nutrients by plants and abates soil nutrient loss via gasification, leaching, or competition with other organisms (Janmohammadi et al. 2016). The bioorganic component (PGPR) of nano-biofertilizer facilitates the restoration of the soil nutrients via nitrogen fixation and phosphate solubilization (Bright et al. 2025), while the NPs assist the slow release of nutrients in a synchronized way as per the crop demands (Morales-Díaz et al. 2017).

5.4 To Protect the Crops

The application of nano-biofertilizer not only enhances crop productivity, shelf life, nutritional quality, and potency of water conservation but also functions as a resistance-inducing agent to escalate the resistance in plants against pests and pathogens (Fig. 1.2). An experiment was conducted by Gatahi et al. (2015) to examine the effect of nano-biofertilizer on tomatoes infected with *Ralstonia solanacearum*, and its pest-resistant role against wilt was determined.

Another study also determines the role of nano-biofertilizer (consisting of PGPR, *B. subtilis*, *P.* fluorescens, *P. putida*, and *Paenibacillus elgii*) against destructive bacterial and fungal pathogens in the rhizosphere of leguminous crops and testifies to their role in crop protection (Gouda et al. 2018).

Mishra and Kumar (2009) observed that the application of titanium-coated nano-biofertilizer on rapeseed mustard enhances root adhesion of beneficial bacteria and protects the crop from various harmful fungal infections. Thakur et al. (2022) also utilized rhizome extract of *Zingiber officinale* for the green synthesis of zinc oxide

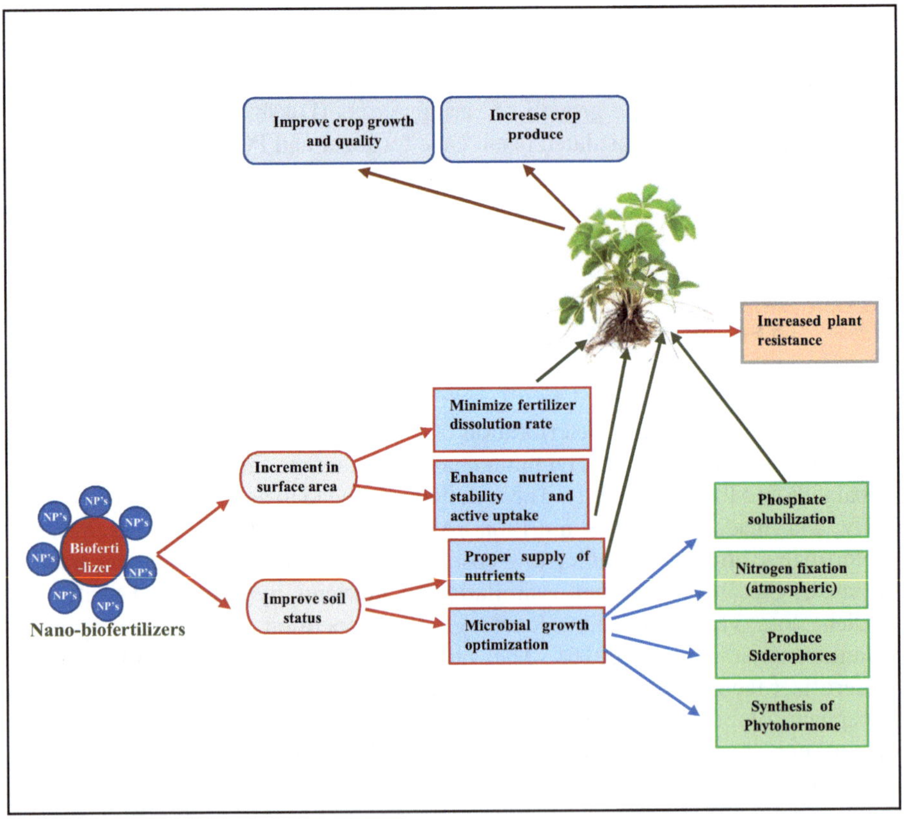

Fig. 1.2 Importance of nano-biofertilizers in improving plant health

nanoparticles (ZnO NPs) against insect pests (*Spodoptera litura* and *Macrosiphum euphorbiae*) associated with tomato crops as an alternative to synthetic insecticide. In another study, Shahbaz et al. (2023) also used the first biocompatible approach to evaluate the potential of green-synthesized selenium nanoparticles (SeNPs) using *Melia azedarach* leaf extracts as growth-promoting substances in wheat under the spot blotch stress and found an effective management strategy to inhibit the fungal growth.

6 Nano-Biofertilizer: Gradual Advancement of Sustainable Agriculture

One of the most common practices that have been going on for several years is the use of chemical fertilizers to enhance agricultural productivity. However, the hazardous effects, such as long residual action and environmental toxicity, linked with the excessive use of chemical fertilizers have had the scientists concerned in the past decades. This impelled the desire to pursue eco-friendly, non-toxic alternatives to attain the coveted aim of enhanced agricultural productivity without any allied side effects. In the current era, to ensure agricultural biosafety, preference is given to biofertilizers and nanofertilizers over chemical fertilizers (Dhir 2017). Biofertilizer is formulated using live formulations of useful microorganisms such as PGPR, i.e., blue-green algae, *Rhizobium*, fungal mycorrhizae, *Azospirillum*, *Azotobacter*, and phosphate-solubilizing bacteria (*Bacillus* and *Pseudomonas*). The application of biofertilizers improves the supply of nutrients to crops by augmenting the nitrogen fixation and breaking the insoluble complex organic into simpler forms through solubilization, which makes it biologically available for plants. It enhances the availability of soil nutrients to plants via enriching the microbial status of the soil. It also boosts the moisture-holding capacity of soil and improves soil aeration and natural fertilization (Itelima et al. 2018). Nevertheless, this compelling tactic has some significant drawbacks, such as on-field stability, consistently poor shelf life, functioning under varying environmental conditions (radiation, temperature, and pH sensitivity), desiccation susceptibility, beneficial bacterial strain scarcity, and above all, the need for high doses to cover large areas (Mishra et al. 2017). Intriguingly, nano-biofertilizers were found superior in terms of encountering all these issues (El-Ghamry et al. 2018). Hence, to combat significant drawbacks of food security, crop production, eco-safety, and sustainability, the development of nano-biofertilizer comes to light by implementing modern agriculture with the espousal of nano-biotechnology (Khan and Rizvi 2017; Teng et al. 2018). The formulation of nano-biofertilizer consists of biofertilizer (consisting of nutrients and PGPR) encapsulated with nanoscale polymers (nanoencapsulation) (Golbashy et al. 2017). This technology can be employed to protect the biofertilizer components comprising PGPR by augmenting their shelf life and dispersion in fertilizer formulation and permitting the restrained discharge of PGPR (Vejan et al. 2016). It assists in the slow and stable release of nutrients to crops without any inadvertent loss (Gouda et al. 2018). The use of nano-biofertilizer significantly benefits the farmers

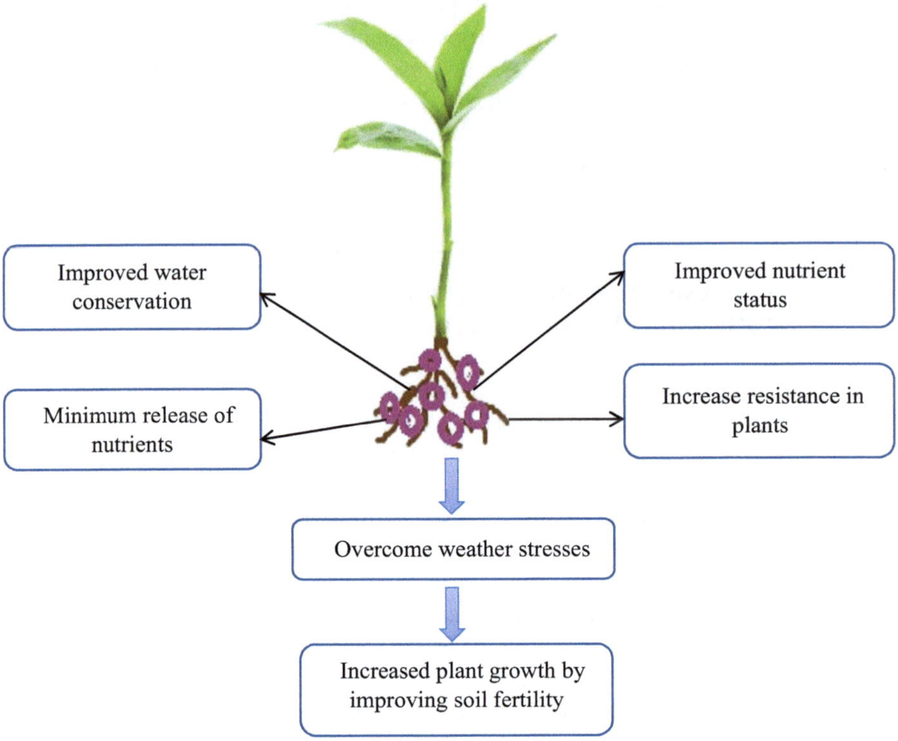

Fig. 1.3 Role of nano-biofertilizers in encouraging plant growth and nutrient status

as they improve nutrient release characteristics and field performance. They also found it economically effective as they reduced the cost and application losses. It is an environmentally sustainable and renewable approach that hastens the proficiency of nutrients (P, N, and K) utilization, improves the related enzyme activity, supplements beneficial microbial populations in the soil, and improves the fertility of the soil. It also aids in enriching the quality of crop products and developing resistance in crops (Thirugnanasambandan 2019). Figure 1.3 represents the role of nanofertilizer in enhancing plant growth and nutritional security in soil or plant systems.

6.1 Current Scenario of Nano-Biofertilizer in Research and Development

6.1.1 Global Scenario

Globally, scientists are emphasizing the promising role of biofertilizers and nanoparticles from agricultural perspectives. Though the role of nano-biofertilizer in improving productivity in agriculture has been documented in past literature. Many Iranian researchers worked on the development of environmentally sustainable nano-biofertilizer (Tahmasbi et al. 2011; Mardalipour et al. 2014; Safaei et al. 2014;

Farnia and Omidi 2015; Ghooshchi 2017). It has been reported that foliar application of Biozar® as nano-biofertilizer in spring wheat significantly enhances the growth and yield components of wheat (Mardalipour et al. 2014). In *Zea mays,* the application of biofertilizer and ZnNPs at 7 days of treatment significantly augments the yield of grain approximately 1 + ½ fold (Farnia and Omidi 2015). Another study in sugar beet shows that nano-biofertilizer application improves the morphological and physiological parameters such as root biomass (42.6%), leaf area (19.6%), sucrose content (1.03%), and net photosynthetic productivity (15.8%), leading to the 19.2% increment in white sugar yield (Jakien et al. 2015).

Another report confirms that the application of nanopharmax and humic acid in *Nigella sativa* increases the level of nutritive ingredients (Safaei et al. 2014). An experiment was conducted to check the efficacy of nanofertilizer + biofertilizer (*Ascophyllum nodosum*) and nanofertilizer alone on the grapevine, and observation revealed that co-application of nano-biofertilizer was found to be more effective in enhancing growth, yield, berry quality, and nutrient level of leaf in grapevine (cv. "Narince") facing alkaline condition compared to nanofertilizer (Sabir et al. 2014). In addition, a study was conducted to analyze the effect of nano-biofertilizer (aztocarvar + phosphobarvar + chelated nanofertilizers) on the nutrient and pigment content of various plants and appraised that it significantly improved the nutrient status and chlorophyll content of the agriculture crop forage sorghum (Speed Feed hybrid) (Mir et al. 2015).

6.1.2 Indian Scenario

Indian researchers also impelled interest in the commercial fertilizer nanoformulation into highly effective nano-biofertilizer by assembling biocompatible, non-toxic nanomaterial with bioorganic constituents. An experiment was conducted by Kumari et al. (2017c) on the extracellular synthesis of silver nanoparticles utilizing bacterial strain (*Pseudomonas aeruginosa*) and biogenic synthesis of AgNP in *Vigna radiata*. A review was presented by Kumari and Singh (2016) on the applicability of AgNPs in agriculture and highlighted that AgNPs could be used as both nanofertilizer and nanoencapsulation agents in formulating nanofertilizer. A report on ZnONP biogenic synthesis utilizing *Pseudomonas aeruginosa* shows that it has broad-spectrum antimicrobial properties, due to which it can be exploited for developing disease resistance in crops (Barsaiya and Singh 2018). Nano-biofertilizer's potential in horticultural crops of the Fabaceae family has been documented in a technical report from Allahabad University to the Ministry of Agriculture. It stated a significant increase in yield and nutritional value of crops by escalating the efficacy of biofertilizer, encapsulating it with nanoparticles, and resulting in the formulation of nano-biofertilizer. That report shows that in India, under limited accessibility of water and land resources, Ag and Au nanoparticles were found to be more effective for nanofertilizer synthesis. However, their application alone as a fertilizer may not be favored for fertilization due to some adverse effects. Nonetheless, when used in combination with biofertilizers (such as *Bacillus subtilis*, *Pseudomonas fluorescens*, and *Paenibacillus elgii*), they act as good plant growth promoters (Shukla et al. 2013). A study was conducted on the role of nanoclay-coated biofertilizer (consisting of *Pseudomonas* and *Trichoderma* species) on crops. It revealed that its application enhances the nutrient use efficiency, water holding capacity, and significant increase in the productivity of rainfed rabi crops (Sarkar et al.2021). Mukhopadhyay and De 2014). In 2014, Celsia and Mala engineered the nanostructured NPK fertilizer system coupled with PGPR and neem cake and demonstrated its substantial role in improving the potency of seed germination in *Vigna radiata*. This experiment was further protracted by Mala et al. (2017) and proves that this treatment progressively enhances the yield attributes of *Vigna radiata* by stimulating the allied biochemical reactions. They affirmed that neem cake acts as an organic carrier or vehicle in efficiently transporting insoluble nanofertilizers and PGPR, whereas PGPR aids in the solubilization of nutrients. Furthermore, the co-application of CuNPs and biofertilizer (PGPR, *Piriformospora indica*) in *Cajanus cajan* improves the growth and vitality of the crop (Rajak et al. 2017). Table 1.3 represents the summarized report of national and international studies on the nano-biofertilizer response to growth and yield attributes of various crops.

7 Constraints and Future Assessment

The application of nano-biofertilizer is supposed to expand for agricultural development. Nevertheless, the formulation should be eco- and human-friendly. Though, there is still not much report on its mechanism of enhancing growth and yield. They consist of two components, i.e., biofertilizer and nanoparticle. Their application is still not common, and they applied based on the trial (Bhari et al. 2021; Zhou et al. 2021). So far, very little research has been executed on its application and development. There is a need to formulate nano-biofertilizer by encapsulating biofertilizer with nanoparticles. To attain nutritional security in the scarcity of land and resources, the prospect of the innovative method of nanotechnology and biotechnology for the sustainable management of crops will always be valued. Our suggestions for the future are as follows:

- For formulating a low-cost and good-quality nano-biofertilizer having a longer shelf life, there is a necessity to use protocols and advanced equipment.
- Detailed scrutiny should be necessary to appraise their effects on human health.
- There is a need to develop multifunctional nano-biofertilizer suitable for many crops.
- Nano-biofertilizers' economic evaluation should be performed, and their compatibility in varied soil and environments should be checked.
- Cooperation of industries to execute the production on a large scale and their field trial should be carried out.
- Need to create cognizance among farmers about the consequences of the application of chemical fertilizer, in addition to the cost-effectiveness and long-lasting effects of nano-biofertilizers.

Table 1.3 Effect of nano-biofertilizers on growth and yield parameters of different crops

Nano-biofertilizers	Crop	Plant reaction	Country	References
Nano Fe + biofertilizer (biochar and compost)	Indian mustard (*Brassica juncea* L.)	Enhanced plant biomass and height also improved total carbon, total nitrogen, pH, etc.	Spain	Baragaño et al. (2020)
Nano Cu and *Piriformospora indica* (plant growth-promoting fungus)	Pigeon pea (*Cajanus cajan* L.)	Increased the vitality, growth, and productivity of the crop	India	Rajak et al. (2017)
Nano Zn + biofertilizers	Maize (*Zea mays* L.)	Improved physiological and morphological characteristics (leaf area, ear length, chlorophyll content, etc.). Also increased yield parameters.	Iran	Farnia et al. (2015)
Nanofertilizer + *Azotobacter* as biofertilizers	Jowar (*Sorghum bicolor* L.)	Boosted carbohydrate, carotenoid, and chlorophyll content (a & b)	Iran	Mir et al. (2015)
Bioorganic nanofertilizer ("Nagro")	Beet (*Beta vulgaris* L.)	Showing ability for proper development of sugar beet also increased productivity and quality of crop	Lithuania	Jakien et al. (2015)
Nano Si + BCAs (*G. mossease*, *T. viride*, and *B. subtilis*)	Tomato (*Lycopersicum esculentum* L.)	Provided resistance against bacterial wilt of tomato (*Ralstonia solanacearum*)	Kenya	Gatahi et al. (2015)
Nano NPK fertilizer + neem cake, PGPR (plant growth-promoting rhizobacteria)	Mung bean (*Vigna radiata* L.)	During germination, improved enzyme activity, and increased seed vigor index	India	Celsia and Mala (2014)
Nanofertilizers (Zn, Fe, and Mn) + Biofertilizers (*Azotobacter*, *Azospirillum*, and *Pseudomonas*)	Wheat (*Triticum aestivum* L.)	Increased plant growth and yield (spike length, spike number, seed number, seed weight, etc.)	Iran	Mardalipour et al. (2014)
Nanofertilizer + seaweed (*Ascophyllum nodosum*)	Grapes (*Vitis vinifera* L.)	Significantly improved plant growth, yield, and berry quality and reduced the adverse effects of abiotic stresses.	Turkey	Sabir et al. (2014)
Nano Ag + nitroxin biofertilizers	Potato (*Solanum tuberosum* L.)	Increased tuber yield (in terms of number, diameter, and weight)	Iran	Tahmasbi et al. (2011)

- Awareness of the biomagnification and biodegradability transfer effects of applications of nano-biofertilizer on plants must be developed to acquire inclusive erudition of its toxicity.
- For validating the admissible and safety limit of nanoparticle dose, a scientific and government-based security appraisal should be commenced.

8 Conclusion

The application of chemical fertilizer causes a disturbance in the ecosystem and resulting adverse impacts on human health. However, nano-biofertilizers enhance growth, shelf life, productivity, nutritional value, and resistance against abiotic and biotic factors. Its application maintains the soil nutrient level and promotes the growth and yield of crops by triggering different mechanisms. They also significantly improve the quality and quantity of products. At a low concentration, they impart good outcomes and have a low probability of bioaccumulation in soil. Compared with a single dose of NPs or biofertilizers, they were found to be more effective in improving soil fertility and exemplify a unique combination of optimistic interactions. This chapter clearly displays that globally many researchers are working on NPs and biofertilizers. However, still, literature is scarce on the formulation, application, and mode of action of nano-biofertilizer at the molecular level of plants. Their amalgamation yields better results than their individual application. Their mutual integration should be augmented through its control delivery and release at the target site. Formulating a multifunctional nano-biofertilizer is necessary, and it should be employed in practice. Its efficiency is linked to the use of novel approaches and detailed experimentations. One must become cognizant of the mechanism of nano-biofertilizer for the development of agriculture. Hence, the future of improving agriculture via nanotechnology is very optimistic. It is an approach that is not only eco-safe but also economical for the development of sustainable agriculture.

References

Abbasifar A, Shahrabadi F, Valizadehk Kaji B (2020) Effects of green synthesized zinc and copper nano-fertilizers on the morphological and biochemical attributes of basil plant. J Plant Nutr 43:1104–1118

Adrees M, Khan ZS, Ali S, Hafeez M, Khalid S, Rehman MZU, Hussain A, Hussain K, Chatha SAS, Rizwan M (2020) Simultaneous mitigation of cadmium and drought stress in wheat by soil application of iron nanoparticles. Chemosphere 238:124681

Ahmadi-Majd M, Nejad AR, Mousavi-Fard S, Fanourakis D (2021) Postharvest application of single, multi-walled carbon nanotubes and nanographene oxide improves rose keeping quality. J Hortic Sci Biotechnol 97(3):1–15

Ahmed RH, Mustafa DE (2020) Green synthesis of silver nanoparticles mediated by traditionally used medicinal plants in Sudan. Int Nano Lett 10:1–14

Ahmed B, Syed A, Rizvi A, Shahid M, Bahkali AH, Khan MS, Musarrat J (2021a) Impact of metal-oxide nanoparticles on growth, physiology and yield of tomato (*Solanum lycopersicum* L.) modulated by *Azotobacter salinestris* strain ASM. Environ Pollut 269:116218

Ahmed S, Choudhury AR, Roy SK, Choi J, Sayyed RZ, Sa T (2021b) Biomolecular painstaking utilization and assimilation of phosphorus under indigent stage in agricultural crops. In: Antioxidants in plant-microbe interaction. Springer, pp 565–588

Ahemad M, Kibret M (2014) Mechanisms and applications of plant growth promoting rhizobacteria: current perspective. J King Saud Univ - Science 26:1–20

Akhavan Hezaveh T, Rahmani F, Alipour H, Pourakbar L (2020) Effects of foliar application of ZnO nanoparticles on secondary metabolite and micro-elements of camelina (*Camelina sativa* L.) under salinity stress. J Stress Physiol Biochem 16:54–69

Akhtar N, Ilyas N, Meraj TA, Aboughadareh AP, Sayyed RZ, Mashwani Z, Poczai P (2022) Improvement of plant responses by nanobiofertilizer: a step towards sustainable agriculture. Nano 12:965. https://www.mdpi.com/2079-4991/12/6/965

Akl MA, Amer MW (2021) Green synthesis of copper nanoparticles by Citrus Limon fruits extract, characterization and antibacterial activity. Chem Int 7:1–8

Ali N, Réthoré E, Yvin J-C, Hosseini SA (2020) The regulatory role of silicon in mitigating plant nutritional stresses. Plants 9:1779

Ali SS, Darwesh OM, Kornaros M, Al-Tohamy R, Manni A, El-Shanshoury AE-RR, Metwally MA (2021) Nano-biofertilizers: synthesis, advantages, and applications. In: Elsamahy T, Sun J (eds) Biofertilizers. Woodhead Publishing, Cambridge, MA, pp 359–370

Anjum S, Anjum I, Hano C, Kousar S (2019) Advances in nanomaterials as novel elicitors of pharmacologically active plant specialized metabolites: current status and future outlooks. RSC Adv 9:40404–40423

Baazaoui N, Sghaier-Hammami B, Hammami SBM, Khefacha R, Chaari S, Elleuch L, Messaoud M, Abdelly C (2021) A handbook guide to better use of nanoparticles in plants. Commun Soil Sci Plant Anal 52:287–321

Baba ZA, Hamid B, Sheikh TA, Alotaibi SH, El Enshasy HA, Ansari MJ, Zuan ATK, Sayyed RZ (2021) Psychrotolerant Mesorhizobium sp. isolated from temperate and cold desert regions solubilizes potassium and produces multiple plant growth promoting metabolites. Molecules 26:5758

Baragaño D, Forján R, Fernandez-Perez B, Ayala J, Afif E, Gallego JLR (2020) Application of biochar, compost and ZVI nanoparticles for the remediation of As, Cu, Pb and Zn polluted soil. Environ Sci Pollut Res 27:33681–33691

Barsaiya M, Singh DP (2018) Green synthesis of zinc oxide nanoparticles by Pseudomonas aeruginosa and their broad spectrum antimicrobial effects. J Pure Appl Microbiol 12:2123–2134

Bastami A, Amirnia R, Sayyed R, Enshasy H (2021) The effect of mycorrhizal fungi and organic fertilizers on quantitative and qualitative traits of two important Satureja species. Agronomy 11:1285

Basu A, Prasad P, Das SN, Kalam S, Sayyed RZ, Reddy MS, El Enshasy H (2021) Plant growth promoting rhizobacteria (PGPR) as green bioinoculants: recent developments, constraints, and prospects. Sustainability 13:1140

Battard N (2012) Convergence and multidisciplinarity in nanotechnology: laboratories as technological hubs. Technovation 32(3–4):234–244

Bhari R, Kaur M, Singh RS (2021) Chicken feather waste hydrolysate as a superior biofertilizer in agroindustry. Curr Microbiol 78:2212–2230

Bharde A, Rautray D, Bansal V, Ahmad A, Sarkar I, Yusuf SM, Sanyal M, Sastry M (2006) Extracellular biosynthesis of magnetite using fungi. Small 2(1):135–141

Bibi F, Ilyas N, Arshad A, Khalid A, Saeed M, Ansar S, Batley J (2022) Formulation and efficacy testing of bio-organic fertilizer produced through solid-state fermentation of agro-waste by *Burkholderia cenocepacia*. Chemosphere 291:132762

Bright JP, Maheshwari HS, Thangappan S, Perveeen K, Bukhari NA, Mitra D, Sayyed RZ, Mantinu A (2025) Biofilmed-PGPR: a next-generation bioinoculant for plant growth promotion in rice (*Oryza sativa* L.) under changing climate. Rice Sci 32. https://doi.org/10.1016/j.rsci.2024.08.008

Castro-Longoria E, Vilchis-Nestor AR, Avalos-Borja M (2011) Biosynthesis of silver, gold and-bimetallic nanoparticles using the filamentous fungus *Neurospora Crassa*. Colloids Surf BBiointerfaces 83:42–48.

Caubet M, Cornu S, Saby NPA, Meunier J-D (2020) Agriculture increases the bioavailability of silicon, a beneficial element for crop, in temperate soils. Sci Rep 10:1999

Celsia ASR, Mala R (2014) Fabrication of nano structured slow release fertilizer system and its influence on germination and biochemical characteristics of *Vigna raidata*. Int J Chem Tech Res 6:4497–4503

Chung I, Rahuman AA, Marimuthu S, Kirthi AV, Anbarasan K, Padmini P, Rajakumar G (2017) Green synthesis of copper nanoparticles using *Eclipta prostrata* leaves extract and their anti-oxidant and cytotoxic activities. Exp Ther Med 14:18–24

Danish Toor M, Adnan M, Javed MS, Habibah UE, Arshad A, Mughees M, Ahmad R, Danish M (2020) Foliar application of Zn: best way to mitigate drought stress in plants. A review Int J Appl Res 6:16–20

Dhir B (2017) Biofertilizers and biopesticides: eco-friendly biological agents. In: Kumar R, Sharma A, Ahluwalia S (eds) Advances in environmental biotechnology. Springer, pp 167–188

Dikshit A, Shukla SK, Mishra RK (2013) Exploring nanomaterials with PGPR in current agricultural scenario. Lap Lambert Academic Publishing, Saarbrücken. ISBN 978-3-659-36774-8 17

Dimkpa CO, Singh U, Bindraban PS, Elmer WH, Gardea-Torresdey JL, White JC (2019) Zinc oxide nanoparticles alleviate drought-induced alterations in sorghum performance, nutrient acquisition, and grain fortification. Sci Total Environ 688:926–934

Du C, Abdullah JJ, Greetham D, Fu D, Yu M, Ren L, Li S, Lu D (2018) Valorization of food waste into biofertiliser and its field application. J Clean Prod 187:273–284

Duhan JS, Kumar R, Kumar N, Kaur P, Nehra K, Duhan S (2017) Nanotechnology: the new perspective in precision agriculture. Biotechnol Rep 15:11–23

Ebrahiminezhad A, Zare-Hoseinabadi A, Sarmah AK, Taghizadeh S, Ghasemi Y, Berenjian A (2017) Plant-mediated synthesis and applications of iron nanoparticles. Mol Biotechnol 60:154–168

El-Desouky HS, Islam KR, Bergefurd B, Gao G, Harker T, Abd-El-Dayem H, Ismail F, Mady M, Zewail RMY (2021) Nano iron fertilization significantly increases tomato yield by increasing plants' vegetable growth and photosynthetic efficiency. J Plant Nutr 44:1649–1663

El-Ghamry AM, Mosa AA, Alshaal TA, El-Ramady HR (2018) Nanofertilizers vs. biofertilizers: new insights. Environ Biodivers Soil Secur 2:51–72

Elsheery NI, Helaly MN, El-Hoseiny HM, Alam-Eldein SM (2020) Zinc oxide and silicone nanoparticles to improve the resistance mechanism and annual productivity of salt-stressed mango trees. Agronomy 10:558

Enrico JM, Piccinetti CF, Barraco MR, Agosti MB, Eclesia RP, Salvagiotti F (2020) Biological nitrogen fixation in field pea and vetch: response to inoculation and residual effect on maize in the Pampean region. Eur J Agron 115:126016

FAO, Ifad, WFP (2015) The state of food insecurity in the world meeting the international hunger targets: taking stock of uneven progress. FAO, Rome

Faridvand S, Amirnia R, Tajbakhsh M, Enshasy HE, Sayyed RZ (2021) The effect of foliar application of magnetic water and nano, organic, and chemical fertilizers on phytochemical and yield characteristics of different landraces of fennel (*Foeniculum vulgare* Mill). Horticulturae 7:475. https://www.mdpi.com/2311-7524/7/11/475

Farnia A, Omidi M (2015) Effect of nano-zinc chelate and nano-biofertilizer on yield and yield components of maize (*Zea mays* L.), under water stress condition. Indian J Nat Sci 5(4614):4707

Fasusi O, Cruz C, Babalola O (2021) Agricultural sustainability: microbial biofertilizers in rhizosphere management. Agriculture 11:163

Fatima A, Singh S, Prasad SM (2020a) Interaction between copperoxide nanoparticles and plants: uptake, accumulation and phytotoxicity. In: Sustainable agriculture reviews. Springer, Cham, pp 143–161

Fatima F, Hashim A, Anees S (2020b) Efficacy of nanoparticles as nanofertilizer production: a review. Environ Sci Pollut Res, 28(2):1292-1303.

Fayaz M, Balaji K, Girilal M, Yadav R, Kalaichelvan PT, Venketesan R (2010) Biogenic synthesis of silver nanoparticles and their synergistic effect with antibiotics: a study against gram-positive and gram-negative bacteria. Nanomed Nanotechnol Biol Med 6:103–109

Ferioun M, Zouitane I, Bouhraoua S, Elouattassi Y, Belahcen D, Errabbani A, Louahlia S, Sayyed R, El-Ghachtouli N (2025) Applying microbial biostimulants and drought-tolerant genotypes to enhance barley growth and yield under drought stress. Front Plant Sci 15:1494987. https://doi.org/10.3389/fpls.2024.1494987

Gatahi DM, Wanyika H, Kihurani AW, Ateka E, Kavoo A (2015) Use of bio-nanocomposites in enhancing bacterial wilt plant resistance, and water conservation in greenhouse farming. In: The 2015 JKUAT scientific conference. Agricultural sciences, technologies and global networking, vol 41, p 52

Gade A, Bonde PP, Ingle AP, Marcato P, Duran N, Rai MK (2008) Exploitation of *Aspergillus niger* for synthesis of silver nanoparticles. J Biobaased Mater Bioenergy 2(3):1–5

Gajbhiye M, Kesharwani J, Ingle A, Gade A, Rai M (2009) Fungus mediated synthesis of silver nanoparticles and their activity against pathogenic fungi in combination with fluconazole. Nanomed Nanotechnol Biol Med 5:382–386

Ghooshchi F (2017) Influence of titanium and bio-fertilizers on some agronomic and physiological attributes of triticale exposed to cadmium stress. Global NEST J 19(3):458–463

Ghorbani F, Sanati AM, Maleki M (2015) Production of silica nanoparticles from rice husk as agricultural waste by environmental friendly technique. Environ Stud Persian Gulf 2:56–65

Golbashy M, Sabahi H, Allahdadi I, Nazokdast H, Hossein M (2017) Synthesis of highly intercalated urea-clay nanocomposite via domestic montmorillonite as eco-friendly slow-release fertilizer. Arch Agron Soil Sci 63:1

Gomez A, Narayan M, Zhao L, Jia X, Bernal RA, Lopez-Moreno ML, Peralta-Videa JR (2021) Effects of nano-enabled agricultural strategies on food quality: current knowledge and future research needs. J Hard Mater 401:123385

Gouda S, Kerry RG, Das G, Paramithiotis S, Shin H-S, Patra JK (2018) Revitalization of plant growth promoting rhizobacteria for sustainable development in agriculture. Microbiol Res 206:131–140

Greger M, Landberg T, Vaculík M (2018) Silicon influences soil availability and accumulation of mineral nutrients in various plant species. Plants 7:41

Grillo R, Fraceto LF, Amorim MJB, Scott-Fordsmand JJ, Schoonjans R, Chaudhry Q (2021) Ecotoxicological and regulatory aspects of environmental sustainability of nanopesticides. J Hard Mater 404:124148

Hafez E, Osman H, Gowayed S, Okasha S, Omara A, Sami R, El-Monem AA, El-Razek UA (2021) Minimizing the adversely impacts of water deficit and soil salinity on maize growth and productivity in response to the application of plant growth promoting rhizobacteria and silica nanoparticles. Agronomy 11:676

Hamid B, Zaman M, Farooq S, Fatima S, Sayyed R, Baba Z, Sheikh T, Reddy M, El Enshasy H, Gafur A (2021) Bacterial plant biostimulants: a sustainable way towards improving growth, productivity, and health of crops. Sustainability 13:2856

Han Y-Q, Wen J-H, Peng Z-P, Zhang D-Y, Hou M-L (2018) Effects of silicon amendment on the occurrence of rice insect pests and diseases in a field test. J Integr Agric 17:2172–2181

Helaly MN, El-Hoseiny H, El-Sheery NI, Rastogi A, Kalaji HM (2017) Regulation and physiological role of silicon in alleviating drought stress of mango. Plant Physiol Biochem 118:31–44

Herlekar M, Barve S, Kumar R (2014) Plant-mediated green synthesis of iron nanoparticles. J Nanoparticles 140614. https://doi.org/10.1155/2014/140614

Itelima JU, Bang WJ, Onyimba IA et al (2018) A review: biofertilizer—a key player in enhancing soil fertility and crop productivity. J Microbiol Biotechnol Rep 2:22–28

Jabeen N, Maqbool Q, Sajjad S, Minhas A, Younas U, Anwaar S, Nazar M, Kausar R, Hussain SZ (2017) Biosynthesis and characterisation of nano-silica as potential system for carrying streptomycin at nano-scale drug delivery. IET Nanobiotechnol 11:557–561

Jakien E, Spruogis V, Romaneckas K, Dautart A, Avižienyt D (2015) The bio-organic nano fertilizer improves sugar beet photosynthesis process and productivity. Zemdirbyste-Agric 102:141–146

Jakinala P, Lingampally N, Hameeda B, Sayyed RZ, Khan MY, Elsayed EA et al (2021) Silver nanoparticles from insect wing extract: biosynthesis and evaluation for antioxidant and antimicrobial potential. PLoS One 16(3):e0241729. https://journals.plos.org/plosone/article?id=10.1371/journal.pone.0241729

Janmohammadi M, Navid A, Segherloo AE, Sabaghnia N (2016) Impact of nano-chelated micronutrients and biological fertilizers on growth performance and grain yield of maize under deficit irrigation condition. Biologija 62:134–147

Jatav GK, De N (2013) Application of nano-technology in soil– plant system. Asian J Soil Sci 8:176–184

Kalam S, Basu A, Ahmad I, Sayyed RZ, El-Enshasy HA, Dailin DJ, Suriani NL (2020) Recent understanding of soil acidobacteria and their ecological significance: a critical review. Front Microbiol 11:580024

Kalteh M, Taj Alipour Z, Ashraf S, Aliabadi MM, Nosratabadi AF (2014) Effect of silica nanoparticles on basil (*Ocimum basili-cum*) under salinity stress. J Chem Health Risks 4:49–55

Khan MR, Rizvi TF (2017) Application of nanofertilizer and nanopesticides for improvements in crop production and protection. In: Ghorbanpour M, Manika K, Varma A (eds) Nanoscience and plant–soil systems. Soil biology, vol 48. Springer, Cham

Khan MN, Mobin M, Abbas ZK, AlMutairi KA, Siddiqui ZH (2017) Role of nanomaterials in plants under challenging environments. Plant Physiol Biochem 110:194–209

Khan ZS, Rizwan M, Hafeez M, Ali S, Adrees M, Qayyum MF, Khalid S, Rehman MZU, Sarwar MA (2020) Effects of silicon nanoparticles on growth and physiology of wheat in cadmium contaminated soil under different soil moisture levels. Environ Sci Pollut Res 27:4958–4968

Kumar V, Vaid K, Bansal SA, Kim KH (2020) Nanomaterial-based immunosensors for ultrasensitive detection of pesticides/herbicides: current status and perspectives. Biosens Bioelectron 165:112382

Kumari R, Singh DP (2016) Silver nanoparticle in agroecosystem: applicability on plant and risk-benefit assessment. In: Singh A, Prasad SM, Singh RP (eds) Plant responses to xenobiotics. ISBN: 978-981-10-2859-5, pp 293–305

Kumari M, Pandey S, Bhattacharya A, Mishra A, Nautiyal C (2017a) Protective role of biosynthesized silver nanoparticles against early blight disease in *Solanum lycopersicum*. Plant Physiol Biochem 121:216–225

Kumari R, Barsainya M, Singh DP (2017b) Biogenic synthesis of silver nanoparticle by using secondary metabolites from *Pseudomonas aeruginosa* DM1 and its anti-algal effect on *Chlorella vulgaris* and *Chlorella pyrenoidosa*. Environ Sci Pollut Res 24:4645–4654

Kumari R, Singh JS, Singh DP (2017c) Biogenic synthesis and spatial distribution of silver nanoparticles in the legume mungbean plant (*Vigna radiata* L.). Plant Physiol Biochem 110:158–166

Kusale S, Attar Y, Sayyed R, El Enshasy H, Hanapi S, Ilyas N, Elgorban A, Bahkali A, Marraiki N (2021a) Inoculation of *Klebsiella variicola* alleviated salt stress and improved growth and nutrients in wheat and maize. Agronomy 11:927

Kusale S, Attar Y, Sayyed R, Malek R, Ilyas N, Suriani N, Khan N, El Enshasy H (2021b) Production of plant beneficial and antioxidants metabolites by *Klebsiella variicola* under salinity stress. Molecules 26:1894

Lafmejani ZN, Jafari AA, Moradi P, Moghadam AL (2018) Impact of foliar application of copper sulphate and copper nanoparticles on some morpho-physiological traits and essential oil composition of peppermint (*Mentha piperita* L.). Herba Pol 64:13–24

Lawre S, Laware SL, Raskar S (2014) Influence of zinc oxide nanoparticles on growth, flowering and seed productivity in onion metal oxide nanomaterials view project influence of zinc and iron nano-partilces on chickpea: nodulation and growth view project flowering and seed productivity in onion. Int J Curr Microbiol Sci 3:874–881

Liu D-Y, Liu Y-M, Zhang W, Chen X-P, Zou C-Q (2019) Zinc uptake, translocation, and remobilization in winter wheat as affected by soil application of Zn fertilizer. Front Plant Sci 10:426

Logeswari P, Silambarasan S, Abraham J (2015) Synthesis of silver nanoparticles using plants extract and analysis of their antimicrobial property. J Saudi Chem Soc 19:311–317

Lopez-Lima D, Mtz-Enriquez A, Carrión G, Basurto-Cereceda S, Pariona N (2021) The bifunctional role of copper nano-particles in tomato: effective treatment for Fusarium wilt and plant growth promoter. Sci Hortic 277:109810

López-Vargas ER, Ortega-Ortíz H, Cadenas-Pliego G, de Alba RK, Cabrera de la Fuente M, Benavides-Mendoza A, Juárez-Maldonado A (2018) Foliar application of copper nanoparticles increases the fruit quality and the content of bioactive compounds in tomatoes. Appl Sci 8:1020

Ma D, Sun D, Wang C, Ding H, Qin H, Hou J, Huang X, Xie Y, Guo T (2017) Physiological responses and yield of wheat plants in zinc-mediated alleviation of drought stress. Front Plant Sci 8:860

Mahanta D, Rai RK, Mishra SD, Raja A, Purakayastha TJ, Varghese E (2014) Influence of phosphorus and biofertilizers on soybean and wheat root growth and properties. Field Crop Res 166:1–9

Mahanty T, Bhattacharjee S, Goswami M, Bhattacharyya P, Das B, Ghosh A, Tribedi P (2017) Biofertilizers: a potential approach for sustainable agriculture development. Environ Sci Pollut Res 24: 3315–3335

Mahawar H, Prasanna R, Gogoi R, Singh SB, Chawla G, Kumar A (2020) Synergistic effects of silver nanoparticles augmented *Calothrix elenkinii* for enhanced biocontrol efficacy against Alternaria blight challenged tomato plants. 3 Biotech 10:102

Mahendran D, Geetha N, Venkatachalam P (2019) Role of silver nitrate and silver nano-particles on tissue culture medium and enhanced the plant growth and development. In: In vitro plant breeding towards novel agronomic traits: biotic and abiotic stress tolerance. Springer, pp 59–74. ISBN 9789813298248

Mala R, Celsia AV, Bharathi SV, Blessina SR, Maheswari U (2017) Evaluation of nano structured slow release fertilizer on the soil fertility, yield and nutritional profile of *Vigna radiata*. Recent Pat Nanotechnol 11:50–62

Malusà E, Pinzari F, Canfora L (2016) Efficacy of biofertilizers: challenges to improve crop production. In: Singh D, Singh H, Prabha R (eds) Microbial inoculants in sustainable agricultural productivity. Springer, New Delhi, pp 17–40

Manjunatha SB, Biradar DP, Aladakatti YR (2016) Nanotechnology and its applications in agriculture: a review. J Farm Sci 29:1–13

Mardalipour M, Zahedi H, Sharghi Y (2014) Valuation of nano-biofertilizer efficiency on agronomic traits of spring wheat at different sowing date. An Int J Biol Forum 6:49–356

McGinnity P (2015) Silicon and its role in crop production. Available online: http://planttuff.com/wp-content/uploads/2015/12/silicon-agriculture-iiterature-rvw-1.pdf. Accessed on 10 Mar 2022

Meunier J-D, Sandhya K, Prakash NB, Borschneck D, Dussouillez P (2018) pH as a proxy for estimating plant-available Si? A case study in rice fields in Karnataka (South India). Plant Soil 432:143–155

Mikhak A, Sohrabi A, Kassaee MZ, Feizian M (2017) Synthetic nanozeolite/nanohydroxyapatite as a phosphorus fertilizer for German chamomile (*Matricaria chamomilla* L.). Ind Crop Prod 95:444–452

Mir S, Sirousmehr A, Shirmohammadi E (2015) Effect of nano and biological fertilizers on carbohydrate and chlorophyll content of forage sorghum (Speedfeed hybrid). Int J Biosci IJB 6:157–164

Mishra VK, Kumar A (2009) Impact of metal nanoparticles on the plant growth promoting rhizobacteria. Digest J Nanomater Biostruct 4:587–592

Mishra S, Keswani C, Abhilash PC, Fraceto LF, Singh HB (2017) Integrated approach of agrinanotechnology: challenges and future trends. Front Plant Sci 8:471

Mohamadipoor R, Mahboub Khomami A (2013) Effect of application of iron fertilizers in two methods "foliar and soil application" on growth characteristics of Spathyphyllum illusion. Eur J Exp Biol 3:241–245

Monreal CM, De Rosa M, Mallubhotla SC, Bindraban PS, Dimkpa C (2016) Nanotechnologies for increasing the crop use efficiency of fertilizer-micronutrients. Biol Fertil Soils 52:423–437

Mozafari AA, Havas F, Ghaderi N (2018) Application of iron nanoparticles and salicylic acid in in vitro culture of strawberries (*Fragaria* × *ananassa* Duch.) to cope with drought stress. Plant Cell Tissue Organ Cult (PCTOC) 132:511–523

Moharrer S, Mohammad B, Gharamohammad RA, Yargol M (2012) Biological synthesis of sil-ver nanoparticles by *Aspergillus flavus*, isolated from soil of Ahar copper mine. Indian J SciTechnol 5:2443–2444

Morales-Díaz AB, Ortega-Ortíz H, Juárez-Maldonado A, Cadenas-Pliego G, González-Morales S, Benavides-Mendoza A (2017) Application of nanoelements in plant nutrition and its impact in ecosystems. Adv Nat Sci: Nanosci Nanotechnol 8: 013001 (13 pp)

Mukhopadhyay R, De N (2014) Nano clay polymer composite: synthesis, characterization, properties and application in rainfed agriculture. Global J Bio Biotechnol 3:133–138

Mushtaq T, Shah A, Akram W, Yasin NA (2020) Synergistic ameliorative effect of iron oxide nanoparticles and *Bacillus subtilis* S4 against arsenic toxicity in *Cucurbita moschata*: Polyamines, antioxidants, and physiochemical studies. Int J Phytoremediation 22:1408–1419

Mustafa G, Hasan M, Yamaguchi H, Hitachi K, Tsuchida K, Komatsu S (2020) A comparative proteomic analysis of engineered and bio synthesized silver nanoparticles on soybean seedlings. J Proteome 224:103833

Naderi MR, Danesh-Shahraki A (2013) Nanofertilizers and their roles in sustainable agriculture. Int J Agric Crop Sci 5:2229–2232

Nagar N, Devra V (2018) Green synthesis and characterization of copper nano-particles using Azadirachta indica leaves. Mater Chem Phys 213:44–51

Nasab BF, Sayyed RZ, Piri R, Rahmani AF (2021) Biopriming and nanopriming: green revolution wings to increase plant yield, growth, and development under stress condition and forward dimensions. In: Singh HB, Anukool V, Sayyed RZ (eds) Antioxidants in plant-microbe interaction. Springer, pp 623–655

Nasrollahzadeh M, Momeni SS, Sajadi SM (2017) Green synthesis of copper nanoparticles using *Plantago asiatica* leaf extract and their application for the cyanation of aldehydes using K4Fe(CN)6. J Colloid Interface Sci 506:471–477

Nayeri FD, Mafakheri S, Mirhosseini M, Sayyed R (2021) Phyto-mediated silver nanoparticles via *Melissa officinalis* aqueous and methanolic extracts: synthesis, characterization and biological properties against infectious bacterial strains. Int Adv Biol Biomed Res 3:270–285. http://www.ijabbr.com/article_245059.html

Nejatzadeh F (2021) Effect of silver nanoparticles on salt tolerance of *Satureja hortensis* l. during in vitro and in vivo germination tests. Heliyon 7:e05981

Nithyapriya S, Lalitha S, Sayyed RZ, Reddy M, Dailin D, El Enshasy H, Suriani NL, Herlambang S (2021) Production, purification, and characterization of bacillibactin siderophore of *Bacillus subtilis* and its application for improvement in plant growth and oil content in sesame. Sustainability 13:5394

Nithya R, Ragunathan R (2009) Synthesis of silver nanoparticle using pleurotus *Pleurotus sajor caju* and its antimicrobial study. Dig J Nanomater Bios 4:623–629

Öktem M, Keleş Y (2018) The role of silver ions in the regulation of the senescence process in Triticum aestivum. Turk J Biol 42:517–526

Panichikkal J, Prathap G, Nair RA, Krishnankutty RE (2021a) Evaluation of plant probiotic performance of *Pseudomonas* sp. encapsulated in alginate supplemented with salicylic acid and zinc oxide nano-particles. Int J Biol Macromol 166:138–143

Panichikkal J, Puthiyattil N, Raveendran A, Nair RA, Krishnankutty RE (2021b) Application of encapsulated Bacillus licheniformis supplemented with chitosan nanoparticles and rice starch for the control of *Sclerotium rolfsii* in *Capsicum annuum* (L.) seedlings. Curr Microbiol 78:911–919

Paschalidis K, Fanourakis D, Tsaniklidis G, Tzanakakis VA, Bilias F, Samara E, Kalogiannakis K, Debouba FJ, Ipsilantis I, Tsoktouridis G (2021) Pilot cultivation of the vulnerable cretan endemic *Verbascum arcturus* L. (scrophulariaceae): effect of fertilization on growth and quality features. Sustainability 13:14030

Pérez-de-Luque A (2017) Interaction of nanomaterials with plants: what do we need for real applications in agriculture? Front Environ Sci 5:12

Pourjamshid SA (2021) Study the effect of iron, zinc and manganese foliar application on morphological and agronomic traits of bread wheat (Chamran cultivar) under different irrigation regimes. Environ Stress Crop Sci 14:109–118

Priyanka N, Geetha N, Ghorbanpour M, Venkatachalam P (2019) Role of engineered zinc and copper oxide nanoparticles in promoting plant growth and yield: present status and future prospects. In: Advances in phytonanotechnology. Academic, Cambridge, MA, pp 183–201

Prasad R, Bhattacharyya A, Nguyen QD (2017a) Nanotechnology in sustainable agriculture: Recent developments, challenges, and perspectives. Front Microbiol 8:1014.

Qureshi A, Singh DK, Dwivedi S (2018) Nano-fertilizers: a novel way for enhancing nutrient use efficiency and crop productivity. Int J Curr Microbiol App Sci 7:3325–3335

Rafiq M, Ali A, Malik MA, Hussain M (2010) Effect of fertilizer levels and plant densities on yield and protein contents of autumn planted maize. Pak J Agric Sci 47:201–208

Rahayu I, Darmawan W, Zaini LH, Prihatini E (2020) Characteristics of fast-growing wood impregnated with nanoparticles. J For Res 31(2):677–685

Rahman KMA, Zhang D (2018) Effects of fertilizer broadcasting on the excessive use of inorganic fertilizers and environmental sustainability. Sustainability. https://doi.org/10.3390/su10030759

Rahman A, Al Jabri AB, Jassim RAH, Jabar AK (2020) The effect of nano nitrogen and bio-fertilizer types on npk concentration in soil and okra plant. Plant Arch 20:4031–4037

Rajak J, Bawaskar M, Rathod D, Agarkar G, Nagaonkar D, Gade A, Rai M (2017) Interaction of copper nanoparticles and an endophytic growth promoter *Piriformospora indica* with *Cajanus cajan*. J Sci Food Agric 97:4562–4570

Rasouli H, Djordjević JP, Sayyed RZ, Zarayneh S, Jafari M, Nasab BF (2019) Nanoparticles: a new threat to crop plants and soil rhizobia? In: Hayat S, Pichtel J, Faizan M, Hayat FQ (eds) Sustainable agriculture reviews 41. Springer, pp 201–214

Saberi-Rise R, Moradi-Pour M (2020) The effect of *Bacillus subtilis* Vru1 encapsulated in alginate-bentonite coating enriched with titanium nanoparticles against *Rhizoctonia solani* on bean. Int J Biol Macromol 152:1089–1097

Sabir A, Yazara K, Sabira F, Karaa Z, Atilla Yazicib M, Goksu N (2014) Vine growth, yield, berry quality attributes and leaf nutrient content of grapevines as influenced by seaweed extract (*Ascophyllum nodosum*) and nanosize fertilizer pulverizations. Sci Hortic 175:1–8

Sadowski Z, Maliszewska GB, Polowczyk I, Kozlecki T (2008) Synthesis of silver nanoparticles using microorganisms. Mater Sci Pol 26:419–425

Singh P, Balaji R (2011) Biological synthesis and characterization of silver nanoparticles using thefungus *Trichoderma harzianum*. Asian J Exp Biol Sci 2:600–605

Sarkar J, Chattopadhyay D, Patra S, Deo SS, Sinha S, Ghosh M, Mukherjee A, Acharya K (2011) *Alternaria alternata* mediated synthesis of protein capped silver nanoparticles and their geno-toxic activity. Dig J Nanomater Biostruct 6:563–573

Sarkar D, Sankar A, Devika OS, Singh S, Shikha, Parihar M, Rakshit A, Sayyed RZ, Gafur A, Ansari MJ, Danish S, Fahad S, Datta R. (2021) Optimizing nutrient use efficiency, productivity, energetics, and economics of red cabbage following mineral fertilization and biopriming with compatible rhizosphere microbes. Sci Rep. 3;11(1):15680. doi: 10.1038/s41598-021-95092-6.

Salunkhe RB, Patil SV, Salunke BK, Patil CD, Sonawane AM (2011) Studies on silver accu-mulation and nanoparticle synthesis by *Cochliobolus lunatus*. Appl Biochem Biotechnol165:221–234

Safaei Z, Azizi M, Davarynejad G, Aroiee H (2014) The effect of foliar application of humic acid and nanofertilizer (pharmks®) on yield and yield components of black cumin (*Nigella sativa* L.). J Med Plants By-Prod 2:133–140

Sagar A, Rai S, Ilyas N, Sayyed RZ, Al-Turki AI, El Enshasy HA, Simarmata T (2022) Halotolerant rhizobacteria for salinitystress mitigation: diversity, mechanisms and molecular approaches. Sustainability 14:490

Savci S (2012) Investigation of effect of chemical fertilizers on environment. APCBEE Procedia 1:287–292

Semida W, Abdelkhalik A, Mohamed G, El-Mageed TA, El-Mageed SA, Rady M, Ali E (2021) Foliar application of zinc oxide nanoparticles promotes drought stress tolerance in eggplant (*Solanum melongena* L.). Plants 10:421

Sethi G, Behera KK, Sayyed R, Adarsh V, Sipra BS, Singh L, Alamro AA, Behera M (2025) Enhancing soil health and crop productivity: the role of zinc-solubilizing bacteria in sustainable agriculture. Plant Growth Regul. https://doi.org/10.1007/s10725-025-01294-7

Shabbir Z, Sardar A, Shabbir A, Abbas G, Shamshad S, Khalid S, Natasha MG, Dumat C, Shahid M (2020) Copper uptake, essentiality, toxicity, detoxification and risk assessment in soil-plant environment. Chemosphere 259:127436

Shah RK, Boruah F, Parween N (2015) Synthesis and characterization of ZnO nano-particles using leaf extract of *Camellia sinesis* and evaluation of their antimicrobial efficacy. Int J Curr Microbiol Appl Sci 4:444–450

Shah AA, Aslam S, Akbar M, Ahmad A, Khan WU, Yasin NA, Ali B, Rizwan M, Ali S (2021) Combined effect of *Bacillus fortis* IAGS 223 and zinc oxide nanoparticles to alleviate cadmium phytotoxicity in *Cucumis melo*. Plant Physiol Biochem 158:1–12

Shahbaz M, Akram A, Mehak A, Haq E, Fatima N, Wareen G, Aamir S, Hassan H, Poczai P, Sayyed RZ, Ilyas N (2023) Evaluation of selenium nanoparticles in inducing disease resistance against spot blotch disease and promoting growth in wheat under biotic stress. Plants 12(4):761. https://doi.org/10.3390/plants12040761

Sharma SB, Sayyed RZ, Trivedi MH, Gobi TA (2013) Phosphate solubilizing microbes: sustainable approach for managing phosphorus deficiency in agricultural soils. Springer Plus 2:587

Sharma P, Pant S, Dave V, Tak K, Sadhu V, Reddy KR (2019) Green synthesis and characterization of copper nano-particles by Tinospora cardifolia to produce nature-friendly copper nano-coated fabric and their antimicrobial evaluation. J Microbiol Methods 160:107–116

Sharma S, Kumari P, Shandilya M, Thakur S, Perveen K, Sheikh I, Ahmed Z, Sayyed R, Mastinu A (2025) The combination of α-Fe_2O_3NP and *Trichoderma* sp. improves antifungal activity against *Fusarium* wilt. J Basic Microbiol:e2400613. https://doi.org/10.1002/jobm.202400613

Shcherbakova EN, Shcherbakov AV, Andronov E, Gonchar LN, Kalenska S, Chebotar VK (2017) Combined pre-seed treatment with microbial inoculants and Mo nanoparticles changes composition of root exudates and rhizosphere microbiome structure of chickpea (*Cicer arietinum* L.) plants. Symbiosis 73:57–69

Shukla SK, Pandey A, Kumar R, Mishra RK, Dikshit A (2013) Prospects of nano-biofertilizer in horticultural crops of Fabaceae. Technical report: agricultural situation in India, Directorate of Economics and Statistics Department of Agriculture and Co-operation Ministry of Agriculture, Government of India C-1, Hut-ments, Dalhousie road, New Delhi-110001, pp 45–50

Shukla SK, Kumar R, Mishra RK, Pandey A, Pathak A, Zaidi M, Srivastava SK, Dikshit A (2015) Prediction and validation of gold nanoparticles (GNPs) on plant growth promoting rhizobacteria (PGPR): a step toward development of nano-biofertilizers. Nanotechnol Rev 4:439–448

Siddiqi KS, Rahman AU, Tajuddin HA (2018) Properties of zinc oxide nanoparticles and their activity against microbes. Nanoscale Res Lett 13:141

Singh P, Ghosh D, Manyapu V, Yadav M, Majumder S (2019) Synergistic impact of iron (iii) oxide nano-particles and organic waste on growth and development of *Solanum lycopersicum* plants: new paradigm in nanobiofertilizer. Plant Arch 19:339–344

Snehal S, Lohani P (2018) Silica nano-particles: its green synthesis and importance in agriculture. J Pharmacogn Phytochem 7:3383–3393

Soliman M, Qari SH, Abu-Elsaoud A, El-Esawi M, Alhaithloul H, Elkelish A (2020) Rapid green synthesis of silver nanoparticles from blue gum augment growth and performance of maize, fenugreek, and onion by modulating plants cellular antioxidant machinery and genes expression. Acta Physiol Plant 42:148

Sreelekshmi R, Siril EA, Muthukrishnan S (2021) Role of biogenic silver nanoparticles on hyperhydricity reversion in *Dianthus chinensis* L. an in vitro model culture. J Plant Growth Regul 41:23–39

Sumbul A, Ansari RA, Rizvi R, Mahmood I (2020) *Azotobacter*: a potential bio-fertilizer for soil and plant health management. Saudi J Biol Sci 27:3634–3640

Tahmasbi D, Zarghami R, Azghandi AV, Chaichi M (2011) Effects of nanosilver and nitroxin biofertilizer on yield and yield components of potato mini tubers. Int J Agric Biol 13:986–990

Talebi R, Drostkar E, Kanouni H (2016) Foliar application of Fe, Zn and NPK nano-fertilizers on seed yield and morphological traits in chickpea under rainfed condition article citation: foliar application of Fe, Zn and NPK nano-fertilizers on seed yield and morphological traits in chickpea under rainfed condition. J Resour Ecol 4:221–228

Teng Q, Zhang D, Niu X, Jiang C (2018) Influences of application of slow-release nano-fertilizer on green pepper growth, soil nutrients and enzyme activity. In: IOP conference series: earth and environmental science, IOP Publishing Ltd, Paris, France vol 208, p 012014

Thakur P, Thakur S, Kumari P, Shandilya M, Sharma S, Poczai P, Alarfaj AA, Sayyed RZ (2022) Nano-insecticide: synthesis, characterization, and evaluation of insecticidal activity of ZnO NPs against *Spodoptera litura* and *Macrosiphum euphorbiae*. Appl NanoSci 12:3835–3850. https://doi.org/10.1007/s13204-022-02530-6

Thirugnanasambandan T (2019) Advances and trends in nano-biofertilizers. SSRN:59. https://doi.org/10.2139/ssrn.3306998

Tripathi DK, Singh S, Singh VP, Prasad SM, Chauhan DK, Dubey NK (2016) Silicon nanoparticles more efficiently alleviate arsenate toxicity than silicon in maize cultiver and hybrid differing in arsenate tolerance. Front Environ Sci 4:46

Tripathi DK, Singh S, Singh VP, Prasad SM, Dubey N, Chauhan DK (2017) Silicon nanoparticles more effectively alleviated UV-B stress than silicon in wheat (*Triticum aestivum*) seedlings. Plant Physiol Biochem 110:70–81

Usman M, Farooq M, Wakeel A, Nawaz A, Cheema SA, Rehman H, Ashraf I, Sanaullah M (2020) Nanotechnology in agriculture: current status, challenges and future opportunities. Sci Total Environ 721:137778

Vafa Z, Sohrabi Y, Sayyed R, Suriani NL, Datta R (2021) Effects of the combinations of rhizobacteria, mycorrhizae, and seaweed, and supplementary irrigation on growth and yield in wheat cultivars. Plants 10:811

Vaghar MS, Sayfzadeh S, Zakerin HR, Kobraee S, Valadabadi SA (2020) Foliar application of iron, zinc, and manganese nanochelates improves physiological indicators and soybean yield under water deficit stress. J Plant Nutr 43:2740–2756

Vahabi K, Ali Mansoori G, Karimi S (2011) Biosynthesis of silver nanoparticles by fungus *Trichoderma reesei* (a route for largescale production of AgNPs). Insci J 1:65–79

Verma VC, Singh SK, Solanki R, Prakash, S (2011) Biofabrication of anisotropic gold nanotriangles using extract of endophytic *Aspergillus clavatus* as a dual functional reductant and stabilizer. Nanoscale Res Lett 6:16–22

Vedamurthy AB, Bhattacharya S, Das A, Shruthi SD (2021) Exploring nanomaterials with rhizobacteria in current agricultural scenario. In: Advances in nano-fertilizers and nano-pesticides in agriculture. Elsevier, Amsterdam, pp 487–503

Vejan P, Abdullah R, Khadiran T, Ismail S, Boyce AN (2016) Role of plant growth promoting rhizobacteria in agricultural sustainability—a review. Molecules 21(573):1–17

Wilson MA, Tran NH, Milev AS, Kannangara GSK, Volk H, Lu GQM (2008) Nanomaterials in soils. Geoderma 146:291–302

Yasmeen F, Razzaq A, Iqbal MN, Jhanzab HM (2015) Effect of silver, copper and iron nanoparticles on wheat germination. Int J Biosci 6:112–117

Yasmin H, Mazher J, Azmat A, Nosheen A, Naz R, Hassan MN, Noureldeen A, Ahmad P (2021) Combined application of zinc oxide nanoparticles and biofertilizer to induce salt resistance in safflower by regulating ion homeostasis and antioxidant defence responses. Ecotoxicol Environ Saf 218:112262

Yorseng K, Siengchin S, Ashok B, Rajulu AV (2020) Nanocomposite egg shell powder with in situ generated silver nanoparticles using inherent collagen as reducing agent. J Bioresour Bioprod 5(2):101–107
Zahedi SM, Moharrami F, Sarikhani S, Padervand M (2020) Selenium and silica nanostructure-based recovery of strawberry plants subjected to drought stress. Sci Rep 10:17672
Zhou P, Adeel M, Shakoor N, Guo M, Hao Y, Azeem I, Li M, Liu M, Rui Y (2021) Application of nano-particles alleviates heavy metals stress and promotes plant growth: an overview. Nano 11:26

Benefits and Environmental Risks of Using Nanopesticides in Agricultural Fields

2

P. Saranraj, Riyaz Sayyed, K. Gayathri, B. Lokeshwari, Abdel Rahman Mohammad Al-Tawaha, and P. Sivasakthivelan

Abstract

Nanopesticides are a future technology to traditional pesticides and have been an emerging innovative technology in pest control. Compared to traditional pesticides, nanopesticides are more effective and specific and hence are a significant innovation in pest control. The formulations enable precise pest control with minimal chemical application by blending nanoscale materials, which enhance the delivery, stability, and bioavailability of active ingredients. This development can reduce such environmental effects as contamination and harm to unforeseen organisms as well as make pest control mechanisms more effective. This evaluates the present status of nanopesticides and their action mechanisms, such as increased penetration, regulated release, and extended action. The necessity for complete risk analyses and safety measures is done by the critical analysis of pivotal issues such as nanoparticle toxicity, persistence in the environment, and regulatory problems. Nanopesticides provide improved pest management, lowered chemical use, and improved yields. They also carry ecological hazards such as soil toxicity, water pollution, and damage to non-target species. Their long-term ecological consequence needs regulation and ecological application in

P. Saranraj (✉) · K. Gayathri · B. Lokeshwari
PG and Research Department of Microbiology, Sacred Heart College (Autonomous), Tirupattur, India

R. Sayyed
Department of Biological Sciences and Chemistry, University of Nizwa, Nizwa, Oman

A. R. M. Al-Tawaha
Department of Biological Sciences, Al-Hussein Bin Talal University, Maan, Jordan

P. Sivasakthivelan
Tamil Nadu Agricultural University, Agricultural Research Station, Bhavanisagar, India

R. Sayyed et al. (eds.), *Nanobiostimulants in Innovative Agriculture*,
https://doi.org/10.1007/978-981-95-2477-8_2

nanopests. Environmental safety is necessary to ensure the proper use of nanopesticides. The balance between effectiveness, safety, and environmental stewardship, and the potential of nanopesticides to become a key component of sustainable agriculture. This chapter analyzes the position of nanopesticides in agriculture, including their environmental advantages, drawbacks, and benefits, such as precise pest control, chemical reduction, and enhanced crop yields, while considering possible risks for sustainable farming.

Keywords

Agricultural sustainability · Ecotoxicology · Soil health · Environmental risks · Nanopesticides · Pest control efficiency

1 Introduction

The increasing world population demands plenty of healthy food sources that can satisfy the needs of human beings on every continent. The application of fertilizers, pesticides, herbicides, and other agrochemicals to increase crop yields has negatively contributed to environmental balance, infection protection, and agrochemical demand (Duro et al. 2020). Pesticide use has decreased biodiversity, increased the defense against disease, stabilized nitrogen, and enabled pesticides to build up in crops of fruits and vegetables, cattle, and aquatic plants (Rasouli et al. 2019). Agriculture uses nanotechnology-based delivery systems and controlled release profiles of agrochemicals and fertilizers to boost crop yields (Watson et al. 2011).

The efficacy of fertilizers, agrochemicals, and nanopesticides is essential to agriculture (Gogos et al. 2012). Because they can have genotoxic and toxic effects, humans are at serious risk of being exposed to toxic agrochemicals that can cross biological barriers. They also reduce application losses, increase stability, increase leaf cover, and utilize fewer formulations or novel nanopesticides. Nanomaterials are intended to protect plants. Among the nanocarriers used as pesticide delivery systems are clay, mesoporous silica nanoparticles, polymers, and porous inorganic materials. The pesticide industry uses nanotechnology to protect crops by improving product performance and environmental stability (Akhtar et al. 2022).

Nanoparticles are tiny pieces of metal, organic molecules, carbon, or metal oxides that range in size from one to one hundred nanometers. They are more reactive and have a larger surface area, a higher surface charge, and better absorption efficiency. The use of nanotechnology in pesticides, sometimes referred to as "nanopesticides," has unique methods for boosting diffusion, efficiency, and selectivity. Nanopesticides are a fantastic agricultural innovation that uses tiny chemicals and novel techniques to kill particular pests. They improve efficiency while lowering environmental impact, and as food demands rise, more agricultural practices are required to avoid food insecurity (Sharma et al. 2025).

2 Nanotechnology in Agriculture

Precision agriculture employs nanotechnology to boost crop yields while managing processes according to environmental conditions (Akhtar et al. 2022). Nanotechnology is used in agriculture to boost crop yield through disease detection, water purification, nutrient balancing, fertilizer supply, gene transfer, pest control, crop growth promotion, seed treatment, germination, lowering hazardous chemical levels, and the use of nanosensors (Sanzari et al. 2019). Nanoparticles have enormous surface areas, mobility, thermal stability, spreading capacity, and spontaneous degradation due to their extremely small size (between 1 and 100 nm). This makes them very useful in agriculture (Fletcher et al. 2020). They enhance nitrogen transport, stabilize fertilizers and herbicides, and safeguard crops (Haller et al. 2019).

Nanofertilizers are far more effective than conventional fertilizers due to their more uniform nutrient distribution and faster plant absorption (Hazra and Purkait 2019). Nanofertilizers boost photosynthesis and agricultural productivity by promoting nutrient absorption by plant roots (Surendhiran et al. 2019). According to studies, because they serve as nutrient carriers, nanoparticles are essential for plant growth (Fang et al. 2015). Zinc oxide The use of nanoparticles as nanofertilizers increased soybean yields and improved their health when grown in soil deficient in zinc (Gunalan et al. 2012). Furthermore, it has been documented that nanomaterials have an impact on seed germination and growth (Dimkpa et al. 2013). One of these studies looked at how spinach seed germination and growth were affected by nano-titanium dioxide (nano-TiO_2) (Jayaseelan et al. 2012). Multi-walled carbon nanotubes (MWCNTs) enhanced tomato seedling germination by enhancing their ability to absorb water (Kantrao et al. 2017) (Fig. 2.1).

3 Criterion for the Selection of Nanomaterials in Agricultural Application

Chen et al. (2017) state that some key factors, such as cost, safety, compatibility with living organisms, and natural degradation, determine the choice of suitable culture materials. The tiny carriers, known as nanocarriers, differ based on whether their targets are weeds, plants, fungi, viruses, or insects. DeSouza et al. (2020) state that nanocarriers have to satisfy the chemical and physical requirements of their targets. The ability to release the nutrient, its rate of delivery, its efficacy in being absorbed by plants, the quantity and quality of the crop, the price, and the environmental residue are some of the requirements of nutrient delivery, according to Hwang et al. (2011). Nanocarriers must have surface, size, and shape properties due to the complexity of plant cell walls (Vaculikova et al. 2015). It is crucial when the polymers naturally degrade, as reported by Wang et al. (2016).

In the use of packaging foods, nanoparticles are chosen given their cost, safety, effect on the environment, as well as their technical functions. Thuesombat et al. (2014) established parameters for the choice of nanocarrier material to be utilized for functional food: low toxicity, solubility, surface tension, electric charge, ability

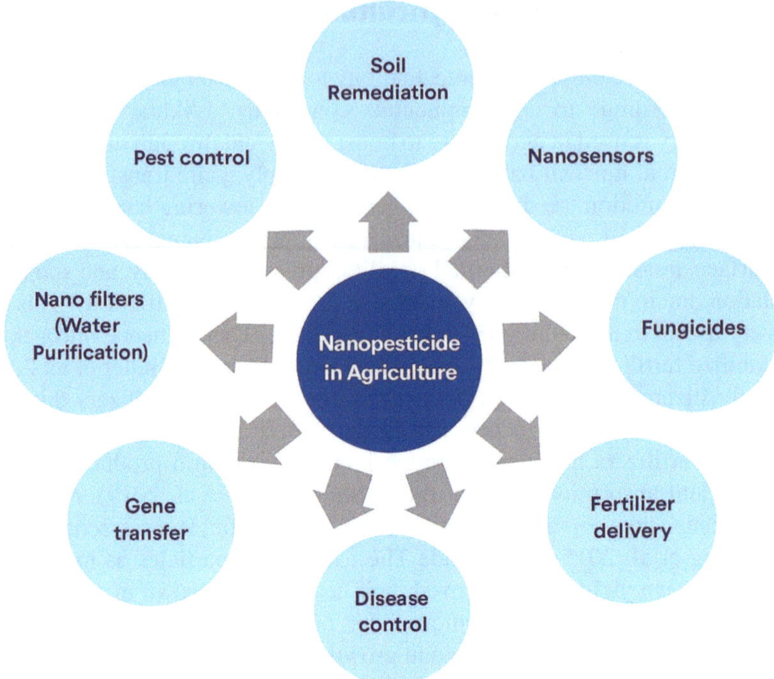

Fig. 2.1 Nanotechnology applications in precision farming

to create gel, source, cost factor, and limitations from sources under cultural as well as dietary limitations. Nanosensors use nanoparticles in food packaging to detect and remove toxic constituents and infections (Gahukar and Das 2020). Some of the benefits of nanosensors are that they are extremely sensitive, economical, compatible, and can detect in real time (Giunti et al. 2019).

4 Nanopesticides

According to Giunti et al. (2021), nanopesticides are pesticides applied in agricultural areas as nanoparticles. According to Hashem and Ramadan (2021), nanopesticides can be embedded into a matrix, functionalized nanocarriers for enzyme stimuli or external stimuli, or directly deposited onto a hybrid substrate. To explore pesticide processes in unconventional nanocarrier formulations, nanoscale materials made of a variety of various materials, including silica, lipids, polymers, copolymers, ceramics, metals, carbon, and others, should be used (Heydarzade et al. 2019). Nanopesticide formulations, as Emamjomeh et al. (2021) reported, have the potential to revolutionize crop management by improving bioavailability and water solubility and protecting agrochemicals from environmental degradation. Nanomaterials are genotoxic and cytotoxic but not toxic (Khoshraftar et al. 2020).

Due to the indiscriminate use of pesticides, nature has been disrupted and the health of all is disturbed. They were ingested unintentionally or intentionally; pesticide residues in water or food may lead to disability-adjusted life years or death (Kamaraj et al. 2018). According to Giongo et al. (2016), kids exposed to pesticides have a greater likelihood of suffering long-term damage to tissues and organs. The study of Jafer and Annon 2018 demonstrated significant peripheral and central neurotoxicity, as well as the impact on blood coagulation capacity. For the application of nanoparticles in agriculture to expand sustainably and safely, a proper analysis of the benefits and limitations influencing the toxicity and efficacy of nanoparticle pesticides must be conducted (Grillo et al. 2021). The formulation affects the efficiency of nanopesticides on farmers, consumers, ecosystems, and all other product segments of the agricultural value chain (Gottardo et al. 2021).

Nanoformulations are critical in increasing the biological availability of active constituents, stabilizing water solubility, and slowing down the rate at which active ingredients degrade. The amount and quality of food and crops produced in agriculture minimize plant damage, native insect pest infestation, and economic loss (Hua et al. 2015). Nanopesticides are transported to the water supply through runoff from industrial and agricultural wastewater during rain, where they leach into the soil, decrease the quality of the water, extend human exposure, and create environmental problems (Kah 2015). Due to their biomimetic characteristics and potential for bioaccumulation and dispersion in foods, water, and soil, as well as in all living organisms, but especially mammals, nanoparticles have been discovered to pose toxicological implications (Kah et al. 2016).

Acute and long-term pathological syndromes in humans, e.g., respiratory, cardiovascular, lymphatic, autoimmune, neurologic, and cancerous ones, may be induced by bioaccumulation and certain features of nanoparticles (Kah et al. 2018). Their appearance might require years, while in other instances, they come immediately. Their adverse impact occurs due to nanoparticles' personal susceptibility and exposure period (Kah et al. 2021). Two types of nanopesticide products are deployed onto farming fields: physically integrated and chemically attached (Table 2.1).

5 Mechanism of Action of Nanopesticides

Despite the potential toxicity of nanopesticides against insect pests, little is known about their exact mechanisms (Stadler et al. 2010). While toxicokinetics includes absorption, distribution, metabolism, and excretion, toxicodynamics describes the physiological, biochemical, and molecular actions of chemicals (Butler et al. 2018). There aren't many studies that look into how silver, metal, gold, alumina, silica, and graphene oxide nanoparticles work against insects (Margulis Goshen and Magdassi 2013). By controlling key insect genes, silver nanopesticides cause developmental and reproductive harm by blocking protein synthesis and gonadotropin release (Krishnaraj et al. 2012).

The phosphorus (P) in nucleic acids and sulphur (S) in proteins and metal nanoparticles reduce membrane permeability, which denatures organelles and

Table 2.1 Interaction types and carrier systems for nano-based pesticide delivery: formulations, pesticides, and references

Interaction type	Carrier system	Formulation	Pesticides loaded	References
Chemical-based systems				
Covalent Bonding	Hybrid nanocomposites	CNT-grafted-PCA	Zineb, Mancozeb	Sánchez-López et al. (2020)
Multifunctional assembly	Peptide-polymer conjugate	Trypsin-PEG hybrid	Oostatic hormone regulator	Wilson et al. (2019)
Ionic bonding	Hollow alginate spheres	Calcium-alginate matrix	Cypermethrin	Severino et al. (2019)
Electrostatic complexation	Polyelectrolyte clusters	Clay-gelatine, pGPMA-dsRNA	MCPA herbicide, dsRNA	Zhao et al. (2018)
Metal cluster integration	Metallic nanoparticles	Cu-TM, Cu^0, Ag^0 nanoclusters	Thiophanate-methyl fungicide	Ranganathan et al. (2019)
Physical-based systems				
Encapsulation	Coprecipitation/vesicles	Polyelectrolyte-surfactant mix	Trichlorfon, Acetochlor, Benzoylurea	Li et al. (2020)
Emulsion systems	Nanoemulsions/Liposomes	mPEG-PLGA, Alginate-Ca^{2+}	Pyrethrin, Citronella oil	Xiang et al. (2020)
Matrix embedding	Hybrid polymer matrix	mPEG-PLGA copolymer	Metolachlor herbicide	Schnoor et al. (2018)
Porous carrier	Sol-gel/biochar composite	Silica-ATP-biochar	Glyphosate, Benzoylurea-Fe_2O_3	Wang et al. (2015)
Polymeric foams	Alkylene-oxide emulsion	Poly(alkylene-oxide) foam	Glyphosate acid, Acetochlor	DeSouza et al. (2020)
Osmotic diffusion	Cellulose-PEG coating	Cellulose ester/PEG/salt	Diazinon insecticide	Fang et al. (2015)

causes cell death (Bryaskova et al. 2011). Trypsin activity is suppressed by gold nanoparticles, which impacts growth and reproduction (Ocsoy et al. 2013). Insects lose water through lipid and wax physical sorption, while silicon dioxide and aluminum oxide nanoparticles physically adhere to their cuticles (Stadler et al. 2010). Tribolium confusum is dehydrated by nanozeolite products that scratch its cuticle (Stadler et al. 2010). Through triboelectric forces, nanostructured alumina adheres to Sitophilus oryzae, absorbs wax coatings, and dehydrates (Butler et al. 2018).

6 Applications of Nanopesticides

Due to their unique characteristics, nanopesticides possess several potential uses in pest control and agriculture. They protect crops from pests such as bacteria, viruses, fungi, and insects, and their nanoscale and controlled release methods enhance pest control efficiency while minimizing pesticide use. They also provide long-term pest control, which minimizes the need for repeated applications and renders bug elimination less hazardous. Nanopesticides, which are produced specifically to act against diseases or pests, have low off-target effects, according to Servin et al. (2015). The prevention of pesticide resistance is also supported by the controlled release of active ingredients. Through their enhanced attachment to plant surfaces, they enhance field effect and plant persistence. Increasing biological pest control entails utilizing essential biological agents, like parasites or predators, to attack pests (Nayeri et al. 2021). In organic agriculture, nanopesticides may be a major alternative to traditional chemical pesticides, protecting organic agriculture methods.

7 Benefits of Using Nanopesticides in Agricultural Fields

7.1 Beneficial Enhancements of Nanopesticides to Control Insect Pests

Some nanomaterials find applications in nanoagriculture to increase crop yields. One of the sub-disciplines of nanotechnology is called nanoencapsulation, which provides various advantageous features, including lower human exposure to active components, controlled release, higher residual levels, exclusion of organic solvents, and increased efficacy. The discovery of nanoscience in pest management in insects has transformed encapsulation products by reducing dosages of pesticides as opposed to regular pesticides, promoting more target-oriented pesticide action and complete effect (Margulis Goshen and Magdassi 2013). The process entails wrapping the primary product (pesticide), placing it inside a shell or encapsulation product, and miniaturizing its size to the nanoscale. All aspects of the production, processing, packaging, storage, and transport of crops benefit significantly from nanotechnology. Compared to traditional pesticide delivery mechanisms, the new slow- or controlled-release insecticides developed due to nanotechnology are extremely efficient.

Controlled release systems are aimed at slowly releasing the expected amount of required and adequate quantities of agrochemicals and optimizing biological competence in a bid to avoid losses and adverse effects. In contrast to conventional pesticide compositions, which cause over 90% of the pesticides to vaporize into the air and adhere to treated food crops, controlled delivery systems release nanoparticles to give agrochemicals to the crops (Shahbaz et al. 2023). The effectiveness of the administration of pesticides to action points, including pests, diseases, and crops, is enhanced through delivery system control, which allows for continuous and extended release of the active agent at the point of action. Besides, by extending the shelf life, enhancing bio-efficacy, reducing chemical applications to crops, and avoiding light-induced photo-degradation of light-sensitive chemicals, fat-soluble chemicals are more dispersible and soluble in water solutions, minimizing the application of pesticides and the number of treatments (Thakur et al. 2022). Carbon nanotubes can efficiently replace traditional methods for crop enhancement, seed germination, and plant growth, as indicated in research (Kashyap et al. 2015).

7.2 Nanopesticides Can Enhance Crop Yield and Nutritional Value

Nanopesticides can improve food quality, nutritional quality, and farm yield (Elmer and White 2016). The protein, amino acid, vitamin, antioxidant, and organic acid composition of crops and their edible parts was greater (Tasnim et al. 2024). Various plants treated with nanopesticides exhibited enhanced sugar, fatty acid, chlorophyll, carotenoid, and major element levels. Potential explanations for these improvements include improved chlorophyll structure and function, improved light absorption and electron transfer efficacy, improved water and carbon dioxide uptake during photosynthesis, effective removal of excess ROS, reduced pathogenic activity, and improved plant immunity through beneficial plant metabolite composition modulation.

7.3 Target Pest Control and the Specificity of Nanopesticides

Pesticides that do not only affect the intended target, the active ingredient in nanopesticides, may be transported to the insect itself. To give the pest a long-lasting effect, the active ingredient can be encapsulated inside nanoparticles that release it gradually. By designing nanoparticles to deliver the active ingredient to a particular pest or stage of development, overall pest control efficiency can be increased (Mishra et al. 2014). Nanopesticides can become specific in a variety of ways, including targeted delivery systems. Most nanopesticides have controlled release mechanisms that allow the active ingredients to be released over an extended period. The controlled release increases pesticide efficacy and reduces the need for treatments. In comparison to traditional pesticides, nanoparticles' minuscule size facilitates efficient pest control by facilitating their entry into plant tissues or the exoskeletons of pests (Peteu et al. 2010). Another advantage of nanopesticides is that they have less of an adverse environmental impact. Nanopesticides reduce the amount of chemicals used and the chance of contaminating water and soil by improving the pesticide's action specificity (Nasab et al. 2021). Nanopesticides: As long as their safe application is ensured by routine evaluations, nanopesticides represent a major advancement in the development of more efficient and ecologically friendly pest control techniques.

7.4 Stress Resilience of Nanopesticides in a Rapidly Changing Climate

Agricultural productivity has been maintained by climate constraints like the increasing frequency of heatwaves, droughts, floods, and other climatic extremes. Up to 50% of the world's food production could be lost as a result of these abiotic stresses. To become more resilient to the quickly changing environment, agricultural practices must adapt (Elmer and White 2016). In the nanopesticides, system resilience and response to biotic and abiotic stresses may be improved. When heat stress is present, responsive nanopesticides can be altered to release active compounds. The total amount of active chemicals released from the nanocarriers increases as the temperature rises. For instance, a thermosensitive poly-dopamine–graphene oxide nanocarrier was used to encapsulate hymexazol. Within 10 min, the nanocarrier's dispersion temperature can rise quickly. This suggests that even at high temperatures, the nanofungicide can release hymenal in a temperature-controlled manner (Grillo et al. 2021). This is significant because the nanofungicide can increase crop growth by releasing active compounds during the summer when target infection susceptibility and competition for resources among competing organisms are at their highest.

Furthermore, in saline environments, responsive nanopesticides can efficiently dispense active compounds. MSN@2,4-D In comparison to pure water, nanoherbicide released 2,4-D more quickly and to a larger extent in an NaCl background solution. The release of 2,4-D is made possible by the electrostatic interaction between the positively charged trimethylammonium groups of MSNs and the negatively charged chloride ions (Servin et al. 2015). MSN@2,4-D nanoherbicide was used to drastically reduce target species without posing a risk of phototoxicity to non-target species, in addition to exhibiting pH- and temperature-dependent release. Additionally, MSNs can promote photosynthesis, which aids in plant growth. MSN-based nanopesticides are therefore regarded as an emerging solution to reduce some of the abiotic stressors since they can maintain or enhance crop growth.

7.5 Reduced Adverse Impacts from Nanopesticides on Non-target Organisms

Studies have demonstrated that nanopesticides are far safer than their non-nanoscale counterparts. The primary cause of this is the decreased capacity for non-target delivery, which lowers exposure to non-target organisms by providing active agents to target organisms and releasing them in response to specific stimuli. For example, essential oil bio-nanopesticides grafted with chitosan showed decreased ecotoxicity to fish species, decreased cytotoxicity to cell lines, and decreased phototoxicity to beans and maize. In both the target soil agroecosystem and the non-target wetland ecosystem, repeated exposures to long-term Cu(OH)2 Nanopesticide did not have any detrimental effects on soil biodiversity (Kashyap et al. 2015). Non-target organisms and unanticipated secondary effects on a variety of terrestrial and aquatic ecosystems must be considered in future evaluations of the environmental hazard posed by nanopesticides.

7.6 Minimized Premature Loss and Enhanced Foliage Adhesion of Active Ingredients

The UV radiation, photolysis, volatilization, and spray spread cause unnecessary loss of plant leaves and target organisms; traditional pesticides become ineffective. Compared to traditional pesticides, the lesser non-target delivery capacity of some compounds can notably decrease the initial loss of active ingredients from UV radiation, photolysis, and volatilization (Servin et al. 2015). The level of UV-blocking and anti-photolysis activity depends on the nanocarrier type. Their efficacy is further constrained by their weak adhesion and holding onto leaf surfaces. The decreased non-target distribution capacity can enhance their tackiness and anti-washing properties, which will enhance the efficacy of the chemical compounds' application. The contact angle on plant leaves is minimized by the active ingredients in nanocarriers. Also, modification of surface chemistry and functional groups of nanocarriers enhances the conjugation of active molecules. Amine groups with positive charges are held by strong hydrogen bonds, covalent bonds, and electrostatic forces to leaves, whereas carboxy groups are held by hydrogen bonds and electrostatic repulsion to leaves.

MSNs enhance the anti-washing activity of active ingredients and promote adherence to plant leaves. Compared to bare cyantraniliprole, MSNs enhanced the anti-washing activity of cyantraniliprole after simulated rainwater washing experiments (Dutta and Kaman 2017). The implications of these results suggest that surface modification of nanocarriers will be more effective in promoting adherence on leaf surfaces and the anti-washing activity of active ingredients against plants cultivated under adverse environmental conditions. Moreover, by avoiding lateral runoff and vertical leaching of nanopesticides in soils in cropland, the minimized non-target delivery potential could minimize collateral harm to surface and groundwater. The leaching and transportability of conventional insecticides in soil can be decreased by nanocarriers. After implantation within MSNs, the high leaching ability of 2,4-D was significantly decreased within soil-packed columns. Peteu et al. (2010) assert that since graphitized carbon immobilized atrazine due to physical contacts, chemical bindings, and π–π interactions, the overall atrazine leaching within the soil after rinses was decreased. These developments have tremendous benefits for sustainable agriculture.

7.7 Advantages of Nanopesticides in Post-harvest Management

The use of agrochemicals at the nanoscale in agriculture, such as nanopesticides, nanofertilizers, nanosensors, and nanoformulations, has completely changed conventional farming. According to Dutta and Kaman (2017), their increased toxicity may also result in a rise in pest penetration. If there are fewer active ingredients per spot, it can deliver more active chemicals that can remain active for longer. Furthermore, a lower dose requirement means a lower application cost. In contrast to conventional pesticide formulations, nanoformulations are designed to increase the solubility of insoluble or weakly soluble active chemicals and to release the biocide in a targeted and controlled manner.

7.8 Nanopesticides and Sustainable Agriculture

The prevention of climate change, the management of natural resources, and global food and nutritional security will all be aided by the integration of cultivation with the chemosphere. It is necessary to address the potential unintended and detrimental effects of nanomaterial use on water, soil microbes, and related ecosystems. The high cost of the input is the second main issue with agri-nanobiotechnology. The production of carbon dots, the most important nanomaterial, from raw biomass is a way to harvest renewable energy and use it in a variety of industries, including agriculture (Servin et al. 2015). The use of sustainable biomass and their photoluminescence mechanism using nanotechnology in agriculture has a lot of potential advantages, but releasing these products into the environment could have toxicological risks. To determine the fate of engineered nanoparticles in terms of their distribution, plant uptake, and the bioavailability of the micronutrients they contain, more research must be done. Because the application of nanoparticles in agriculture is more complicated than in other domains, these problems necessitate a longer procedure.

8 Advantages of Nanopesticides over Conventional Pesticides

Nanopesticides will overcome the shortcomings of current strategies and produce new formulations that are less expensive, lessen pest resistance, and are still effective against the target insect while being innocuous to organisms that are not the target. The benefits of using nanotechnology-based pesticide formulations have been confirmed by several studies (Ulrichs et al. 2006). Less volatility, increased stability of the nanoformulation, removal of hazardous organic solvents, gradual release of active ingredients, increased bioavailability, UV protection, increased mobility, increased insecticidal action, low dosage levels, and reduced environmental residues and runoff are all benefits of this approach. Insecticides made with nanotechnology are very effective, use little, and produce hydrophobic pesticides with better solubility (Timmer et al. 2008).

9 Impact of Nanopesticides and Their Environmental Risks (Fig. 2.2)

9.1 Adverse Effect of Nanopesticides in Plant Models

The effectiveness of nanoparticle absorption determines how it affects plant growth and metabolic processes. Through a variety of mechanisms, including phagocytosis, endocytosis, and pinocytosis, the plants absorb the nanoparticles. They react differently, but they can tolerate unpleasant situations like nanomaterial toxicity. Zein nanoparticles treated with neem oil have been shown to reduce the DNA damage index in Allium cepa (Kumari et al. 2017). In low-anthocyanin basil varieties, copper (II) hydroxide nanowires increased certain acids without stopping manganese deposition. Citric acid-capped cerium oxide nanoparticles, however, did not affect the tomato tissues' nutrient ratio. Moreover, β-cyclodextrin-grafted chitosan nanoparticles co-loaded with linalool and carvacrol did not cause phytotoxicity when applied to maize plants. Nanoparticles are harmful to crops because of their intrinsic characteristics, even though they are useful on farms. Numerous studies have confirmed the phytotoxicity of formulations used in commercial food crop nano-pest control products (Wen et al. 2005). However, nanopesticide reduced the amount of chlorophyll and biomass in maize leaves, as well as the antioxidant capacity of spinach and lettuce.

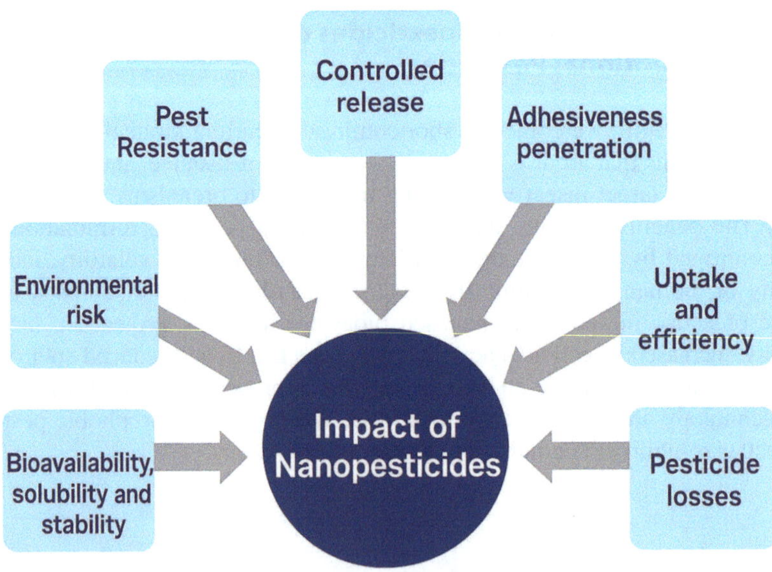

Fig. 2.2 Environmental interactions of nanopesticides

9.2 Adverse Effect of Nanopesticides in Non-target Animal Models

The major advances in agriculture have a significant impact on many non-target animals, populations, communities, and ecosystems. Their characteristics must be taken into account for the nano-enabled herbicides since they have the potential to significantly influence the exposure to non-target species. Atrazine nanocapsules altered the rate of Ca^{2+}-ATPase, raised the rate of DNA damage in erythrocytes, and dramatically decreased the hemoglobin content in fish after exposure. The harmful effects of pyrethrum extract-encapsulated nanoparticles in bullfrog tadpoles, which result in significant DNA damage and nuclear abnormalities, have been studied in a few studies (Peteu et al. 2010). Numerous findings, including a high mortality rate at the highest concentration of silver nanoparticles, have been reported in the majority of studies on the effects of silver nanoparticles in certain freshwater fish species. A higher proportion of protein and albumin and a decrease in white blood cells, red blood cells, hematocrit, and serum glucose concentration are the reproductive and developmental parameters that change.

9.3 Interaction of Nanopesticides with Other Environmental Substances/Pollutants

The engineered nanoparticles are becoming more common in the environment as their production and use increase. The use of nanopesticides in agriculture will increase (Servin et al. 2015). Due to their manufacturing, accumulation through transportation and bioaccumulation, and mixing with other environmental contaminants and dissolved organic compounds to further contaminate the environment, these substances are toxic to organisms that are not their intended target. Various nanoparticles have been discovered in a variety of environmental matrices, including soil, sediment, groundwater, and surface water (Zabrieski et al. 2015). Because of their higher surface area-to-volume ratio and capacity to alter their surface characteristics after being released into the environment, nanoparticles may become extremely dynamic in environmental systems. Nanoparticles can quickly accumulate in the food chain and have an impact on plant growth. This could put people and animals at risk for long-term effects from nanoparticle buildup. Pure active chemicals are used as active ingredients in environmental remediation nanoformulations, which alter their conventional formulations and nanoformulations, despite studies demonstrating the environmental risks posed by nanoparticles.

9.4 Impact of Nanopesticides on the Human

Pesticides can enter the body through a variety of routes, including inhalation of pesticide-containing aerosols, dust, vapor, food, water, and direct dermal contact through oral exposure. On the other hand, prolonged and careless use has detrimental impacts on well-being (Dutta and Kaman 2017). Because pesticides are nonspecific and poorly applied, humans—especially babies and children—are extremely vulnerable to their negative effects.

9.5 Nanopesticide-Induced Phytotoxicity

Through the suppression of their hydraulic conductivity, nanopesticides may limit the roots' ability to absorb and supply water, resulting in stomatal closure, decreased rates of transpiration and photosynthesis, wilting plants, and ultimately desiccation-induced death (Kashyap et al. 2015). According to experiments, after being exposed to certain nanoparticles, the transpiration process drastically decreased from the control values. At certain concentrations, nanoparticles also reduced the hydraulic conductivities of roots.

9.6 Nanopesticide Inhibits the Ectomycorrhiza and Nitrification Process in Soil

In the absence of ectomycorrhizal fungi in the roots of AgNP-treated plants, research revealed that the fine root growth, root biomass, root length, and root weight of pine trees were diminished as a result of the AgNPs' treatment of the soil. In contrast to the control plant roots, which had multiple ectomycorrhizal genera, the roots of pine plants in AgNP-contaminated soil had only one genus and no ectomycorrhizae. Lateral root production and mycorrhiza helper bacterial density were also decreased. Researchers also looked into how AgNPs affected biological nitrification (Margulis Goshen and Magdassi 2013). AgNPs have been shown to inhibit the nitrification process and bacteria that oxidize ammonia. The degree of nitrification suppression rose in tandem with AgNP levels and incubation duration. At a relatively low dose, nitrification also showed a hormetic effect because of the growth of the silver-tolerant bacterium (Gogos et al. 2012).

9.7 Impact of Nanopesticides on Soil Invertebrates

Enchytraeus crypticus is an invertebrate used as an indicator species in toxicological testing. The effect of the atrazine nanoformulation on this species was examined. The commercial atrazine formulation and the nanoformulation were identical. To assess the toxicity endpoints over a range of dosages, the organism's entire life cycle was used. In suppression experiments, the organism avoided the commercial formulation and pure atrazine, but there was no obvious avoidance of the nanoformulations (Dutta and Kaman 2017). In these experiments, the nanoformulation resulted in a decrease in both adults and juveniles. Pure atrazine reduced the number of juveniles in a dose-dependent manner and did not affect survival. Commercial production affected survival or reproduction.

9.8 Nanopesticides Restrict Degradation of Thiacloprid and Organic Pollutants

Thiacloprid degradation in $Cu(OH)_2$ NPs-contaminated soil was studied. $Cu(OH)_2$ Nanopesticides at low concentrations had a minimal impact on thiacloprid breakdown efficiency when compared to the control (Peteu et al. 2010). For example, after incubation with nanopesticides, residual thiacloprid percentages in the soil were greater than in the control group. This meant that since $Cu(OH)_2$ nanopesticides possess greater adsorption capacity and enhance the immobilization of organic pollutants, their use on soil minimized the degradation of thiacloprid.

9.9 Impact of Nanopesticides on Aquatic Fauna

The impact of lambda-cyhalothrin-loaded nanocapsules on Danio rerio zebrafish was shown. Though larger-sized microcapsules deposit at the bottom and release slowly, constituting a chronic and long-term risk to benthic organisms, nanoscale microcapsules exhibited excellent dispersity and the most rapid release profiles in aquatic bodies, showing acute toxicity towards a range of aquatic organism species. To compare the surviving larvae's morphological defects, certain responses were investigated. The characteristic notochord bending was evident for nanoparticles, and the malformation percentage was greater.

9.10 Impact of Nanopesticides on Industrial Insects and Pollinators

Experiments showed how CuO and ZnO nanoparticles influenced silkworms. NPs treatment decreased growth, body weight, survival, and cocoon yield; changed gene expression and antioxidant enzyme activity; inhibited food metabolism; and led to gut microbial dysbiosis. Production of peroxides, oxidative stress, cell and DNA damage, overexpression of genes involved in antioxidant biosynthesis, and down-regulation of ovarian and vitellogenin development-related genes were all detected following graphene oxide nanoparticle exposure. Wilson et al. (2019) also noted the reduction in reproduction rate, the number of unfertilized eggs produced, and the destruction of the ovarian tissues. Likewise, in honey bees, the nanopesticides resulted in enhanced expulsion of digestive cells into the lumen and enhanced cell death in the midgut.

10 Current Approaches and Strategies for Assessing the Environmental Risks of Nanopesticides

There are various degrees of complexity in the environment, including air, soil, and water. Consequently, as with traditional pesticides, nanopesticides may find their way into the environment unintentionally or intentionally, and caution is therefore essential when evaluating and controlling the risks they present. Issues regarding nanopesticides fate, toxicity, and biodegradability, together with environmental risk evaluation methods, are raised through growing interest in their application (Severino et al. 2019). As they were exposed to lower concentrations, engineered nanomaterials, including carbon nanotubes, silver, and zinc oxide, typically manifested relatively low risks. Nevertheless, additional applications of nanomaterials in agricultural activities or potential food incorporation might increase exposure rates and change the balance of benefits to risks. Nanostructures are multifaceted, including reactivity, size, shape, and electric charge. This renders it difficult to describe the biological safety of nanopesticides as well as to assess and estimate the required elements of cytotoxicity and genotoxicity.

The environmental hazard of a pesticide could be quantified in four ways: it entails simple exposure models, complex exposure models, indirect effects, biomagnification, and recovery, and it encompasses field monitoring of the pesticide concentration and its impacts. The ecotoxicological impacts of nanopesticides are unfortunately poorly reported, especially concerning the behavior and environmental fate of nanoformulations (Zhao et al. 2018). The toxicity, behavior, and fate of nanopesticide formulations, for example, have been investigated in various habitats and organ systems. Because of their variation in release kinetics, active ingredient solubility, and interactions with plants and soil, nanopesticides can sometimes not be tested using the same standardized ecotoxicological tests as traditional pesticides. Traditional pesticides are also likely to play a vital role in determining the fate of nanopesticides in the environment since they can interact with the organic or environmental pollutants already present. A variety of environmental and physicochemical parameters, including pH, temperature, salinity, soil type, porosity, water flow, ionic strength of water, mineral composition, microbial community, quantity and type of organic matter, and their dynamic interaction, may influence the fate of nanomaterials. These factors can also change the properties of the particles. Dispersion of traditional pesticides in water, soil, and air; stability of pesticides in water, soil, and sediment; environmental concentrations; bioaccumulation; and environmental toxicity are the pillars of the existing methods for evaluating their environmental hazard (Ranganathan et al. 2019). In addition, the uptake route of nanopesticides into the organism can be very different from that of the traditional pesticides in terms of sorption coefficients, half-life or half-dissipation time, octanol/water partition coefficient, bioconcentration factor, and hydrophobicity. Regulatory authorities apply a variety of standards for nanoparticles with varying size ranges because there is not yet a universally recognized standard for nanopesticides. The European Food Safety Authority and the Organisation for Economic Co-operation and Development are two of the bodies that have been engaged to address the constraints and bridge these gaps (Severino et al. 2019). Severe environmental issues such as neonicotinoids and DDT were previously endured because there was no organized assessment of the risks to the ecosystem.

11 Limitations and Challenges in the Use of Nanopesticides

Nano-based pesticides minimize environmental pollution through reduced rates of pesticide application and losses, which is an argument in their favor compared to traditional pesticides. Despite the possibility that nano-based pesticides could prove to be effective, there remain many unsettled issues. They include potential short- and long-term impacts on non-target species, biodiversity, soil and groundwater quality, and human and animal health. Nanodroplets evaporate upon emission because their droplet size is small. Interactions between nanopesticides with a wide variety of animals, plants, and microbes at diverse trophic levels are an essential area of investigation (Schnoor et al. 2018).

Commercialization of nano-based pesticides faces three fundamental challenges: acceptability in the market, legislative acceptability, and practical utilization. The characteristics of the nanocarriers and how the active ingredient diffuses within the nanoformulation matrix determine the release of active ingredients into the environment. The long-term release of nanoparticles has been found to impact non-target organisms. The use of non-biodegradable nanocarriers, like metal and metal oxides, has not been given much attention since most nanocarriers utilized in nanopesticides are natural polymers, polysaccharides, or short-duration degrading lipids (Khodakovskaya et al. 2009). Conventional nano-based pesticide formulations require the application of toxic organic solvents to produce pesticidal nanoformulations, and these solvents are of special concern to the environment. Large-scale applications of nano-based pesticide formulations have raised alarms among the scientific community regarding their possible ecological consequences and toxicity.

12 Conclusion

The development and formulation of nanopesticides represent a significant leap forward in modern agricultural practices. Through a diverse array of systems, including nanoemulsions, colloidal carriers, nanosuspensions, and various polymer-, lipid-, clay-, and metal-based nanostructures, scientists have been able to tailor pesticide delivery mechanisms with enhanced precision and efficiency. Each formulation type offers unique advantages, such as improved solubility, controlled release, better target specificity, and compatibility with different active ingredients. Nanogels, nanofibers, micelles, and chitosan nanoparticles, for instance, have also stretched the functional capacity of polymer-based delivery systems. Lipid-based nanoliposomes and solid lipid nanoparticles give biocompatible options for housing hydrophobic pesticides. Smart, sustainable crop protection solutions amenable to precision agriculture objectives result from these higher-order formulation tools. Still, while the advantages of nanopesticides like increased pest control effectiveness, less environmental pollution, and increased crop resistance, are becoming more apparent, the environmental and health risks posed need to be extensively analyzed. The difficulty still exists in guaranteeing compliance with regulation, identifying long-term ecological effects, and determining standardized risk assessment procedures. Thus, while nanopesticide technology is full of potential to revolutionize agriculture, its introduction needs to be steered by strong scientific, environmental, and ethical principles to ensure safe and sustainable application.

References

Akhtar N, Ilyas N, Meraj TA, Aboughadareh AP, Sayyed RZ, Mashwani Z, Poczai P (2022) Improvement of plant responses by Nanobiofertilizer: a step towards sustainable agriculture. Nano 12:965

Bryaskova R, Pencheva D, Dubois P, Georgieva N (2011) Synthesis and comparative study on the antimicrobial activity of hybrid materials based on silver nanoparticles (AgNPs) stabilized by polyvinylpyrrolidone (PVP). J Chem Biol 4(4):185–191

Butler M, Lopez Garcia G, Stadler T (2018) The potential of nanostructured alumina for leaf-cutting ants *Acromyrmex lobicornis* (Hymenoptera: Formicidae) management. Austral Entomol 57(3):292–296

Chen Y, Hong S, Fu CW, Hoang T, Li X, Valencia V, Zhang Z, Perman JA, Ma S (2017) Investigation of the mesoporous metal-organic framework as a new platform to study the transport phenomena of biomolecules. ACS Appl Mater Interf 9:10874–10881

DeSouza JF, da Silva PK, Alves TFR, Torqueti de Barros C, Amaral VA, de Moura Crescencio KM, Rios AC, Batain F, Souto EB, Severino P (2020) Structural comparison, physicochemical properties, and in vitro release profile of curcumin-loaded lyotropic liquid crystalline nanoparticle: influence of hydrotrope as interface stabilizers. J Mol Liq 306:112861

Dimkpa CO, McLean JE, Britt DW, Johnson WP, Arey B, Anderson AJ (2013) Antifungal activity of ZnO nanoparticles and their interactive effect with a biocontrol bacterium on growth antagonism of the plant pathogen *Fusarium graminearum*. Biometals 26(6):913–924

Duro JA, Lauk C, Kastner T, Erb KH, Haberl H (2020) Global inequalities in food consumption, cropland demand and land-use efficiency: a decomposition analysis. Glob Environ Chang 64:102124

Dutta P, Kaman P (2017) Nanocentric plant health management with special reference to silver. Int J Curr Microbiol Appl Sci 6(6):2821–2830

Elmer WH, White JC (2016) The use of metallic oxide nanoparticles to enhance the growth of tomatoes and eggplants in disease-infested soil or soilless medium. Environ Sci Nano 3(5):1072–1079

Emamjomeh L, Imani S, Talebi Jahromi K, Moharramipour S (2021) Nanoencapsulation enhances the contact toxicity of *Eucalyptus globulus* Labill and *Zataria multiflora* Bois's essential oils against the third instar larvae of *Ephestia kuehniella* (Lepidoptera: Pyralidae). Int J Pest Manag 17:1–9

Fang Z, Yang D, Gao Y, Li H (2015) pH-responsible Pickering emulsion and its catalytic application for reaction at water-oil interface. Colloid Polym Sci 293(6):1505–1513

Fletcher SJ, Reeves PT, Hoang BT, Mitter N (2020) A perspective on RNAi-based biopesticides. Front Plant Sci 11:51–59

Gahukar RT, Das RK (2020) Plant-derived nanopesticides for agricultural pest control: challenges and prospects. Nanotechnol Environ Eng 5(1):3–10

Giongo AMM, Vendramim JD, Forim MR (2016) Evaluation of neem-based nanoformulations as an alternative to control fall armyworm. Cienc Agrotec 40(1):26–36

Giunti G, Palermo D, Laudani F, Algeri GM, Campolo O, Palmeri V (2019) Repellence and acute toxicity of a nano-emulsion of sweet orange essential oil toward two major stored grain insect pests. Ind Crop Prod 142:111869

Giunti G, Campolo O, Laudani F, Zappalà L, Palmeri V (2021) Bioactivity of essential oil-based nano biopesticides toward *Rhyzopertha dominica* (Coleoptera: Bostrichidae). Ind Crop Prod 162:113257

Gogos A, Knauer K, Bucheli TD (2012) Nanomaterials in plant protection and fertilization: current state, foreseen applications, and research priorities. J Agric Food Chem 60(39):9781–9792

Gottardo S, Mech A, Drbohlavová J, Małyska A, Bowadt S, Sintes JR, Rauscher H (2021) Towards safe and sustainable innovation in nanotechnology: state-of-play for smart nanomaterials. NanoImpact 21:100297

Grillo R, Fraceto LF, Amorim MJ, Scott-Fordsmand JJ, Schoonjans R, Chaudhry Q (2021) Ecotoxicological and regulatory aspects of environmental sustainability of nanopesticides. J Hazard Mater 404:124148

Gunalan S, Sivaraj R, Rajendran V (2012) Green synthesized ZnO nanoparticles against bacterial and fungal pathogens. Prog Nat Sci Mater Int 22(6):693–700

Haller S, Widmer F, Siegfried BD, Zhuo X, Romeis J (2019) Responses of two ladybird beetle species (Coleoptera: Coccinellidae) to dietary RNAi. Pest Manag Sci 75(10):2652–2662

Hashem AS, Ramadan MM (2021) Nanoemulsions of chamomile and cumin essential oils as an alternative bio rational control approach against the red flour beetle, *Tribolium castaneum.* J Plant Prot Pathol 12(1):11–17

Hazra DK, Purkait A (2019) Role of pesticide formulations for sustainable crop protection and environment management: a review. J Pharmacogn Phytochem 8(3):686–693

Heydarzade A, Valizadegan O, Negahban M, Mehrkhou F (2019) Efficacy of *Mentha spicata* and *Mentha pulegium* essential oil nanoformulation on mortality and physiology of *Tribolium castaneum* (Col.: Tenebrionidae). J Crop Prot 8(4):501–520

Hua KH, Wang HC, Chung RS, Hsu JC (2015) Calcium carbonate nanoparticles can enhance plant nutrition and insect pest tolerance. J Pestic Sci 40(4):208–213

Hwang IC, Kim TH, Bang SH, Kim KS, Kwon HR, Seo MJ, Youn YN, Park HJ, Yasunaga-Aoki C, Yu YM (2011) Insecticidal effect of controlled release formulations of etofenprox based on nano-bio technique. J Fac Agric Kyushu Univ 56:33–40

Jafer FS, Annon MR (2018) Larvicidal effect of pure and green-synthesized silver nanoparticles against *Tribolium castaneum* (herb.) and *Callosobruchus maculatus* (fab.). J Glob Pharma Technol 10(3):448–454

Jayaseelan C, Rahuman AA, Kirthi AV, Venkatesan J, Kim SK (2012) Novel microbial route to synthesize ZnO nanoparticles using *Aeromonas hydrophila* and their activity against pathogenic bacteria and fungi. Spectrochim Acta A Mol Biomol Spectrosc 90:78–84

Kah M (2015) Nanopesticides and nanofertilizers: emerging contaminants or opportunities for risk mitigation? Front Chem 3:64–70

Kah M, Weniger AK, Hofmann T (2016) Impacts of (nano)formulations on the fate of an insecticide in soil and consequences for environmental exposure assessment. Environ Sci Technol 50(20):10960–10967

Kah M, Kookana RS, Gogos A, Bucheli TD (2018) A critical evaluation of nanopesticides and nanofertilizers against their conventional analogs. Nat Nanotechnol 13:677–684

Kah M, Johnston LJ, Kookana RS, Bruce W, Haase A, Ritz V, Dinglasan J, Doak S, Garelick H, Gubala V (2021) Comprehensive framework for human health risk assessment of nano pesticides. Nat Nanotechnol 16(9):955–964

Kamaraj C, Gandhi PR, Elango G, Karthi S, Chung IM, Rajakumar G (2018) Novel and environmentally friendly approach; the impact of neem (*Azadirachta indica*) gum nanoformulation (NGNF) on *Helicoverpa armigera* (hub.) and *Spodoptera litura* (fab.). Int J Biol Macromol 107:59–69

Kantrao S, Ravindra MA, Akbar SMD, Jayanthi PK, Venkataraman A (2017) Effect of biosynthesized silver nanoparticles on growth and development of *Helicoverpa armigera* (Lepidoptera: Noctuidae): interaction with midgut protease. J Asia Pac Entomol 20(2):583–589
Kashyap PL, Xiang X, Heiden P (2015) Chitosan nanoparticle-based delivery systems for sustainable agriculture. Int J Biol Macromol 77:36–51
Khodakovskaya M, Dervishi E, Mahmood M, Xu Y, Li Z, Watanabe F, Biris AS (2009) Carbon nanotubes can penetrate plant seed coat and dramatically affect seed germination and plant growth. ACS Nano 3(10):3221–3227
Khoshraftar Z, Safekordi AA, Shamel A, Zaefizadeh M (2020) Evaluation of the insecticidal activity of nanoformulation of *Melia azedarach* (leaf) extract as a safe environmental insecticide. Int J Environ Sci Technol 17(2):1159–1170
Krishnaraj C, Jagan EG, Rajasekar S, Selvakumar P, Kalaichelvan PT, Mohan N (2012) Effect of biologically synthesized silver nanoparticles on *Bacopa monnieri* L. Wettst Plant growth metabolism. Process Biochem 47(4):651–658
Kumari M, Pandey S, Bhattacharya A, Mishra A, Nautiyal CS, Singh HB (2017) Protective role of biosynthesized silver nanoparticles against early blight disease in *Solanum lycopersicum*. Plant Physiol Biochem 121:216–225
Li Z, Yu T, Paul R, Fan J, Yang Y, Wei Q (2020) Agricultural nanodiagnostics for plant diseases: recent advances and challenges. Nanoscale Adv 2(7):3083–3094
Margulis Goshen K, Magdassi S (2013) Nanotechnology: an advanced approach to the development of potent insecticides. In: Advanced technologies for managing insect pests. Springer, pp 295–314
Mishra S, Singh BR, Singh A, Keswani C, Naqvi AH, Singh HB (2014) Biofabricated silver nanoparticles act as a strong fungicide against *Bipolaris sorokiniana* causing spot blotch disease in wheat. PLoS One 9(5):e97881
Nasab BF, Sayyed RZ, Piri R, Rahmani AF (2021) Biopriming and nanopriming: green revolution wings to increase plant yield, growth, and development under stress condition and forward dimensions. In: Singh HB, Anukool V, Sayyed RZ (eds) Antioxidants in plant-microbe interaction. Springer, pp 623–655
Nayeri FD, Mafakheri S, Mirhosseini M, Sayyed R (2021) Phyto-mediated silver nanoparticles via *Melissa officinalis* aqueous and methanolic extracts: synthesis, characterization and biological properties against infectious bacterial strains. Int Adv Biol Biomed Res 3:270–285
Ocsoy I, Paret ML, Ocsoy MA, Kunwar S, Chen T, You M, Tan W (2013) Nanotechnology in plant disease management: DNA-directed silver nanoparticles on graphene oxide as an antibacterial against *Xanthomonas perforans*. ACS Nano 7(10):8972–8980
Peteu SF, Oancea F, Sicuia OA, Constantinescu F, Dinu S (2010) Responsive polymers for crop protection. Polymers 2(3):229–251
Ranganathan P, Mutharani B, Chen SM, Sireesha P (2019) Biocompatible chitosan-pectin polyelectrolyte complex for simultaneous electrochemical determination of metronidazole and metribuzin. Carbohydr Polym 214:317–327
Rasouli H, Djordjević JP, Sayyed RZ, Zarayneh S, Jafari M, Fazeli-Nasab B (2019) Nanoparticles: a new threat to crop plants and soil rhizobia. In: Hayat S, Pichtel J, Faizan M, Hayat FQ (eds) Sustainable agriculture reviews 41. Springer, pp 201–214
Sánchez-López E, Gomes D, Esteruelas G, Bonilla L, Lopez-Machado AL, Galindo R, Cano A, Espina M, Ettcheto M, Camins A (2020) Metal-based nanoparticles as antimicrobial agents: an overview. Nano 10(2):292–300
Sanzari I, Leone A, Ambrosone A (2019) Nanotechnology in plant science: to make a long story short. Front Bioeng Biotechnol 7:1–12
Schnoor B, Elhendawy A, Joseph S, Putman M, Chacón-Cerdas R, Flores-Mora D, Bravo-Moraga F, Gonzalez-Nilo F, Salvador-Morales C (2018) Engineering atrazine loaded poly (lactic-co-glycolic acid) nanoparticles to ameliorate environmental challenges. J Agric Food Chem 66:7889–7898

Servin A, Elmer W, Mukherjee A, De la Torre-Roche R, Hamdi H, White JC, Bindraban P (2015) A review of the use of engineered nanomaterials to suppress plant disease and enhance crop yield. J Nanopart Res 17(2):1–21

Severino P, da Silva CF, Andrade LN, de Lima OD, Campos J, Souto EB (2019) Alginate nanoparticles for drug delivery and targeting. Curr Pharm Des 25(11):1312–1334

Shahbaz M, Akram A, Mehak A, Haq E, Fatima N, Wareen G, Aamir S, Hassan H, Poczai P, Sayyed RZ, Ilyas N (2023) Evaluation of selenium nanoparticles in inducing disease resistance against spot blotch disease and promoting growth in wheat under biotic stress. Plants 12(4):761

Sharma S, Kumari P, Shandilya M, Thakur S, Perveen K, Sheikh I, Ahmed Z, Sayyed R, Mastinu A (2025) The combination of α-Fe2O3NP and *Trichoderma sp.* improves antifungal activity against Fusarium wilt. J Basic Microbiol 65:e2400613

Stadler T, Buteler M, Weaver DK (2010) Novel use of nanostructured alumina as an insecticide. Pest Manag Sci 66(6):577–579

Surendhiran M, Raja K, Jerlin R, Marimuthu S, Srivignesh S (2019) Nanoemulsion seed invigoration for improvement. Int J Agric 9(4):333–340

Tasnim A, Roy A, Akash SR, Ali H, Habib MR, Barasarathi J, Muthukumaran M, Sayyed RZ, Yeasmin T (2024) *Hibiscus sabdariffa* L. petal biomass: a green source of nanoparticles of multifarious potential. Open Agric 9:20220332

Thakur P, Thakur S, Kumari P, Shandilya M, Sharma S, Poczai P, Alarfaj AA, Sayyed RZ (2022) Nano-insecticide: synthesis, characterization, and evaluation of insecticidal activity of ZnO NPs against Spodoptera litura and Macrosiphum euphorbiae. Appl NanoSci 12:3835–3850

Thuesombat P, Hannongbua S, Akasit S, Chadchawan S (2014) Effect of silver nanoparticles on rice (*Oryza sativa* L. cv. KDML 105) seed germination and seedling growth. Ecotoxicol Environ Safe 104:302–309

Timmer LW, Dewdney MM, Chung KR (2008) 2009 Florida citrus pest management guide: melanose. Coop Ext Serv, Inst Food Agric Sci, Univ Florida, Gainesville

Ulrichs C, Goswami A, Mewis I, Schreiner M (2006) Electrostatic application of inert silica dust based insecticides onto plant surfaces. Commun Agric Appl Biol Sci 71(2):171–178

Vaculikova E, Placha D, Jampilek J (2015) Toxicology of drug nanocarriers. Chem List 109:346–352

Wang M, Chen Y, Zhang R, Wang W, Zhao X, Du Y, Yin H (2015) Effects of chitosan oligosaccharides on the yield components and production quality of different wheat cultivars (*Triticum aestivum* L.) in Northwest China. Field Crop Res 172:11–20

Wang T, Ren J, Qu G, Liang D, Hu S (2016) Glyphosate contaminated soil remediation by atmospheric pressure dielectric barrier discharge plasma and its residual toxicity evaluation. J Hazard Mater 320:539–546

Watson SB, Gergely A, Janus ER (2011) Where is agro nanotechnology heading in the United States and European Union. Nat Resour Environ 26:8–12

Wen LX, Chen JF, Li XL, Lu Y, Wang YH (2005) Controlled release of avermectin from porous hollow silica nanoparticles. Pest Manag Sci 61(6):583–590

Wilson DR, Suprenant MP, Michel JH, Wang EB, Tzeng SY, Green JJ (2019) The role of assembly parameters on polyplex poly (beta-amino ester) nanoparticle transfections. Biotechnol Bioeng 116(5):1220–1230

Xiang Y, Lu X, Yue J, Zhang Y, Sun X, Zhang G, Cai D, Wu Z (2020) Stimuli-responsive hydrogel as carrier for controlling the release and leaching behavior of hydrophilic pesticide. Sci Total Environ 722:137811

Zabrieski Z, Roberts DP, Dery PD, Meyer SLF, Vinyard B, Buyer JS (2015) Pesticidal activity of metal oxide nanoparticles on plant pathogenic isolates of *Pythium*. Ecotoxicol 24(6):1305–1314

Zhao X, Li J, Feng Y, Yu G, Zhou Q, He F, Xiao D, Chen K, Zhang L (2018) Self-aggregation behavior of hydrophobic sodium alginate derivatives in aqueous solution and their application in the nanoencapsulation of acetamiprid. Int J Biol Macromol 106:418–424

Response of Crop Microbiome to Nanoparticles

3

Nosheen Akhtar, Noshin Ilyas, Hina Mustafa, Muhammad Azhar Jameel, and Muhammad Ahsan Altaf

Abstract

Crop microbiomes are the key factors that decide plants' health and productivity. The association of microbiomes with the plants significantly affects the end product. Nanoparticles enhanced agricultural productivity, but their interaction with the crop microbiomes and with the host plant directly or indirectly modulates the microbial composition and their activities. They also play their role in changing the soil's physical and chemical properties. NPs are biphasic in nature, which means their role as either beneficial or phytotoxic is linked with the type of NPs, the concentration of NPs, the type of host species, soil type, and other environmental factors. NPs contain reactive surfaces, and due to their small size, they are used as fungicides, bactericides, and nanofertilizers. Different products, like nanofertilizers, nanopesticides, and encapsulated nanocarriers, have been developed. They are effective for the removal of weeds, pathogens, and insects and enhance the production of crops but sometimes create toxicities in the environment. This chapter summarizes the effects of commonly used NPs on plant microbiomes, including bacteria, viruses, and mycorrhiza. They change the

N. Akhtar
Department of Agricultural Sciences, Texas State University, San Marcos, TX, USA

Department of Botany, PMAS Arid Agriculture University, Rawalpindi, Pakistan

N. Ilyas (✉) · H. Mustafa
Department of Botany, PMAS Arid Agriculture University, Rawalpindi, Pakistan

M. A. Jameel
Department of Zoology, Wildlife & Fisheries, PMAS Arid Agriculture University, Rawalpindi, Pakistan

M. A. Altaf
Key Laboratory for Quality Regulation of Tropical Horticultural Crops of Hainan Province, Sanya Nanfan Research Institute, Hainan University, Sanya, China

R. Sayyed et al. (eds.), *Nanobiostimulants in Innovative Agriculture*,
https://doi.org/10.1007/978-981-95-2477-8_3

microbial composition, diversity, and soil properties as well. The positive and negative interactions of NPs with the microbes have been discussed in detail.

Keywords

AMF-plant symbiosis · Microbiome · Microbial diversity · Soil properties · Nanoparticles

1 Introduction

1.1 Overview of Crop Microbiomes

The microbes that live inside or outside the plant tissues are called plant microbiomes. They are essential to maintain plant growth. Generally, plant microbiomes include endophytes, epiphytes, plant pathogens, oomycetes, viruses, archaea, and arbuscular mycorrhizal fungi (Fig. 3.1). They are involved in various biological functions like nutrient acquisition, stress tolerance, and immune modulation. They play a significant role in fighting against biotic and abiotic stresses. For instance, plants develop a mutualistic relationship with microbes to fulfill the demands of nutrients, i.e., legume-rhizobium symbiosis and mycorrhiza (Coşkun et al. 2023).

Plant growth-promoting rhizobacteria are used to maintain sustainable agriculture (Sethi et al. 2025; Ferioun et al. 2025; Bright et al. 2025). They colonize the

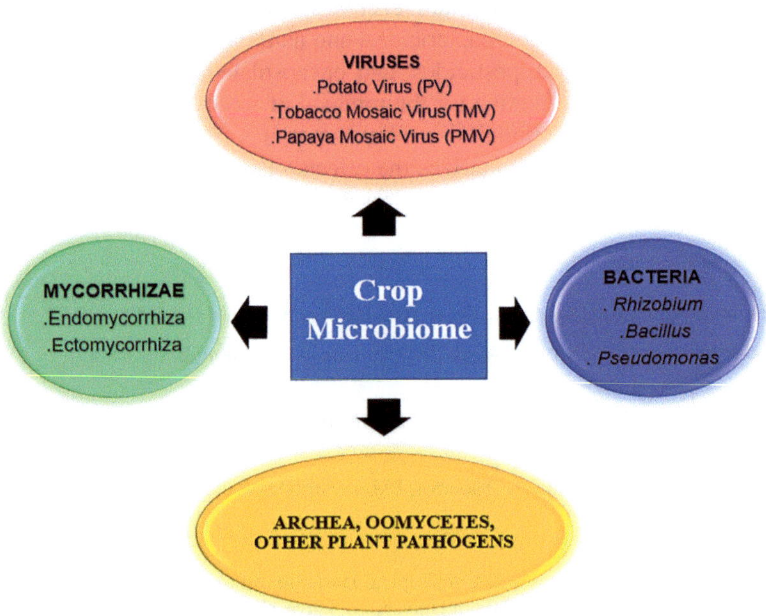

Fig. 3.1 Crop microbiome

rhizosphere and improve plant growth by various known mechanisms. A wide variety of bacterial strains were identified as PGPR, including *Aeromonas*, *Allorhizobium*, *Agrobacterium*, *Arthrobacter*, *Azospirillum*, *Azotobacter*, *Azoarcus*, *Azorhizobium*, *Bradyrhizobium*, *Caulobacter*, *Burkholderia*, *Chromobacterium*, *Enterobacter*, *Flavobacterium*, *Klebsiella*, *Mesorhizobium*, *Paenibacillus*, *Micrococcus*, *Pantoea*, *Pseudomonas*, *Rhizobium*, and *Streptomyces*. They synthesize auxin, gibberellins, cytokinin, osmolytes, siderophores, and other secondary metabolites that improve plant growth and tolerance (Pérez-García et al. 2023).

A large number of legumes (approximately 88% of species) and non-legume plants have symbiotic associations with nitrogen-fixing bacteria. The symbiotic association of legume plants with nitrogen-fixing bacterial strains evolved around 58 million years ago and annually converts 60 million tons of dinitrogen to ammonia. It reduces the requirement for synthetic nitrogen fertilizer. Various strains like Actinomycetes, Frankia, and alpha- or beta-proteobacteria develop a symbiotic relationship with the plant and fix atmospheric nitrogen. Legume plants develop nodules in their roots (Chen et al. 2021).

Mycorrhizas is the association between fungi and plant roots. In this relationship, fungi provide mineral nutrients (like phosphorus), water, and protection to plant roots and, in turn, get sugar. Approximately 80% of plant species have this association. This association creates resistance in plants against pathogen attacks. If aphids attack plants, they use the underground network for signal transfer and warn the nearby plants about the attack. The host plant synthesizes volatile organic compounds that attack insect predators. They also relocate the carbon compounds in case of herbivore attack. They are helpful in the colonization of plants in barren soil and also protect against heavy metal toxicity. They create resistance in plants against soil-borne pathogens. Mycorrhiza releases enzymes that are toxic to pathogens and activate the defense system in plants by modulating their immune system (Budamagunta et al. 2022; Liu et al. 2023; Wang et al. 2023).

Mycorrhizae increase the absorption area of plants through physical and chemical methods. In physical methods, the mycorrhizal mycelia are smaller in diameter than root hairs and penetrate to cover a larger area of soil for absorption than plant roots (Dave et al. 2024; Vafa et al. 2024). Chemically the fungi secrete organic acids that chelate ions and make them available to plants by ion exchange. In this association, fungi get a constant supply of glucose, sucrose, and carbohydrates. The interaction of soil, roots, and mycorrhiza creates a microenvironment. They play a significant role in improving soil physiochemical properties, plant health, and fitness. The external mycelium creates water-stable soil aggregates, which is a property of good soil tilth. They create changes in the root exudates that were responsible for maintaining the diversity and quality of microbes in the rhizosphere and are termed "mycorrhizosphere" (Fig. 3.2). They change hormonal balance, mineral nutrient composition, carbon allocation patterns, and some other features of plant physiology (Barea 2000).

Mycorrhizal associations have been classified under two categories, i.e., ectomycorrhizas and endomycorrhizas. Ectomycorrhiza occurs in angiosperms and some families of gymnosperms. It is estimated that ectomycorrhizal association

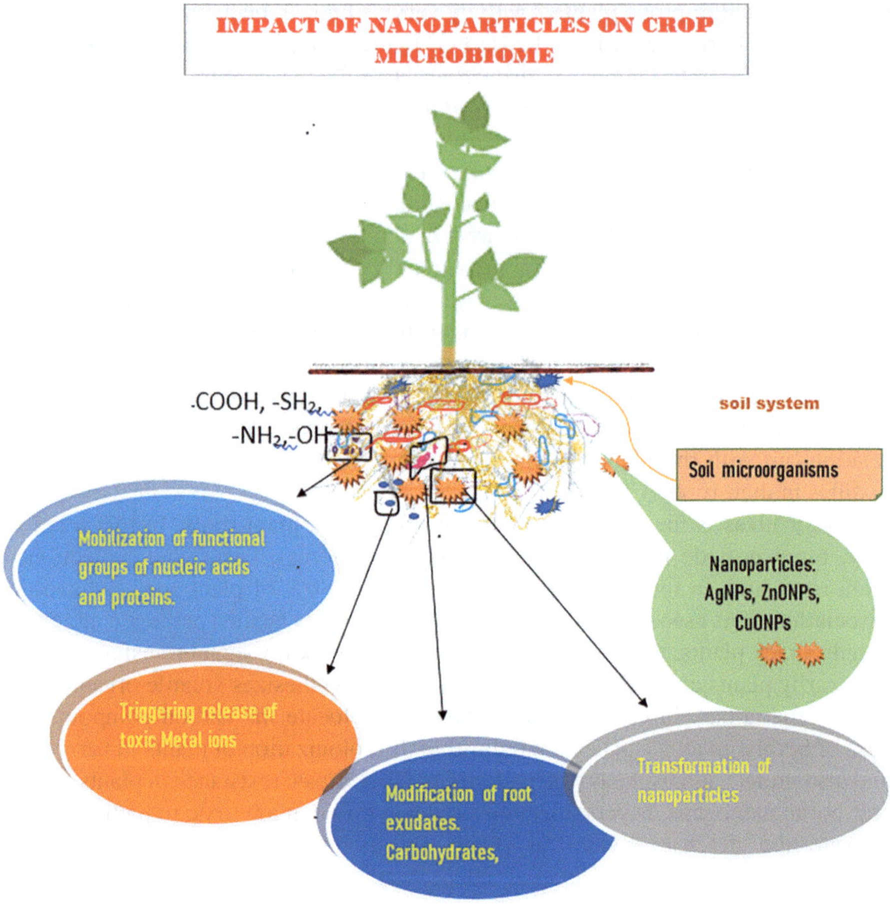

Fig. 3.2 Nanoparticles and soil microbial activities

developed nearly 130 million years ago. They form a sheath that encloses the rootlet, and hyphae form an interconnected network and penetrate between the root cells. This network is called the "Hartig net"; they did not penetrate intracellularly (Mohammadi et al. 2011). Endomycorrhizal association exists in 80% of plant species, including vegetables, crops, fruit trees, and grasses. The hyphae in endomycorrhiza penetrate the root cortex intracellularly and intercellularly. One of the major tasks of endomycorrhiza is nutrient acquisition. They improved the uptake of nitrogen, phosphorus, amino acids, and water. Endomycorrhiza has been categorized into five groups: ericoid, arbuscular, monotropoid, orchid, and arbutoid mycorrhizae (Feng et al. 2023; Natesan et al. 2023).

1.2 Definition and Concepts of NPs/Plants

Nanoparticles are defined as particles having a size between 1 and 100 nm. The first use of NPs was done in the nineteenth century in Mesopotamia to create glittering effects on the pots. The Ag and Cu NPs create shiny effects by homogeneous dispersal on the surface of pots (Sudha et al. 2018). Nanotechnology was applied at industrial levels initially, and then it was recently introduced in agriculture to enhance nutrient availability by providing an effective fertilizer. They are obtained from biopolymers like carbohydrates and proteins that are less toxic to the ecosystem. Plant extracts contain phenols, alkaloids, flavonoids, alcohols, sugar, and terpenoids, so they are used for the green/biosynthesis of NPs (Akhtar et al. 2022; Hassan et al. 2020; Jakinala et al. 2021). They contain reactive surfaces, and due to their small size, they can be used effectively as fungicides, bactericides, and nanofertilizers. Nanoparticles were introduced in agriculture for the proper development of agrosystems. Different products, like nanofertilizers, nanopesticides, and encapsulated nanocarriers, have been developed. They are effective for the removal of weeds, pathogens, and insects and enhance the production of crops. They are eco-friendly and cost-effective as well (Nasab and Sayyed 2021; Nayeri et al. 2021; Singh et al. 2021). NPs are also used for the transfer of DNA to create transgenic plants. They also produce transgenic seeds by transferring DNA into the pollen. Researchers are also trying to develop insect-resistant varieties by genetic transformation using NPs as a carrier of genes/DNA (Zhang et al. 2018).

There is not enough literature to indicate the effects of NPs on the ecosystem. Sometimes NPs create toxicities in the environment that affect plants, animals, and other living biotas; for example, they produce reactive oxygen species, or free radicals, that cause oxidative damage in living organisms. Nanopesticides are causing environmental pollution. Although they are very effective in the removal of pests, even at low concentrations they block the stomata in the leaves, make a layer on the stigma that inhibits pollen germination, enter the conductive tissues, and block the transport of food, water, and assimilation products (Altindal and Altindal 2020; Shahbaz et al. 2023; Thakur et al. 2022).

2 Interaction of Plant Microbiome with Nanoparticles

2.1 Bacteria

Plant growth-promoting (PGP) bacterial strains can sustain plant growth in unfavorable conditions, while nanoparticles augment plant growth in the presence of PGPR. The bacterial strains convert the nutrients from bound to available form in the rhizosphere, and NPs enhance the uptake of these nutrients. Bacterial strains and NPs act in a collaborative way to improve plant growth (Ma et al. 2022). Some nanoparticles have antibacterial properties; they act as pesticides and have very good potential against pathogens (Fig. 3.3) (Olga et al. 2022). Ferric oxide NPs have toxic effects on soil microbes. They have high mobility for electrons due to their

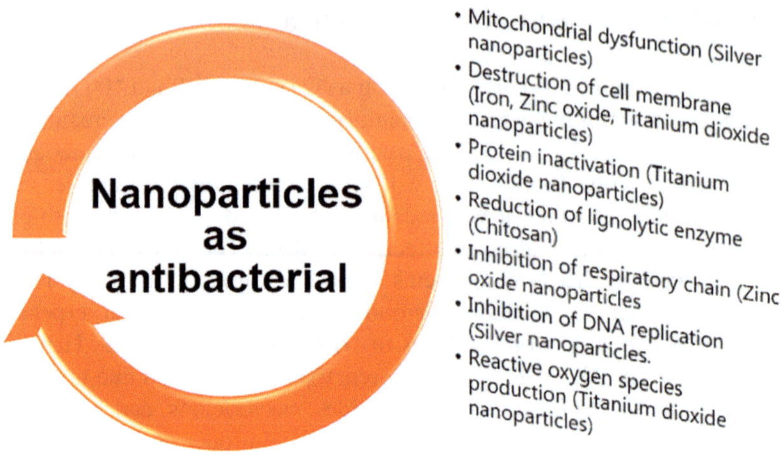

Fig. 3.3 Antibacterial activities of nanoparticles

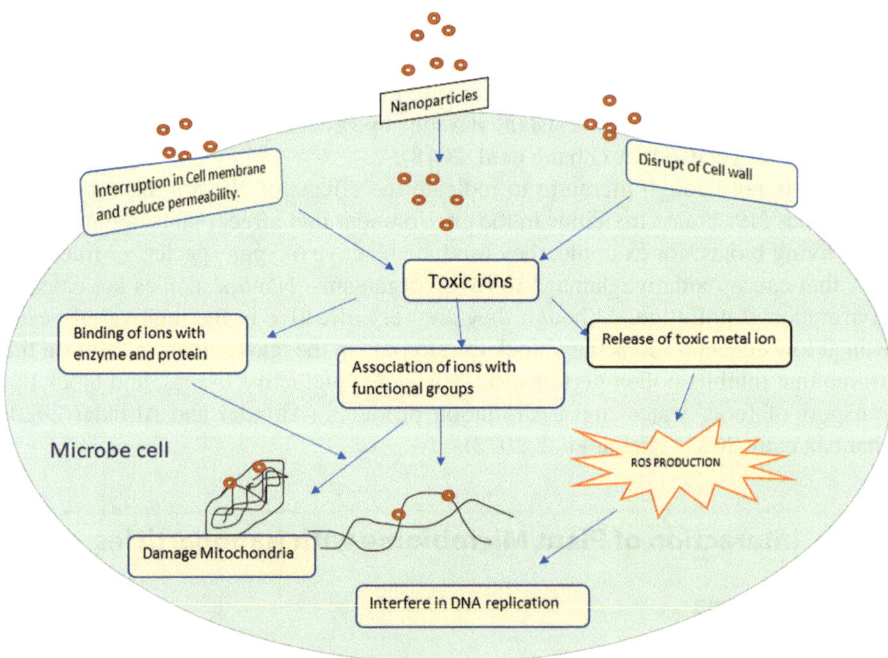

Fig. 3.4 Nanoparticle toxicity on microbial cells

improved surface-area-to-volume ratio. They release ferrous and ferric ions that are good at producing reactive oxygen species. These reactive oxygen species cause oxidative stress for microbes that deteriorate DNA, proteins, carbohydrates, and lipids, and in the end cause cell apoptosis (Fig. 3.4). Ferric oxide NPs decreased the amount of decaying organic matter in the soil (which is the main substrate for bacterial growth), which changed the soil biodiversity. For example, the growth of Acidobacteria decreased in low organic carbon, so the reduction in organic carbon by ferric oxide NPs may affect the composition and colonization of this strain (Hackl et al. 2004).

2.1.1 Rhizobium

Nanoparticles play their role in legume formation and biological nitrogen fixation. Rhizobium colonizes the rhizosphere and increases its colony number. They improve nitrogen fixation by nodule formation, while the inoculation of bacterial strains depends on the efficiency of the *Rhizobium* strain, host plant, and other climatic factors. The application of iron NPs enhanced the nodulation by *Rhizobium*, increasing nitrogenase activity and leghemoglobin content in *Cicer arietinum* (Abd-Alla et al. 2019). Iron NPs improved the symbiotic performance of *Rhizobium* with plants. They increase the number of nodules and size of nodules and improve nitrogenase activity, iron, and leghemoglobin content in plants. Analysis of nitrogen content shows that the nitrogen content in root and shoot was high in treated plants as compared to the control. Iron nanoparticles activate nodule formation. Gene expression in the *Rhizobium* leads to enhancing the number of nodules and nitrogen fixation (De Souza-Torres et al. 2021). An experiment was designed to analyze the alone and combined effect of zinc oxide nanoparticles and *Rhizobium leguminosarum* on plant pathogens like *Pseudomonas syringae* and *Meloidogyne incognita.* In alone application, the inoculation of *R. leguminosarum* and seed priming with zinc oxide nanoparticles improved plant growth and their photosynthetic activity. *R. leguminosarum* increases nodulation in plants, but zinc oxide NPs decrease the number of nodules and have negative effects on nodulation. The co-application of *R. leguminosarum* and zinc oxide nanoparticles was more effective in reducing the effect of disease and the number of nematodes (Kashyap and Siddiqui 2022). The application of zinc oxide NPs has negative effects on legume-*Rhizobium* symbiosis. They decreased the development of roots in *Pisum sativum* and also affected the rate of transpiration, stem growth, and surface area of leaves. They damage the cell surface of bacterial species, cause early senescence of root nodules, and disturb the process of nitrogen fixation (Huang et al. 2014). Silicon dioxide NPs and *R. leguminosarum* have some synergistic effects on plants against pathogens. Their combined application reduced the multiplication and damaging effects of *Pseudomonas syringae* and *Meloidogyne incognita* on *Pisum sativum.* They enhanced the number of nodules, carotenoid content, chlorophyll content, and growth of plants (Kashyap and Siddiqui 2021). The application of zinc oxide NPs also affects nodulation in *Lens culinaris* in the presence and absence of *Rhizobium* sp. The use of zinc oxide NPs and *Rhizobium* sp. fights against plant pathogens and improves the growth and development of plants (Siddiqui et al. 2018). The application of silver NPs supports the development of legume-*Rhizobium* symbiosis. In a 2-year field study, it was observed that low doses of silver NPs increased the development of root nodules and enhanced the growth and yield of *Glycine max.* This improvement can be linked with reduced ethylene production, a decrease in oxidative stress, and the development of systematic acquired resistance in plants (Krutyakov et al. 2022).

The application of metal oxide NPs modulates the synthesis of plant growth-promoting substances in nodule-forming *Rhizobium* sp. The effect of nanoparticles on *Rhizobium* sp. depends on various factors, but the most important are the type of nanoparticles and dose of nanoparticles. The low concentration of NPs did not affect the activities of *Rhizobium* sp., but the high concentration reduced the synthesis of hydrogen cyanide, indole acetic acid, ammonia, and siderophore. Exopolysaccharides are the polymers secreted by bacterial strains that provide them protection from unfavorable conditions. The application of metal oxide NPs enhanced the synthesis of exopolysaccharides (EPS) in rhizobia species. In a comparative study, it was observed that copper oxide nanoparticles effectively increased the production of EPS in bacteria as compared to zinc oxide and iron oxide NPs, i.e., 58%, 29%, and 46% improvement in the synthesis of EPS, respectively. It can be concluded that NPs interact with the *Rhizobium* sp. and modulate their physiological expression but need further research to explore the hidden knowledge (Oves et al. 2014).

Green synthesis of nanoparticles was also performed by using bacterial strains. Green synthesis involves the use of bacteria, fungi, and plant extract to reduce the size of salts at the nanoscale. It is a safe, inexpensive, and environmentally friendly technique. *R. leguminosarum* was used to synthesize silver NPs. Their synthesis was confirmed by Fourier transform infrared spectroscopy, dynamic light scattering, particle size analyzer, scanning electron microscopy, and energy dispersive X-rays. These characterization techniques revealed that the biosynthesized silver NPs have rough and spherical shapes and a size range from 13 to 40 nm. They have various functional groups that show their stability and capping. So, *Rhizobium* sp. is a good candidate for the synthesis of green nanoparticles (Abdelkhalek et al. 2022).

2.1.2 Bacillus

Bacillus subtilis produces various substances like proteins, peptides, and other molecules that are important for plant growth. The exogenous application of copper oxide NPs and *Bacillus subtilis* improved the germination, growth, and biomass of *Triticum aestivum* and *Zea mays*. They have synergistic effects on plants to support the growth of crops (Haider et al. 2022). Nanozeolite improves soil characteristics and has a good effect on *Bacillus* sp. The assessment of soil parameters like organic carbon, potassium, nitrate nitrogen, phosphorus, alkaline phosphate activity, and dehydrogenase fluorescein diacetate hydrolysis activity shows that nanozeolite and *Bacillus* sp. in combination improve soil health and growth of plants. They enhanced shoot length, photosynthesis, and total organic compounds (ketones, aldehydes, and phenols). GC-MS analysis of plants shows a high number of alcohols and acid esters in treated plants; they are responsible for stress tolerance in plants (Khati et al. 2018).

Bacillus pumilus, titanium, and zinc oxide NPs have a synergetic effect on plants. They create heavy metal tolerance-enhanced biomass, phytohormones, osmolytes, and antioxidant activities in plants. It was reported that titanium and zinc oxide NPs have an interaction with *Bacillus pumilus*. Titanium oxide NPs upregulate katX1, gls, KatX2, and glsA, while zinc oxide NPs upregulate treA, BPUM_071, and yqgT proteins. Some components like BPUM_0725, ykuU, ytbE, sodf, BPUM_1020yvgN, soda, BPUM_3004, and prkc did not show any association with the titanium and zinc oxide network. These protein network components are mainly involved in D-glutamate and D-glutamine metabolism, microbial metabolism, alanine-aspartate-arginine-proline-tryptophan metabolism, glyoxylate, and dicarboxylate metabolism pathways. These pathways play their role in the formation of carbohydrates from fatty acids, which results in better growth and development of *Bacillus pumilus* under unfavorable conditions. The titanium and zinc oxide NP network shows interaction with the proteins involved in reactive oxygen metabolism, oxidative stress, hydrogen peroxide metabolism, and various other pathways in plants. These pathways create stability and make the plant responsive in a stressful situation (Shafiq et al. 2022).

2.1.3 Pseudomonas

Pseudomonas sp. falls in the category of plant growth-promoting rhizobacteria that have the potential to improve plant growth during biotic and abiotic stress. Zinc oxide NPs in the presence of *Pseudomonas* sp. produce high amounts of antioxidants, including ascorbate peroxidase, peroxidase, superoxide dismutase, catalase, dehydroascorbate reductase, and glutathione reductase, in plants facing heat and drought stress (Eswaran et al. 2024). A significant reduction in the level of hydrogen peroxide, malondialdehyde content, and electrolyte leakage also confirmed the synergistic action of *Pseudomonas* sp. and zinc oxide NPs (Azmat et al. 2022). The application of silicon NPs *Pseudomonas putida* and *Pseudomonas fluorescence* increases the synthesis of primary and secondary metabolites that indicate their positive interaction. They improve plant tolerance by scavenging reactive oxygen species, enhancing nutrient uptake, improving the root system, and releasing root exudates that support the growth of microbial populations. Seed priming with this combination is also a good option that augments germination and supports crop establishment (Hatami et al. 2021).

2.2 Bacterial Consortium and Nanoparticles

Plant growth-promoting bacterial species have a synergistic effect on each other; they are more effective in improving plant growth as compared to the single strains. To analyze the effect of nanoparticles on microbes' population in the rhizosphere and to check their interaction, researchers performed various experiments. It was documented that the application of NPs with a bacterial consortium works together for the better growth of plants. Silver NPs, when applied in combination with *Bacillus cereus*, *Pseudomonas fluorescens*, and *Planomicrobium chinense*, enhanced the antioxidant production and reduced nickel, cobalt, manganese, and lead accumulation in plants. They enhance the uptake of nutrients and modulate the phytohormones (Faridvand et al. 2021; Khan and Bano 2016). The application of silicon NPs improves the physiochemical properties of soil and supports growth in unfavorable conditions. They increase soil water content by maintaining soil water holding capacity and improve osmoregulation, soil chemistry, and root exudates that further enhance the growth of microbes in the rhizosphere. They create a balance between calcium, potassium, and sodium ions, which are required for nutrient and water uptake in plants. The exopolysaccharides produced by PGPR in the presence of silicon NPs bound sodium ions in the saline soil and reduced their leaching and uptake in plants. They work together and support the development of plants in drought and salinity stress (Hafez et al. 2021). The application of silicon NPs has a positive interaction with *Azospirillum brasilense*, *Bacillus* sp., and *Azospirillum lipoferum*. They work together to improve the biomass, antioxidants, phytohormones, nutrients, and yield of *Triticum aestivum*. It was due to the increment in photosynthetic pigments, the rate of photosynthesis, and the activation of enzymes involved in sugar metabolism. They maintain plant turgor through osmolytes production and resume metabolic activities (Akhtar et al. 2021). The use of *Pseudomonas koreensis* and *Azotobacter chroococcum* with silica NPs improves microbial count (i.e., *Bacillus*, *Azotobacter*, etc.) and soil enzyme activity, including dehydrogenase and urease (Alharbi et al. 2022).

PGPR was also used to prepare nano-encapsulated biofertilizers that have some edge over the commonly used biofertilizers. They encapsulate PGPR and nanoparticles by using some other substances that protect PGPR from contamination and are helpful in the slow release of nanoparticles, making them more effective. They improve root length and proliferation in the soil. The encapsulation of *Bacillus subtilis* and *Pseudomonas fluorescence* with silica NPs and carbon nanotubes develops an effective nano-biofertilizer (Moradipour et al. 2019). Nanosilicab is also a nano-biofertilizer developed by using silicon NPs, *Azospirillum brasilense*, *Bacillus* sp., and *Azospirillum lipoferum*. It is a cost-effective, environment-friendly, and sustainable approach. They improved the germination attributes and morphological, physiological, and biochemical attributes of crops (Akhtar and Ilyas 2022).

2.3 Viruses

Approximately 50% of plant diseases were due to viruses, which are very difficult to cure. It is important to develop some strategies for the proper management of viral infections that damage economically important crops. Metal nanoparticles have the potential to work as antivirals due to their specific physiochemical properties (Vargas-Hernandez et al. 2020). It was reported that more than 700 species of crops have been affected by viruses (900 species). The rate of viral infection depends on various factors like virus concentration, rate of replication, stage of infection, type of viral species, and host species. They cause necrosis, drying of leaves, molting, malformation of fruits, and even death (Hill and Whitham 2014).

The application of silver NPs decreases the activity of the bean yellow mosaic virus. The NPs attach to the glycoproteins of the virus envelope and form a bond with the disulfide region. They restrict the fusion of the viral membrane with the host cell, so reducing the rate of infection. Apart from the interaction with the envelope, NPs interact with the viral nucleic acid and inhibit their activity in the host.

The synthesis of viral RNA and extracellular virion was also locked (Singh et al. 2022). Table 3.1 shows the effect of nanoparticle application on different plant viruses.

Plant viruses and nanoparticles have a very interesting interaction that is in favor of both humans and plants. It was documented that plant viruses accumulate on tumor cells and create an immune response in the surrounding cells, so plant virus-based nanoparticles were developed and are under examination to explore their effect on T cell activation against cancer cells. Virus-based nanoparticles are immunogenic and did not cause negative reactions (Peruzzi and Chiocca 2016). Plant viruses, including Hibiscus chlorotic ringspot virus, Brome Mosaic virus, Cowpea Mosaic Virus, Red clover necrotic virus, Physalis Mottle Virus, Johnson Grass Chlorotic Stripe Mosaic Virus, Potato Virus, Tobacco Mosaic Virus, and Papaya Mosaic Virus, were used for the virus-based nanoparticles that act as a carrier in the field of biomedicine for vaccine development (epitope) or targeted delivery for therapy (or diagnosis). Plant viruses are suitable for nanoscale application, as they are small in size and easy to produce, and they are structurally uniform, so the cargo molecules can be loaded into their cavity, or their outer surface can be modified (Lico et al. 2013).

2.4 Mycorrhiza and Plant Symbiosis

The effect of NPs on arbuscular mycorrhizal (AM) association with plants depends on various factors like the type of NPs, the concentration of NPs, the type of host species, AM species, and environmental conditions. Silicon dioxide NPs improve the colonization of AM in the rhizosphere of *Solanum tuberosum*. The application of iron NPs increases the solubility of phosphorus in the soil, which internally improved the nutritional status of AM fungi and plants (Wang et al. 2021). Zinc ferrite NPs and AMF have a positive interaction. They augmented plant growth by improving nutrient availability in the roots of plants. Zinc ferrite NPs slightly reduced AMF colonization, but their positive effect on plant growth was more pronounced; i.e., they enhanced the root fresh weight and shoot fresh weight by 91% and 74% in *Pisum sativum*, respectively (Abdelhameed et al. 2021).

The presence of NPs reduced the colonization intensity and richness and deteriorated the structure of mycorrhiza. They reduced the diversity and affected the community composition of the AMF-plant system. The effect of NPs on the ecosystem is more complicated; it needs detailed research to explore the hidden facts. NPs have antifungal properties; they attach to the cell wall of mycelium and damage the cell wall and cell membrane; they increase membrane permeability for NP entry and deposition while restricting the water channels. They are bound with the spores by van der Waals forces and inhibit the spores' germination, impairing the antioxidant defense system that results in the accumulation of free radicals (Tian et al. 2019). Iron oxide NPs and silver oxide NPs have a negative effect on AMF at the rate of 3.2 and 0.01 mg/kg, respectively, i.e., they reduced the nutrient availability and

Table 3.1 Pathogenic plant viruses and nanoparticles

Plant viruses	Treatment	Results	References
Potato virus Y (PVY) and tomato mosaic virus (ToMV)	Silver nanoparticles	Induce systematic acquired resistance Enhance total soluble protein, photosynthetic pigments, and antioxidants (polyphenol oxidase and peroxidase)	Noha et al. (2018)
Bean yellow mosaic virus	Silver nanoparticles	Inactivation of viruses by damaging their genomic material	Elbeshehy et al. (2015)
Anti-tomato bushy stunt virus (TBSV)	Graphene-based silver nanocomposites	Inhibit the replication and activity of viruses	Elazzazy et al. (2017)
Tobacco mosaic virus (TMV)	Zinc oxide nanoparticles and silica nanoparticles	Decreased viral colonization Inhibit viral replication Activate plant immune systems Upregulation of systemic resistance-related genes (like *PR1* and *PR2*) Enhance catalase and peroxidase activity Increase abscisic acid and salicylic acid hormones by 517% and 162%	Cai et al. (2019)
Tomato mosaic virus (ToMV)	Zinc oxide nanoparticles	Improved plant resistance against ToMV Enhanced plant growth Improve antioxidant activities Decrease hydrogen peroxide, oxygen-free radicals, and malondialdehyde content	Sofy et al. (2021)
Cucumber mosaic virus (CMV)	Nickel oxide nanostructures	Activation of defense-related genes Decrease in disease severity Reduction in viral accumulation and colonization Enhanced plant fresh and dry biomass	Derbalah and Elsharkawy (2019)
Turnip mosaic virus (TuMV)	Titanium and iron oxide nanoparticles	15% to 60% decrease in viral coat proteins inside the infected cells 40% increase in phytohormones that activate the defense system and improve plant growth 50% increase in shoot biomass	Hao et al. (2018)
Potato virus Y	Silver nanoparticles	Produce resistance in plants against the virus Improved plant growth	El-Shazly et al. (2017)

decreased the mycorrhizal growth (Feng et al. 2013). The effect of silver NPs is biphasic, i.e., their role as either beneficial or phytotoxic is linked with their dose. AMF reduced the damaging effects of silver NPs and augmented silver accumulation in plants; it (AMF) acts as a phytoremediator for the soil. AMF sequesters heavy metals in the extraradical mycelium and has glomalin proteins that perform metal chelation. NPs easily move inside the root cells and are translocated to aerial parts by the vascular system. AMF enhances metabolic activity and flow that supports the streamlined transport of silver NPs from root to shoot. They also control silver content and its allocation in the aerial parts of plants (Yang et al. 2022). Ferric oxide NPs modulate the composition of AMF, i.e., they enhance the proportion of Glomus and decrease the portion of Gigaspotacease and Acaulosporaceae, so they disturb the AMF and plant symbiosis. The low dose of ferric oxide NPs is less damaging for soil microbes and AMF. The AMF association combats the damaging effects of NPs on bacterial communities (Cao et al. 2016).

3 Conclusion and Recommendations

The positive effects of NPs on crops imply their importance for viable agriculture, but their effects on the ecosystem should be taken into consideration for maintaining a sustainable and safe environment. The interaction of NPs with soil microbiomes is complex and demands extensive and specific (species-based, dose, and type of NPs) research. It is a critical factor to determine the effect of NPs on soil microbes, as highly toxic NPs gave good responses at low concentrations, while significantly beneficial NPs impart damaging effects at high doses. The following suggestion will be helpful for future plans:

- Analyze the ecotoxicological effects and other aspects of NPs on large-scale field trials with comprehensive assessment (including the interaction of NPs with soil, microbes, and plants).
- Evaluate the effect of NPs on the structure and functionality of plant-microbe interaction.
- The role of naturally existing NPs on soil microbiota and experiments based on realistic agronomic conditions, including the use of chemical/biofertilizers.
- Accumulation and transformation of NPs in the fields with the passage of time and their role in the growth of soil microbes, soil fertility, and its physical and chemical properties.
- Proper rules and regulations should be developed and implemented for the application of nanoparticles as nanofertilizer or nanopesticides to control the release of NPs in the natural ecosystem.

References

Abd-Alla MH, Nafady NA, Bashandy SR, Hassan AA (2019) Mitigation of effect of salt stress on the nodulation, nitrogen fixation and growth of chickpea (*Cicer arietinum* L.) by triple microbial inoculation. Rhizosphere 10:100148. https://doi.org/10.1016/j.rhisph.2019.100148

Abdelhameed RE, Abu-Elsaad NI, Abdel Latef AA, Metwally RA (2021) Tracking of zinc ferrite nanoparticle effects on pea (*Pisum sativum* L.) plant growth, pigments, mineral content and arbuscular mycorrhizal colonization. Plants 10(3):583. https://doi.org/10.3390/plants10030583

Abdelkhalek A, Yassin Y, Abdel-Megeed A, Abd-Elsalam KA, Moawad H, Behiry SI (2022) *Rhizobium leguminosarum* bv. viciae-mediated silver nanoparticles for controlling bean yellow mosaic virus (BYMV) infection in faba bean plants. Plants 12(1):45. https://doi.org/10.3390/plants12010045

Akhtar N, Ilyas N (2022) Role of nanosilicab to boost the activities of metabolites in *Triticum aestivum* facing drought stress. Plant Soil:1–7. https://doi.org/10.1007/s11104-021-05285-1

Akhtar N, Ilyas N, Hayat R, Yasmin H, Noureldeen A, Ahmad P (2021) Synergistic effects of plant growth promoting rhizobacteria and silicon dioxide nano-particles for amelioration of drought stress in wheat. Plant Physiol Biochem 166:160–176. https://doi.org/10.1016/j.plaphy.2021.05.039

Akhtar N, Ilyas N, Meraj TA, Pour-Aboughadareh A, Sayyed RZ, Mashwani ZU, Poczai P (2022) Improvement of plant responses by nanobiofertilizer: a step towards sustainable agriculture. Nano 12(6):965. https://doi.org/10.3390/nano12060965

Alharbi K, Rashwan E, Mohamed HH, Awadalla A, Omara AE, Hafez EM, Alshaal T (2022) Application of silica nanoparticles in combination with two bacterial strains improves the growth, antioxidant capacity and production of barley irrigated with saline water in salt-affected soil. Plants 11(15):2026. https://doi.org/10.3390/plants11152026

Altindal N, Altindal D (2020) Agriculture and nanoparticles. In: Biogenic nano-particles and their use in agro-ecosystems. Springer, pp 59–66. https://doi.org/10.1007/978-981-15-2985-6_4

Azmat A, Tanveer Y, Yasmin H, Hassan MN, Shahzad A, Reddy M, Ahmad A (2022) Coactive role of zinc oxide nanoparticles and plant growth promoting rhizobacteria for mitigation of synchronized effects of heat and drought stress in wheat plants. Chemosphere 297:133982. https://doi.org/10.1016/j.chemosphere.2022.133982

Barea J (2000) Rhizosphere and mycorrhiza of field crops. In: Biological resource management connecting science and policy. pp 81–92. https://doi.org/10.1007/978-3-662-04033-1_7

Bright JP, Maheshwari HS, Thangappan S, Perveeen K, Bukhari NA, Mitra D, Sayyed RZ, Mantinu A (2025) Biofilmed-PGPR: a next-generation bioinoculant for plant growth promotion in rice (*Oryza sativa* L.) under changing climate. Rice Sci 2025(32). https://doi.org/10.1016/j.rsci.2024.08.008

Budamagunta V, Shameem N, Irusappan S, Parray JA, Thomas M, Marimuthu S, Kirubakaran R, Jothi KNA, Sayyed RZ, Lim HR (2022) Microbial nanovesicle and extracellular polymeric substances mediate nickel tolerance and remediate heavy metal ions from soil. Environ Res 219(15):114997. https://doi.org/10.1016/j.envres.2022.114997

Cai L, Liu C, Fan G, Liu C, Sun X (2019) Preventing viral disease by ZnONPs through directly deactivating TMV and activating plant immunity in *Nicotiana benthamiana*. Environ Sci Nano 6(12):3653–3669

Cao J, Feng Y, Lin X, Wang J (2016) Arbuscular mycorrhizal fungi alleviate the negative effects of iron oxide nanoparticles on bacterial community in rhizospheric soils. Front Environ Sci 4:10. https://doi.org/10.3389/fenvs.2016.00010

Chen QL, Hu HW, He ZY, Cui L, Zhu YG, He JZ (2021) Potential of indigenous crop microbiomes for sustainable agriculture. Nat Food 2(4):233–240. https://doi.org/10.1038/s43016-021-00253-5

Coşkun F, Alptekin Y, Demir S (2023) Effects of arbuscular mycorrhizal fungi and salicylic acid on plant growth and the activity of antioxidative enzymes against wilt disease caused by *Verticillium dahliae* in pepper. Eur J Plant Pathol 165(1):163–177. https://doi.org/10.1007/s10658-022-02596-6

Dave A, Ingle S, Perveen K, Bukhari NA, Sayyed R, Mastinu A (2024) Harnessing plant growth-promoting and wilt-controlling biopotential of a consortium of actinomycetes and mycorrhizae in pigeon pea. J Phytopathol 172:e13399. https://doi.org/10.1111/jph.13399

De Souza-Torres A, Govea-Alcaide E, Gómez-Padilla E, Masunaga SH, Effenberger FB, Rossi LM, López-Sánchez R, Jardim RF (2021) Fe3O4 nanoparticles and rhizobium inoculation enhance nodulation, nitrogen fixation and growth of common bean plants grown in soil. Rhizosphere 17:100275. https://doi.org/10.1016/j.rhisph.2020.100275

Derbalah AS, Elsharkawy MM (2019) A new strategy to control cucumber mosaic virus using fabricated NiO-nanostructures. J Biotechnol 306:134–141. https://doi.org/10.1016/j.jbiotec.2019.10.003

Elazzazy AM, Elbeshehy EK, Betiha MA (2017) In vitro assessment of activity of graphene silver composite sheets against multidrug-resistant bacteria and Tomato Bushy Stunt Virus. Trop J Pharm Res 16(11):2705–2711

Elbeshehy EK, Elazzazy AM, Aggelis G (2015) Silver nanoparticles synthesis mediated by new isolates of Bacillus spp., nanoparticle characterization and their activity against Bean Yellow Mosaic Virus and human pathogens. Front Microbiol 6:453. https://doi.org/10.3389/fmicb.2015.00453

El-Shazly MA, Attia YA, Kabil FF, Anis E, Hazman M (2017) Inhibitory effects of salicylic acid and silver nanoparticles on potato virus Y-infected potato plants in Egypt. Middle East J Agric Res 6:835–848

Eswaran SUD, Sundaram L, Perveen K, Bukhari NA, Sayyed RZ (2024) Osmolyte-producing microbial biostimulants regulate the growth of *Arachis hypogaea* L. under drought stress. BMC Microbiol 4(165). https://doi.org/10.1186/s12866-024-03320-6

Faridvand S, Amirnia R, Tajbakhsh M, Enshasy HE, Sayyed RZ (2021) The effect of foliar application of magnetic water and nano, organic, and chemical fertilizers on phytochemical and yield characteristics of different landraces of fennel (*Foeniculum vulgare* Mill). Horticulturae 7:475. https://www.mdpi.com/2311-7524/7/11/475

Feng Y, Cui X, He S, Dong G, Chen M, Wang J, Lin X (2013) The role of metal nanoparticles in influencing arbuscular mycorrhizal fungi effects on plant growth. Environ Sci Technol 47(16):9496–9504

Feng Z, Zhu H, Yao Q (2023) Multi-disciplinary nature of microbes in agricultural research: current status. In: Microbes in agri-forestry biotechnology, pp 1–34. CRC Press, Taylor and Francis Group, https://doi.org/10.1201/9781003110477

Ferioun M, Zouitane I, Bouhraoua S, Elouattassi Y, Belahcen D, Errabbani A, Louahlia S, Sayyed R, El-Ghachtouli N (2025) Applying microbial biostimulants and drought-tolerant genotypes to enhance barley growth and yield under drought stress. Front Plant Sci 15:1494987. https://doi.org/10.3389/fpls.2024.1494987

Hackl E, Zechmeister-Boltenstern S, Bodrossy L, Sessitsch A (2004) Comparison of diversities and compositions of bacterial populations inhabiting natural forest soils. Appl Environ Microb 70:5057–5065. https://doi.org/10.1128/AEM.70.9.5057-5065.2004

Hafez EM, Osman HS, Gowayed SM, Okasha SA, Omara AE, Sami R, Abd El-Monem AM, Abd El-Razek UA (2021) Minimizing the adversely impacts of water deficit and soil salinity on maize growth and productivity in response to the application of plant growth-promoting rhizobacteria and silica nanoparticles. Agronomy 11(4):676. https://doi.org/10.3390/agronomy11040676

Haider HI, Zafar I, Ain QU, Noreen A, Nazir A, Javed R, Sehgal SA, Khan AA, Rahman MM, Rashid S, Garai S (2022) Synthesis and characterization of copper oxide nanoparticles: its influence on corn (*Z. mays*) and wheat (*Triticum aestivum*) plants by inoculation of Bacillus subtilis. Environ Sci Pollut Res:1–6. https://doi.org/10.1007/s11356-022-24877-7

Hao Y, Yuan W, Ma C, White JC, Zhang Z, Adeel M, Zhou T, Rui Y, Xing B (2018) Engineered nanomaterials suppress turnip mosaic virus infection in tobacco (*Nicotiana benthamiana*). Environ Sci Nano 5(7):1685–1693

Hassan R, Jelena PD, Sayyed RZ, Simin Z, Majid J, Bahman FN (2020) Nanoparticles: a new threat to crop plants and soil rhizobia? In: Hayat S, Pichtel J, Faizan M, Hayat FQ (eds) Sustainable agriculture reviews 41. Springer, pp 201–214

Hatami M, Khanizadeh P, Bovand F, Aghaee A (2021) Silicon nanoparticle-mediated seed priming and *Pseudomonas* spp. inoculation augment growth, physiology and antioxidant metabolic status in *Melissa officinalis* L. plants. Ind Crop Prod 162:113238. https://doi.org/10.1016/j.indcrop.2021.113238

Hill J, Whitham S (2014) Control of virus diseases in soybeans. Adv Virus Res 90:355–390

Huang YC, Fan R, Grusak MA, Sherrier JD, Huang CP (2014) Effects of nano-ZnO on the agronomically relevant rhizobium–legume symbiosis. Sci Total Environ 497:78–90

Jakinala P, Lingampally N, Hameeda B, Sayyed RZ, Khan MY, Elsayed EA, Elenshasy H (2021) Silver nanoparticles from insect wing extract: biosynthesis and evaluation for antioxidant and antimicrobial potential. PLoS One 16(3):e0241729. https://journals.plos.org/plosone/article?id=10.1371/journal.pone.0241729

Kashyap D, Siddiqui ZA (2021) Effect of silicon dioxide nanoparticles and *Rhizobium leguminosarum* alone and in combination on the growth and bacterial blight disease complex of pea caused by Meloidogyne incognita and *Pseudomonas syringae* pv. pisi. Arch Phytopathol Plant Protect 54(9–10):499–515. https://doi.org/10.1080/03235408.2020.1843306

Kashyap D, Siddiqui ZA (2022) Effect of zinc oxide nanoparticles and *Rhizobium leguminosarum* on growth, photosynthetic pigments and blight disease complex of pea. Gesunde Pflanzen 74:29–40. https://doi.org/10.1007/s10343-021-00586-y

Khan N, Bano A (2016) Modulation of phytoremediation and plant growth by the treatment with PGPR, Ag nanoparticle and untreated municipal wastewater. Int J phytoremed 18(12):1258–1269. https://doi.org/10.1080/15226514.2016.1203287

Khati P, Bhatt P, Kumar R, Sharma A (2018) Effect of nanozeolite and plant growth promoting rhizobacteria on maize. Biotech 8(3):1–2. https://doi.org/10.1007/s13205-018-1142-1

Krutyakov YA, Mukhina MT, Shapoval OA, Zargar M (2022) Effect of foliar treatment with aqueous dispersions of silver nanoparticles on legume-Rhizobium symbiosis and yield of soybean (*Glycine max* L. Merr). Agronomy 12(6):1473. https://doi.org/10.3390/agronomy12061473

Lico C, Schoubben A, Baschieri S, Blasi P, Santi L (2013) Nanoparticles in biomedicine: new insights from plant viruses. Curr Med Chem 20(28):3471–3487

Liu Y, Lu J, Cui L, Tang Z, Ci D, Zou X, Zhang X, Yu X, Wang Y, Si T (2023) The multifaceted roles of Arbuscular Mycorrhizal Fungi in peanut responses to salt, drought, and cold stress. BMC Plant Biol 23(1):1–9. https://doi.org/10.1186/s12870-023-04053-w

Ma J, Alshaya H, Okla MK, Alwasel YA, Chen F, Adrees M, Hussain A, Hameed S, Shahid MJ (2022) Application of cerium dioxide nanoparticles and chromium-resistant bacteria reduced chromium toxicity in sunflower plants. Front Plant Sci 13. https://doi.org/10.3389/fpls.2022.876119

Mohammadi K, Khalesro S, Sohrabi Y, Heidari G (2011) A review: beneficial effects of the mycorrhizal fungi for plant growth. J Appl Environ Biol Sci 1(9):310–319

Moradipour M, Saberi-Riseh R, Mohammadinejad R, Hosseini A (2019) Nano-encapsulation of plant growth-promoting rhizobacteria and their metabolites using alginate-silica nanoparticles and carbon nanotube improves ucb1 pistachio micropropagation. J Microbiol Biotechnol 29(7):1096–1103

Nasab BF, Sayyed RZ, Piri R, Rahmani AF (2021) Biopriming and Nanopriming: green revolution wings to increase plant yield, growth, and development under stress condition and forward dimensions. In: Singh HB, Anukool V, Sayyed RZ (eds) . Springer, pp 623–655

Natesan S, Rajaram S, Manoharan D, Ramachandran T (2023) The beneficial plant microbial association for sustainable agriculture. In: Plant microbiome for plant productivity and sustainable agriculture. Springer Nature Singapore, pp 137–210. https://doi.org/10.1007/978-981-19-5029-2_7

Nayeri FD, Mafakheri S, Mirhosseini M, Sayyed R (2021) Phyto-mediated silver nanoparticles via *Melissa officinalis* aqueous and methanolic extracts: synthesis, characterization and biological properties against infectious bacterial strains. Int J Adv Biol Biomed Res 3:270–285. http://www.ijabbr.com/article_245059.html

Noha K, Bondok AM, El-Dougdoug KA (2018) Evaluation of silver nanoparticles as antiviral agent against ToMV and PVY in tomato plants. Sciences 8(01):100–111

Olga M, Jana M, Anna M, Irena K, Jan M, Alena Č (2022) Antimicrobial properties and applications of metal nanoparticles biosynthesized by green methods. Biotechnol Adv:107905. https://doi.org/10.1016/j.biotechadv.2022.107905

Oves M, Khan MS, Zaidi A, Ahmed AS, Azam A (2014) Production of plant-growth promoting substances by nodule forming symbiotic bacterium rhizobium sp. os1 is influenced by cuo, zno and Fe nanoparticles. IIOAB J 5(4):1

Pérez-García LA, Sáenz-Mata J, Fortis-Hernández M, Navarro-Muñoz CE, Palacio-Rodríguez R, Preciado-Rangel P (2023) Plant-growth-promoting rhizobacteria improve germination and bioactive compounds in cucumber seedlings. Agronomy 13(2):315. https://doi.org/10.3390/agronomy13020315

Peruzzi PP, Chiocca EA (2016) Cancer immunotherapy: a vaccine from plant virus proteins. Nat Nanotechnol 11:214–215

Sethi G, Behera KK, Sayyed R, Adarsh V, Sipra BS, Singh L, Alamro AA, Behera M (2025) Enhancing soil health and crop productivity: the role of zinc-solubilizing bacteria in sustainable agriculture. Plant Growth Regul. https://doi.org/10.1007/s10725-025-01294-7

Shafiq T, Yasmin H, Shah ZA, Nosheen A, Ahmad P, Kaushik P, Ahmad A (2022) Titanium oxide and zinc oxide nanoparticles in combination with Cadmium tolerant *Bacillus pumilus* ameliorates the Cadmium toxicity in maize. Antioxidants 11(11):2156. https://doi.org/10.3390/antiox11112156

Shahbaz M, Akram A, Mehak A, Haq E, Fatima N, Wareen G, Aamir S, Hassan H, Poczai P, Sayyed RZ, Ilyas N (2023) Evaluation of selenium nanoparticles in inducing disease resistance against spot blotch disease and promoting growth in wheat under biotic stress. Plants 12(4):76. https://doi.org/10.3390/plants12040761

Siddiqui ZA, Khan A, Khan MR, Abd-Allah EF (2018) Effects of zinc oxide nanoparticles (ZnO NPs) and some plant pathogens on the growth and nodulation of lentil (*Lens culinaris* Medik.). Acta Phytopathol Entomol Hung 53(2):195–211. https://doi.org/10.1556/038.53.2018.012

Singh R, Kuddus M, Singh PK, Choden D (2022) Nanotechnology for nanophytopathogens: from detection to the management of plant viruses. Biomed Res Int 3:2022. https://doi.org/10.1155/2022/8688584

Singh RP, Handa R, Manchanda G (2021) Nanoparticles in sustainable agriculture: an emerging opportunity. J Control Release 329:1234–1248. https://doi.org/10.1016/j.jconrel.2020.10.051

Sofy AR, Sofy MR, Hmed AA, Dawoud RA, Alnaggar AE, Soliman AM, El-Dougdoug NK (2021) Ameliorating the adverse effects of tomato mosaic tobamovirus infecting tomato plants in Egypt by boosting immunity in tomato plants using zinc oxide nanoparticles. Molecules 26(5):1337. https://doi.org/10.3390/molecules26051337

Sudha PN, Sangeetha K, Vijayalakshmi K, Barhoum A (2018) Nanomaterials history, classification, unique properties, production and market. In: Emerging applications of nanoparticles and architecture nanostructures. Elsevier, pp 341–384. https://www.researchgate.net/publication/309733691

Thakur P, Thakur S, Kumari P, Shandilya M, Sharma S, Poczai P, Alarfaj AA, Sayyed RZ (2022) Nano-insecticide: synthesis, characterization, and evaluation of insecticidal activity of ZnO NPs against *Spodoptera litura* and *Macrosiphum euphorbiae*. Appl Nanosci 12:3835–3850. https://doi.org/10.1007/s13204-022-02530-6

Tian H, Kah M, Kariman K (2019) Are nanoparticles a threat to mycorrhizal and rhizobial symbioses? A critical review. Front Microbiol 10:1660. https://doi.org/10.3389/fmicb.2019.01660

Vafa ZN, Sohrabi Y, Heidari G, Rizwan M, Sayyed RZ (2024) Wheat growth and yield response are regulated by mycorrhizae application and supplemental irrigation. Chemosphere 364:143068. https://doi.org/10.1016/j.chemosphere.2024.143068

Vargas-Hernandez M, Macias-Bobadilla I, Guevara-Gonzalez RG, Rico-Garcia E, Ocampo-Velazquez RV, Avila-Juarez L, Torres-Pacheco I (2020) Nanoparticles as potential antivirals in agriculture. Agriculture 10(10):444. https://doi.org/10.3390/agriculture10100444

Wang F, Li K, Shi Z (2021) Phosphorus fertilization and mycorrhizal colonization change silver nanoparticle impacts on maize. Ecotoxicology 30(1):118–129. https://doi.org/10.1016/j.chemosphere.2021.131644

Wang Y, Zou YN, Shu B, Wu QS (2023) Deciphering molecular mechanisms regarding enhanced drought tolerance in plants by arbuscular mycorrhizal fungi. Sci Hortic 308:111591. https://doi.org/10.1016/j.scienta.2022.111591

Yang D, Wang L, Ma F, Wang G, You Y (2022) Effects of Ag nanoparticles on plant growth, Ag bioaccumulation, and antioxidant enzyme activities in Phragmites australis as influenced by an arbuscular mycorrhizal fungus. Environ Sci Pollut Res 1–1. https://doi.org/10.1007/s11356-022-22540-9

Zhang R, Meng Z, Abid MA, Zhao X (2018) Novel pollen magnetofection system for transformation of cotton plant with magnetic nanoparticles as gene carriers. In: Zhang B (ed) Transgenic cotton. Methods in molecular biology, vol 1902. Humana Press, New York. https://doi.org/10.1007/978-1-4939-8952-2_4

Nanopesticides: Unleashing the Power of Tiny Warriors in Plant Protection

4

Tanmaya Kumar Bhoi, Vipula Vyas, Ipsita Samal, Deepak Kumar Mahanta, J. Komal, Sangeeta Singh, and Kuldeep Sharma

Abstract

The fast-developing sector of agricultural technology known as nanopesticides has the power to completely transform plant protection methods. Nanopesticides, which use nanoscale materials to their full potential, have several advantages over traditional pesticides, including increased effectiveness, controlled release, targeted delivery, and less environmental impact. The idea of nanopesticides is examined in this book chapter along with their traits and many forms, including nanoparticles, nanocapsules, nanofibers, and nanoemulsions. In this book chapter, the improved effectiveness of nanopesticides in pest management is highlighted along with their modes of action and techniques of manufacture. In addition, the difficulties and prospects for nanopesticides are discussed, including regulatory frameworks and safety issues.

Keywords

Nanocapsules · Nanoemulsions · Nanofibers · Nanoparticles · Plant protection

T. K. Bhoi · V. Vyas · S. Singh (✉) · K. Sharma
Forest Protection Division, ICFRE-Arid Forest Research Institute (ICFRE-AFRI), Jodhpur, India

I. Samal
ICAR-National Research Centre for Litchi, Muzaffarpur, Bihar, India

D. K. Mahanta
Forest Entomology Discipline, Forest Protection Division, Forest Research Institute (ICFRE-FRI), Dehradun, Uttarakhand, India

J. Komal
Basic Tasar Silkworm Seed Organization (BTSSO), Central Silk Board, Bilaspur, Chhattisgarh, India

R. Sayyed et al. (eds.), *Nanobiostimulants in Innovative Agriculture*,
https://doi.org/10.1007/978-981-95-2477-8_4

1 Introduction

As the world's population grows, so does the demand for food, and it is predicted that by 2050, there will be 9.7 billion people on the planet, necessitating a considerable rise in agricultural output (Bhoi et al. 2023; Bhoi et al. 2022). It is essential to safeguard crops from pests, diseases, and weeds that might result in large yield losses in order to satisfy this need (Mahanta et al. 2023; Samal et al. 2023). A variety of tactics are included in plant protection measures, such as the application of pesticides, integrated pest management (IPM) procedures, and the creation of disease-resistant varieties (Mahanta et al. 2022). Farmers may maintain healthy and abundant crop yields by putting into practice efficient plant protection techniques, which will support both economic stability and food security (Shaila and Begum 2021). The mainstay of plant protection initiatives has historically been conventional insecticides (Rana et al. 2022). These chemical substances are made to specifically target and eradicate certain weeds, illnesses, and pests that endanger crops (Ala et al. 2020). According to Mehrwar et al. (2021), they have made a substantial contribution to raising agricultural output and decreasing crop losses. Traditional pesticides' shortcomings, on the other hand, have prompted questions about both their long-term viability and environmental impact (Rathore et al. 2021). The non-selectiveness of conventional insecticides is one of their main drawbacks. They frequently kill beneficial creatures, including pollinators, pests' natural enemies, and soil microorganisms, even while they can efficiently target and eradicate pests (Naharki and Jaishi 2020). This unintentional harm throws off the sensitive ecosystems' equilibrium, which has unforeseen effects like decreased biodiversity and ecological imbalance (Deka et al. 2021). Additionally, pest resistance has developed as a result of the excessive and careless use of pesticides, making some chemicals less effective over time (Hajam and Kumar 2022). Moreover, conventional pesticides have the potential to linger in the environment, contaminating soil, water, and air (Endshaw and Hiruy 2021). Concerns regarding possible health risks to consumers are raised by the buildup of pesticide residues in food crops (Bernaola and Holt 2021). Pesticide discharge into waterways can also taint drinking water supplies and destroy aquatic ecosystems. These environmental worries have increased the demand for more sustainable and eco-friendly plant protection techniques (Yan et al. 2021). The development of nanotechnology in recent years has created new opportunities for overcoming the constraints of conventional pesticides in agriculture. According to Melanie et al. (2022), nanotechnology entails the manipulation and control of matter at the nanoscale, often at length scales between 1 and 100 nm. Nanotechnology is a viable area for innovation in many industries, including agriculture, due to the special characteristics and behavior of materials at this scale. Nanotechnology has a number of potential benefits for plant protection (Constantinescu and Sicuia 2020). For example, nanopesticides can improve the effectiveness and precision of insect management. Targeted delivery to pests is made possible by encapsulating or attaching pesticide molecules to nanoscale carriers, such as nanoparticles or nanocapsules, which lowers the amount of pesticide required and minimizes off-target effects (Zobir et al. 2021). Due to the more efficient use of lower doses, targeted administration can also prevent the emergence of pesticide resistance in pests. Additionally, pesticides' durability and controlled release can be improved by nanotechnology-based formulations, extending their efficacy and lowering the frequency of administration (Shivashakarappa et al. 2022). This increased effectiveness can help farmers produce their crops more cheaply and with less of an impact on the environment. Nanomaterials, such as nanoparticles and nanoclays, can also be used to create smart crop protection materials, such as self-healing materials and sensors for early pest and disease detection (Shahbaz et al. 2023; Pramanik et al. 2020). Nanotechnology use in agriculture is not without difficulties, though. Nanomaterials may be hazardous to humans, animals, and the environment, raising concerns (Fabiyi 2023). Prior to their widespread usage, it is essential to carefully investigate the safety aspects of nanopesticides and provide effective regulation and risk assessment (Maroufpour et al. 2020). Additionally, small-scale farmers in underdeveloped countries may not be able to afford nanotechnology-based products due to their high production costs (Ingle and Gupta 2021). This book chapter provides a comprehensive exploration of nanopesticides, covering their concept, characteristics, and types, mechanisms of action, synthesis methods, challenges, and future perspectives. It serves as a valuable resource for researchers, practitioners, and policymakers seeking to understand and harness the power of nanotechnology in the field of plant protection.

2 Nanotechnology in Agriculture

A promising area of research in agriculture is nanotechnology, which provides creative ways to boost plant protection and boost crop output (Mujtaba et al. 2021). To create cutting-edge instruments and methods for agricultural applications, materials exhibit distinctive characteristics and behaviors at the nanoscale (Bindra and Singh 2021). The use of nanotechnology in agriculture, also known as nanoagricultural or nanotechnology in agriculture, has the potential to transform plant protection methods and overcome the drawbacks of conventional pesticides (Hofmann et al. 2020). The creation of nanopesticides is one important area where nanotechnology is changing things. In order to accurately deliver the active chemicals to the intended pests or diseases, nanopesticide formulations use nanoscale carriers, such as nanoparticles or nanocapsules (Farooq et al. 2021). The effectiveness and efficiency of pest management can be significantly increased by encapsulating or attaching the pesticide molecules to these nanocarriers. With less pesticide needed due to the focused distribution, there are fewer off-target impacts and less environmental contamination (Yadav et al. 2020). Additionally, nanocarriers' controlled release capabilities can increase a pesticide's shelf life, lessening the need for repeated applications and the expense of production for farmers (Gondal 2022).

Nanotechnology presents possibilities for the creation of intelligent plant protection materials. Crop pests, illnesses, and environmental challenges can all be detected early using nanosensors (Wang et al. 2021). With the help of these sensors, plant health metrics may be monitored in real time, allowing for prompt mitigation of threats. Additionally, self-healing nanomaterials can be created, enabling the repair of plant tissues harmed by pathogens or pests. Crop resilience and yield losses can both be reduced by this self-healing ability (Tipu et al. 2021). Nanotechnology also offers the potential to deal with the problem of pesticide resistance. Traditional insecticides may become ineffective over time due to the development of pest and pathogen resistance (Pérez-de-Luque 2020). However, nanopesticides can be made to target many physiological processes in pests through a variety of modes of action. This multifaceted strategy improves long-term pest control efficiency while lowering the risk of resistance development (Vijayakumar et al. 2022).

Despite the many advantages that nanotechnology in agriculture offers, there are still issues that need to be resolved. The potential toxicity of nanomaterials to people, animals, and the environment is a major worry (Al-Hadede et al. 2020). However, nanopesticides can be made to target many physiological processes in pests through a variety of modes of action. This multifaceted strategy improves long-term pest control efficiency while lowering the risk of resistance development (Vijayakumar et al. 2022). Despite the many advantages that nanotechnology in agriculture offers, there are still issues that need to be resolved. The potential toxicity of nanomaterials to people, animals, and the environment is a major worry (Al-Hadede et al. 2020).

To ensure the responsible use of nanotechnology in agriculture, certain rules and regulations need to be created (Kalita et al. 2021). Accessibility and affordability of nanotechnology-based solutions for small-scale farmers, especially in developing regions, provide another problem (Usman et al. 2020). Farmers with little resources may find it difficult to obtain nanomaterials and nanoscale technologies due to their high manufacturing costs (Saxena et al. 2020). A variety of farming groups should be given the opportunity to use nanotechnology by making it more accessible and inexpensive (Neme et al. 2021). To encourage the development and distribution of practical nano-agricultural solutions, cooperation between researchers, industry stakeholders, and policymakers is essential.

3 Nanotechnology Definition and Principle

The control, creation, and manipulation of matter at the nanoscale are all part of the multidisciplinary discipline of nanotechnology (Soni et al. 2022). It deals with objects that are typically between 1 and 100 nm (one billionth of a meter) in size and have peculiar features and functions (Kokotovich et al. 2021). Materials have distinct physical, chemical, and biological properties that set them apart from their bulk counterparts at this scale. The fields of physics, chemistry, biology, materials science, and engineering are all included in the field of nanotechnology, which has applications in a variety of fields, including electronics, medicine, energy, and agriculture (Rennick et al. 2021). The fundamental idea behind nanotechnology is that matter behaves differently at the nanoscale than it does at the macroscopic level (Svendsen et al. 2020). The creation, use, and application of nanomaterials and nanoscale technologies are guided by these concepts. The following key ideas underlie nanotechnology:

3.1 Size and Surface Area

Materials have a very high surface area-to-volume ratio at the nanoscale. The surface area of particles or structures grows dramatically in relation to their volume as their size decreases (Sindhu et al. 2021). This huge surface area offers improved features, including increased catalytic activity or better drug delivery efficiency, as well as enhanced reactivity and increased interactions with the environment (Cummings et al. 2021).

3.2 Quantum Effects

At the nanoscale, quantum mechanical phenomena are increasingly obvious. Quantum confinement effects can influence the behavior of electrons in nanosized particles, changing their optical, electrical, and magnetic properties. Due to quantum confinement, quantum dots, for instance, display distinctive fluorescence characteristics (Khan 2020).

3.3 Self-Assembly

Nanoscale parts have the ability to spontaneously self-assemble or organize into well-defined structures and patterns. It is possible to precisely arrange nanoparticles or molecules into desired configurations by self-assembly processes, which are fueled by fundamental forces like Van der Waals, electrostatic, or magnetic interactions. According to Furxhi et al. (2021), this characteristic is essential for the creation of nanoscale materials and technologies.

3.4 Novel Properties

Nanomaterials frequently have special qualities and functions that set them apart from their bulk counterparts. These characteristics may result from surface, size, or quantum effects alone or in combination (Hou et al. 2021). For instance, nanoparticles may differ from their larger counterparts in terms of their optical, electrical, mechanical, or thermal properties, opening up novel capabilities and applications.

3.5 Bottom-Up and Top-Down Approaches

Nanomaterials commonly differ from their bulk counterparts by having unique properties and capabilities. These traits could be the result of surface, size, or quantum processes acting alone or in concert (Hou et al. 2021). For instance, the optical, electrical, mechanical, or thermal properties of nanoparticles may be different from those of their larger counterparts, creating new possibilities and uses.

3.6 Multifunctionality

A single nanoscale substance or device can incorporate many functionality thanks to nanotechnology. Multifunctional nanodevices can be made by combining multiple nanomaterials or engineering particular features at the nanoscale. For instance, nanocarriers can be made to transport medications, imaging agents, and targeting molecules all at once (Buruk et al. 2023; Engelmann et al. 2020).

These nanotechnology principles serve as the cornerstone for the creation of novel nanomaterials, nanodevices, and nanoscale systems in a variety of industries (Acharya and Pal 2020). By enabling improvements in electronics, health, energy storage, environmental cleanup, and agriculture, among other fields, nanotechnology has the potential to transform entire sectors. Researchers and scientists continue to investigate fresh applications and push the limits of scientific knowledge in nanotechnology by taking advantage of the special characteristics and capabilities of nanomaterials.

4 Application Area of Nanotechnology in Agriculture

From crop production and protection to soil management and water purification, nanotechnology has the potential to revolutionize several facets of agriculture (Saritha et al. 2022). The uses of nanotechnology in agriculture are numerous, provide creative answers to problems faced by farmers, and support the use of sustainable agricultural methods (Ioannou et al. 2020). The creation of nanopesticides is a well-known use of nanotechnology in agriculture. In order to accurately transport active ingredients to desired pests or pathogens, nanopesticide formulations use nanoscale carriers, such as nanoparticles or nanocapsules (Zhang et al. 2021). Targeted distribution is made possible by the use of nanopesticides, which also reduces the amount of pesticide required and lessens off-target effects. Additionally, nanocarriers can improve the stability and controlled release of pesticides, extending their duration of action and decreasing the frequency of administration (Spanos et al. 2021) (Fig. 4.1). In addition to enhancing pest management, this also reduces environmental contamination and lowers farmers' production expenses (Kale et al. 2021). Additionally, nanotechnology presents opportunities for better nutrient management in agriculture. For instance, nanofertilizers can improve the effectiveness of plants absorbing nutrients. The slow release of nutrients from nano-sized nutrient

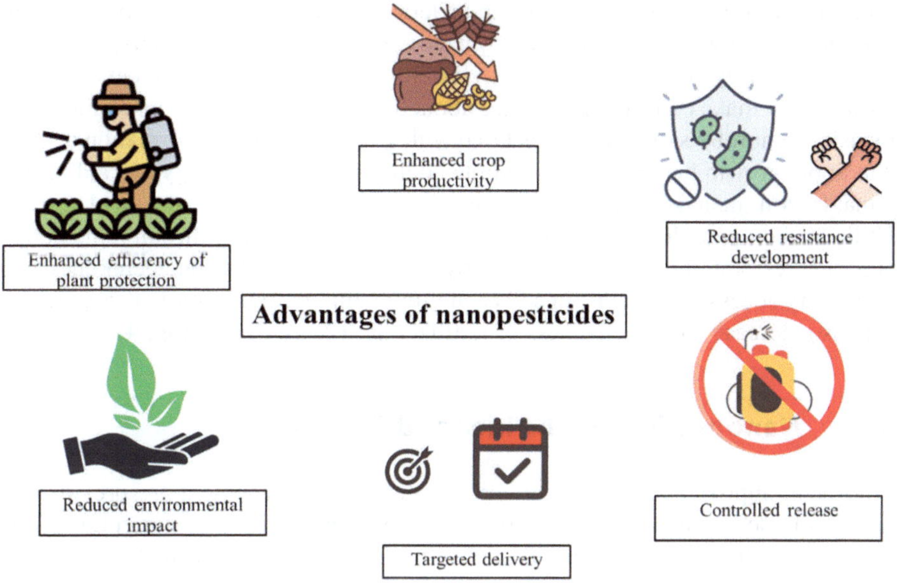

Fig. 4.1 Advantages of nanopesticide over conventional pesticide

particles can guarantee their availability to plants for a long time (Akhtar et al. 2022; Faridvand et al. 2021; Garza-Garca et al. 2022). With this regulated release mechanism, nutrient leaching is decreased, environmental contamination is reduced, and crop nutrient uptake is maximized (Deshmukh et al. 2022). In addition, real-time monitoring of soil nutrient levels using nanosensors enables precise and targeted fertilizer delivery depending on the individual requirements of crops. Nanotechnology offers prospects for early identification and detection of plant illnesses and insect infestations in the field of crop protection (Saini and Ledwani 2022). Nanosensors and nanobiosensors are able to identify particular infections or disease markers in plants at an early stage, enabling farmers to take prompt and precise preventative steps (Rehmanullah et al. 2020). To ensure the safety and purity of food, nanobiosensors can also find pesticide residues in crops. Additionally, antibacterial capabilities of nanoparticles can be created, offering a natural and environmentally benign substitute to synthetic pesticide (Saqib et al. 2022). Nano-based antimicrobial coatings on seeds or other surfaces have the ability to stop the spread of infections and lower loss after harvesting (Avşar 2022). The management of water in agriculture could be improved via nanotechnology. Nanosensors can measure the amount of soil moisture, enabling precise irrigation planning and effective water utilization. The retention of water in soil can be achieved with the help of nanomaterials such as hydrogels, which also improve plant water availability (Thangadurai et al. 2020). Additionally, water treatment and desalination, addressing difficulties with water scarcity in dry places, and providing a reliable source of water for agriculture can be accomplished using nanofiltration and nanodevices. Nanotechnology provides instruments specific to the site and data-driven farming methods in the realm of precision farming. Environmental parameters, including moisture, light, and temperature intensity, can be monitored by nanosensors and nanobiosensors, which can provide useful data for enhancing crop development factors (Hazarika et al. 2022). Farmers can make educated judgments about the cultivation and resource allocation by using nanorobots or nanodrones fitted with sensors to gather information on crop condition, growth patterns, and nutritional deficits. Nanotechnology can aid in sustainable soil management in addition to crop-related uses (Kale et al. 2021). Utilizing nanoparticles to break down or bind contaminants in soil, soil remediation and site restoration are made easier by using nanoremediation techniques. By increasing access to nutrients in soils and minimizing nutrient losses, nanoparticles can also increase the effectiveness of fertilizer use (Avşar 2022) (Fig. 4.1).

5 Nanopesticides: The Idea and Features

Innovative approaches to efficient and targeted pest management are provided by nanopesticides, which are a potential usage of nanoparticles in agriculture (Pho et al. 2020). To accurately transport active ingredients to the intended pests or diseases, these pesticide formulations use nanocarriers, which include nanoparticles or nanocapsules (Rani et al. 2023). Compared to conventional pesticide formulations, nanopesticides have a number of benefits, such as increased sustainability, a smaller ecological footprint, and greater efficacy (Vasseghian et al. 2022). The unique characteristics and behaviors of nanoparticles at the nanoscale are the basis for the notion of nanopesticides (Fraceto et al. 2020). Nanocarriers can increase the effectiveness and specificity of pest management by taking advantage of these characteristics. Additionally, because nanoparticles have a high surface area to volume ratio, they are easier to deliver just where they are needed, and a lesser amount of pesticide is required for effective control. The off-target impacts of typical pesticides, such as harm to beneficial creatures and environmental contamination, are reduced by this focused administration (Gahukar and Das 2020). Nanopesticides have a number of distinguishing qualities that make them very appealing for pest control. Their capacity for controlled release is one of their essential traits (López-Cabeza 2020). A continuous and extended effect on pests or diseases can be achieved by engineering nanocarriers that are used in nanopesticides to dissipate the active compounds gradually (Grillo et al. 2021). In addition to extending the pesticide's effectiveness, this controlled dispersion mechanism lowers the amount and duration of application, saving farmers money and reducing environmental exposure (Djiwanti and Kaushik 2019). The stability and bioavailability of the active components are enhanced by nanopesticides. According to Zaheer et al. (2022), many traditional pesticides are susceptible to deterioration or have limited solubility in water, which can reduce their efficacy. The active components can be encapsulated or bound by nanocarriers, enhancing their stability under a variety of environmental factors and preventing them from degrading. Nanoparticles' hydrophilic or lipophilic properties can make poorly soluble pesticides more soluble, increasing their dispersion and accessibility (Ikawati et al. 2021). Another noteworthy quality of nanopesticides is their versatility. Nanocarriers may be functionalized or altered to have different characteristics or functions. To improve the targeting and adherence of the pesticides to pests or pathogens, the outermost layer of nanoparticles can be altered to increase adherence to plant surfaces or specialized pest receptors. A pesticide can be combined with other bioactive substances to create a pesticide that has a synergistic effect, and nanocarriers can be made to release numerous active components at once. Additionally, nanopesticides provide ways to combat pesticide resistance in vermin (de Oliveira 2021). Traditional insecticides can lose their efficacy over time when viruses and pests develop resistance to them. Different physiological systems in pests or diseases can be targeted by nanocarriers that have diverse modes of action (Santos et al. 2021). This multifaceted strategy improves long-term pest management efficacy and decreases the probability of resistance development. The possibility for a decreased ecological effect of nanopesticides is another aspect of them (Ashitha and Mathew 2020). The chance of soil, bodies of water, and non-target organisms becoming contaminated is decreased by the focused delivery and controlled release of active substances. Reduced environmental exposure and associated dangers are also a result of the lower pesticide dosage required for efficient control. Additionally, using nanopesticides can lessen the requirement for repeated applications, lowering the overall pesticide load in the ecosystem and encouraging sustainable pest control techniques (Raj et al. 2021). Despite the many benefits and possibilities of nanopesticides, there are issues and things to think about while developing and using them (Kookana et al. 2014). The safety of nanopesticides to humans, animals, and the environment is a crucial factor. Nanomaterials' special qualities, such as their small size and higher reactivity, raise questions regarding their possible toxicity. Thorough risk assessments and toxicity tests must be conducted in order to fully analyze the safety aspects of nanopesticides (Balaure et al. 2017). To ensure their responsible usage and reduce any potential dangers, proper legislation and guidelines should be put in place (Schiavi et al. 2022). Additionally, it can be difficult to produce nanopesticides in an economical manner. Nanocarriers must be synthesized and manufactured using specific methods and tools, which can be expensive (Stadler et al. 2018). It is important to consider how affordable and available nanopesticides are to small-scale farmers, especially in developing nations. To make nanopesticides more widely accessible, research and development activities should concentrate on discovering affordable and scalable techniques for their manufacturing.

6 Types of Nanopesticides

The term "nanopesticides" refers to a group of formulations that deliver active chemicals for efficient pest control in agriculture using nanoscale carriers. The most widely used nanopesticides include nanocapsules, nanoemulsions, nanofibers, and nanoparticles (Chaud et al. 2021). Regarding formulation support, controlled dispersion, and targeted distribution, each of these classes has particular qualities and benefits.

6.1 Nanoemulsions

Colloidal dispersions called nanoemulsions are made up of tiny droplets of one liquid mixed with another liquid. According to Demir et al. (2022), the active ingredient in nanopesticides is often dissolved or disseminated in an oil phase before being emulsified in water or another liquid phase. A substantial interfacial area is provided by the tiny droplet dimension of the emulsion particles, which typically ranges from 20 to 200 nm, allowing for effective interactions with pathogens or pests. The benefits of nanoemulsions include greater stability, longer release of the active component, and increased solubility. They are adaptable for a variety of applications since they can incorporate both hydrophobic and hydrophilic insecticides (Shahavi et al. 2019).

6.2 Nanocapsules

Nanocapsules are tiny structures with an inner core that houses the active component and an outside, thin coating. Nanocapsules' core-shell design enables the protective covering and controlled release of the active component. The material of the shell can be selected to offer controlled release capabilities and protection against deterioration (Tiwari et al. 2020). Improved stability, decreased volatility, targeted distribution, and precisely controlled release kinetics are just a few benefits of nanocapsules. To maximize the effectiveness of encapsulation, release profile, and longevity of the active component, nanocapsule size and composition can be customized (Yin et al. 2023).

6.3 Nanoparticles

Solid particles known as nanoparticles typically have a size range of fewer than 100 nm. Nanoparticles operate as carriers for active chemicals in nanopesticides, enabling controlled release and targeted delivery. Metals, metal oxides, polymers, and lipids are just a few of the components that can be used to create nanoparticles (Bratovcic et al. 2021). The material for the nanoparticles is chosen based on the intended release kinetics, stability, biocompatibility, and compatibility with the active component. To improve their interaction with pathogens or pests, increase their adherence to plant surfaces, or enable targeted delivery to particular plant tissues, nanoparticles can be modified or have their surfaces altered (Arora et al. 2022).

6.4 Nanofibers

Ultrafine fibers known as nanofibers typically have sizes that fall within the nanometer range. They can be made from polymers that are organic or synthetic utilizing methods like electrospinning. Nanofibers can be used as carriers for the regulated release of active substances in the context of nanopesticides. A sustained release of the active component is made possible by the high surface area and porosity of nanofibers (Wang et al. 2022). Nanofibers can be designed to have particular properties like controlled release profiles, biodegradability, and biocompatibility. They can be sprayed on plant surfaces as mats or coatings, which prolongs the pesticide's release and encourages adhesion to plant fibers (Sun et al. 2019).

According to the unique pest control requirements and intended applications, each form of nanopesticide has distinct benefits and considerations (Okey-Onyesolu et al. 2021). The selection of a nanopesticide formulation is influenced by a variety of elements, including the active ingredient's physicochemical qualities, desired release profile, compatibility with the pests or pathogens that it is intended to control, environmental considerations, and ease of manufacture and administration. The possible adverse effects of these nanopesticide compositions are being investigated, and great effort is being made to improve their effectiveness, strength, and environmental safety for long-term pest management in agriculture.

7 Mode of Action of Nanopesticide

In order to effectively manage diseases and pests in agriculture, nanopesticides, which use nanoscale carriers to deliver active chemicals, use a variety of modes of action (Kannan et al. 2023). The chemical, biological, and physical mechanisms that underlie these methods of action can be categorized. The specific method varies depending on the sort of nanopesticide used and the pathogen or pest that it is intended to kill (Lima et al. 2021).

7.1 Physical Mechanisms

(a) Adhesion and Penetration: Because of their small size and distinct surface characteristics, nanopesticides can stick to the surface of pests or diseases. This stickiness makes it easier for them to enter the target species. Insect cuticles and other protective layers can be penetrated by nanoparticles or nanocapsules, which may then communicate with physiological systems (Pradhan and Mailapalli 2020).

(b) Disruption of Cellular Structures: Nanoparticles can damage cell membranes, cause structural integrity to be lost, or interfere with crucial processes by disrupting the cellular structures of diseases, pests, or their physical defenses. This disruption may be caused by harmful compounds released from the formulation of the nanopesticide or by direct interactions between nanoparticles and cellular components (Adisa et al. 2019).

(c) Targeted Delivery: Active chemicals can be delivered specifically to pests or pathogen-infected locations using nanoscale carriers. The ability to selectively attach to target cells or tissues by the functioning of nanoparticles or nanocapsules with ligands or particular receptors improves the delivery of the active component to the targeted site of action (Zhang et al. 2023a, b).

7.2 Chemical Mechanisms

(a) Controlled Release: Active ingredients in nanopesticides can be formulated to release gradually over time. The active component is encapsulated or bound within nanocarriers, allowing for a slow release that prolongs the time that pests or diseases are exposed to the pesticide. The effectiveness and longevity of the active component are improved by this regulated release mechanism (Devi et al. 2019).

(b) Improved Solubility: By making insoluble active compounds more soluble, nanoparticles can increase the availability and bioavailability of such substances. Increased solubility guarantees that the pesticide is effectively dispersed and distributed throughout the target species, maximizing its ability to kill pests or pathogens (Santiago et al. 2020).

7.3 Biological Mechanisms

7.3.1 Synergistic Effects

When several active chemicals or components are integrated within the nanoscale carrier, nanopesticides may demonstrate synergistic effects. By concurrently attacking several targets or several physiological pathways, these synergistic effects can increase the effectiveness of pest management. A more effective control method, for instance, may be achieved by combining pesticides with a living thing or plant extracts within a nanocarrier (Takeshita et al. 2023; Sharma et al. 2025).

7.3.2 Increased Stability

By shielding the active ingredient from deterioration, nanocarriers can increase both the stability and bioactivity of the substance. With improved stability, pests or pathogens are controlled for a longer period of time, and the pesticide stays active during application (Santiago et al. 2020).

7.3.3 Lessened Development of Resistance

Nanopesticides can aid in lessening the evolution of resistance in pathogens or pests. In comparison to conventional pesticides, nanopesticides can reduce the risk of resistance development by leveraging numerous modes of action or combining many pesticides within a nanocarrier (Adetunji and Ugbenyen 2019).

7.3.4 Targeted Bioactivity

Nanopesticides can be designed to selectively target infections or pests while causing the least amount of damage to unintended organisms. By creating nanocarriers with certain surface functionalizations or alterations, it is possible to achieve this targeted bioactivity and enable selective conversations with the organisms of interest (Graily-Moradi and Asgari Lajayer 2021).

8 Synthesis and Formulation of Nanopesticides

Making nanoscale carriers and incorporating active chemicals are key steps in the design and synthesis of nanopesticides (Priyanka et al. 2019). Nanoparticles or nanocapsules are produced throughout the synthesis process using techniques like chemical reduction, sol-gel procedures, and green synthesis. There are also physical techniques used, such as vapor deposition and high-energy ball milling. According to Roseline et al. (2019), biological techniques use microbial or plant-mediated synthesis. Among the formulation strategies are matrix formation, surface modification, and encapsulation. Electrostatic assembly, emulsion/solvent evaporation, and coacervation are three techniques for encapsulation (Raj et al. 2021). Coating nanocarriers or adding ligands to them allows for surface modification. Nanoscale carriers are incorporated into a hydrogel system or polymeric matrix during matrix formation. According to Lade and Gogle (2019), these procedures improve the stability, controlled release, and targeted delivery of the active components in nanopesticides.

8.1 Nanomaterials for Nanopesticide Development

By acting as carriers for active ingredients and offering special qualities that improve their efficacy, stability, and targeted distribution, nanomaterials play a critical role in the fabrication of nanopesticides (Jasrotia et al. 2022). Nanopesticide formulations use a variety of nanomaterial kinds, each of which has unique benefits and functions. Due to their tiny size, high surface area-to-volume ratio, and diverse features, nanoparticles are frequently used as carriers in nanopesticides (Shao et al. 2022). The intrinsic antibacterial capabilities of metal-based nanoparticles like zinc oxide, copper, and silver make them efficient against microorganism diseases and pests. These nanoparticles have the ability to damage cell membranes, prevent the action of enzymes, and cause oxidative stress in diseases or pests (Jasrotia et al. 2022). Metal oxide nanoparticles with photocatalytic characteristics, such as zinc oxide and titanium dioxide, can be used to kill pests. These nanoparticles generate reactive oxygen species (ROS) when exposed to light, which harms pests and pathogens (Yin et al. 2023). Polymer nanoparticles are versatile carriers of different kinds of active substances since they are frequently made from biodegradable polymers. Due to their biocompatibility and controlled release qualities, polymers like poly(lactic-co-glycolic acid) (PLGA) and chitosan are frequently employed in pharmaceutical products. Nanoscale structures called nanocapsules have a core ingredient that is enclosed by a shell or covering. According to Schiavi et al. (2022), these nanocapsules offer the active component protection, controlled release, and targeted delivery. Lipid-based nanocapsules, such as solid lipid nanoparticles (SLNs) and nanostructured lipid carriers (NLCs), have superior stability, controlled release, and increased bioavailability of the active component since they are made of lipids or lipid-like materials. By adjusting the shell characteristics, polymer-based nanocapsules like dendrimers and polymeric micelles offer regulated release of the active ingredient. For a targeted interaction with pathogens or pests, these nanocapsules can be functionalized with ligands or targeting molecules. Ultrafine fibers known as nanofibers typically have sizes in the nanometer range. Using methods like electrospinning, they can be made from organic or synthetic polymers (Peng et al. 2022). The vast surface area and porosity of nanofibers in nanopesticides enable high loading of active chemicals and regulated release. Polymeric nanofibers can be filled with active substances and offer prolonged release when produced from polymers like polyvinyl alcohol (PVA), poly(lactic acid), or poly(-caprolactone). Due to the flexibility of the electrospinning technique, formulation design can incorporate both hydrophilic and hydrophobic active substances (Mali et al. 2020). Another class of nanomaterials employed in nanopesticides is called nanocomposites, which are hybrid materials made of nanoscale components embedded within a matrix material. Nanoparticles or nanocapsules can be added to a polymeric matrix to create nanocomposite particles. These particles enable controlled release and increase the active ingredient's stability (Del Prado-Audelo et al. 2022). The synergistic benefits of each property's individual properties are made possible by the combination of nanoparticles and polymers. The active ingredient's characteristics, the desired release kinetics, the stability requirements, and the target pest or pathogen all have a role in the choice of nanomaterials. Each nanomaterial has distinct benefits and abilities that can be customized to meet particular pest management requirements (Kah et al. 2021). The progress and optimization of nanopesticide formulations for efficient and long-lasting pest control in agriculture will be facilitated by ongoing research and development activities in nanomaterial synthesis and characterization (Grillo et al. 2021).

8.2 Methods of Nanopesticide Synthesis

To create nanoscale carriers and include active components, a variety of methodologies and procedures are used in the synthesis of nanopesticides (Sangeetha et al. 2021). These techniques can be broadly divided into three categories: chemical synthesis, physical techniques, and biological methods.

To create nanoscale carriers, chemical synthesis techniques use controlled chemical reactions. Chemical reduction is a popular technique that creates metallic nanoparticles by reducing metal salts in the presence of reducing chemicals (Ahmadi et al. 2021). For instance, using this technique, one can create nanoparticles of zinc, copper, or silver oxide. By hydrolyzing and condensing precursor materials, the procedure known as sol-gel is another chemical synthesis method used to produce metallic oxide nanoparticles, such as titanium oxide or silica nanoparticles (Shekhar et al. 2021). Green synthesis techniques use natural resources like plant extracts, microbes, or macromolecules as reduction and stabilization agents to create nanopesticides in an eco-friendly manner.

To create nanoscale carriers, physical procedures entail the manipulation of physical attributes. Through blending and grinding, high-energy ball milling mechanically decreases the particle size of bulk materials. According to Zou et al. (2022), this technique is used to create nanoparticles with regulated size and morphology. Thin films or coatings are produced on a substrate using vapor deposition processes, such as chemical vapor deposition (CVD) and physical vapor deposition (PVD), which deposit vaporized precursor materials. According to Ogunyemi et al. (2023), these techniques are frequently used to produce nanoscale thin films or nanocomposites for nanopesticide applications.

For the manufacturing of nanopesticides, biological techniques use living organisms or biological systems. Microbes that can gather or produce nanoparticles enzymatically are used in microbial synthesis (Tran et al. 2022). These microorganisms function as stabilizing and reducing agents, allowing for the precise creation of nanoparticles with particular characteristics. Plant extracts that contain phytochemicals that function as reducing agents are used in plant-mediated synthesis to transform metal ions into nanoparticles. This technology for manufacturing nanopesticides is economical and environmentally friendly (Zhang et al. 2023a, b).

8.3 Formulation Techniques for Nanopesticide

As they affect the stability, controlled release, and targeted administration of active ingredients, formulation techniques are crucial to the development of nanopesticides (Bocca et al. 2020). The inclusion of nanoscale carriers and active chemicals into an appropriate matrix or system is a key component of nanopesticide formulation approaches (Deka et al. 2021).

8.3.1 Encapsulation

Active compounds are enclosed within nanoscale carriers using the commonly used technique of encapsulation (Ahmed et al. 2019). This method offers greater stability, controlled release, and protection for the active ingredient. Various encapsulation techniques can be used, including:

Coacervation: Coacervation is the process of separating a polymeric phase from a liquid phase that contains the active component. Droplets created by the liquid phase are enclosed by the polymeric phase to create microcapsules or nanoparticles.

Emulsion/Solvent Evaporation: In this technique, the active component is dispersed inside an emulsion or solvent solution. The active ingredient-containing nanocapsules or nanoparticles are created as a result of the evaporation of the emulsion.

Electrostatic Assembly: Using electrostatic interactions, active substances are encapsulated inside nanocarriers during electrostatic assembly. This method creates a multilayered structure on top of the active ingredient by layer-by-layer depositing elements or polymers with opposing charges.

8.3.2 Surface Modification

In order to improve stability, controlled release, and targeted distribution, surface modification strategies concentrate on changing the surface properties of nanoscale carriers (Manna et al. 2023). These methods consist of the following:

Surface Coating: The nanocarriers' stability, capacity to avoid aggregation, and ability to deliver controlled release qualities can all be enhanced by coating them with substances like polymers, lipids, or surfactants. The coating ingredients may also promote particular interactions with the viruses or pests that are the focus.

Ligand Functionalization: Adding particular ligands or targeting molecules to the surface of nanoscale carriers is known as ligand functionalization. These ligands have the ability to preferentially attach to biomarkers or receptors on pests or pathogens, enabling targeted administration and enhancing the effectiveness of the nanopesticide.

8.3.3 Matrix Formation

In order to generate nanopesticide formulations, matrix formation procedures entail adding nanoscale carriers into an appropriate matrix or system (Pascoli et al. 2019). The following are typical matrix creation methods:

Polymer Matrix: A formulation for a nanocomposite can be made by dispersing nanoscale carriers inside a polymeric matrix. The polymer matrix gives the active ingredient stability, controlled dissolution, and security.

Hydrogel Formation: Active substances and nanoscale carriers can be enclosed in hydrogels, made from three-dimensional networks of polymers that are hydrophilic. Hydrogels offer controlled dissolution in response to external or physiological circumstances and can be made to react to specific stimuli.

9 Efficacy of Nanopesticides

In comparison to conventional pesticides, nanopesticides are more effective for a number of reasons. These elements help them perform better in managing and controlling pests.

9.1 Greater Surface Area and Better Contact

Compared to traditional pesticides, nanopesticides have a much bigger surface area since they are made by nanoscale particles or carriers. Increased close interaction between the pesticide and the intended pests or diseases is made possible by the larger surface area. Increased contact results in better adsorption, penetration, and interaction with the cellular structures of the pest or disease, increasing efficacy (Ding et al. 2023).

9.2 Targeted Delivery

Nanopesticides can be made to distribute the active chemicals just to the infections or pests that they are intended to kill. Nanopesticides can preferentially stick or bind to the target species through surface modification or functionalization, enhancing the likelihood of contact and uptake. This method of targeted distribution reduces off-target effects and makes sure that the active components effectively kill the targeted infections or pests (Campos 2021).

9.3 Controlled Release

Nanopesticides have controlled release mechanisms, which is a big benefit. The active substances can be released from the nanoscale carriers in a steady, controlled manner. This controlled release makes sure that pests or diseases are exposed to the pesticide continuously and for an extended period of time, improving its efficacy. Nanopesticides offer longer-lasting pest control when compared with conventional pesticides with quick-release characteristics, lowering the frequency of treatment and total pesticide usage (Devi et al. 2019).

9.4 Synergistic Effects

Synergistic effects can result from the incorporation of numerous active chemicals or components into nanoscale carriers in nanopesticides. This mixture of many active chemicals can attack pests or pathogens through a variety of mechanisms, increasing the overall effectiveness of pest management. Synergistic formulations can strengthen the effectiveness against different pest species or strains, increase the range of control, and overcome resistance difficulties (Zobir et al. 2021).

9.5 Improved Solubility and Dispersion

Active chemicals in nanopesticides that aren't very soluble are frequently made more soluble by the introduction of nanoparticles. The pesticide can now be distributed and dispersed more effectively among the target species because of its enhanced solubility. A greater amount of interaction and contact between the pesticide and pathogens or pests results in more effective control (Devi et al. 2019).

9.6 Protection from Degradation

Nanopesticides have the ability to shield their active components from environmental deterioration, including chemical and photochemical deterioration. The nanoscale carriers serve as a shield, keeping the active components from coming into contact with outside elements. This defense improves the pesticide's stability and absorption, maintaining its efficacy for longer periods of time (Agathokleous et al. 2020).

9.7 Minimized Development of Resistance

The multi-modal action and synergistic effects of nanopesticides can aid in reducing the growth of resistance in pathogens or pests. Nanopesticides can target numerous physiological pathways by mixing different active ingredients or including components with alternative modes of action, which decreases the possibility that pests or pathogens would develop resistance to a single pesticide. This multifaceted strategy enhances pest management's sustainability and long-term efficacy (Jogaiah et al. 2020).

10 Environmental Implication of Nanopesticides

When assessing the potential of nanopesticides for pest management, it is essential to take into account their effects on the environment. Although nanopesticides have advantages, including increased efficacy and targeted distribution, they also raise some environmental issues.

10.1 Ecotoxicity

Nanopesticides may possess special qualities that affect how they interact with organisms and ecosystems that are not their intended targets. Beneficial creatures, including pollinators, aquatic life, and soil microorganisms, may become hazardous as a result of some nanoparticles. To make sure nanopesticides don't affect unintended targets or upset the ecological balance, it is essential to evaluate their ecotoxicological effects (Asif et al. 2021).

10.2 Persistence and Accumulation

Due to their stability and resistance to degradation, several nanomaterials utilized in nanopesticides may linger in the environment. The possibility of their buildup in soil, water bodies, or sediments as a result of this persistence raises questions regarding their long-term impacts. Nanomaterial buildup can have an effect on ecosystems and put aquatic and terrestrial creatures at risk (Agathokleous et al. 2020).

10.3 Fate and Transport

Knowledge of the fate and movement of nanomaterials in the environment is crucial for assessing the likelihood of their dispersion and distribution in the environment. The behavior of nanopesticides and their potential exposure to non-target organisms can change as a result of transformation, aggregation, or interactions with environmental matrices. In order to predict potential effects and create effective risk management measures, it is helpful to evaluate the mobility and destiny of nanomaterials (Castillo-Henríquez et al. 2020).

10.4 Risk to Aquatic Systems

Aquatic creatures may be at risk if nanopesticides are released into waterways. Nanoparticles' large surface area and reactivity may improve their bioavailability and uptake by aquatic animals, which could have an impact on their survival. To maintain the protection of aquatic biodiversity and ecosystem services, including freshwater and marine ecosystems, it is essential to assess the threats to aquatic systems (Gahukar and Das 2020).

10.5 Interaction with Soil

Nanopesticides can interact with soil elements, potentially changing the soil's characteristics, the availability of nutrients, and microbial populations. These interactions may have an impact on ecosystem services, crop growth, and soil fertility. Sustainable agricultural practices require an understanding of how nanopesticides affect soil health and interactions between plants and soil (Dangi and Verma 2021).

10.6 The Potential for Bioaccumulation

Some of the nanoparticles employed in nanopesticides have the capacity to build up inside of living things. Bioaccumulation, which may result in biomagnification of food chains, may happen if nanomaterials are ingested by plants or other creatures and remain in their tissues. To avoid negative impacts on higher trophic levels, it is essential to evaluate the potential for bioaccumulation and the dangers associated with it (Gahukar and Das 2020).

10.7 Risk Assessment and Regulation

Because nanopesticides have particular features, risk assessment and regulation can be difficult. The unique traits and behaviors of nanomaterials might not be adequately captured by conventional regulatory systems. To ensure the safe and responsible use of nanopesticides and reduce their possible environmental repercussions, adequate risk assessment methodologies and regulatory requirements must be established (Galhardi et al. 2022).

11 Challenges Associated with Nanopesticide

For its successful use and widespread adoption, nanopesticide must overcome a number of obstacles (Singh et al. 2021a, b). These difficulties cover a wide range of topics, such as public perception, scalability, safety, legal and regulatory frameworks, and cost-effectiveness.

11.1 Safety and Environmental Issues

The safety and environmental effects of nanopesticides are one of the main issues. Unique physicochemical characteristics that nanoparticles may have give rise to questions about their potential toxicity to people, non-target species, and the environment. To make sure that nanopesticides are used safely, thorough risk analyses and toxicity studies must be carried out. To reduce any potential negative consequences, it is essential to understand the destiny and behavior of nanomaterials in the environment, particularly their potential accumulation and persistence (do Espirito Santo Pereira et al. 2021).

11.2 Regulatory Environment for Nanopesticides

It is continually evolving and differs depending on the region. To guarantee the safe and responsible use of nanopesticides, there is a need for clear, uniform rules and criteria. The proper protocols for assessing the efficacy, effectiveness, and environmental effects of nanopesticides must be established by regulatory agencies. To facilitate the commercialization and widespread acceptance of nanopesticides, uniformity in regulations and international cooperation are required (Ur Rahim et al. 2021).

11.3 Scale-Up and Manufacturing

It is difficult to increase the manufacturing of nanopesticides without sacrificing their effectiveness, reliability, or affordability. Large-scale production is complicated by the need for specific tools and processes in the synthesis and formulation of nanopesticides. For nanopesticides to be widely used, it is essential to develop scalable and affordable manufacturing procedures that can match the demand for commercial applications (Sangeetha et al. 2021).

11.4 Cost-Effectiveness

Nanopesticides can be more expensive to make and formulate than traditional pesticides. Cost-effectiveness of nanopesticides is a problem, especially for small-scale farmers. Their availability and adoption may be constrained by high production costs combined with the requirement for specialized tools and materials. Making nanopesticides more efficient, effective, and economically viable can be accomplished by finding ways to lower production costs, boost productivity, and optimize formulations (Kumar et al. 2019).

11.5 Stability and Formulation Issues

Agglomeration sedimentation, or deterioration of the active component or nanomaterials as time passes, are stability issues that nanopesticides may experience. For nanopesticide formulations to remain effective and have a long shelf life, stability is essential. To overcome these difficulties and guarantee constant performance throughout storage and application, formulation optimization, such as the use of stabilizers or encapsulating techniques, is required (Gaur et al. 2024).

11.6 Public Perception and Acceptance

How people perceive and embrace nanotechnology, particularly nanopesticides, can have an impact on whether people use it and whether the market accepts it. The use of nanomaterials in agriculture may raise issues of safety, ethics, and the environment for some parties. To promote acceptance and overcome any opposition, it is essential to establish public trust through effective communication, transparency, and education regarding the advantages, safety precautions, and risk management strategies related to nanopesticides (He et al. 2019).

11.7 Risk Association

Effective risk communication and education measures are needed because of the complexity and distinctive qualities of nanopesticides. Farmers, legislators, regulatory agencies, and the general public should all be informed about the advantages, dangers, and appropriate usage of nanopesticides. To close knowledge gaps and promote responsible and informed decision-making, transparent communication channels, reliable information dissemination, and training programs can all be beneficial (Camara et al. 2019).

11.8 Resistance Control

Similar to traditional pesticides, nanopesticides raise concerns about the emergence of pest and pathogen resistance. Use of nanopesticides with a single mechanism of action over time can lead to the emergence of resistance. To prevent the emergence of resistance and guarantee the long-term efficacy of nanopesticides, it is essential to incorporate resistance management techniques, such as combination formulations, rotating active ingredients, or putting IPM practices into practice (Zhai et al. 2020).

12 Future Direction

For nanopesticides to be successfully implemented, it is crucial to address safety concerns, regulatory hurdles, cost-effectiveness, and scalability. Nanopesticides offer enormous potential to revolutionize pest control in agriculture. Future research and development initiatives, in conjunction with strong regulatory frameworks, can assist in overcoming these difficulties and unlocking the full potential of nanopesticides, paving the way for environmentally friendly and productive pest management techniques.

Improved Selectivity and Efficacy: Nanopesticides have the potential to improve the selectivity and effectiveness of pest management. The development of nanopesticides with improved pest-specific targeting, decreased off-target effects, and improved control of pest resistance can be the subject of future study.

Controlled Release Systems: Creating nanopesticides with sophisticated controlled release mechanisms can enhance their effectiveness and lower the number of times they need to be applied. This may result in less pesticide use, more environmentally friendly pest management techniques, and fewer negative effects on the environment.

Synergistic Formulations: Multiple active ingredients or components can be combined to create nanopesticides with synergistic action against infections or pests. This strategy offers a wider range of pest management options while improving efficacy and overcoming resistance.

Smart Delivery Systems: By combining nanosensors, nanotechnology, and nanopesticides, it is possible to monitor insect populations, environmental conditions, and the efficacy of pesticides in real-time. This may result in focused application that improves pest control methods.

Combination with Biological Agents: Biological agents, such as biopesticides or biocontrol agents, can be coupled with nanopesticides to provide integrated pest management strategies. This combination can increase sustainability while decreasing the usage of chemical pesticides and improving pest control effectiveness.

Biodegradable and Eco-Friendly Nanomaterials: Long-term effects and environmental persistence are two issues that can be addressed by the creation of biodegradable and eco-friendly nanomaterials. Nanopesticide formulations that are greener and more ecologically friendly can benefit from the use of biodegradable and sustainable nanomaterials.

Nanopesticides for Vector Control: By focusing on disease-carrying insects like ticks and mosquitoes, nanopesticides can play a significant part in vector control. The management of vector-borne diseases can be aided by the development of nanopesticides intended specifically for vector control. Precision agriculture and data integration technologies can help with real-time monitoring, analysis, and decision-making for the best pesticide application, leading to more effective and sustainable farming practices (Thakur et al. 2022).

13 Conclusion

The use of nanopesticides in agriculture holds considerable promise for transforming pest management. Nanopesticides have a number of benefits over conventional pesticide formulations, including increased efficacy, targeted distribution, controlled release, and less environmental impact, thanks to the use of nanoscale carriers. Nanomaterials' distinctive characteristics enable improved stability, solubility, and adherence to pathogens or pests, resulting in more successful control methods. Nanopesticides administer active chemicals precisely and strategically, limiting side effects and lowering the overall pesticide load in the environment. Their controlled release mechanisms extend the pesticides' efficiency and decrease the frequency of treatment, saving farmers money and fostering environmentally friendly pest control techniques. Additionally, nanopesticides provide chances to beat pesticide resistance because of their multi-modal activity and synergistic results. However, there are obstacles that must be overcome in the creation and application of nanopesticides. For the responsible use of nanopesticides to be ensured, thorough risk assessments and toxicity studies are required. Establishing regulatory frameworks is necessary to direct the examination and approval procedure. To ensure that farmers, especially those in developing nations, have access to nanopesticides, it is also necessary to consider their cost-effectiveness of manufacturing and scalability.

References

Acharya A, Pal PK (2020) Agriculture nanotechnology: translating research outcome to field applications by influencing environmental sustainability. Nano Impact 19:100232. https://doi.org/10.1016/j.impact.2020.100232

Ahmadi S, Fazilati M, Nazem H, & Mousavi SM (2021) Green synthesis of magnetic nanoparticles using Satureja hortensis essential oil toward superior antibacterial/fungal and anticancer performance. BioMed Res Int (1):8822645

Adetunji CO, & Ugbenyen MA (2019) Mechanism of action of nanopesticide derived from microorganism for the alleviation of abiotic and biotic stress affecting crop productivity. In Nanotechnology for Agriculture: Crop Prod & Prot (119–142)

Adisa IO, Pullagurala VLR, Peralta-Videa JR, Dimkpa CO, Elmer WH, Gardea-Torresdey JL et al (2019) Recent advances in nano-enabled fertilizers and pesticides: a critical review of mechanisms of action. Environ Sci Nano 6(7):2002–2030. https://doi.org/10.1039/C9EN00265K

Agathokleous E, Feng Z, Iavicoli I, Calabrese EJ (2020) Nano-pesticides: a great challenge for biodiversity? The need for a broader perspective. Nano Today 30:100808. https://doi.org/10.1016/j.nantod.2019.100808

Ahmed K, Mikhail WZ, Sobhy HM, Radwan EM, El Din TS, Youssef A (2019) Effect of lambda-cyahalothrin as nanopesticide on cotton leafworm, *Spodoptera littoralis* (Boisd.). Egypt J Chem 62(7):1263–1275. https://doi.org/10.21608/ejchem.2019.6871.1581

Akhtar N, Ilyas N, Meraj TA, Aboughadareh AP, Sayyed RZ, Mashwani Z, Poczai P (2022) Improvement of plant responses by nanobiofertilizer: a step towards sustainable agriculture. Nanomaterials 12:965. https://doi.org/10.3390/nano12060965

Al-Hadede LT, Khaleel SA, Hasan SK (2020) Some applications of nanotechnology in agriculture. Biochem Cell Arch 20:1447

Ala AA, Otitodun GO, Ajao SK, Nwaubani SI, Opit GP, Ogundare MO & Ileleji KE (2020) A comparison of integrated pest management and traditional practices for bagged maize stored in Nigeria. J Stored Prod Res 86:101571

Arora S, Murmu G, Mukherjee K, Saha S, Maity D (2022) A comprehensive overview of nanotechnology in sustainable agriculture. J Biotechnol 355:21–41. https://doi.org/10.1016/j.jbiotec.2022.06.007

Ashitha A, Mathew J (2020) Characteristics and types of slow/controlled release of pesticides. In: Controlled release of pesticides for sustainable agriculture. Springer International Publishing, pp 141–153

Asif M, Islam S, Malik MA, Dar ZM, Masood A, Shafi S et al (2021) Nano pesticides application in agriculture and their impact on environment. In: Agricultural development in Asia: potential use of nano-materials and nano-technology. IntechOpen

Avşar C (2022) A novel assessment strategy for nanotechnology in agriculture: evaluation of nanohydroxyapatite as an alternative phosphorus fertiliser. Kem Ind 71(5–6). https://doi.org/10.15255/KUI.2021.057

Balaure PC, Gudovan D, Gudovan I (2017) Nanopesticides: a new paradigm in crop protection. In: New pesticides and soil sensors. Academic, pp 129–192

Bernaola L, Holt JR (2021) Incorporating sustainable and technological approaches in pest management of invasive arthropod species. Ann Entomol Soc Am 114(6):673–685. https://doi.org/10.1093/aesa/saab041

Bhoi TK, Samal I, Majhi PK, Komal J, Mahanta DK, Pradhan AK et al (2022) Insight into aphid mediated *Potato virus Y* transmission: a molecular to bioinformatics prospective. Front Microbiol 13:1001454. https://doi.org/10.3389/fmicb.2022.1001454

Bhoi TK, Samal I, Mahanta DK, Komal J, Jinger D, Sahoo MR et al (2023) Understanding how silicon fertilization impacts chemical ecology and multitrophic interactions among plants, insects, and beneficial arthropods. Silicon 15(6):2529–2549. https://doi.org/10.1007/s12633-022-02220-6

Bindra HS, Singh B (2021) Nanofertilizers and nanopesticides: future of plant protection. In: Advances in nano-fertilizers and nano-pesticides in agriculture. Woodhead Publishing, pp 57–84

Bocca B, Barone F, Petrucci F, Benetti F, Picardo V, Prota V et al (2020) Nanopesticides: physico-chemical characterization by a combination of advanced analytical techniques. Food Chem Toxicol 146:111816. https://doi.org/10.1016/j.fct.2020.111816

Bratovcic A, Hikal WM, Said-Al Ahl HAH, Tkachenko KG, Baeshen RS, Sabra AS et al (2021) Nanopesticides and nanofertilizers and agricultural development: scopes, advances and applications. Open J Ecol 11(4):301–316. https://doi.org/10.4236/oje.2021.114022

Buruk B, Ekmekci PE, Arda B (2023) An ethical analysis of a prospective new paradigm of life: nanotechnology-enabled human beings within the framework of principlism. Dev World Bioeth. https://doi.org/10.1111/dewb.12394

Camara MC, Campos EV, Monteiro RA, do Espirito Santo Pereira A, de Freitas Proença PL, Fraceto LF (2019) Development of stimuli-responsive nano-based pesticides: emerging opportunities for agriculture. J Nano Biotechnol 17(1):1–9

Campos EV (2021) Commercial nanoproducts available in world market and its economic viability. In: Advances in nano-fertilizers and nano-pesticides in agriculture. Woodhead Publishing, pp 561–593

Castillo-Henríquez L, Alfaro-Aguilar K, Ugalde-Álvarez J, Vega-Fernández L, Montes de Oca-Vásquez G, Vega-Baudrit JR (2020) Green synthesis of gold and silver nanoparticles from plant extracts and their possible applications as antimicrobial agents in the agricultural area. Nanomaterials (Basel) 10(9):1763. https://doi.org/10.3390/nano10091763

Chaud M, Souto EB, Zielinska A, Severino P, Batain F, Oliveira-Junior J et al (2021) Nanopesticides in agriculture: benefits and challenge in agricultural productivity, toxicological risks to human health and environment. Toxics 9(6):131. https://doi.org/10.3390/toxics9060131

Constantinescu F, Sicuia OA (2020) Phytonanotechnology and plant protection. In: Phytonanotechnology. Elsevier, pp 245–287

Cummings CL, Kuzma J, Kokotovich A, Glas D, Grieger K (2021) Barriers to responsible innovation of nanotechnology applications in food and agriculture: a study of US experts and developers. Nano Impact 23:100326. https://doi.org/10.1016/j.impact.2021.100326

Dangi K, Verma AK (2021) Efficient and eco-friendly smart nano-pesticides: emerging prospects for agriculture. Mater Today Proc 45:3819–3824

Deka B, Babu A, Baruah C, Barthakur M (2021) Nanopesticides: a systematic review of their prospects with special reference to tea pest management. Front Nutr 8:686131. https://doi.org/10.3389/fnut.2021.686131

Del Prado-Audelo ML, Bernal-Chávez SA, Gutiérrez-Ruíz SC, Hernández-Parra H, Kerdan IG, Reyna-González JM et al (2022) Stability phenomena associated with the development of polymer-based nanopesticides. Oxid Med Cell Longev:5766199. https://doi.org/10.1155/2022/5766199

Demir E, Kansız S, Doğan M, Topel Ö, Akkoyunlu G, Kandur MY et al (2022) Hazard assessment of the effects of acute and chronic exposure to permethrin, copper hydroxide, acephate, and validamycin nanopesticides on the physiology of Drosophila: novel insights into the cellular internalization and biological effects. Int J Mol Sci 23(16):9121. https://doi.org/10.3390/ijms23169121

Deshmukh SK, Kochar M, Kaur P, Singh PP (2022) Nanotechnology in agriculture and environmental science. CRC

Devi PV, Duraimurugan P, Chandrika KS (2019) *Bacillus thuringiensis*-based nanopesticides for crop protection. In: Nano-biopesticides today and future perspectives. Academic, pp 249–260

Ding Y, Xiao Z, Chen F, Yue L, Wang C, Fan N et al (2023) A mesoporous silica nanocarrier pesticide delivery system for loading acetamiprid: effectively manage aphids and reduce plant pesticide residue. Sci Total Environ 863:160900. https://doi.org/10.1016/j.scitotenv.2022.160900

Djiwanti SR, Kaushik S (2019) Nanopesticide: future application of nanomaterials in plant protection. In: Plant nanobionics approaches nanoparticles biosynthesis toxicity, vol 2, pp 255–298

Endshaw W, Hiruy B (2021) Farmers' pest management practices of stored faba bean and their implication to food security in Farta District, North West Ethiopia. Turkish JAF Sci Tech 9(4):665–674. https://doi.org/10.24925/turjaf.v9i4.665-674.3860

Engelmann W, Hupffer HM, Von Hohendorff R (2020) Potential legal avenues for managing the environmental risks of nanotechnology. In: Sustainable consumption: the right to a healthy environment, pp 479–494

do Espirito Santo Pereira A, Caixeta Oliveira H, Fernandes Fraceto L, Santaella C (2021) Nanotechnology potential in seed priming for sustainable agriculture. Nanomaterials (Basel) 11(2):267. https://doi.org/10.3390/nano11020267

Fabiyi OA (2023) The perspective of nanotechnology in plant protection. In: Nanophytopathology. CRC, pp 23–38

Faridvand S, Amirnia R, Tajbakhsh M, Enshasy HE, Sayyed RZ (2021) The effect of foliar application of magnetic water and nano, organic, and chemical fertilizers on phytochemical and yield characteristics of different landraces of *Foeniculum vulgare* mill. Horticulturae 7:475. https://www.mdpi.com/2311-7524/7/11/475

Farooq T, Adeel M, He Z, Umar M, Shakoor N, da Silva W et al (2021) Nanotechnology and plant viruses: an emerging disease management approach for resistant pathogens. ACS Nano 15(4):6030–6037. https://doi.org/10.1021/acsnano.0c10910

Fraceto LF, De Castro VL, Grillo R, Ávila D, Oliveira HC, Lima R (2020) Nanopesticides. Springer International Publishing

Furxhi I, Murphy F, Poland CA, Cunneen M, Mullins M (2021) Precaution as a risk in data gaps and sustainable nanotechnology decision support systems: a case study of nano-enabled textiles production. NanoEthics 3:245–270. https://doi.org/10.1007/s11569-021-00400-z

Gahukar RT, Das RK (2020) Plant-derived nanopesticides for agricultural pest control: challenges and prospects. Nanotechnol Environ Eng 5:1–9

Gaur J, Kumar S, Kaur H, Pal M, Bala K, Batoo KM & Hussain S (2024) Eco-friendly innovation: Harnessing nature's blueprint for enhanced photocatalysis and antimicrobial potential in multi-structured PN/ZnO nanoparticles. Funct Compos Struct 6(1):015005

Galhardi JA, Wang P, Bueno V, Ghoshal S, Gravel V, Wilkinson KJ et al (2022) Field evaluation of the potential effects of polymer and silica-based nanopesticides on strawberries and agricultural soils. Environ Sci Nano 9(10):3833–3843. https://doi.org/10.1039/D2EN00329E

Garza-García JJ, Hernández-Díaz JA, Zamudio-Ojeda A, León-Morales JM, Guerrero-Guzmán A, Sánchez-Chiprés DR, ... & García-Morales S (2022) The role of selenium nanoparticles in agriculture and food technology. Biol Trace Elem Res 200(5):2528–2548

Gondal AH, & Tayyiba L (2022) Prospects of using nanotechnology in agricultural growth, environment and industrial food products. Rev Agric Sci 10:68–81

Graily-Moradi F, Asgari Lajayer B (2021) Nanoinsecticides: preparation [application], and mode of action. In: Nanotechnology applications in health and environment science. Springer Nature Switzerland AG, pp 385–405

Grillo R, Fraceto LF, Amorim MJB, Scott-Fordsmand JJ, Schoonjans R, Chaudhry Q (2021) Ecotoxicological and regulatory aspects of environmental sustainability of nanopesticides. J Hazard Mater 404(A):124148. https://doi.org/10.1016/j.jhazmat.2020.124148

Hajam YA, & Kumar R (2022) Management of stored grain pest with special reference to Callosobruchus maculatus, a major pest of cowpea: A review. Heliyon 8(1)

Hazarika A, Yadav M, Yadav DK, Yadav HS (2022) An overview of the role of nanoparticles in sustainable agriculture. Biocatal Agric Biotechnol 43:102399. https://doi.org/10.1016/j.bcab.2022.102399

He X, Deng H, Hwang HM (2019) The current application of nanotechnology in food and agriculture. J Food Drug Anal 27(1):1–21. https://doi.org/10.1016/j.jfda.2018.12.002

Hofmann T, Lowry GV, Ghoshal S, Tufenkji N, Brambilla D, Dutcher JR et al (2020) Technology readiness and overcoming barriers to sustainably implement nanotechnology-enabled plant agriculture. Nat Food 1(7):416–425. https://doi.org/10.1038/s43016-020-0110-1

Hou J, Hu C, White JC, Yang K, Zhu L, Lin D (2021) Nano–zoo interfacial interaction as a design principle for hybrid soil remediation technology. ACS Nano 15(9):14954–14964. https://doi.org/10.1021/acsnano.1c05180

Ikawati S, Himawan T, Abadi AL, Tarno H (2021) Toxicity nanoinsecticide based on clove essential oil against *Tribolium castaneum* (Herbst). J Pestic Sci 46(2):222–228. https://doi.org/10.1584/jpestics.D20-059

Ingle AP, Gupta I (2021) Role of metal-based nanoparticles in plant protection. In: Nanotechnology in plant growth promotion and protection, vol 6, pp 220–238

Ioannou A, Gohari G, Papaphilippou P, Panahirad S, Akbari A, Dadpour MR et al (2020) Advanced nanomaterials in agriculture under a changing climate: the way to the future? Environ Exp Bot 176:104048. https://doi.org/10.1016/j.envexpbot.2020.104048

Jasrotia P, Nagpal M, Mishra CN, Sharma AK, Kumar S, Kamble U et al (2022) Nanomaterials for postharvest management of insect pests: current state and future perspectives. Front Nanotechnol 3. https://doi.org/10.3389/fnano.2021.811056

Jogaiah S, Singh HB, Fraceto LF, De Lima R (2020) Advances in nano-fertilizers and nano-pesticides in agriculture: a smart delivery system for crop improvement. Woodhead Publishing

Kah M, Johnston LJ, Kookana RS, Bruce W, Haase A, Ritz V et al (2021) Comprehensive framework for human health risk assessment of nanopesticides. Nat Nanotechnol 16(9):955–964. https://doi.org/10.1038/s41565-021-00964-7

Kale SK, Parishwad GV, Husainy ASN, Patil AS (2021) Emerging agriculture applications of silver nanoparticles. ES Food Agrofor 3:17–22. https://doi.org/10.30919/esfaf438

Kalita R, Saha O, Rahman N, Tiwari S, Phukon M (2021) Nanotechnology in agriculture. In: Nanobiotechnology mitigation abiotic stress plants, pp 101–116

Kannan M, Bojan N, Swaminathan J, Zicarelli G, Hemalatha D, Zhang Y et al (2023) Nanopesticides in agricultural pest management and their environmental risks: a review. Int J Environ Sci Technol 7:1–26. https://doi.org/10.1007/s13762-023-04795-y

Khan SH (2020) Green nanotechnology for the environment and sustainable development. In: Green materials for wastewater treatment, pp 13–46. https://doi.org/10.1007/978-3-030-17724-9_2

Kokotovich AE, Kuzma J, Cummings CL, Grieger K (2021) Responsible innovation definitions, practices, and motivations from nanotechnology researchers in food and agriculture. NanoEthics 15(3):229–243. https://doi.org/10.1007/s11569-021-00404-9

Kookana RS, Boxall AB, Reeves PT, Ashauer R, Beulke S, Chaudhry Q et al (2014) Nanopesticides: guiding principles for regulatory evaluation of environmental risks. J Agric Food Chem 62(19):4227–4240. https://doi.org/10.1021/jf500232f

Kumar S, Nehra M, Dilbaghi N, Marrazza G, Hassan AA, Kim KH (2019) Nano-based smart pesticide formulations: emerging opportunities for agriculture. J Control Release 294:131–153. https://doi.org/10.1016/j.jconrel.2018.12.012

Lade BD, Gogle DP (2019) Nano-biopesticides: synthesis and applications in plant safety. In: Nanobiotechnology applications in plant protection, vol 2, pp 169–189. https://doi.org/10.1007/978-3-030-13296-5_9

Lima PH, Antunes DR, Forini MM, Pontes MD, Mattos BD, Grillo R (2021) Recent advances on lignocellulosic-based nanopesticides for agricultural applications. Front Microbiol 11:580024. https://doi.org/10.3389/fnano.2021.809329

López-Cabeza R (2020) Methods for understanding the fate of nanopesticides in soil and water. Nanopesticides Res Dev Mech Action Sustain Use Agric 111–136

Mahanta DK, Jangra S, Priti A, Ghosh A, Sharma PK, Iquebal MA et al (2022) Groundnut bud necrosis virus modulates the expression of innate immune, endocytosis, and cuticle development-associated genes to circulate and propagate in its vector. Thrips Palmi Front Microbiol 11:580024. https://doi.org/10.3389/fmicb.2022.773238

Mahanta DK, Komal J, Samal I, Bhoi TK, Dubey VK, Pradhan K et al (2023) Nutritional aspects and dietary benefits of "Silkworms": current scenario and future outlook. Front Microbiol 11:580024

Manna S, Roy S, Dolai A, Ravula AR, Perumal V, Das A (2023) Current and future prospects of "all-organic" nanoinsecticides for agricultural insect pest management. Front Microbiol 11:580024. https://doi.org/10.3389/fnano.2022.1082128

Mali SC, Dhaka A, Githala CK, & Trivedi R (2020) Green synthesis of copper nanoparticles using Celastrus paniculatus Willd. leaf extract and their photocatalytic and antifungal properties. Biotechnol Rep 27, e00518

Maroufpour N, Mousavi M, Abbasi M, Ghorbanpour M (2020) Biogenic nanoparticles as novel sustainable approach for plant protection. In: Biogenic nano-particles and their use in agro-ecosystems, pp 161–172

Mehrwar V, & Uniyal VP (2021) Insect pest diversity of standing crops and traditional pest management in agricultural areas of the Mandakini Valley, Garhwal Himalaya, Uttarakhand, India. Intl J Hortic Agric Food Sci 5(4):1–9

Melanie M, Miranti M, Kasmara H, Malini DM, Husodo T, Panatarani C et al (2022) Nanotechnology-based bioactive antifeedant for plant protection. Nanomaterials (Basel) 12(4):630. https://doi.org/10.3390/nano12040630

Mujtaba M, Wang D, Carvalho LB, Oliveira JL, Espirito Santo Pereira AD, Sharif R, Jogaiah S, Paidi MK, Wang L, Ali Q, Fraceto LF (2021) Nanocarrier-mediated delivery of *miRNA*, *RNAi*, and *CRISPR-Cas* for plant protection: current trends and future directions. ACS Agric Sci Technol 1(5):417–435

Naharki K, Jaishi M (2020) Documentation of indigenous technical knowledge and their application in pest management in western mid hill of Nepal. SAARC J Agric 18(1):251–261. https://doi.org/10.3329/sja.v18i1.48397

Neme K, Nafady A, Uddin S, Tola YB (2021) Application of nanotechnology in agriculture, postharvest loss reduction and food processing: food security implication and challenges. Heliyon 7(12). https://doi.org/10.1016/j.heliyon.2021.e08539

Ogunyemi SO, Xu X, Xu L, Abdallah Y, Rizwan M, Lv L et al (2023) Cobalt oxide nanoparticles: an effective growth promoter of *Arabidopsis* plants and nano-pesticide against bacterial leaf blight pathogen in rice. Ecotoxicol Environ Saf 257:114935. https://doi.org/10.1016/j.ecoenv.2023.114935

Okey-Onyesolu CF, Hassanisaadi M, Bilal M, Barani M, Rahdar A, Iqbal J et al (2021) Nanomaterials as nanofertilizers and nanopesticides: an overview. ChemistrySelect 6(33):8645–8663. https://doi.org/10.1002/slct.202102379

de Oliveira JL (2021) Nano-biopesticides: present concepts and future perspectives in integrated pest management. In: Advances in nano-fertilizers and nano-pesticides in agriculture, pp 1–27

Pascoli M, Jacques MT, Agarrayua DA, Avila DS, Lima R, Fraceto LF (2019) Neem oil based nanopesticide as an environmentally friendly formulation for applications in sustainable agriculture: an ecotoxicological perspective. Sci Total Environ 677:57–67. https://doi.org/10.1016/j.scitotenv.2019.04.345

Peng F, Wang X, Zhang W, Shi X, Cheng C, Hou W et al (2022) Nanopesticide formulation from pyraclostrobin and graphene oxide as a nanocarrier and application in controlling plant fungal pathogens. Nanomaterials (Basel) 12(7):1112. https://doi.org/10.3390/nano12071112

Pérez-de-Luque A (2020) Guest edited collection: nanotechnology in agriculture. Sci Rep 10(1):15738. https://doi.org/10.1038/s41598-020-73198-7

Pho QH, Losic D, Ostrikov KK, Tran NN, Hessel V (2020) Perspectives on plasma-assisted synthesis of N-doped nanoparticles as nanopesticides for pest control in crops. React Chem Eng 5(8):1374–1396. https://doi.org/10.1039/D0RE00069H

Pradhan S, Mailapalli DR (2020) Nanopesticides for pest control. Sustain Agric Rev 40:43–74

Pramanik P, Krishnan P, Maity A, Mridha N, Mukherjee A, Rai V (2020) Application of nanotechnology in agriculture. In: Environmental nanotechnology, vol 4, pp 317–348. https://doi.org/10.1007/978-3-030-26668-4_9

Priyanka P, Kumar D, Yadav K, Yadav A (2019) Nanopesticides: synthesis, formulation and application in agriculture. In: Nanobiotechnology applications in plant protection, vol 2, pp 129–143. https://doi.org/10.1007/978-3-030-13296-5_7

Raj S, Trivedi R, & Soni V (2021) Biogenic synthesis of silver nanoparticles, characterization and their applications—a review. SurfS 5(1):67–90

Rana R, Samal I, & Bhoi TK (2022, November 21–23). Traditional methods of insect pest management in India. In Proceedings of the VIIth International Conference on Global Research Initiatives for Sustainable Agriculture & Allied Sciences (GRISAAS-2022)

Rani N, Duhan A, Pal A, Kumari P, Beniwal RK, Verma D et al (2023) Are nano-pesticides really meant for cleaner production? An overview on recent developments, benefits, environmental hazards and future prospectives. J Clean Prod 411:137232. https://doi.org/10.1016/j.jclepro.2023.137232

Rathore S, Chandola M, Raghuvanshi R, Kaur M, Singh KV (2021) Indigenous pest management practices of Indian hill farmers: introspecting their rationale and communication pattern for secure ecosystems. Sustainability 13(21):11608. https://doi.org/10.3390/su132111608

Rehmanullah MZ, Inayat N, Majeed A (2020) Application of nanoparticles in agriculture as fertilizers and pesticides: challenges and opportunities. In: New frontiers in stress management for durable agriculture, pp 281–293

Rennick JJ, Johnston APR, Parton RG (2021) Key principles and methods for studying the endocytosis of biological and nanoparticle therapeutics. Nat Nanotechnol 16(3):266–276. https://doi.org/10.1038/s41565-021-00858-8

Roseline TA, Murugan M, Sudhakar MP, Arunkumar K (2019) Nanopesticidal potential of silver nanocomposites synthesized from the aqueous extracts of red seaweeds. Environ Technol Innov 13:82–93. https://doi.org/10.1016/j.eti.2018.10.005

Saini N, Ledwani L (2022) Potential applications of nanotechnology in agriculture: conceptions, characteristics, prospects, and limitations – a review. Nano World J 8(Suppl 1):S147

Samal I, Bhoi TK, Majhi PK, Murmu S, Pradhan AK, Kumar D et al (2023) Combatting insects mediated biotic stress through plant associated endophytic entomopathogenic fungi in horticultural crops. Front Plant Sci 13:1098673

Sangeetha J, Hospet R, Thangadurai D, Adetunji CO, Islam S, Pujari N et al (2021) Nanopesticides, nanoherbicides, and nanofertilizers: the greener aspects of agrochemical synthesis using nanotools and nanoprocesses toward sustainable agriculture. In: Handbook of nanomaterials and nanocomposites for energy and environmental applications. Springer, Cham, pp 1663–1677

Santiago EF, Pontes MS, Arruda GJ, Caires AR, Colbeck I, Maldonado-Rodriguez R et al (2020) Understanding the interaction of nanopesticides with plants. In: Nanopesticides. Springer, pp 69–109

Santos TS, Silva TM, Cardoso JC, Albuquerque-Júnior RLC, Zielinska A, Souto EB et al (2021) Biosynthesis of silver nanoparticles mediated by entomopathogenic fungi: antimicrobial resistance, nanopesticides, and toxicity. Antibiotics (Basel) 10(7):852. https://doi.org/10.3390/antibiotics10070852

Saqib S, Nazeer A, Ali M, Zaman W, Younas M, Shahzad A et al (2022) Catalytic potential of endophytes facilitates synthesis of biometallic zinc oxide nanoparticles for agricultural application. BioMetals 35(5):967–985. https://doi.org/10.1007/s10534-022-00417-1

Saritha GNG, Anju T, Kumar A (2022) Nanotechnology-big impact: how nanotechnology is changing the future of agriculture? J Agric Food Res 10:100457. https://doi.org/10.1016/j.jafr.2022.100457

Saxena R, Kumar M, Tomar RS (2020) Implementation of nanotechnology in agriculture system: a current perspective. Nanobiotechnol Concepts Appl Health Agric Environ 121

Schiavi D, Ronchetti R, Di Lorenzo V, Salustri M, Petrucci C, Vivani R et al (2022) Circular hazelnut protection by lignocellulosic waste valorization for nanopesticides development. Appl Sci 12(5):2604. https://doi.org/10.3390/app12052604

Shahavi MH, Hosseini M, Jahanshahi M, Meyer RL, Darzi GN (2019) Evaluation of critical parameters for preparation of stable clove oil nanoemulsion. Arab J Chem 12(8):3225–3230. https://doi.org/10.1016/j.arabjc.2015.08.024

Shahbaz M, Akram A, Mehak A, Haq E, Fatima N, Wareen G, Aamir S, Hassan H, Poczai P, Sayyed RZ, Ilyas N (2023) Evaluation of selenium nanoparticles in inducing disease resistance against spot blotch disease and promoting growth in *Triticum aestivum* under biotic stress. Plants 12(4):761. https://doi.org/10.3390/plants12040761

Shao C, Zhao H, Wang P (2022) Recent development in functional nanomaterials for sustainable and smart agricultural chemical technologies. Nano Converg 9(1):11. https://doi.org/10.1186/s40580-022-00302-0

Shaila M, & Begum N (2021) Ancient farming methods of seed storage and pest management practices in India—A Review. Plant Arch 21:499–509

Sharma S, Kumari P, Shandilya M, Thakur S, Perveen K, Sheikh I, Ahmed Z, Sayyed R, Mastinu A (2025) The combination of α-Fe_2O_3NP and *Trichoderma* sp. improves antifungal activity against *Fusarium* wilt. J Basic Microbiol:e2400613. https://doi.org/10.1002/jobm.202400613

Shekhar S, Sharma S, Kumar A, Taneja A, Sharma B (2021) The framework of nanopesticides: a paradigm in biodiversity. Mater Adv 2(20):6569–6588

Shivashakarappa K, Reddy V, Tupakula VK, Farnian A, Vuppula A, Gunnaiah R (2022) Nanotechnology for the detection of plant pathogens. Plant Nano Biol 2:100018. https://doi.org/10.1016/j.plana.2022.100018

Sindhu RK, Chitkara M, Sandhu IS (2021) Nanotechnology: principles and applications. CRC

Singh S, Sangwan S, Sharma P, Devi P, Moond M (2021a) Nanotechnology for sustainable agriculture: an emerging perspective. J Nanosci Nanotechnol 21(6):3453–3465. https://doi.org/10.1166/jnn.2021.19012

Singh H, Sharma A, Bhardwaj SK, Arya SK, Bhardwaj N, Khatri M (2021b) Recent advances in the applications of nano-agrochemicals for sustainable agricultural development. Environ Sci Process Impacts 23(2):213–239. https://doi.org/10.1039/d0em00404a

Soni RA, Rizwan MA, Singh S (2022) Opportunities and potential of green chemistry in nanotechnology. Nanotechnol Environ Eng 7(3):661–673. https://doi.org/10.1007/s41204-022-00233-5

Spanos A, Athanasiou K, Ioannou A, Fotopoulos V, Krasia-Christoforou T (2021) Functionalized magnetic nanomaterials in agricultural applications. Nanomaterials (Basel) 11(11):3106. https://doi.org/10.3390/nano11113106

Stadler T, Buteler M, Valdez SR, Gitto JG (2018) Particulate nanoinsecticides: a new concept in insect pest management. Insectic Agric Toxicol 23:83

Sun Y, Liang J, Tang L, Li H, Zhu Y, Jiang D et al (2019) Nano-pesticides: a great challenge for biodiversity? Nano Today 28:100757. https://doi.org/10.1016/j.nantod.2019.06.003

Svendsen C, Walker LA, Matzke M, Lahive E, Harrison S, Crossley A et al (2020) Key principles and operational practices for improved nanotechnology environmental exposure assessment. Nat Nanotechnol 15(9):731–742. https://doi.org/10.1038/s41565-020-0742-1

Takeshita V, Munhoz-Garcia GV, Pereira AES, Tornisielo VL, Fraceto LF (2023) Radiometric strategy to track nanopesticides: an important approach to understand the fate, mechanisms of action and toxicity. TrAC Trends Anal Chem 165:117156. https://doi.org/10.1016/j.trac.2023.117156

Thakur P, Thakur S, Kumari P, Shandilya M, Sharma S, Poczai P, Alarfaj AA, Sayyed RZ (2022) Nano-insecticide: synthesis, characterization, and evaluation of insecticidal activity of ZnO NPs against Spodoptera litura and Macrosiphum euphorbiae. Appl Nanosci 12:3835–3850. https://doi.org/10.1007/s13204-022-02530-6

Thangadurai D, Sangeetha J, Prasad R (2020) Nanotechnology for food. Agric Environ

Tipu MM, Baroi A, Rana J, Islam S, Jahan R, Miah MS et al (2021) Potential applications of nanotechnology in agriculture: a smart tool for sustainable agriculture. In: Agricultural development in Asia—potential use of nanomaterials and nanotechnology. IntechOpen

Tiwari E, Mondal M, Singh N, Khandelwal N, Monikh FA, Darbha GK (2020) Effect of irrigation water type and other environmental parameters on CeO_2 nanopesticide–clay colloid interactions. Environ Sci Process Impacts 22(1):84–94. https://doi.org/10.1039/c9em00428a

Tran NN, Le TN, Pho HQ, Tran TT, Hessel V (2022) Nanofertilizers and nanopesticides for crop growth. In: Plant and nanoparticles. Springer Nature Singapore, pp 367–394

Ur Rahim H, Qaswar M, Uddin M, Giannini C, Herrera ML, Rea G (2021) Nano-enabled materials promoting sustainability and resilience in modern agriculture. Nanomaterials 11(8):2068. https://doi.org/10.3390/nano11082068

Usman M, Farooq M, Wakeel A, Nawaz A, Cheema SA, Rehman HU et al (2020) Nanotechnology in agriculture: current status, challenges and future opportunities. Sci Total Environ 721:137778. https://doi.org/10.1016/j.scitotenv.2020.137778

Vasseghian Y, Arunkumar P, Joo SW, Gnanasekaran L, Kamyab H, Rajendran S et al (2022) Metal-organic framework-enabled pesticides: an emerging tool for sustainable cleaner production and environmental hazard reduction. J Clean Prod 373:133966. https://doi.org/10.1016/j.jclepro.2022.133966

Vijayakumar MD, Surendhar GJ, Natrayan L, Patil PP, Ram PMB, Paramasivam P (2022) Evolution and recent scenario of nanotechnology in agriculture and food industries. J Nanomater 2022:1–17. https://doi.org/10.1155/2022/1280411

Wang Y, Deng C, Rawat S, Cota-Ruiz K, Medina-Velo I, Gardea-Torresdey JL (2021) Evaluation of the effects of nanomaterials on Oryza sativa responses: underlining the benefits of nanotechnology for agricultural applications. ACS Agric Sci Technol 1(2):44–54. https://doi.org/10.1021/acsagscitech.1c00030

Wang D, Saleh NB, Byro A, Zepp R, Sahle-Demessie E, Luxton TP et al (2022) Nano-enabled pesticides for sustainable agriculture and global food security. Nat Nanotechnol 17(4):347–360. https://doi.org/10.1038/s41565-022-01082-8

Yadav RK, Singh NB, Singh A, Yadav V, Bano C, Khare S et al (2020) Expanding the horizons of nanotechnology in agriculture: recent advances, challenges and future perspectives. Vegetos 33(2):203–221. https://doi.org/10.1007/s42535-019-00090-9

Yan S, Ren BY, Shen J (2021) Nanoparticle-mediated double-stranded RNA delivery system: a promising approach for sustainable pest management. Insect Sci 28(1):21–34. https://doi.org/10.1111/1744-7917.12822

Yin J, Su X, Yan S, Shen J (2023) Multifunctional nanoparticles and nanopesticides in agricultural application. Nanomaterials 13(7):1255. https://doi.org/10.3390/nano13071255

Zaheer T, Ali MM, Abbas RZ, Atta K, Amjad I, Suleman A et al (2022) Insights into nanopesticides for ticks: the superbugs of livestock. Oxid Med Cell Longev 2022:7411481. https://doi.org/10.1155/2022/7411481

Zhai Y, Abdolahpur Monikh FA, Wu J, Grillo R, Arenas-Lago D, Darbha GK et al (2020) Interaction between a nano-formulation of atrazine and rhizosphere bacterial communities: atrazine degradation and bacterial community alterations. Environ Sci Nano 7(11):3372–3384. https://doi.org/10.1039/D0EN00638F

Zhang P, Guo Z, Ullah S, Melagraki G, Afantitis A, Lynch I (2021) Nanotechnology and artificial intelligence to enable sustainable and precision agriculture. Nat Plants 7(7):864–876. https://doi.org/10.1038/s41477-021-00946-6

Zhang X, He Y, Yuan Z, Shen G, Zhang Z, Niu J et al (2023a) A pH- and enzymatic-responsive nanopesticide to control pea aphids and reduce toxicity for earthworms. Sci Total Environ 861:160610. https://doi.org/10.1016/j.scitotenv.2022.160610

Zhang J, Kothalawala S, Yu C (2023b) Engineered silica nanomaterials in pesticide delivery: challenges and perspectives. Environ Pollut 320:121045. https://doi.org/10.1016/j.envpol.2023.121045

Zobir SAM, Ali A, Adzmi F, Sulaiman MR, Ahmad K (2021) A review on nanopesticides for plant protection synthesized using the supramolecular chemistry of layered hydroxide hosts. Biology 10(11):1077. https://doi.org/10.3390/biology10111077

Zou R, Li B, Duan W, Lin G, Cui Y (2022) Synthesis of 3-carene-derived nanocellulose/1,3,4-thiadiazole-amide complexes with antifungal activity for plant protection. Pest Manag Sci 78(8):3277–3286. https://doi.org/10.1002/ps.6952

Neoformulations in Crop and Soil Nutrition

5

Maimona Saeed, Noshin Ilyas, Fatima Bibi, Saman Sarfraz, and Saoli Chanda

Abstract

The formation of new soil and nutrient availability became a recent approach in sustainable agriculture. Neoformation consists of various processes such as the addition of organic matter, microbial activity, symbiotic interaction, and nutrient cycling. These processes are the important part in the formation of nutrients in soil. For this purpose, microbiomes play an important role in the maintenance of soil health. Microbiomes are the collection of microorganisms found in the soil environment. These helpful microorganisms engage in a variety of plant growth-promoting activities, like fixing, mineralizing, solubilizing, and mobilizing nutrients; producing siderophores, antagonistic substances, and antibiotics; and releasing substances that stimulate plant growth, like auxin and gibberellin hormones, through interactions with the roots of the host plant. Some species of plant make symbiotic associations with microbes and profit from the availability of minerals and nutrients while using little energy. The distribution of plant species determines the nature and amount of root exudates, as well as the crop species and cultivars. On the other side, it is possible to isolate, formulate, and create biofertilizers from microbes that play crucial functions in the microbiome. In agriculture, significant microorganisms with Fe- and Zn-solubilizing properties can be employed to biofortify micronutrients in various cereal crops. A safe envi-

M. Saeed (✉)
Department of Botany, Rawalpindi Women University, Rawalpindi, Pakistan

N. Ilyas · F. Bibi
Department of Botany, PMAS Arid Agriculture University, Rawalpindi, Pakistan

S. Sarfraz
Government College Women University, Sialkot, Pakistan

S. Chanda
Earth and Environment Department, Florida International University, Miami, FL, USA

R. Sayyed et al. (eds.), *Nanobiostimulants in Innovative Agriculture*,
https://doi.org/10.1007/978-981-95-2477-8_5

ronment is achieved through sustainable agriculture, which in turn has a good impact on soil health when it is used effectively. The subject of future research will be how to promote crop growth by shaping soil characteristics with the help of plant microbial communities and organic amendments in the soil. This chapter illustrates how microbiomes help in the neoforming of crops and the integration of nutrients into the soil.

Keywords

Microbiome · Phytoremediation · Plant growth promoting traits · Root exudates · Soil-integrated nutrient

1 Introduction

Neoformation is the process of soil mineral formation and its transfer into organic matter. It occurs during the weathering of soil and the addition of new minerals into the soil with time. It helps to control water cycling, maintain biodiversity, and especially plant health. When new soil and minerals are added into the soil for the growth and development of plants, it is called neoformation in crops (Lavelle and Spain 2001). Different mechanisms take part in the modification and addition of minerals into soil. For this purpose, we will understand the process of soil formation.

2 Soil Formation

Soil is a delicate combination of organic matter and tiny mineral particles that is gradually generated from rocks that outcrop at the surface and other surficial parent materials that are typically out of balance with the atmosphere and biosphere. As sub-components of ecosystems, soils are open thermodynamic systems that allow energy to move through and dissipate, as evidenced by the variety of biological and physical structures (Parmar 2014).

The intricate relationships between the lithosphere, atmosphere, and biosphere result in soils, which are results of these interactions and represent the distinct effects of each. Although soils are primarily composed of organic matter and lithosphere-derived minerals, their pore spaces also have unique hydrological surroundings and atmospheres. Through the development and penetration of their roots, plants contribute energy to the soil system. It takes tens to hundreds of thousands of years for completely developed soils to form, as soil development is a gradual process. Erosion and leaching are the main degradation processes, whereas aggravating activities like weathering and translocation are balanced to produce it (Singh et al. 2017).

Most significantly, organisms, e.g., microorganisms, also provide energy to the soil through partially decomposed byproducts of their digestive and metabolic activities, including metabolites like earthworm mucus, microbial polysaccharides, leachates, root exudates, and, ultimately, their dead biomasses (Rani et al. 2017).

2.1 Soil Nutrient

For the growth of plants, the major source of nutrients is soil. Nutrient management provides a comprehensive and sophisticated approach to nutrient supply, making it a shining example of sustainable agriculture. In addition to optimizing current crop yields, soil nutrient also guarantees the long-term health of soils and ecosystems by balancing organic and inorganic sources, taking crop rotation into account, and introducing beneficial bacteria. The synergy created by combining several nutrition sources is the core of integrated nutrient management (INM). Composted from manure, compost, or crop leftovers, organic fertilizers improve soil structure, microbial activity, and water retention in addition to providing vital nutrients. As with inorganic fertilizers, they provide accuracy in treating certain nutrient deficits found by soil analysis (Wu and Ma 2015).

2.2 Role of Crop in Neoformation

The practice of using plants to improve soil fertility has been in existence for millennia, tracing back to the Greek and Roman periods, wherein legumes, specifically fava beans, were utilized to cover soil and enrich vineyards with nitrogen. During the 1700s, lupine cultivation was implemented in northern Europe as a means of enriching nutrient-deficient sandy soils through nitrogen fixation and root turnover, which helped introduce organic carbon and nutrients into the soil (Venturi and Keel 2016). These were perhaps the first plants developed for phytoremediation (the cleaning up of land). However, until recently, the ability of plants to clean up toxic waste and damaged soil remained unknown. Plants cover the majority of the planet's surface; it makes sense that they have a significant impact on the destiny and transportation of chemicals and xenobiotic substances. Land makes up 29% of the surface of the planet, and 71% of it is livable (the rest of the area is either covered in glaciers or is barren). Plants cover 98% of the area that is livable; 50% of it is used for agriculture, 37% is forested, and 11% is covered in shrubs and grassland (Briggs et al. 2021). About half of all primary output and carbon sequestration is contributed by trees and forests. In addition, due to their massive biomass, plants are the most powerful oxidative enzyme systems on Earth. Their catabolic enzyme systems have developed for respiration, plant defense, and detoxification, and as a byproduct, they incidentally break down several detrimental organic compounds through biodegradation (Rezapour et al. 2021; Patel et al. 2016). The use of biomass for bioenergy, maintaining biological diversity, stabilizing soil, and many other ecosystem services are further advantages of phytoremediation. By stimulating plant-associated microbes or using secondary metabolism, contaminants that the plant has absorbed can be detoxified (Budamagunta et al. 2022; Coleman et al. 2021). Due to its sustainability and affordability, this technology is viewed as an alternative to conventional physical and chemical cleanup techniques. The use of specific plant species for their capacity to absorb and accumulate contaminants, the presence of a robust and diverse soil microbiome. The success of phytoremediation depends on various

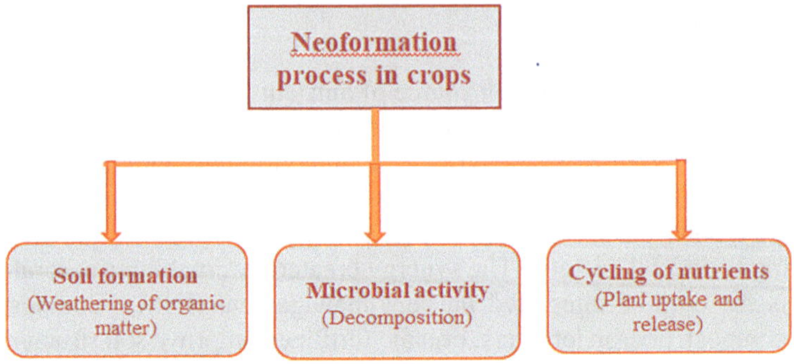

Fig. 5.1 Process of neoformation in crop

factors, including environmental elements such as contaminant concentration, soil pH, nutrient status, and its ability to reduce contaminants. Additionally, the presence and activity of soil- and plant-associated microorganisms also play a crucial role in this process (Kour and Sayyed 2019). Phytoremediation uses plants to stabilize, filter, volatilize, degrade, and remove pollutants (Fig. 5.2). Absorption, translocation, and pollutant metabolism by the plant, as well as contaminant degradation mediated by root exudates, are three components of the breakdown of contaminants via plant-produced enzymes released into the soil or through metabolic processes. Phytovolatilization, as described by Farrell and Germida (Kamath et al. 2004), is the process by which pollutants travel from the soil or groundwater through the plants and ultimately into the atmosphere. This occurred as a result of the pollution being taken up by the roots, moving to the plant, and volatilizing into the atmosphere. Phytostabilization, the term used by plants to restrict the pollutant in the soil or groundwater, can be attributed to three different mechanisms: (1) absorption, (2) root accumulation, and (3) surface adsorption by root and entry into humic matter in the plant rhizosphere. As previously stated, the three methods are all reliant on the value of Known. Utilizing plants' innate capacities to absorb, digest, and collect hazardous chemicals from the environment is a key component of this process Fig 5.1 describe the process of neoformation in crops. Neoformation process needs various essential elements which is enlisted in Table 5.1.

2.3 Role of Microorganisms in Neoformation of Crops

Soil is made up of four basic components: a mineral component, an organic component, water, and air. The organic component contains living Microorganisms that add minerals into soil by decomposition. The term "microbiome" was initially introduced by Lederberg and McCray to refer to a group of microorganisms that reside in a particular host or environment. The various populations of microorganisms that live in and around crops are referred to as crop microbiomes. This includes microorganisms that interact intricately with crops and one another, such as bacteria, fungi, viruses, and others. The interaction between the plant and its associated

Table 5.1 Essential elements of crop neoformation

Element	Characterization	Impact on soil
Organic matter	Decomposition of plant residues and root exudates	Enhance soil fertility
Weathering	The breakdown of rocks and minerals into smaller pieces	Enhance soil minerals. Raises the mineral content of the soil
Microbial activity	The participation of bacteria, fungi, and microorganisms in the processes present on soil.	Restore nutrient availability and cycling
Symbiotic interaction	Associations with mycorrhizal fungi and nitrogen-fixing bacteria.	Improves nutrient absorption and plant health.
Nutrient cycling	Recycling and mineralization of nutrients	Increase soil fertility

microbiome, which consists of bacteria, fungus, and other microorganisms living in and around the plant roots, affects the efficacy of phytoremediation (Huang et al. 2014; Bahman and Sayyed 2019).

2.4 Phytoremediation in Crop Neoformation

Depending on how they affect plants, bacteria that are linked with plants can be categorized as harmful, useful, or neutral. Therefore, the ongoing release of root exudates, or rhizodeposition, acts as a mechanism that demands high levels of carbon (C) and nitrogen (N), and it is a crucial factor in determining the composition of microbiological communities (Chaparro et al. 2012). The nature and makeup of root exudates can draw in beneficial bacteria and drive away potentially harmful ones (Novo et al. 2018; Venturi and Keel 2016). Biochemical compounds or metabolites are just a few of the substances that may be discovered in the root exudates of many plant species that are present in the rhizosphere. They might make up as much as a third of the C. Thus, plants offer nutrients, hormones, toxins, and substances that serve as antagonists to promote the proliferation of plant-beneficial bacteria species (Alves et al. 2021). Otherwise, bacteria can encourage growth and development of plants by providing. Plant-available nutrients help alleviate both biotic and abiotic stressors and provide a protective barrier against infections in plants. Bacteria are essential for nutrient absorption and transport, driving the cycle of critical nutrients (Saeed et al. 2023). These microorganisms alter the growth patterns of roots and decompose root exudates that are subsequently reabsorbed by the plants (Benizri and Kidd 2018). By dissolving and eliminating pollutants from the soil, some bacteria and fungi in the soil can be extremely important in the phytoremediation process. Fig 5.2 describe the mechanisms used in process of phytoremediation. For instance, rhizosphere bacteria like Pseudomonas putida and *Sphingomonas* can aid in the degradation of contaminants like hydrocarbons and heavy metals. Similar to this, it has been demonstrated that fungal species like *Trichoderma viridae* and *Fusarium oxysporum* promote plant development and increase the

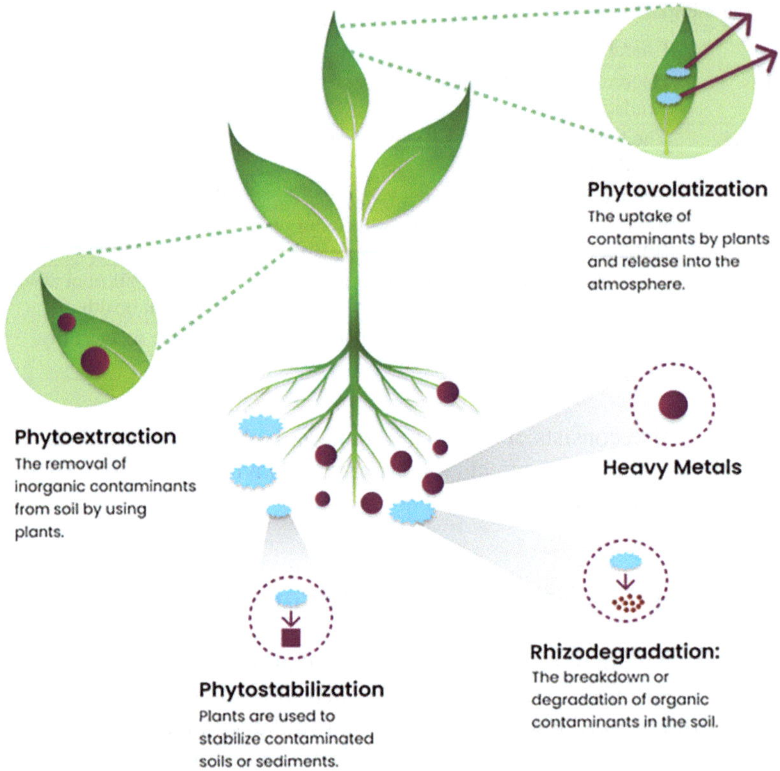

Fig. 5.2 Mechanism of phytoremediation

absorption of pollutants during the phytoremediation process (Novo et al. 2018). For instance, some bacteria can assist the plant in detoxifying hazardous compounds by converting them into less toxic forms, such as heavy metals and organic pollutants. This can make it simpler for the plant to absorb the toxins and store them in its tissues, where they can later be removed from the area and appropriately disposed of. The microbiome can aid in enhancing the development and survival of the plants employed in phytoremediation in addition to helping to break down pollutants. This can be done by giving the plant the vital nutrients and water it needs or by making the plant more resilient to environmental stresses like drought or extremely high or low temperatures (Rashid et al. 2016). One recent study examined the function of the microbiome in phytoremediation utilizing a variety of crops, including maize, wheat, and sunflowers. It was published in the journal *Frontiers in Microbiology* in 2020. The ability of the plant to absorb and retain pollutants like heavy metals was found to be enhanced by the microbiome, and the plant's growth and survival in polluted conditions were also improved (Table 5.2).

Recent research has shown that using crop microbiomes in phytoremediation has promise. One study revealed that *Brassica juncea* (Indian mustard) plants absorbed more cadmium and lead when a particular strain of bacteria, *Bacillus subtilis*, was

Table 5.2 Microbiomes and their association with crops

Crop	Microbiome	Mechanism	Reference
Wheat	*Bacillus thuringiensis* AZP2	Improve stress tolerance	Doty et al. (2018)
Soybean	*Pseudomonas putida*	Enhance tolerance against abiotic stress	Zhang et al. (2019)
Tomato	*Streptomyces, Luteimonas,* and *Enterobacter*	Improve plant growth	Pilon-Smits and LeDuc (2009)
Potato	*Methylobacterium* sp.	Change endophytic bacterial community	Weyens et al. (2010)
Chili	*Pseudomonas aeruginosa*	Improve iron availability by siderophores	Hunter (2016)
Maize	*Bacillus subtilis*	Improve abiotic stress tolerance	Saeed et al. (2022a, b)

added to the soil. This suggested that the phytoremediation process can be improved by the presence of certain bacteria in the soil. Another study revealed that the phytoremediation ability of *Brassica napus* (rape) plants in polluted soils was enhanced by the co-inoculation of several rhizosphere bacteria. The scientists discovered that the interaction of several bacteria had a synergistic impact that boosted the absorption of heavy metals by the plants (Sagar et al. 2020a, b). A different investigation, which was presented in the 2019 issue of the journal *Applied Microbiology and Biotechnology*, suggested that the microbiome may be crucial to the breakdown of polycyclic aromatic hydrocarbons (PAHs) in polluted soils. Researchers found that plants containing particular bacteria were better able to break down PAHs than plants without the bacteria. Saeed et al. (2021) reported the role of the microbiome in the remediation of crude oil contaminated soil. *Pseudoarthrobacter phenanthrenivorans* and *Azospirillum oryzae* degrade the hydrocarbons and improve the growth of maize in crude oil contaminated soil.

3 Current Challenges in Exploiting Crop and Microbiome Interaction

3.1 Limitation of Phytoremediation in Neoforming

At the moment, phytoremediation falls short of becoming a highly effective, reliable, and quick cleanup solution. This is demonstrated by the sometimes-inconsistent field results, the sluggish and incomplete disintegration, and the prolonged cleanse (Gowtham et al. 2022; Hosini et al 2022; Ilyas et al. 2022; Ilyas et al. 2021). Numerous investigations have concentrated on improving the state of the soil, making contaminants more accessible and bioavailable, creating favorable conditions for plant growth, and examining and utilizing the microorganisms that are associated with the soil and plants in the process of phytoremediation (Jabborova et al. 2022; Jabborova et al. 2021; Kapadia et al. 2022; Khan et al 2020; Kour et al. 2021). From an ecological perspective, the methodologies used for a considerable period were centered on individual organisms rather than an integrated meta-organism approach. In agriculture, improving the plant and the accompanying microorganisms, or the meta-organism, has been beneficial in increasing plant growth and disease resistance (Kusale et al. 2021a; Mawar et al 2023; Mendes et al. 2013; Rafique et al. 2023). Nevertheless, a holistic meta-organism technique that facilitates the discovery of interactions between the plant host and potentially hundreds of naturally occurring degradative microbial taxa that inhabit the host has not been widely implemented. Compared to microbial communities, it is crucial to the health of the carrier by producing vital vitamins, enhancing nutrient solubility, and defending the host against pathogens. It is asserted that the plant microbiome increases the host's practical capability. Additionally, the microbiome has the power to control how certain plant features are expressed, which enhances the physiological condition of the plant. The effectiveness of phytoremediation is highly influenced by the activities of the plant microbiome, according to recent studies. It's possible that the influence has been overestimated. A deeper comprehension of microbial and plant interactions and reactions to pollutants is crucial for improving phytoremediation. One specific issue that has to be addressed is how a carrier or host creates a healthy microbiome and how it functions. Molecular data and ecological models have already provided insights into the process of how insects establish and sustain a beneficial cuticular microbiome (Tarfeen et al. 2023). Microbiome establishment and functioning have also been closely examined in gut microbiome research (Trivedi et al. 2017), which has already led to the translation of these findings into helpful therapies. However, neither polluted soil nor a complex rhizosphere environment has had the concepts ever thoroughly investigated. We firmly think that a more focused stimulation of phytoremediation is achievable by knowing how the host communicates, forms, and maintains its microbiome capable of promoting growth of plant and pollutant breakdown.

4 Conclusion

The crop microbiomes are essential to the process of neoforming and integrating nutrients into soil. The development and health of the plants can be improved by the presence of certain bacteria and fungi in the soil, which can help in decomposition and add organic nutrients into the soil. This latest study shows the value of using microbiomes in crop neoforming and emphasizes the necessity of a diversified and healthy soil microbiome for this procedure. Despite trade-offs between bacteria that actively promote plant growth and health and those that break down pollutants, the plant host established a healthy microbiome. In order to function optimally for pollutant degradation, it may need more focus on human interventions because the plant-microbiome is not functioning at its best. To sum up, phytoremediation is difficult, but figuring out how plants influence their associated microbiota and vice versa will create new opportunities for improving the dependability and effectiveness of phytoremediation. It's essential to know how to capture and maintain the advantages of natural systems. Challenge for the present and the future, one that we must support in order to decrease the consequences of human activities on the environment and move toward a more sustainable society.

References

Alves ARA, Oliveira RS, Silva EF, Novo LAB (2021) Role of bacteria and mycorrhizal fungi in phytomining: status and future perspectives. Microbe Med Remed Environ Cont Elsevier 11(24):15–26

Bahman F-N, Sayyed RZ (2019) Plant growth promoting rhizobacteria and salinity stress: a journey into the soil. In: Sayyed, Arora, Reddy (eds) Plant growth promoting rhizobacteria for sustainable stress management, Abiotic stress management, vol 1, Springer, pp 21–34

Benizri E, Kidd PS (2018) The role of the rhizosphere and microbes associated with hyperaccumulator plants in metal accumulation. Springer Int Pub 11(786):157–188

Briggs GG, Bromilow RH, Evans AA (2021) Relationships between lipophilicity and root uptake and translocation of non-ionized chemicals by barley. Pestic PS 13(5):495–504

Budamagunta V, Shameem N, Irusappan S, Parray JA, Thomas M, Marimuthu S, Kirubakaran R, Jothi KNA, Sayyed RZ, Lim HR (2022) Microbial nanovesicle and extracellular polymeric substances mediate nickel tolerance and remediate heavy metal ions from soil. Environ Res 219:114997. https://doi.org/10.1016/j.envres.2022.114997

Chaparro JM, Sheflin AM, Manter DK, Vivanco JM (2012) Manipulating the soil microbiome to increase soil health and plant fertility. Biol Fertil Soils 48(5):489–499. https://doi.org/10.1007/s00374-012-0691-4

Coleman JOD, BlakeKalff MMA, Davies TGE (2021) Detoxification of xenobiotics by plants: chemical modification and vacuolar compartmentation. Trends Plant Sci 2(4):144–151

Kour D, Sayyed RZ (2019) Drought tolerant phosphorus solubilizing microbes: biodiversity and biotechnological applications for alleviation of drought stress in plant. In: Sayyed, Arora, Reddy (eds) Plant growth promoting rhizobacteria for sustainable stress management, Abiotic stress management, vol 1, Springer, pp 255–308

Doty SL, James CA, Moore AL, Vajzovic A, Singleton GL, Ma C, Khan Z, Xin G, Kang JW, Park JY, Meilan R, Strauss SH, Wilkerson J, Farin F, Strand SE (2018) Enhanced phytoremediation of volatile environmental pollutants with transgenic trees. Proc Natl Acad Sci USA 104(43):16816–16821

Gowtham HG, Singh SB, Shilpa N, Aiyaz M, Nataraj K, Udayashankar AC, Amruthesh KN, Murali M, Poczai P, Gafur A, Almalki WH, Sayyed RZ (2022) Insight into recent progress and perspectives in improvement of antioxidant machinery upon PGPR augmentation in plants under drought stress: a review. Antioxidants 11(9):1763. https://doi.org/10.3390/antiox11091763

Huang XF, Chaparro JM, Reardon KF, Zhang R, Shen Q, Vivanco JM (2014) Rhizosphere interactions: root exudates, microbes, and microbial communities. Botany 92(4):267–275. https://doi.org/10.1139/cjb-2013-0225

Hunter P (2016) Plant microbiomes and sustainable agriculture. EMBO Rep 17(11):1696–1699. https://doi.org/10.15252/embr.201643476

Ilyas N, Akhtar N, Naseem A, Qureshi R, Majeed A, Sayyed RZ (2022) The potential of *Bacillus subtilis* and phosphorus in improving the growth of wheat under chromium stress. J Appl Microbiol 00:1–15. https://sfamjournals.onlinelibrary.wiley.com/doi/10.1111/jam.15676

Ilyas N, Mumtaz K, Akhtar N, Yasmin H, Sayyed RZ, Khan W, Enshasy HE, Dailin DJ, Elsayed EA, Ali Z (2021) Exo-polysaccharides producing bacteria for the amelioration of drought stress in wheat. Sustainability 12:8876. https://www.mdpi.com/2071-1050/12/21/8876

Jabborova D, Annapurna K, Azimov A, Tyagi S, Pengani KR, Sharma S, Vikram K, Poczai P, Nasif O, Ansari MJ, Sayyed RZ (2022) Co-inoculation of biochar and arbuscular mycorrhizae for growth promotion and nutrient fortification in soybean under drought conditions. Front Plant Sci 13:947547. https://doi.org/10.3389/fpls.2022.947547

Jabborova D, Kannepalli A, Davranov K, Narimanov A, Enakiev Y, Syed A, Elgorban AM, Bahkali AH, Wirth S, Sayyed RZ, Gafur A (2021) Co-inoculation of rhizobacteria promotes growth, yield, and nutrient contents in soybean and improves soil enzymes and nutrients under drought conditions. Sci Rep 11:22081. https://doi.org/10.1038/s41598-021-01337-9

Kamath R, Schnoor JL, Alvarez PJ (2004) Effect of root-derived substrates on the expression of nah-lux genes in Pseudomonas fluorescens HK44: implications for PAH biodegradation in the rhizosphere. Environ Sci Technol 38(6):1740–1745

Kapadia C, Patel N, Rana A, Vaidya H, Alfarraj A, Ansari MJ, Gafur A, Poczai P, Sayyed RZ (2022) Evaluation of plant growth promoting and salinity ameliorating potential of halophillic bacteria isolated from saline soil. Front Plant Sci 13:946217. https://doi.org/10.3389/fpls.2022.946217

Khan I, Awan SA, Ikram R, Rizwan M, Akhtar N, Yasmin H, Sayyed RZ, Ali S, Ilyas N (2020) Effects of 24-epibrassinolide on plant growth, antioxidants defense system, and endogenous hormones in two wheat varieties under drought stress. Physiol Plant:1–11. https://onlinelibrary.wiley.com/doi/abs/10.1111/ppl.13237

Kour D, Kaur T, Devi R, Yadav A, Singh M, Joshi D, Singh J, Suyal DC, Kumar A, Rajput VD, Yadav AN, Singh K, Singh J, Sayyed RZ, Arora NK, Saxena AK (2021) Beneficial microbiomes for bioremediation of diverse contaminated environments for environmental sustainability: present status and future challenges. Environ Sci Pollut Res 28:24917–24939. https://doi.org/10.1007/s11356-021-13252-7

Kusale SP, Attar YC, Sayyed RZ, Malek RA, Ilyas N, Suriani NL, Khan N, Enshasy HE (2021a) Production of plant beneficial and antioxidants metabolites by *Klebsiella variicola* under salinity stress. Molecules 26:1894. https://doi.org/10.3390/molecules26071894

Lavelle P, Spain AV (2001) Soil formation. Soil Ecol: vol (1)143–200. https://doi.org/10.1007/0-306-48162-6

Mawar R, Ranawat M, Ram L, Sayyed RZ (2023) Harnessing drought tolerant PGPM in arid agro ecosystem for plant disease management and soil amelioration. In: Mawar R, Sayyed RZ, Sharma SK, Sattiraju KS (eds) Plant growth promoting microorganisms of arid region: status and prospects. Springer, pp 27–43

Mendes R, Garbeva P, Raaijmakers JM (2013) The rhizosphere microbiome: significance of plant beneficial, plant pathogenic, and human pathogenic microorganisms. FEMS Microbiol 37(5):634–663. https://doi.org/10.1111/1574-6976.12028

Novo LAB, Castro PML, Alvarenga P, da Silva EF (2018) Plant growth-promoting rhizobacteria-assisted phytoremediation of mine soils. Bio-geotechnologies for mine site rehabilitation Elsevier 89(67):281–295. https://doi.org/10.1016/B978-0-12-812986-9.00016-6

Patel PR, Shaikh SS, Sayyed RZ (2016) Dynamism of PGPR in bioremediation and plant growth promotion in heavy metal contaminated soil. Indian J Exp Biol 54(4):286–290

Pilon-Smits EA, LeDuc DL (2009) Phytoremediation of selenium using transgenic plants. Curr Opin Biotechnol 20(2):207–212

Parmar DK (2014) Yield, produce quality and soil health under vegetable cropping systems as influenced by integrated nutrient management in mid-hill zone of Himachal Pradesh. J Indian Soc Soil Sci 62(1):45–51

Rafique N, Aqeel M, Raja NI, Shabbir G, Ajaib M, Sayyed RZ, Alharbi SA, Ansari MJ (2023) Interactive effects of melatonin and salicylic acid on *Brassica napus* under drought condition. In: Plant and Soil. https://doi.org/10.1007/s11104-023-05942-7

Rani YS, Triveni U, Patro TSSK (2017) Integrated nutrient management for enhancing the soil health, yield and quality of little millet (Panicum sumatrense). Int J Bioresour Stress Manag 8(1):26–32

Rashid MI, Mujawar LH, Shahzad T, Almeelbi T, Ismail IMI, Oves M (2016) Bacteria and fungi can contribute to nutrients bioavailability and aggregate formation in degraded soils. Microbiol Res 183:26–41. https://doi.org/10.1016/j.micres.2015.11.007

Rezapour S, Moghaddam SS, Jalil HM, Sayyed R (2021) An analysis of bioaccumulation, phyto-translocation, and health risk potential of soil cadmium released from waste leachate on a calcareous-semi-arid transect. Int J Environ Sci Tech. https://doi.org/10.1007/s13762-021-03777-2

Saeed M, Ilyas N, Arshad M, Sheeraz M, Ahmed I, Bhattacharya A (2021) Development of plant microbiome bioremediation system for crude oil contamination. J Environ Chem Eng 9:105401

Saeed M, Ilyas N, Bibi F, , Shabir S, Jayachandran K, Sayyedd RZ, Showe PL, A A Shati, M Y Alfaifi (2023) Adsorption kinetics of a bioengineered microbiome based biochar system for insitu remediation of total petroleum hydrocarbons (TPHs) contaminated soil. Chemosphere 324(2023):138311. https://doi.org/10.1016/j.chemosphere.2023.138311

Saeed M, Ilyas N, Bibi F, Jayachandran K, Dattamudi S, Elgorban MA (2022a) Biodegradation potential of PAHs by *Bacillus marsiflavi*, genome analysis and its plant growth promoting potential. Environ Pollut 292:118343

Saeed M, Ilyas N, Jayachandran K, Shabir S, Akhtar N, Shahzad A, Sayyed RZ, Bano A (2022b) Advances in biochar and PGPR engineering system for hydrocarbon degradation: a promising strategy for environmental remediation. Environ Pollut 2022(35):199282. https://doi.org/10.1016/j.envpol.2022.119282

Sagar A, Sayyed RZ, Ramteke PW, Sharma S, Marraiki N, Elgorban AM, Syed A (2020a) ACC deaminase and antioxidant enzymes producing halophilic *Enterobacter* sp. PR14 promotes the growth of rice and millets under salinity stress. Physiol Mol Biol Plants 26:1847–1854. https://doi.org/10.1007/s12298-020-00852-9

Sagar A, Riyazuddin R, Shukla PK, Ramteke PW, Sayyed RZ (2020b) Heavy metal stress tolerance in *Enterobacter* sp. PR14 is mediated by plasmid. Indian J Exp Biol 58(2):115–121

Singh R, Kumar S, Kumar H, Kumar M, Kumar A, Kumar D (2017) Effect of irrigation and integrated nutrient management on growth and yield of chickpea (*Cicer arietinum* L.). Plant Arch 17(2):1319–1323

Tarfeen N, Ul-Nisa K, Hamid H, Bashir Z, Yatoo AM, Dar MA, Mohiddin FA, Amin Z, Ahmad R, Sayyed RZ (2023) Microbial remediation: a promising tool for reclamation of contaminated sites with special emphasis on heavy metal and pesticide pollution: a review. Processes 10:1358. https://doi.org/10.3390/pr10071358

Trivedi P, Delgado-Baquerizo M, Trivedi C, Hamonts K, Anderson IC, Ingh B (2017) Keystone microbial taxa regulate the invasion of a fungal pathogen in agro-ecosystems. Soil Biol Biochem 98(23):10–14. https://doi.org/10.1016/j.soilbio.2017.03.013

Venturi V, Keel C (2016) Signaling in the rhizosphere. Trends Plant Sci 21(3):187–198. https://doi.org/10.1016/j.tplants.2016.01.005

Weyens N, Truyens S, Dupae J, Newman L, Taghavi S, van der Lelie D, Carleer R, Vangronsveld J (2010) Potential of the TCEdegrading endophyte Pseudomonas putida W619-TCE to improve plant growth and reduce TCE phytotoxicity an evapotranspiration in poplar cuttings. Environ Pollut 158(9):2915–2919

Wu W, Ma B (2015) Integrated nutrient management (INM) for sustaining crop productivity and reducing environmental impact: a review. Sci Total Environ 512:415–427

Zhang L, Rylott EL, Bruce NC, Strand SE (2019) Genetic modification of western wheatgrass (*Pascopyrum smithii*) for the phytoremediation of RDX and TNT. Planta 24(4):1007–1015

Nanoparticles as Growth-Stimulating Elements for Sustainable Agriculture

6

Hesham Ali El Enshasy, Rozaliana Abdul Karim, Intan Baizura Mohamed, Mohamed Rashad, Lisnawita, L. Lee San, and Johnny Meddy

Abstract

The formation of new soil and nutrient availability became a recent approach in sustainable agriculture. Neoformation consist of various process such as addition of organic matter, microbial activity, symbiotic interaction, and Nutrient Cycling. These processes are the important part in the formation nutrients into of soil. For this purpose, microbiome play important role in maintenance of soil health. Microbiomes are the collection of microorganisms found in the soil environment. These helpful microorganisms engage in a variety of plant growth-promoting activities, like fixing, mineralizing, solubilizing, and mobilizing nutrients, producing siderophores, antagonistic substances, antibiotics, and

H. Ali El Enshasy (✉)
Innovation Centre in Agritechnology for Advanced Bioprocessing (ICA), Universiti Teknologi Malaysia (UTM), Pagoh, Johor, Malaysia

Faculty of Chemical Engineering, Universiti Teknologi Malaysia (UTM), Skudai, Johor, Malaysia

City of Scientific Research and Technology Applications, Alexandria, Egypt
e-mail: hesham@utm.my

R. A. Karim · I. B. Mohamed
Innovation Centre in Agritechnology for Advanced Bioprocessing (ICA), Universiti Teknologi Malaysia (UTM), Pagoh, Johor, Malaysia

M. Rashad
City of Scientific Research and Technology Applications, Alexandria, Egypt

Lisnawita
Program Study of Agrotechnology, Faculty of Agriculture, Universitas Sumatera Utara, Medan, Indonesia

L. Lee San · J. Meddy
Department of Agriculture Sarawak, Kuching, Sarawak, Malaysia

R. Sayyed et al. (eds.), *Nanobiostimulants in Innovative Agriculture*,
https://doi.org/10.1007/978-981-95-2477-8_6

releasing substances that stimulate plant growth, like auxin and gibberellin hormones, through interactions with the roots of the host plant. Some species of plant make symbiotic associations with microbes and profit from the availability of minerals and nutrients while using little energy. The distribution of plant species determines the nature and amount of root exudates, as well as the crop species and cultivars. On the other side, it is possible to isolate, formulate, and create biofertilizers from microbes that play crucial functions in the microbiome. In agriculture, significant microorganisms with Fe- and Zn-solubilizing properties can be employed to biofortify micronutrients in various cereal crops. A safe environment is achieved through sustainable agriculture, which in turn has a good impact on soil health when it is used effectively. The subject of future research will be how to promote crop growth by shaping soil characteristics with the help of plant microbial communities and organic amendments in the soil. This chapter illustrates how microbiomes help in the neoforming of crops and the integration of nutrient into the soil.

Keywords

Microbiome · Phytoremediation · Plant growth promoting traits · Root exudates · Soil integrated nutrient

1 Introduction

With the increased population worldwide, there has been pressure on increasing the food production to achieve sustainable food security. In addition, farming faces many challenges, such as climate change, soil depletion, and the extensive utilization of chemical fertilizers and pesticides, which causes negative impact on soil quality. With so much demand put on growers to produce even more food, there is a great demand for efficient and environmentally friendly methods (Sethi et al. 2025; Ferioun et al. 2025). Of such promising approaches being considered, the nanoparticle application in agriculture can significantly address these issues and enhance farming. Nanoparticles are very tiny particles whose size is less than 100 nm. They have properties hat differ from those of larger particles. Consequently, nanoparticles can interact with the target elements, especially plants and the environment, unlike the larger particles. As such, the nanoparticles are proving to be more efficient and selective in terms of the effects that they produce. Hence, they can revolutionize farming practices by improving plant growth, controlling pests, and making fertilizers and pesticides more effective (Ghidan and Antary 2019).

2 Nanoparticle Characteristics and Production

Nanotechnology showed promising research subjects with wide applications in many technological fields (Selmani et al. 2022; Taha et al. 2024). This technology is based on the development of functional ultrasmall-scale particles of size less than 100 nm. This small particle has the capacity to provide more functionality compared to using the materials in a larger size. In addition, these nanoparticles can be easily encapsulated to support the targeted delivery of the active compounds to the targeted plant part (in the case of nanofertilizers) to improve nutrient uptake with the exact needed concentration without waste production. On the other hand, the application of nanoparticles in pest control can improve the specificity and efficacy and reduce the potential accumulation of toxic compounds in the environment. In addition, the application of nanoparticles in fertilizers can be designed to produce a sustainable release system for a continuous supply of nutrients with an optimal plant uptake rate to reduce the cost and frequent application. Nanofertilitzers can be designed for the nutrient supply for both macro- and micro-nutrients in mono form or in combined form to ease the field application. However, nanofertilizers can be classified under three main categories: nano-sized particles of the conventional macro- or micronutrient fertilizers, nanostructured materials, and nanoencapsulated nutrients with sustainable release properties (Arora et al. 2024).

Production of nanoparticles (NPs) in agriculture applications, either for chemical fertilizers or biological control agents, can be achieved either using physicochemical methods or biological methods (green synthesis). The applications of different methods for nanoparticle production have been reviewed recently by many authors (Kumari et al. 2023; Szczyglewska et al. 2023). In the case of physicochemical methods, it can be achieved using two main approaches as follows:

2.1 Top-Down Synthesis

In this production method, larger-sized materials are broken down to smaller non-scaled size particles using different methods of physical/chemical or a combination thereof. The common methods used for the production of nanoparticles using this approach include pure mechanical treatment, thermal or laser extirpation, sputtering, and pure chemical treatment. In the mechanical approach, large chemical particles are micronized and then nano-sized down to a size of 100 nm, commonly using ball milling. Whereas, the thermal method is carried out by carbon vapor deposition with many hours' treatment in a particular gaseous environment. On the other hand, chemical methods (which are mainly called wet methods) are based on the utilization of different types of oxidizing or reducing agents such as sodium bromide, sodium citrate, or hydrazine.

2.2 Bottom-Up Synthesis

In this method, the nanoparticles are synthesized from smaller molecules, which act as building blocks for the development of nano-sized particles. This involves the use of physical/chemical methods and forces to help in assembling the molecular building blocks into the desired size and structure with a nano-sized range. This method involves different processes such as molecular condensation, laser pyrolysis, spray pyrolysis, vapor deposition, and the sol-gel process in addition to simple precipitation processes. However, this method is considered easier with less expensive equipment compared to the top-down synthesis and easier in controlling the size distribution and shape compared to other methods (Eid et al. 2013).

In addition to these main methods for nanoparticle biosynthesis, biological synthesis methods are widely used as an alternative clean/green method for the production. This includes the utilization of biomaterials such as enzymes or whole-cell catalysis using plants, microorganisms (either prokaryotes or eukaryotes), and insect cells for the production of nano-sized particles (Jakinata et al. 2021; Ying et al. 2022). However, the method used for nanoparticle production needs to be carefully selected to ensure that the residue produced during the production process does not affect the product quality and safety. This is especially true when using nanoparticles for direct release to the environment or in applications related to direct exposure to animals and humans.

3 Nanoparticles in Agriculture

Applications of nanoparticles in agriculture span from enhanced growth of soil and adequate water capacity to increased production or crop productivity and disease and pest control affecting plants (Faridvand et al. 2021). It has been reported that the incorporation of nanotechnology in fertilizer formulation enhances nutrient release and absorption efficiency (Elsabagh et al. 2024). They aimed to improve the effectiveness of fertilizers while reducing the quantity needed. These nanoparticles can directly transport nutrients to plants without using regular fertilizers, destroying the environment. Therefore, the application of nanotechnology in agriculture is perceived as a way of developing a new sustainable, efficient, and environmentally friendly food production system. Some of the features and applications of nanoparticles are appealing; however, practically, there are still many unknowns regarding the impact of nanoparticles on ecosystems, human health, and the environment in the long run. However, there is an increasing interest in this field of study, and scientists have continued to seek ways to utilize nanotechnology in modern farming society as power reserves (Table 6.1).

Despite the availability of diverse types of nanoparticles, some types have attracted interest because of their properties and potential uses in agriculture. Such nanoparticles have promised benefits in increasing crop yields by helping fend off pests and diseases while making fertilizers and pesticides more effective. Some of the most famous and widely researched nanoparticles in agriculture are metallic

Table 6.1 Beneficial impacts of utilizations of nanotechnology in agriculture

Benefit	Description	Reference
Enhances nutrient delivery	Nanomaterials can encapsulate and deliver nutrients like nitrogen, phosphorus, and potassium more efficiently to plants.	Balusamy et al. (2023), Al-darwesh et al. (2024)
Increases pesticide and herbicide efficacy	Nanoencapsulation of pesticides and herbicides allows for controlled release and targeted application, minimizing the amount needed and reducing potential harm to non-target organisms and the environment.	Ghidan and Antary (2019), Balusamy et al. (2023)
Improves water management	Enhances water retention in soil, reducing water consumption and improving drought tolerance in crops.	Methela et al. (2023)
Enhances seed germination and growth	Improves seed germination rates and seedling vigor.	Del Buono et al. (2022), Sharon et al. (2020)

nanoparticles, including silver (Ag), copper (Cu), zinc oxide (ZnO), and gold (Au), which are very effective in agriculture (Balusamy et al. 2023). These metallic nanoparticles have such characteristics because of their small size, which impacts plants and soil they interact with in a manner that more extensive materials cannot. For instance, while silver nanoparticles have remarkably high antibacterial properties, as highlighted by Draghiceanu et al. (2021), plants sprayed with them usually can be safeguarded from many bacterium attacks and fungi. Therefore, it is confirmed that silver nanoparticles possess very high antimicrobial activity that can be utilized to protect crops against attack by bacteria and fungi (Ghidan and Antary 2019). A combination of iron nanoparticles with *Trichoderma* sp. has been reported to offer potent antifungal activity (Sharma et al. 2025).

In general, as reported recently, copper and zinc oxide nanoparticles help better absorption of nutrients and increase plant growth efficiency, while gold shows potential to treat plant disease and give drug delivery (Wang et al. 2023). Therefore, these metallic nanoparticles are gradually being used as critical factors that can enhance crop production and counter other problems experienced in modern farming systems. Elsabagh et al. (2024) used nanoparticles obtained from nanostructured date palm pits (nDDP) to study the performance of the nanoparticle-based fertilizers in terms of nutrient release kinetics and their impact on plant growth. These nanoparticles were produced via a planetary ball milling process, which reduced material derived from the date palm pit to the nanoscale. The resulting nanostructured material is then incorporated with macronutrients, specifically potassium dihydrogen phosphate (KH_2PO_4) and magnesium oxide (MgO), to form the eco-friendly nano-enabled fertilizers (nDPF1 and nDPF2). They also include agricultural waste into their process and improve the efficiency of fertilizers to help plant nutrients release and retain to support plant growth and increase sustainability at the same time. They concluded that the nanofertilizer demonstrated a controlled release of phosphorus, potassium, and magnesium, minimizing leaching losses compared to traditional fertilizers. Using these nanofertilizers resulted in a sustained release of nutrients, preventing leaching and nutrient loss. The fertilizers also could maximize nutrient use efficiency and reduce environmental impact.

Del Buono et al. (2022) conducted a study to determine the effectiveness of nanoparticles in maize seed priming. They utilized hybrid nanoparticles composed of lignin and zinc oxide. The influence of the treatment on germination and vigor of seeds, biomass, and the biochemical changes in the growth and development of the maize plant was also determined. The results showed that the hybrid nanoparticle treatment positively influenced several aspects of maize development, particularly in enhancing the seed germination rate. This concluded that it promotes early maize development and enhances plant growth and biochemical properties. A similar experiment was also conducted by Essa et al. (2021). CuNPs synthesized from biogenic sources were administered to investigate the impact on the growth and development of wheat seedlings. The study also revealed that seedling growth rates could be enhanced with biogenic CuNPs, and the physiological activity of the seedlings would also be beneficial. However, higher concentrations can induce stress responses and inhibit growth, highlighting the importance of dose optimization for potential agricultural applications.

Nanoparticles also play a significant role in water management, increasing water availability, decreasing evaporation, and increasing water uptake by roots. Nanoparticles can efficiently manage water by increasing the water-holding capacity and irrigation efficiency. Methela et al. (2023) describe synthesizing and using chitosan-GSNO nanoparticles to improve drought stress tolerance in soybean plants. The study also notes an enhanced physiological and biochemical alteration of the functional nanoparticles for better functioning under drought stress for better yield. Combating drought has been one of the main problems facing the water management of agriculture since it causes a water deficit and, therefore, reduces crop productivity. These nanoparticles may increase root vigor and general plant resistance. Thus, soybean varieties better adapted to periods of limited water availability may be produced. In another study by Ghidan and Antary (2019), they reported that the nanoparticles can be used in wastewater treatment and hence can be used in the agricultural sector. They can assist in the water treatment system, making it useful for irrigation. These technologies remove water waste and assist farmers in supplementing water at the right time for crops. Because of climate change and generally fluctuating rainfall systems, other measures like water management with nanoparticles should be sought to enhance agricultural production and food security. Despite the potential benefits of nanotechnology in agriculture and food processing, there are still significant challenges, including health risks, public skepticism, the need for further research, and evolving regulatory frameworks (Neme et al. 2021).

4 Mechanisms of Action of Nanoparticles in Plant Growth

Nanoparticles have emerged as innovative carriers for the delivery of essential nutrients to plants, significantly enhancing nutrient uptake efficiency. This capability is particularly crucial in modern agriculture, where traditional fertilization methods often lead to nutrient loss and suboptimal plant growth. The mechanisms of action of nanoparticles in plant growth represent an expanding field of research that integrates nanotechnology with agricultural science. Nutrient delivery systems, enhancement of soil properties, interaction of plant hormones, and the roles of nanoparticles in photosynthesis and respiration are the four main mechanisms that will be investigated in this chapter. Each of these mechanisms is supported by recent studies that elucidate the multifaceted roles of nanoparticles in promoting plant growth and resilience.

4.1 Nutrient Delivery Systems

The transfer of vital nutrients to plants is made possible by nanoparticles, which greatly increase the efficiency of nutrient uptake. According to research by Karny et al. (2018), therapeutic nanoparticles can efficiently infiltrate leaf tissues, enabling direct nutrient administration to crops. This direct delivery method circumvents the limitations of conventional fertilization techniques, which often struggle with nutrient leaching and poor absorption rates. The ability of nanoparticles to enhance biological targeting and facilitate intra-plant mobility is crucial for improving nutrient absorption, as highlighted in studies focusing on the interaction of nanoparticles with plant physiological processes in crops like *Lycopersicon esculentum*, also known as tomato (Karny et al. 2018).

Moreover, the use of magnetite nanoparticles has been shown to promote growth and increase chlorophyll content in various plant species, including maize and *Pseudostellaria heterophylla* (Racuciu et al. 2022). These nanoparticles not only enhance nutrient delivery but also stimulate physiological responses that contribute to improved plant health. For instance, the application of zinc oxide nanoparticles has been linked to increased growth and yield in peanut plants, demonstrating the potential of nanoparticles to act as effective nutrient carriers (Prasad et al. 2022). A recent study also highlights the role of metal nanoparticles in improving nutrient delivery and stress tolerance in plants. This study emphasizes the intricate processes by which nanoparticles penetrate plant cells and then move within them, which is essential for maximizing root-shoot translocation and nutrition distribution throughout the body (Nawaz et al. 2024). Similarly, Adams et al. (2020) discussed how mesoporous silica nanoparticles (MSNPs) can stabilize nutrients for plant uptake by immobilizing and releasing them in a regulated manner. This controlled release mechanism not only enhances nutrient availability but also minimizes nutrient loss, which is a common issue in conventional fertilization practices.

4.2 Nanoparticles as Soil Property Enhancers

Besides aiding in the delivery of nutrients, nanoparticles can also improve the nutrients' bioavailability in the soil. Zhang et al. (2022) pointed out that although the knowledge of how nanoparticles move through plant tissues is still developing, the benefits of delivering nutrients using nanoparticles are obvious since they can greatly increase the effectiveness of nutrient absorption in agricultural systems. This is particularly important in the context of sustainable agriculture, where maximizing nutrient use efficiency is critical for reducing environmental impacts.

Enhancing the qualities of the soil is another way that nanoparticles help to support healthy plant growth. The physical and chemical properties of the soil can be enhanced by adding nanoparticles, which will improve nutrient availability, aeration, and water retention. For instance, it has been demonstrated that applying nanomaterials to soil improves its stability and structure, which can reduce erosion and increase root penetration (Zahra et al. 2022). This is due to nanoparticle-induced changes to soil characteristics that may result in higher microbial activity, which is advantageous for the cycling of nutrients and the general health of the soil. Besides, Tarafder et al. (2020) also discussed the formulation of hybrid nanofertilizers designed for the slow and sustainable release of micronutrients, which can significantly improve nutrient use efficiency in agricultural practices. The slow-release mechanism minimizes nutrient loss and ensures that plants have a consistent supply of essential elements.

Furthermore, the interaction of nanoparticles with soil microorganisms can enhance nutrient cycling and availability. The presence of nanoparticles can stimulate microbial activity, which plays an essential role in breaking down organic matter and releasing nutrients into the soil (Adams et al. 2020). This synergistic effect between nanoparticles and soil biology underscores the potential of nanotechnology to create more sustainable agricultural systems.

Research also indicates that nanoparticles can alter the soil chemical composition, which will indirectly enhance its capacity to retain nutrients and water. This is particularly relevant in arid and semi-arid regions where water scarcity poses a significant challenge to agriculture. By improving soil moisture retention, nanoparticles can help sustain plant growth during periods of drought, thereby contributing to food security (Zahra et al. 2022). Furthermore, the application of nanotechnology in soil management can lead to reduced fertilizer requirements as nanoparticles can enhance the efficiency of nutrient uptake by plants (Nawaz et al. 2024).

4.3 Interaction with Plant Hormones

The interaction of nanoparticles with plant hormones is another critical mechanism through which they influence plant growth. Nanoparticles can modulate the levels of phytohormones, which are vital for regulating various physiological processes in plants, including growth, development, and stress responses. For example, studies have shown that nanoparticles can enhance the synthesis of auxins, which are crucial for root development and overall plant growth (Pereira et al. 2021). Nanoparticles also influence plant growth by interacting with plant hormones, which are critical for regulating various physiological processes. For example, the application of titanium dioxide (TiO_2) nanoparticles has been shown to enhance the activation of Rubisco, which is a key enzyme in the photosynthesis process, thereby increasing the photosynthesis rate during critical growth stages (Yagız and Çalışkan 2024).

This hormonal modulation can lead to improved seed germination, root elongation, and biomass accumulation.

In addition, the use of kinetin-capped zinc oxide nanoparticles has been reported to improve plant growth and enhance resistance to drought stress in mung beans (Ajmal et al. 2023). The zinc provided by these nanoparticles plays a crucial role in the synthesis of carbonic anhydrase enzymes, which facilitate carbon dioxide transport during photosynthesis. This highlights the multifaceted role of nanoparticles in not only delivering nutrients but also enhancing the physiological resilience of plants against environmental stressors. Moreover, the research conducted by Garg and Maheshwari (2023) indicates that seed priming with nanoparticles can alter the balance of reactive oxygen species (ROS) and plant growth regulators (PGRs), resulting in enhanced stress resistance and improved plant health. This interaction between nanoparticles and plant hormones underscores the potential of nanotechnology to not only facilitate nutrient delivery but also to enhance the physiological resilience of plants against biotic and abiotic stresses.

4.4 Role in Photosynthesis and Respiration

Nanoparticles also play a significant role in enhancing photosynthesis and respiration, which are two fundamental processes that underpin plant growth and productivity. The introduction of iron oxide nanoparticles has been shown to promote photosynthetic performance and improve the overall growth of plants like *Pseudostellaria heterophylla* (Li et al. 2021). These nanoparticles can stimulate the degradation of pectin in plant cell walls, facilitating root growth and enhancing nutrient uptake. Besides, the coupling of photosynthesis and respiration is critical for optimizing energy use in plants. Research by Gain et al. (2021) suggests that nanoparticles can influence the energetic coupling between these two processes, thereby enhancing overall metabolic efficiency. This is particularly important in the context of global warming, where the balance between photosynthesis and respiration can be disrupted, leading to reduced crop yields. Furthermore, the study by Yudina et al. (2020) also highlights the impact of nanoparticles on the photosynthetic machinery, suggesting that they can enhance the efficiency of photosynthesis by improving the function of chloroplasts and the electron transport chain. This enhancement can lead to increased biomass production and improved crop yields, which are essential for meeting the growing global food demand.

In addition to enhancing photosynthesis, nanoparticles can also influence respiration rates in plants. Research indicates that nanoparticles can modulate mitochondrial respiration, which is crucial for energy production and overall plant health (Xu et al. 2018). By optimizing respiration, nanoparticles can help plants adapt to environmental stressors, thereby improving their resilience and productivity. Moreover, studies have indicated that nanoparticles can modulate mitochondrial respiration, which is essential for energy production in plants. The interaction of nanoparticles with chloroplasts and other cellular structures involved in metabolic processes can optimize energy use and enhance overall plant health (Pietrzak et al. 2024). This is particularly important in the context of global warming, where the balance between photosynthesis and respiration can be disrupted, which will then lead to reduced crop yields. The mechanisms of action of nanoparticles in plant growth encompass a wide range of processes, including nutrient delivery systems, enhancement of soil properties, interaction with plant hormones, and roles in photosynthesis and respiration. The integration of nanotechnology into agricultural practices holds significant promise for improving crop productivity and sustainability. As research in this field continues to evolve, it is essential to further explore the complex interactions between nanoparticles and plant systems to fully harness their potential for enhancing agricultural outcomes.

5 Conclusion

The positive impact of nanotechnology application in agriculture has been extensively reported during the last two decades. Nanofertilizers as shown provides many advantages to the plantation system and can reduce the cost and efficacy of fertilization applied. In addition, I can reduce the overall negative impact of extensive uses of chemical agriculture in the whole system and thus support sustainable good agriculture practice. In addition, application of fertilizers and chemical pesticides in nanoscale formulations can reduce the negative toxic impact on the soil microorganisms. Thus, it reduces the risk of chemical applications to soil with less impact on microbial diversity and sustainable soil fertility. However, more research is needed to optimize the overall applications of nanoparticles in fertilization and biological control systems. The process scalability for the production of nanoparticles on an industrial scale needs to be available at a reasonable cost, as the fertilizer/biological control business is high–volume, low-value business compared to other sectors using nanoparticles, such as the medical, cosmetic, and food industries. In addition, precise determination of chemical concentration is needed when used in nanoparticle form to understand in depth the potential risk of the application of nanoparticles in the whole ecosystem and to support this risk analysis study with many field studies.

References

Adams CB, Erickson JE, Bunderson L (2020) A mesoporous silica nanoparticle technology applied in dilute nutrient solution accelerated establishment of zoysiagrass. Agrosyst Geosci Environ 3(1):e20006

Ajmal M, Ullah R, Muhammad Z, Khan MN, Kakar HA, Kaplan A, Okla MK, Saleh IA, Kamal A, Abdullah A, Abdul Razak S (2023) Kinetin capped zinc oxide nanoparticles improve plant growth and ameliorate resistivity to polyethylene glycol (PEG)-induced drought stress in *Vigna radiata* (L.) R. Wilczek (Mung Bean). Molecules 28(13):5059

Al-darwesh MY, Ibrahim SS, Mohammed MA (2024) Enhanced radical scavenging activity of green-synthesized ZnO nanoparticles: implications for pharmacological applications. Results Chem 7:101368. https://doi.org/10.1016/j.rechem.2024.101368

Arora PK, Tripathi S, Omar RA, Chauhan P, Sinhal VK, Singh A, Srivastava A, Garg SK, Singh VP (2024) Next-generation fertilizers: the impact of bionanofertilizers on sustainable agriculture. Microb Cell Fact 23:254

Balusamy SR, Joshi AS, Perumalsamy H, Mijakovic I, Singh P (2023) Advancing sustainable agriculture: a critical review of smart and eco-friendly nanomaterial applications. J Nanobiotechnol 21:372. https://doi.org/10.1186/s12951-023-02135-3

Del Buono D, Luzi F, Tolisano C, Puglia D, Di Michele A (2022) Synthesis of a lignin/zinc oxide hybrid nanoparticles system and its application by nano-priming in maize. Nanomaterials 12(3):568. https://doi.org/10.3390/nano12030568

Draghiceanu O, Şuţan AN, Dobrescu CM, Batut-Andrei ND, Soare LC, Topala CM (2021) Characterization in terms of composition and phytotoxicity of aqueous spores extract. Curr Trends Nat Sci 10(20):53–60. https://doi.org/10.47068/ctns.2021.v10i20.008

Eid AM, Elmarzugi NA, El Enshasy HA (2013) Preparation and evaluation of olive oil nanoemulsion using sucrose monoester. Int J Pharm Pharmaceut Sci 5:434–440

Elsabagh SS, Elkhatib EA, Rashad M (2024) Eco-friendly nano-enabled fertilizers derived from date industry waste for sustainable and controlled-release of P, K and Mg nutrients: sorption mechanisms, controlled-release performance and kinetics. Bioresour Bioprocess 11(1). https://doi.org/10.1186/s40643-023-00716-6

Essa HL, Abdelfattah MS, Marzouk A, Shedeed Z, Guirguis HA, El-Sayed MMH (2021) Biogenic copper nanoparticles from Avicennia marina leaves: impact on seed germination, detoxification enzymes, chlorophyll content and uptake by wheat seedlings. PLoS One 16(4):e0249764. https://doi.org/10.1371/journal.pone.0249764

Faridvand S, Amirnia R, Tajbakhsh M, El Enshasy HA, Sayyed RZ (2021) The effect of foliate application of magnetic water and nano-fertilizers on phytochemical and yield characteristics of fennel. Horticulturae 7(11):475. https://doi.org/10.3390/horticulturae7110475

Ferioun M, Zouitane I, Bouhraoua S, Elouattassi Y, Belahcen D, Errabbani A, Louahlia S, Sayyed R, El-Ghachtouli N (2025) Applying microbial biostimulants and drought-tolerant genotypes to enhance barley growth and yield under drought stress. Front Plant Sci 15:1494987. https://doi.org/10.3389/fpls.2024.1494987

Gain G, Vega de Luna F, Cordoba J, Perez E, Degand H, Morsomme P, Thiry M, Baurain D, Pierangelini M, Cardol P (2021) Trophic state alters the mechanism whereby energetic coupling between photosynthesis and respiration occurs in *Euglena gracilis*. New Phytol 232(4):1603–1617

Garg R, Maheshwari S (2023) Seed priming and fortification of seeds using nanotechnology: a review. EPH-Int J Agric Environ Res 9(1):11–17

Ghidan AY, Antary TMA (2019) Applications of nanotechnology in agriculture. In: IntechOpen eBooks. https://doi.org/10.5772/intechopen.88390

Jakinata P, Lingampally N, Hameeda B, Sayyed RZ, Kan YM, Elsayed EA, El Enshasy H (2021) Silver nanoparticles from insect wing extract: biosynthesis and evaluation for antioxidant and antimicrobial potential. PLoS One 16(3):e0241729. https://doi.org/10.1371/journal.pone.0241729

Karny A, Zinger A, Kajal A, Shainsky-Roitman J, Schroeder A (2018) Therapeutic nanoparticles penetrate leaves and deliver nutrients to agricultural crops. Sci Rep 8(1):1–10

Kumari P, Raturi S, Kulshrestha S, Chauhan K, Shingra S, Andras K, Thu K, Khargotra R, Singh T (2023) A comprehensive review on various techniques used for synthesizing nanoparticles. J Mat Res Technol 27:1739–1763

Li J, Ma Y, Xie Y (2021) Stimulatory effect of Fe_3O_4 nanoparticles on the growth and yield of *Pseudostellaria heterophylla* via improved photosynthetic performance. HortScience 56(7):753–761

Methela NJ, Pande A, Islam MS, Rahim W, Hussain A, Lee D, Mun B, Raj NPMJ, Kim S, Kim Y, Yun B (2023) Chitosan-GSNO nanoparticles: a positive modulator of drought stress tolerance in soybean. BMC Plant Biol 23:639. https://doi.org/10.1186/s12870-023-04640-x

Nawaz T, Gu L, Fahad S, Saud S, Bleakley B, Zhou R (2024) Exploring sustainable agriculture with nitrogen-fixing cyanobacteria and nanotechnology. Molecules 29(11):2534

Neme K, Nafady A, Uddin S, Tola YB (2021) Application of nanotechnology in agriculture, post-harvest loss reduction and food processing: food security implication and challenges. Heliyon 7:e08539. https://doi.org/10.1016/j.heliyon.2021.e08539

Pereira ADES, Oliveira HC, Fraceto LF, Santaella C (2021) Nanotechnology potential in seed priming for sustainable agriculture. Nanomaterials 11(2):267

Pietrzak M, Skiba E, Wolf WM (2024) Root-applied cerium oxide nanoparticles and their specific effects on plants: a review. Int J Mol Sci 25(7):4018

Prasad TNVKV, Satisha GC, Kumar AN, Swethasree M, Girish BP, Sudhakar P, Reddy BR, Saritha M, Sabitha N, Bhaskar Reddy BV, Rajasekhar P (2022) Particulate nanoscale silica induced novel morphological and biochemical stimulus effects in chilli (*Capsicum annuum* L.). ACS Agric Sci Technol 2(3):555–563

Racuciu M, Tecucianu A, Oancea S (2022) Impact of magnetite nanoparticles coated with aspartic acid on the growth, antioxidant enzymes activity and chlorophyll content of maize. Antioxidants 11(6):1193

Selmani A, Kovačević D, Bohinc K (2022) Nanoparticles: from synthesis to applications and beyond. Adv Colloid Interface Sci 303:102640–102652

Sethi G, Behera KK, Sayyed R, Adarsh V, Sipra BS, Singh L, Alamro AA, Behera M (2025) Enhancing soil health and crop productivity: the role of zinc-solubilizing bacteria in sustainable agriculture. Plant Growth Regul. https://doi.org/10.1007/s10725-025-01294-7

Sharma S, Kumari P, Shandilya M, Thakur S, Perveen K, Sheikh I, Ahmed Z, Sayyed R, Mastinu A (2025) The combination of α-Fe_2O_3NP and *Trichoderma* sp. improves antifungal activity against *Fusarium* wilt. J Basic Microbiol 2025:e2400613. https://doi.org/10.1002/jobm.202400613

Sharon FB, Sumithra D, Sangeetha G (2020) Impact of silver nanoparticles on germination and growth features in mung bean plant (Vigna radiata). Int J Eng Appl Sci Technol 5(2):274–281. https://doi.org/10.33564/ijeast.2020.v05i02.043

Szczyglewska P, Feliczak-Guzik A, Nowak I (2023) Nanotechnology–general aspects: a chemical reduction approach to the synthesis of nanoparticles. Molecules 28(13):4932. https://doi.org/10.3390/molecules28134932

Taha HAIM, Agamy NFM, Soliman TN, Younes NM, El Enshasy HA, Darwish AMG (2024) Preparation and characterization of vitamin E/calcium/soy protein isolate nanoparticles for soybean milk beverage fortification. Peer J 12:e17007. https://doi.org/10.7717/peerj.17007

Tarafder C, Daizy M, Alam MM, Ali MR, Islam MJ, Islam R, Ahommed MS, Aly Saad Aly M, Khan MZH (2020) Formulation of a hybrid nanofertilizer for slow and sustainable release of micronutrients. ACS Omega 5(37):23960–23966

Wang Q, Xu S, Zhong L, Zhao X, Wang L (2023) Effects of zinc oxide nanoparticles on growth, development, and flavonoid synthesis in Ginkgo biloba. Int J Mol Sci 24(15):15775. https://doi.org/10.3390/ijms242115775

Xu Y, Ibrahim IM, Wosu CI, Ben-Amotz A, Harvey PJ (2018) Potential of new isolates of *Dunaliella salina* for natural β-carotene production. Biology 7(1):14

Yagız AK, Çalışkan ME (2024) Effects of TiO_2 nano-priming on tomato seed germination and plant development. J Anim Plant Sci 34(1):62–72

Ying S, Guan Z, Ofoegbu PC, Clubb P, Rico C, He F, Hong J (2022) Green synthesis of nanoparticles: current developments and limitations. Environ Technol Innov 26:102336–102356

Yudina L, Sherstneva O, Sukhova E, Grinberg M, Mysyagin S, Vodeneev V, Sukhov V (2020) Inactivation of H+-ATPase participates in the influence of variation potential on photosynthesis and respiration in peas. Plants 9(11):1585

Zahra Z, Habib Z, Hyun H, Shahzad HMA (2022) Overview on recent developments in the design, application, and impacts of nanofertilizers in agriculture. Sustainability 14(15):9397

Zhang H, Goh NS, Wang JW, Pinals RL, González-Grandío E, Demirer GS, Butrus S, Fakra SC, Del Rio FA, Zhai R, Zhao B (2022) Nanoparticle cellular internalization is not required for RNA delivery to mature plant leaves. Nat Nanotechnol 17(2):197–205

Microbial Nano-biostimulants for Improved Crop Productivity

7

Neelum Naheed, Asim Shahzad, Aqsa Jabeen, Muhammad Umer, and Noshin Ilyas

Abstract

The rising population causing a significant threat to meeting the world food demand results in increased use of chemical fertilizers. Excessive use of chemical fertilizers causes challenges like soil degradation, fertility loss, mineral deficiency, pest resistance, climate change, and chemical pollution. These emerging issues impose a considerable burden on existing agricultural problems. Biostimulants are a good initiative to overcome agricultural issues by improving overall production, sustainability, metabolic reactions, and physiochemical response of plants against various biotic and abiotic stresses. Seaweeds, plant extracts, hydrolyzed proteins, nitrogen-containing compounds, humic substances, and microorganisms are various biostimulants. Microbial biostimulants are helpful to lower the use of chemical fertilizers, but their efficacy is greater when combined with nanostimulants. Nanotechnology is an advanced field that creates nano-sized particles and is used in every field of life. In agriculture, nanoparticles improve crop production and remove agro-waste from water bodies. Nanoparticles (both organic and inorganic) are synthesized by biological, chemical, and mechanical processes. Green synthesized nanoparticles (NPs) are considered environmentally friendly as compared to mechanical and chemical approaches due to less production of toxic chemicals. Nanomaterial integration with microbial biostimulants into agricultural practices boosts crop productivity. Integrating microbial nanotechnology into agricultural practices and biomedicine is promising, but it must overcome synthesis, standardization, and scalability challenges.

N. Naheed · A. Jabeen · M. Umer · N. Ilyas
Department of Botany, Pir Mehr Ali Shah Arid Agriculture University, Rawalpindi, Pakistan

A. Shahzad (✉)
The College of Geography and Environmental Sciences, Henan University, Kaifeng, China

R. Sayyed et al. (eds.), *Nanobiostimulants in Innovative Agriculture*,
https://doi.org/10.1007/978-981-95-2477-8_7

Keywords

Hormones · Humic acids · Microorganisms · Plant extracts · Seaweeds

1 Introduction

The rapidly growing population and substantial changes in climate and ecosystem pose significant threats to global crop production and food demand. Increased food demand leads to increased land use for cultivation and chemical fertilizers to meet food requirements (Akhtar et al. 2022). Scientists and agriculturists are trying to improve the crop yield as per demand, but they are facing several challenges like soil degradation, fertility loss, deficiency of essential minerals, pest resistance, and most importantly climate change. Soil degradation refers to reduced soil fertility and ability to accomplish various ecosystem functions, particularly food and fiber production. This deprivation is due to numerous degradation processes involving plant yield and quality reduction (Rather et al. 2024). The climate changes also contribute largely to the world food scarcity.

Moreover, the greenhouse effect causes a rise in temperature due to increased CO_2 emissions. This rise in temperature causes changes in weather patterns, raised crop evapotranspiration rate, increased insect attack, shorter crop life, and change in weed flora. The climate changes also affect soil's microbial diversity and their enzymatic activities, plant growth, production, and quality. Another prominent agricultural challenge is soil degradation, reducing soil capacity to perform various ecological functions like food and fiber production (Desmet and Rossi-Hansberg 2024). The major causes of soil degradation and erosion are overgrazing, poor agriculture practices, deforestation, compaction, salinization, and excessive use of agrochemicals. Soil degradation adversely affects the food quantity and quality, which in turn affects human health and nutrition. Furthermore, soil degradation increases crop vulnerability to numerous abiotic as well as abiotic stresses such as water deficit stress and nutrient imbalance, ultimately reducing the crop yield. The third most challenging threat to agriculture is agro-pollution caused by excessive use of inorganic fertilizers, pesticides, and herbicides. The chemical fertilizer application not only enhanced the crop's overall yield but also threatens the soil's biological and physiochemical properties.

Excessive use of chemical fertilizers reduces SOM (soil organic matter) content, changes soil pH, increases pests, and alters soil structure and soil acidification. Additionally, it disturbs the microbial diversity and activities in rhizospheric soil, especially plant growth-promoting bacteria (PGPR), which decreases crop growth and development (Zhang et al. 2024). The long-term persistence of chemical fertilizers in the soil also causes food contamination. Its overuse is not only a cause of soil chemical pollution but also water and air pollution. All these emergent issues impose a considerable burden on the existing agricultural problems that challenge the global food demand. Plant biostimulants are a good initiative to overcome agricultural issues. Moreover, it improves crop production, sustainability, plant metabolic, and physiological parameters in response to numerous biotic as well as abiotic stresses. Microbial biostimulants are helpful to lower the use of chemical fertilizers, but their efficacy is greater when combined with nanostimulants (Chen et al. 2024). In this book chapter, nanotechnology, microbes, and their mechanism of action and synergistic effect on plants and ecology will be briefly discussed.

2 Nanotechnology

Nanotechnology is an emergent and auspicious technology with great potential to modernize the agricultural sector and livestock. Nanotechnology is an advanced field that creates nano-sized particles of less than 100 nm. This field applies to both basic and advanced research, including disease diagnosis and as a therapeutic agent. In this modern world, nanoparticles are used in every field of life; one of the most important uses of nanoparticles is in the agricultural field (Nayeri et al. 2021). In the agricultural field, nanoparticles are used to improve crop production and remove agro-waste from water bodies (Karnwal et al. 2024; Wei et al. 2024). Furthermore, nanoparticles enhance the effectiveness of agrochemicals, including fertilizers, weedicides, and pesticides that reduce crop loss. Nanomaterials ensure the efficient targeted delivery of nutrients to various plant parts for improved crop productivity and defense. Furthermore, nano-biosensor helps to develop advanced agricultural technologies that record the crop and soil nutrient demand as well as enable crop management at the time. Nanotechnology is a leading-edge technology that allows agricultural development and ecosystem safety. It allows the synchronization of various strategies for agriculture management and yield improvement.

2.1 Nanoparticles Synthesis

Nanoparticles (NPs) are primarily produced by using two approaches, either bottom-up or top-down, either chemically or through natural sources. Mainly nanoparticle synthesis is categorized into the following methods: biological, chemical, and mechanical methods (Rasouli et al. 2020). Biologically synthesized nanoparticles are considered safer than chemical and mechanical methods due to the lack of toxic chemical production through this method (Jakinala et al. 2021). Among biologically prepared nanoparticles, the most effective and widely used are green synthesized nanoparticles that revolutionize the existing fields, especially the agriculture sector (Sembada and Lenggoro 2024).

2.1.1 Bottom-Up Approach

The bottom-up approach involves the synthesis of material from small elements to aggregates, then nanoparticles (NPs). The sol-gel, pyrolysis, chemical vapor deposition (CVD), spinning, and green synthesis are the main techniques employed in the bottom-up approach for nanoparticle (NP) fabrication (Sembada and Lenggoro 2024).

2.1.2 Sol-Gel

The sol-gel is the most promising bottom-up approach because of its simplicity; additionally, it comprises the sol, which is a solid colloidal solution dispersed in a liquid phase. In this process, the metal oxides plus chloride precursors are generally used that are dissolved in a liquid by continuous agitation, sonication, and shaking. This sol-gel system is separated by various methods like filtration, centrifugation, or sedimentation, then finally removal of water by drying (Abidin et al. 2024).

2.1.3 Biosynthesis

Nanoparticle synthesis through plants as well as microorganisms like algae, bacteria, fungi, and actinomycetes is a green and environment-friendly technology (Tasnim et al. 2024). Various microorganisms are utilized for the green synthesis of metallic NPs like platinum, gold, silver, cadmium, iron, and zirconium, as well as metal oxides, specifically zinc oxide, titanium oxide etc. (Wu and Zou 2009). Nanoparticles may be synthesized intracellularly and extracellularly, according to the site of nanoparticles.

2.1.4 Intracellular Synthesis of Nanoparticles (NPs) by Fungi

In nanoparticle (NP) synthesis process, ions are transferred to microbial cellular structures manifested with enzymes. Nanoparticles synthesized inside the microbes are usually smaller as compared to those nanoparticles synthesized outside the cells. This size limitation is likely due to the nucleation process occurring within the organisms (Yildirim et al. 2006).

2.1.5 Extracellular Synthesis of Nanoparticles (NPs) by Fungi

Nanoparticles (NPs) synthesized outside the microbe's body have more applications in various fields of science because they avoid several intracellular adjoining components that are unnecessary. Most fungi efficiently produce nanoparticles outside the body due to their numerous secretory components. These components are really helpful in nanoparticle capping, reduction, and stability (Hasan 2015).

Top-Down Approach

The top-down approach involves the breakdown of large-scale substances into nanosized particles. Nanolithography, laser ablation, mechanical milling, thermal decomposition, and sputtering are the most commonly applied techniques to synthesize nanoparticles (Nazir et al. 2024).

Mechanical Milling

The mechanical milling is the best approach as compared to other top-down processes for NP synthesis. This method involves milling and post-annealing of nanoparticles in an inert atmosphere during nanoparticle production. The major factor in the process of mechanical milling involved plastic deformation as a major factor that influences the particle shape; any break caused a reduction in particle size, resulting in a size increase (Abidin et al. 2024).

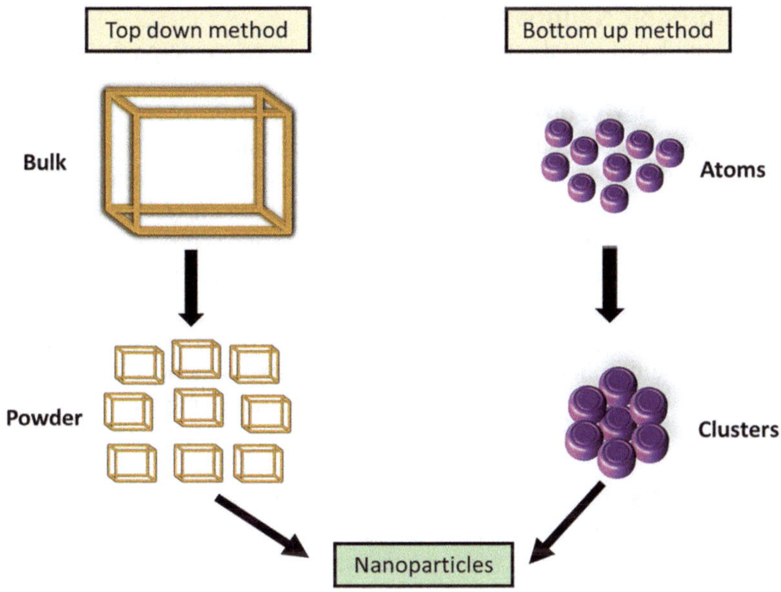

Fig. 7.1 Nanoparticle synthesis approaches

Nanolithography

Lithography is a method of producing a desired shape and structure on a light-sensitive substance by removing a part of the material for manufacturing the required morphology. Nanolithography involves the nanometric-scale fabrication of structures in the range of 1 to 100 nm with the required size and shape. Nanolithography involves processes including electron-beam, multiphoton, instance optical, nanoimprint, and scanning probe (Fig. 7.1).

2.2 Types of Nanomaterials

2.2.1 Organic Nanoparticles

The most generally known organic NPs are micelles, liposomes, ferritin, dendrimers, etc. Among these nanoparticles, some particles, such as liposomes and micelles, consist of hollow core, are eco-friendly, are biodegradable, and are sensitive to heat and electromagnetic radiation. Nanoparticle size, composition, and surface morphology—all these unique properties make them good options for targeted drug delivery. The nanoparticle's stability, delivery system, and drug-carrying capacity determine their applications in various fields. The organic NPs are commonly applied in the field of biomedicine; for the targeted drug delivery, these particles are precisely injected into plant parts.

2.2.2 Inorganic Nanoparticles

Metals and oxides of metals are commonly categorized as inorganic NPs.

2.2.3 Metal-Derived NPs

Inorganic NPs are produced by either destructive or constructive approach from the metals to nanosizes. Approximately all metals are used in the synthesis of nanoparticles, but the most commonly used metals are aluminum (Al), silver (Ag), zinc (Zn), cobalt (Co), copper (Cu), gold (Au), iron (Fe), lead (Pb), and cadmium (Cd). Nanoparticles' efficiency and their use in various fields depend on their unique features, including size ranges between 10 and 100 nm, pore size, surface charge, and surface charge density. The surface properties involve surface area/volume ratio, crystallization, amorphous morphology, shapes (like spherical and cylindrical), color, reactivity, and sensitivity to environment (including moisture, heat, sunlight, air, etc.)

2.2.4 Metal Oxide-Derived Nanoparticles (NPs)

Metallic NPs' properties are modified to get oxides of metal NPs, e.g., Mg nanoparticles are promptly oxidized to magnesium oxide nanoparticles in the presence oxygen. The efficiency and reactivity of these nanoparticles, including cerium oxide (CeO_2), magnetite (Fe_3O_4), iron oxide (Fe_2O_3), silicon dioxide (SiO_2), zinc oxide (ZnO), and titanium oxide (TiO_2), can be enhanced through the oxidation process (Ealia and Saravanakumar 2017).

3 Biostimulants

Biostimulants, as advanced agronomic tools, are bioactive substances containing microorganisms that stimulate both crop growth and productivity through nutrient enhancement, absorption, and assimilation efficiency. These biostimulants are applied to the plant through foliar or rhizosphere application, which assists plants to resist abiotic stresses (Bulgari et al. 2019; Rouphael and Colla 2020). The word "biostimulants" is mentioned in the most recent European fertilizer regulation involving microbes and nanotechnology. The chemical nature of organic material, hence the constituents of the resulting product, makes it challenging to determine which molecules are mainly effective and chiefly responsible for the biological activity. Minerals like selenium, silicon, and vitamins; amino acids, chitin, chitosan; poly- and oligosaccharides; and traces of natural plant hormones can all function as biostimulants. The agro-industrial waste, including food residues and agrochemicals, can also be used to formulate biostimulants (Xu and Geelen 2018). Using agricultural by-products as raw materials to produce biostimulants is a component of the circular economy plan, which aims to make agriculture more sustainable. Biostimulants can either increase photosynthetic activity and primarily affect metabolism or activate certain biochemical pathways to increase secondary metabolites content (Bulgari et al. 2019; Bulgari et al. 2015). The synergistic activities of many bioactive compounds in biostimulants are helpful in various stress tolerances in plants. Biostimulants have different types based on their source, primary ingredient, or action mechanism (Bulgari et al. 2015).

3.1 Biostimulants Classification

Biostimulants are categorized into five major groups, as mentioned below:

3.1.1 Seaweeds and Plant Extracts

Seaweeds, including both multicellular and macroscopic aquatic algae, including green, brown, and red, are diversified groups (Bulgari et al. 2019; Bulgari et al. 2015). They are a rich source of bioactive chemicals, essential nutrients, organic materials, and fertilizers. Algae as fertilizers have been utilized in agricultural science since ancient times because of their beneficial features for crop productivity. The extract from them is now used in agricultural land as soil conditioners and crop biostimulants (Battacharyya et al. 2015). Seaweeds and plant extracts as biostimulants significantly improve plant growth and development, yield quantity and quality, and provide resistance to abiotic stress (Ali et al. 2021). Furthermore, algae-based biostimulants include a variety of mineral and bioactive components such as polysaccharides like laminarine, alginates, and fucoidan (Bulgari et al. 2019). Plants enriched in secondary metabolites are the primary source of biostimulants, a type of bioactive chemical that has been linked to the activation of the physiological reactions of plants. Extracts from borage leaves or flowers have been seen to enhance lettuce development as well as quality (Moreno-Hernández et al. 2020).

3.1.2 Hydrolyzed Proteins and Nitrogen-Containing Compounds

Protein hydrolysates are a combination of substances, including amino acids, peptides, and polypeptides derived from complex biological matrices via chemical, enzymatic, or thermal hydrolysis. Extensive research on horticultural crops showed that protein hydrolysates of animals and plants significantly increase crop growth and production besides resistance to a variety of environmental stresses (Colla et al. 2015).

3.1.3 Humic Substance

Humic compounds are the combination of humic and fulvic acids. It is actually soil organic matter that is naturally formed by partially decomposed residues of plants, animals, and microorganisms. Humic compounds get fully decomposed by the enzymatic activity in soil that in return provides essential nutrients to plants. These nutrients improve food and water uptake, regulate plant metabolism, and improve growth and development, production, and stress adaptability (Zamljen et al. 2024).

3.1.4 Microorganisms

Soil is one of the major habitats for both types of microorganisms, harmful and beneficial, that are essential to maintain and regulate soil processes. Some are soil-borne phytopathogens, but on the other hand, the majority are plant growth-promoting microbial communities (Sharma et al. 2025). These tiny creatures determine soil fertility by playing a vital role in organic matter decomposition. Microorganism composition is an indicator to determine and regulate the biological process in soil. In addition, microbes' composition determines the percentage of soil organic content and defines the soil type as well. Numerous microorganisms act as plant biostimulants, which activate various natural processes to enhance nutrient uptake efficiency, crop quantity, and biotic and abiotic resistance (Del Buono et al. 2021). Mycorrhizal fungi and bacterial strains like *Rhizobium* spp., *Azospirillum* spp., or *Azotobacter* spp. and mycorrhizal fungi (Bastami et al. 2021). These can be applied to plants as foliar spray, root tips, soil improvement, or seed treatment (Ganugi et al. 2021). However, foliar spray is more favorable because it can provide nutrients to plants at the seedling stage to increase plant development (Preininger et al. 2018). The micro-organisms increase plant resistance against pathogens, stimulating the plant's defense system by producing defense compounds (Heil and Bostock 2002). When a plant receives a stress stimulus, the microbes colonize its rhizosphere and increase the oxidative species in plants, in return producing antioxidant compounds (ascorbic acid, anthocyanins, phenolic compounds, terpenoids, etc.) (Rouphael and Colla 2020). Microbe-derived biostimulants enhance plant tolerance against heat stress; they possess antioxidant activity, in addition to reducing ethylene release (Hamid et al. 2021). The beneficial microbes influence food quality by the accumulation or modulation of functional components (Ganugi et al. 2021).

3.1.5 Nanoparticles with Microbial Biostimulants

Nanoparticles, especially selenium, silicon, cobalt, aluminum, silver, and zinc oxides, as biostimulants, are getting popular in the field of agriculture. These elements are found in soils as well as in plants as inorganic salts, including amorphous silica. The beneficial effects of these compounds can be either permanent (e.g., reinforcing cell walls with silica deposits) or temporary (e.g., after abiotic stress exposure). Beneficial components are defined by their chemical properties and the circumstances in which they positively impact plant development and stress response. Nano-biostimulants are the combination of biostimulants and nanoparticles (NPs) and are an emerging domain in agriculture (Ramírez-Rodríguez et al. 2024). This emerging technology aims to enhance overall crop productivity and regulate plant life under environmental stresses. Chemical fertilizers are proven to be dangerous for plants as well as human health. These nano-biostimulants showed the potency of strengthening plant development with increased nutrition, improving overall productivity, and prolonged shelf life. On the other hand, microbial nano-biostimulants also face challenges (Vijayaram et al. 2024). The limited knowledge about the synergistic approach of microbes and nanotechnology also poses a limitation on its utilization in agriculture. This gap needs extensive research and exploration to make this technology widely acceptable (Pukale et al. 2024).

3.1.6 Mechanism of Action of Microbial Nano-niostimulants

The microbial nano-biostimulants initiate several processes under the mixed application of nanoparticles (NPs) and microbes. Nanoparticles (NPs) protect the microbes, including plant growth-promoting rhizobacteria as well as mycorrhizal fungi (Chaudhary et al. 2024). These microbes contribute significantly to maintaining soil health, nutrient uptake, recycling, pathogenic microorganism elimination, and stable aggregate formation that increases soil aeration, water-holding capacity, and overall ecosystem sustainability (Sethi et al. 2025). NPs are another important component of microbial nano-biostimulants that are applied to plants in two ways: foliar application or soil application. Nanoparticles enter plants primarily via root hairs and leaf epidermis or stomata through apoplastic or symplastic pathways; NP transport is also facilitated by vascular bundles (Hong et al. 2021). These nanoparticles assist in essential nutrient uptake from the soil and ensure controlled release and translocation to various plant parts (Faridvand et al. 2021). Additionally, NPs application improves photosynthetic rate and increases chlorophyll content, phytohormone production, and disease resistance (Ali et al. 2024). Therefore, NPs help to enhance plant biomass and yield and improve crop quality. Microbial nano-biostimulant exposure triggers considerable changes in metabolic pathways and gene expression by promoting specific metabolite synthesis and increasing levels of carotenoids, total sugar, indole, phenols, and amino acids in various plant species (Basu et al. 2021). In addition, it stimulates plant defense mechanisms against abiotic stress by antioxidant enzyme activation and osmotic adjustment (Lephatsi et al. 2022). Several factors, including humidity, temperature, soil pH, NP size and shape, and plants' physiological characteristics, affect the nanoparticles' absorption, transport, and accumulation (Chen et al. 2023).

4 Role of Microbial Nano-biostimulant in Stress Tolerance

Microbial nano-biostimulant-treated plants are capable of producing more metabolites and antioxidants under the stimulant of the below-mentioned stresses (Fazeli-Nasab et al. 2021).

4.1 Drought

Water stress is a common abiotic stress for crops throughout regions. The reduction in water availability has a direct effect on photosynthetic activity, changing leaf gas exchanges. This can negatively impact product output and quality (Ferioun et al. 2025). Water stress causes observable signs such as loss of turgor and yellowing/browning of leaves from chlorophyll breakdown (Shahbaz et al. 2023). A decrease in chlorophyll content is a sign of ionic imbalance in a crop in water deficit conditions. Microbial biostimulants derived from seaweed extracts can buffer fluctuations in plant water status by increasing the accumulation of osmotically active molecules (Johnson et al. 2024).

4.2 Salinity Stress

Crops grown along coastal areas may experience saline stress due to an increase in the amount of soluble salts in irrigated water and soil. Salinity inhibits plant growth and thus leads to cellular damage, affecting crop vitality and production. The crop's response is measured by duration of exposure. Therefore, high salt content creates osmotic stress because of the soil's high ionic concentration, which might reduce root water uptake. Thus, stressed plants exhibit signs of water deficiency. The biostimulants can help crops tolerate salt better. Plants adopt the same endogenous mechanisms to tolerate abiotic stresses (like salinity and drought) (Franzoni et al. 2022). During both stresses, biostimulants may produce a rise in the content of proline, simple sugars, alcohols, abscisic acid, and antioxidant molecules capable of preventing damage caused by free radicals. The use of biostimulants improves stress tolerance and reduces production losses. Biostimulant actions depend on species as well as salinity conditions. Protein hydrolysates of vegetable origin have shown promising effects in enhancing salt tolerance in a variety of horticultural crops, including lettuce, consequently enhancing yield and dry matter accumulation. Plants treated with microbial biostimulants have more yield and utilize radiation more effectively. Microbial biostimulants can also provide comparable outcomes (Johnson et al. 2024) (Table 7.1).

4.3 High- and Low-Temperature Stress

Temperature (T) is a major abiotic environmental factor that influences plant metabolic processes by altering chemical reactions, enzyme activity, membrane properties, cytoplasm, vacuoles, xylem, and phloem solution viscosity. Mainly plants experience low or high temperature stress, and both cause a reduction in plant growth and development. Microbial nano-biostimulants can produce protection in response to low temperatures by accumulating cryoprotective chemicals and/or activating membrane repair mechanisms and have shown excellent results. Likewise, high-temperature stress can also destroy crops, which disrupts cell membranes and enzyme activity (Thakur et al. 2022). This stress also damages cell membrane phospholipids, which cause cellular compartmentalization, resulting in metabolic issues (Martínez-Lorente et al. 2024). The high-temperature response in plants involves increasing transpiration, which aids in thermoregulation. High temperatures above 37 °C can be harmful, particularly in areas with inadequate water supply. Microbial biostimulants that improve crop tolerance may work by improving crops' water absorption capability and enhancing accumulation of compounds that protect cell membranes (Niu et al. 2022).

Table 7.1 Biostimulant mode of action and its role in abiotic stress tolerance

Abiotic stress	Crop involved	Microorganisms as biostimulants	Beneficial effects/mode of action	References
Drought stress	*L. lycopersicum*	*A. brasilense*	Increased plant biomass Increased leaf surface area	Romero et al. (2014)
	C. sinensis	*A. nodosum*	Involved in upregulation of stress tolerance genes	Spann and Little (2011)
	P. trifoliata	*G. versiforme*	Increased morphological attributes of plants by uptake of K, P, and ca	Wu and Zou (2009)
	C. annuum	*Glomus claroides*	Provide tolerance to drought stress by maintenance of water potential	Davies et al. (2002)
	P. sativum	*P. putida* and *P. fluorescens*	Involved in plant overall growth and development	Arshad et al. (2008)
	T. aestivum	*A. brasilense*	Improved crop yield and tolerance to stress	Pereyra et al. (2006, 2012)
	C. melo	*R. intraradices* and *G. versiforme*	Involved in enhancing antioxidant activity and net photosynthetic ratio	Huang et al. (2011)
	S. lycopersicum	*R. intraradices*	Increased overall yield by 12–25%	Subramanian et al. (2006)
Salinity stress	*E. japonica*	*G. iranicum*	Improved mineral uptake	Gómez-Bellot (2015)
	V. faba	*A. brasilense*	Enhanced photosynthetic pigments Increased overall plant biomass	Hamaoui et al. (2001)
	C. cajan	*P. fluorescens*	Involved in tolerance to salt stresses by expression of CcWRKY genes	Kumar et al. (2019)
	S. lycopersicum	*U. lactuca*	Decrease the negative impact of salinity on crop	El Boukhari et al. (2021)
	C. arietinum *T. aestivum*	*A. brasilense*	Enhance product quality and storage period	Hamaoui et al. (2001), Fasciglione et al. (2015)
	H. vulgare	*H. diazotrophicus*	Enhanced root water content Lessen Na^+ uptake	Suarez et al. (2015)
	Cucurbita spp.	*Bacillus* spp., *P. azotoformans*	Improves fresh plant biomass Increased $K^+:Na^+$ ratio	Yildirim et al. (2006)
	Z. mays	*A. chrococcum*	Increased chlorophyll and polyphenol content	Rojas-Tapias et al. (2012)
	V. faba/P. sativum	*Rhizobium leguminosarum*	Plant growth and stress tolerance	Cordovilla et al. (1999)

(continued)

Table 7.1 (continued)

Heat stress	*T. aestivum*	*A. chrococcum*	Increase primary plant nutrient content	Chaudhary et al. (2013), Egamberdiyeva and Höflich (2003)
	S. bicolor	*P. putida*	Improved membrane injury Enhanced metabolite content	Ali et al. (2009)
	L. esculentum	*Glomus* sp.	Reduction in lipid peroxidation	Duc et al. (2018)
	S. lycopersicum	PGPB + humic acid	Improved salicylic acid content Enhanced antioxidant enzyme content	Khan et al. (2020)
	S. tuberosum	*Paraburkholderiaphytofirmans*	Decreased H_2O_2 Reduced production of ACC	Sangiorgio (2022)
Chilling stress	*S. lycopersicum*	*P. frederiksbergensis*	Improved seed germination	Subramanian et al. (2016)
	T. aestivum	*P. dispersa*	Enhanced tolerance mechanism	Selvakumar et al. (2008)
	K. alvarezii	*A. nodosum*	Increased tolerance to cold	Loureiro et al. (2014)
	P. pyrifolia *M. domestica*	*Pseudomonas fluorescens*	Competition with bacteria having INA^+	Lindow and Brandl (2003)
	V. vinifera	*Paraburkholderiaphytofirmans*	1-Aminocyclopropane-1-carboxylic acid (ACC)	Theocharis et al. (2012)
	K. Alvarezii	*A. Nodosum*	Enhanced cold tolerance	Loureiro et al. (2014)

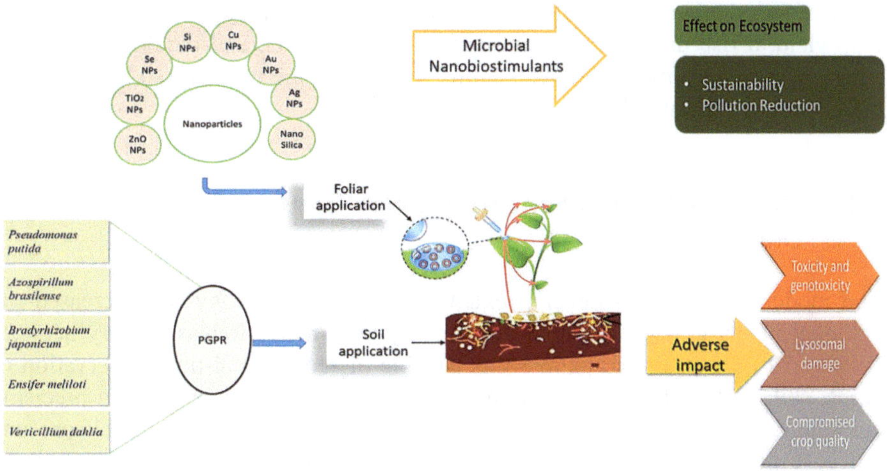

Fig. 7.2 Nano-biostimulant formulation involves various PGPR strains and NPs, its application via foliar spray or soil application, and its effect on ecosystem maintenance

5 Role of Microbial Nano-biostimulants in Agriculture

The agriculture sector plays an important role in the country's economy by providing food and raw materials to nourish the continuously increasing world's population, leading to the excessive use of agrochemicals, especially chemical fertilizers and pesticides. However, this overuse of chemicals in agriculture poses serious issues to environmental sustainability and biodiversity (Singh et al. 2024). The integration of nanotechnology into agriculture possesses several advantages over chemical fertilizers, and it reduces the use of agrochemicals. Farmers choose nanoparticles over other conventional ways because of their compact size, ease of transport, simplicity of handling, extended shelf life, and great efficiency (Singh et al. 2024). Nano-biotechnology may benefit agriculture and the ecosystem by developing nanoinsecticides and regulating plant diseases. This can be accomplished by utilizing environmentally friendly nanoparticles to enhance the efficacy of fertilizers at lower doses. Agricultural nanotechnology and associated processes improve agricultural production while reducing disease impact (Babu et al. 2024) (Fig. 7.2).

6 Precision Agriculture Applications

Precision agriculture incorporates conventional agriculture practices and modern agriculture techniques that employ information and communication, which has modernized agriculture as smart agriculture (Cisternas et al. 2020). Precise agriculture is also integrated with sensory devices by employing wireless sensory networks capable of real-time crop monitoring. It has led to a considerable decrease in food insecurity threads caused by the drastic population growth (Bright et al. 2025). The application of precise agriculture has provided many great benefits, such as high crop yield, a rise in the quality and quantity of products, and the development of better crop management strategies while preserving the climate (Akhter and Sofi 2022; Bhat and Huang 2021).

6.1 Development of Sensors and Real-Time Monitoring System

The modernized way to develop extensive sensor technology has increased the potential of wireless communication, broadly known as wireless sensory network (WSN) (Placidi et al. 2021). WSN is capable of remote real-time observation of crops, soil, and ecological conditions through the integrated sensors. In precision agriculture, real-time monitoring is employed to make appropriate decisions and quick responses based on the analyzed data for better development of crops (Zervopoulos et al. 2020). The customization capabilities of Wireless Sensor Networks (WSN) have been widely adopted in precision agriculture, supporting various applications such as pest detection, disease management, innovative underground planting techniques, ensuring food security, enhancing crop quality, modern drone spraying, and maintaining the privacy and security of sensed data. These advancements aim to address the growing demands of future agriculture (Lin et al. 2018; Ali et al. 2018). While the integration of WSN in precision agriculture has unlocked numerous opportunities, it also presents significant challenges that hinder its effective implementation (Sanjeevi et al. 2020). The biggest challenge in WSN installation is synchronization. The remote sensor nodes require network synchronization to provide correct time-correlated measurements to process individual data to compose new information. Successful synchronization ensures the accurate interpretation of data and its utilization in intelligent decision-making processes (Tavares Bruscato et al. 2017). Similarly, its proper functioning depends on other factors such as the utilized hardware and software, harsh environmental conditions, prevention of external intervention, appropriate installation procedure, and the system's immaculate operation management (Giordano et al. 2018).

6.2 Integration of Microbial and Nanotechnology for Optimized Crop Management

As conventional agriculture techniques are failing to fulfill the increasing demand of food production, it has become necessary to adopt modern agriculture techniques. Microbial and nanotechnology hold a conspicuous position among the most recently rising technical innovations in agriculture (Fu et al. 2020). Modern nanotechnology has widespread applications in agriculture and optimized crop management. Engineered nanoparticles play a vital part in crop genetic manipulation, disease regulation, pest resistance, seed germination, and overall disease protection (Guleria et al. 2023). Microbial communities in plants are essential for effective crop management. They help regulate plant growth, improve nutrient absorption from the soil through the roots, enhance resistance to pests, reduce the impact of harmful pathogens, strengthen the plant's immunity against diseases, and even suppress the growth of competing plants (Mutale-Joan et al. 2023).

The microbial incorporation into nanotechnology has shown great potential for optimized crop management. This multidisciplinary approach can enhance crop productivity significantly and protect the crops from pathogen attacks (Hussain et al. 2023). The synthesis of nanoparticles by microorganisms such as bacteria, viruses, fungi, and algae is known as microbial nanotechnology (Koul et al. 2021). Chemically synthesized nanoparticles can release toxic chemicals and have complex manufacturing protocols. In contrast, microbially synthesized nanoparticles are a green and sustainable approach, have increased efficacy and biocompatibility, and can develop stress-resistant crops (Konappa et al. 2022). Nanoparticle synthesis via microbes utilizes biological components to minimize environmental harms (Singhal et al. 2023). This process involves microbial enzymes and bioactive substances acting as natural reducing agents, facilitating the nanoparticle synthesis without the use of toxic chemicals (Menon et al. 2023). These nanoparticles enhance agricultural productivity by improving nutrient delivery, managing pests more effectively, and increasing crop resilience against abiotic stresses such as extreme temperatures and salinity (Ijaz et al. 2023; Maqsood et al. 2023). Furthermore, they reduce reactive oxygen species (ROS), harmful byproducts of plant stress, thereby promoting healthier and more resilient crops (Maqsood et al. 2023). Microbial nanoparticles have proven to mitigate environmental stress and ensure the regulation of plant survival under harsh conditions. Various types of metal, metal oxide, and carbon-based nanoparticles synthesized by microbes are developed to produce stress-smart crops that can thrive under stress conditions without much reduction in their yield. Literature supports the successful combined applications of nanoparticles and microbes against plant stress factors (Shaikh et al. 2024). A study done by Akhtar et al. (2021) explains the role of silicon nanoparticles in combination with PGPR for the amelioration of drought stress. The results have shown that the combined application of PGPR and nanosilica has increased crop resilience against drought stress. It is reported that nanomaterials can modify the plant rhizospheric microbial community. A study done by Ge et al. (2014) shows that the application of engineered nanoparticles has improved the concentration of some of the rhizospheric microorganisms like *Rhizobium* and *Sphingomonas,* while decreasing others like *Ensifer, Clostridium, Rhodospirillaceae,* and *Azotobacter*. Plants' microbes can absorb the nanomaterial that can affect plant-microbe and microbe-microbe interactions (Sillen et al. 2020). Nanomaterials can also be used to engineer the microbes to achieve targeted beneficial traits in the crops. Nanomaterials such as nanochitosan, iron, zinc, and copper are being reported to reduce crop diseases and regulate microbiomes (Ahmed et al. 2022; Jiang et al. 2021).

7 Sustainable and Environmental Impact

The major elements that are involved in the deterioration of the environment involve either the natural or the manmade. The process of industrialization, the increase in population, and the accumulation of various toxic elements in the environment have made the environment polluted. Therefore, there is an increase in demand for clean and green technology, which is now a global challenge. The increase in nanotechnology is very impressive and a great source of controlling the toxicity-causing element of the environment (Gehrke et al. 2015). The synthesis of nanoparticles by using plant material exhibits great potential for environmental remediations. These particles have shown some impressive dealings against herbicides, heavy metals, wastes, and dyes. The NPs, which are synthesized using plant material or by the use of microbes, show some great benefits to the environment, as this technique is safe for the environment (Muskan et al. 2024). Nano-biofertilizers take crop nutrition to the next level. With their slow-release property, they minimize environmental loss, while their tiny, nano-sized particles ensure crops absorb nutrients more effectively. It's a smarter, more sustainable way to nourish plants. These nano-biofertilizers are the perfect way to improve the yield of crops without making any disturbance to the environment. The delivery of nutrients in a controlled manner is possible by the use of smart nanocarriers, which can be adaptable to the ever-changing soil microbial activity, pH, and soil moisture (Oluwaseyi et al. 2024).

8 Reduction in Use of Chemical Fertilizers

The revolution of industry and the green revolution that remained very fruitful as it fulfilled the demand of an ever-growing population has led to an increase in the total production of yield in a specified area. This revolution has increased production, but it has brought up many challenges, like that of less fertile soil (Ayoub 1999). The use of synthetic fertilizer was for the cause of fulfilling the demand for food for the people, but it then came with many problems related to health and the ecosystem (Jote 2023). Keeney's research suggests that the global population could double by 2050. If developing countries start consuming food and meat at the same level as developed nations, the demand for food and nutrients could increase, potentially tripling current levels. All this should be analyzed by keeping in mind a scenario of increasing urbanization, which caused a decrease in the land use for the production purpose of food (Greidinger 1997). Therefore, all these equations sum up to a finding that nutrient load per unit area is increasing; thus, the production of food should be done through more sustainable techniques than ever used before.

There is a dire need to eliminate synthetic fertilizers' impacts on the environment and human health. Thus, to counter this, a novel agricultural technique is under practice, which is known as sustainable or organic agriculture. This practice is often named ecological agriculture (Keeney and Follett 1991). Organic fertilizers compared to synthetic fertilizers are cheaper and easy to access (Aksoy 2001). Organic matter is the fertility base of soil, and the inoculation of microbial fertilizers plays an important role in plant nutrition. It is very cost-effective, friendly to the environment, and a non-bulk fertilizer (Solomon et al. 2012; Aboudrare 2009). Whereas the inorganic fertilizers are in total contrast to nano-biofertilizers for being very expensive and having harmful effects on the ecosystem (Mahajan Gupta and Sharma 2008). All the events raise the issues of reduction in crop yield, degradation of soil, and imbalance of nutrients (Hijbeek et al. 2018). The intensive use of synthetic fertilizer for a greater period poses a significant impact on the soil, causing damage to the soil microbes, degradation of soil, and many others that are already mentioned above (Vitousek et al. 1997). Thus, to make the soil fertile and of good use, we should opt for controlled-release fertilizers, manures, and biofertilizers (Pandiselvi et al. 2017; Rai et al. 2014). Thereby, with the use of nano-biofertilizer, we can reduce the use of chemical fertilizers, as some promising results were shown by nano-biofertilizer.

9 Promotes Biodiversity

The toxicity in the soil is mainly caused by synthetic fertilizers, although they make a significant contribution to the soil nutrients and crop yield increase. Therefore, a nano-biofertilizer comes up with the improved nutrition status of the crop and with a reduced rate of application. The benefit of nano-biofertilizer is its slow rate of release, and this minimizes the loss to the environment, whereas the use of nano-sized particles increases the absorption of the nutrients. If the use of nano-biofertilizers is effectively done, then they can help increase yield without disturbing the masses. The best consideration for the nano-biofertilizer was given in recent years. The use of the plant extracts and the microbes is done for green synthesis to produce environmentally friendly nanoparticles. The nutrient release as per the crop-specific can be managed by the use of modified surfaces of nanoparticles. The system of controlled release of nutrients is managed by using the smart system of nanocarriers, which immediately adapts to every altering pH, moisture, and activity of microbes in the soil. The increase in crop growth has been reported with the synergistic treatments of plant growth-promoting rhizobacteria (Abioye et al. 2024). A significant impact has been put forward by the nano-biofertilizer in healing and increasing the productivity of agriculture. Nanotechnology and biofertilizers both work in synergy to enhance their microbial role within the coating of nanomaterials, which offers great soil moisture and provides vital nutrients to the plants. This is so because of the presence of PGPR that directly or indirectly provides resistance to the disease, phosphorous solubilization, rhizoremediation, and many more (Shanware and Taiwade 2022). In finding an alternative to synthetic chemical fertilizers, scientists have gone through much hustle, and now it is almost possible to replace synthetic fertilizers with nanotechnology. This approach, which uses nanoparticles and biofertilizers, has shown massive impact and lowered the use of synthetic fertilizers; thus, eventually, good crop productivity has resulted, and better crop management is validated (Mir et al. 2015). Nowadays, nano-biofertilizers are viewed as the agricultural industry's future and are a perfect substitute for current technologies, which have a number of shortcomings (Duhan et al. 2017). The soil benefits from the bioorganic components of a nano-biofertilizer in several ways, including promoting biodiversity and beneficial microflora and regulating the soil's nutrient efficacy.

10 Challenges and Future Directions

Microbial nano-biostimulants have considerable potential in agriculture and medicine, but they confront several hurdles that must be overcome in order to make further progress. Integrating microbial nanotechnology into agricultural practices and biomedicine is promising, but it must overcome synthesis, standardization, and scalability challenges. Microorganisms typically produce polydisperse nanomaterials, hindering their use (Grasso et al. 2021). Furthermore, they found that current approaches for purifying and standardizing nanomaterials lack consistency. High manufacturing costs and low yields limit the economic feasibility of microbial nano-biostimulants (Grasso et al. 2021). Optimizing culture conditions and scaling up production can increase yield and lower costs (Bloch et al. 2021). Synthetic biology can enhance the efficiency and selectivity of nanomaterial production (Grasso et al. 2021). Expanding the application of microbial nano-biostimulants in other areas, such as food safety and environmental remediation, might increase their influence. While microbial nano-biostimulants show considerable potential, their effective deployment will be dependent on overcoming these hurdles and finding novel solutions. The potential for these technologies to revolutionize agriculture and medicine remains considerable, justifying ongoing research and development (Malik and Cinelli 2024).

References

Abidin MZU, Ikram M, Moeen S, Nazir G, Kanoun MB, Goumri-Said S (2024) A comprehensive review on the synthesis of ferrite nanomaterials via bottom-up and top-down approaches advantages, disadvantages, characterizations and computational insights. Coord Chem Rev 520:216158

Abioye OM, Okunola AA, Amodu MF, Olaseheinde DA, Yusuf KO (2024) NanoBiofertilizer and its application in sustainable agriculture, crop specific nutrients delivery and environmental sustainability: a review. Applied Science and Engineering Progress 17(3):7339–7339

Aboudrare A (2009) Agronomie Durable, Principes ET Pratiques. Rapport de Formation Continue. FAO

Ahmed T, Noman M, Jiang H, Shahid M, Ma C, Wu Z et al (2022) Bioengineered chitosan-iron nanocomposite controls bacterial leaf blight disease by modulating plant defense response and nutritional status of rice (Oryza sativa L.). Nano Today 45:101547

Akhtar N, Ilyas N, Hayat R, Yasmin H, Noureldeen A, Ahmad P (2021) Synergistic effects of plant growth promoting rhizobacteria and silicon dioxide nano-particles for amelioration of drought stress in wheat. Plant Physiol Biochem 166:160–176

Akhtar N, Ilyas N, Meraj TA, Pour-Aboughadareh A, Sayyed RZ, Mashwani ZUR, Poczai P (2022) Improvement of plant responses by nanobiofertilizer: a step towards sustainable agriculture. Nano 12(6):965

Akhter R, Sofi SA (2022) Precision agriculture using IoT data analytics and machine learning. Journal of King Saud University-Computer and Information Sciences 34(8):5602–5618

Aksoy U (2001) Ecological farming. II. Ecological farming symposium in Turkey, Antalya, pp 14–16

Ali SKZ, Sandhya V, Grover M, Kishore N, Rao LV, Venkateswarlu B (2009) Pseudomonas sp. strain AKM-P6 enhances tolerance of sorghum seedlings to elevated temperatures. Biol Fertil Soils 46:45–55

Ali R, Pal AK, Kumari S, Karuppiah M, Conti M (2018) A secure user authentication and key-agreement scheme using wireless sensor networks for agriculture monitoring. Futur Gener Comput Syst 84:200–215

Ali O, Ramsubhag A, Jayaraman J (2021) Biostimulant properties of seaweed extracts in plants: Implications towards sustainable crop production. Plant 10:531

Ali S, Akhtar MS, Siraj M, Zaman W (2024) Molecular communication of microbial plant biostimulants in the rhizosphere under abiotic stress conditions. Int J Mol Sci 25(22):12424

Arshad M, Shaharoona B, Mahmood T (2008) Inoculation with pseudomonas spp. containing ACC-deaminase partially eliminates the effects of drought stress on growth, yield, and ripening of pea (Pisum sativum L.). Pedosphere 18:611–620

Ayoub AT (1999) Fertilizers and the environment. Nutr Cycl Agroecosyst 55:117–121. https://doi.org/10.1023/A:1009808118692

Babu GR, Gokuldhev M, Brahmanandam PS (2024) Integrating IoT for soil monitoring and hybrid machine learning in predicting tomato crop disease in a typical South India Station. Sensors 24(19):6177

Bastami A, Amirnia R, Sayyed RZ, Enshasy HAE (2021) The effect of mycorrhizal fungi and organic fertilizers on quantitative and qualitative traits of two important Satureja species. Agronomy 11(7):1285

Basu A, Prasad P, Das SN, Kalam S, Sayyed RZ, Reddy MS, El Enshasy H (2021) Plant growth promoting rhizobacteria (PGPR) as green bioinoculants: recent developments, constraints, and prospects. Sustainability 13:1140

Battacharyya D, Babgohari MZ, Rathor P, Prithiviraj B (2015) Seaweed extracts as biostimulants in horticulture. Sci Hortic 196:39–48

Bhat SA, Huang NF (2021) Big data and ai revolution in precision agriculture: survey and challenges. IEEE Access 9:110209–110222

Bloch K, Pardesi K, Satriano C, Ghosh S (2021) Bacteriogenic platinum nanoparticles for application in nanomedicine. Front Chem 9:624344

Bright JP, Maheshwari HS, Thangappan S, Perveeen K, Bukhari NA, Mitra D, Sayyed RZ, Mantinu A (2025) Biofilmed-PGPR: a next-generation bioinoculant for plant growth promotion in rice (Oryza sativa L.) under changing climate. Rice Sci 32(1):94–106

Bulgari R, Cocetta G, Trivellini A, Vernieri P, Ferrante A (2015) Biostimulants and crop responses: a review. Biol Agric Hortic 31:1–17

Bulgari R, Franzoni G, Ferrante A (2019) Biostimulants application in horticultural crops under abiotic stress conditions. Agronomy 9:306

Chaudhary D, Narula N, Sindhu SS, Behl RK (2013) Plant growth stimulation of wheat (Triticum aestivum L.) by inoculation of salinity tolerant Azotobac ter strains. Physiol Mol Biol Plants 19(4):515–519

Chaudhary A, Poudyal S, Kaundal A (2024) Role of Mycorrhizal fungi in maintaining sustainable agroecosystem

Chen S, Pan Z, Zhao W, Zhou Y, Rui Y, Jiang C et al (2023) Engineering climate-resilient rice using a nanobiostimulant-based "stress training" strategy. ACS Nano 17(11):10760–10773

Chen J, Jia Y, Sun Y, Liu K, Zhou C, Liu C et al (2024) Global marine microbial diversity and its potential in bioprospecting. Nature 633(8029):371–379

Cisternas I, Velásquez I, Caro A, Rodríguez A (2020) Systematic literature review of implementations of precision agriculture. Comput Electron Agric 176:105626

Colla G, Nardi S, Cardarelli M, Ertani A, Lucini L, Canaguier R, Rouphael Y (2015) Protein hydrolysates as biostimulants in horticulture. Sci Hortic 196:28–38

Cordovilla MDP, Berrido SI, Ligero F, Lluch C (1999) Rhizobium strain effects on the growth and nitrogen assimilation in Pisum sativum and Vicia faba plant growth under salt stress. J Plant Physiol 154:127–131

Davies FT, Olalde-Portugal V, Aguilera-Gomez L, Alvarado MJ, Ferrera-Cerrato RC, Boutton TW (2002) Alleviation of drought stress of Chile ancho pepper (Capsicum annuum L. cv. San Luis) with arbuscular mycorrhiza indigenous to Mexico. Sci Hortic 92:347–359

Del Buono D, Regni L, Del Pino AM, Bartucca ML, Palmerini CA, Proietti P (2021) Effects of megafol on the olive cultivar 'Arbequina'grown under severe saline stress in terms of physi-

ological traits, oxidative stress, antioxidant defenses, and cytosolic Ca2+. Front Plant Sci 11:603576
Desmet K, Rossi-Hansberg E (2024) Climate change economics over time and space. Annual Review of Economics 16
Duc NH, Csintalan Z, Posta K (2018) Arbuscular mycorrhizal fungi mitigate negative effects of combined drought and heat stress on tomato plants. Plant Physiol Biochem 132:297–307
Duhan JS, Kumar R, Kumar N, Kaur P, Nehra K, Duhan S (2017) Nanotechnology: the new perspective in precision agriculture. Biotechnol Reports 15:11–23
Ealia SAM, Saravanakumar MP (2017) A review on the classification, characterisation, synthesis of nanoparticles and their application. In: IOP conference series: materials science and engineering, vol 263. IOP Publishing, p 032019
Egamberdiyeva D, Höflich G (2003) Influence of growth-promoting bacteria on the growth of wheat in different soils and temperatures. Soil Biol Biochem 35(7):973–978
El Boukhari MEM, Barakate M, Choumani N, Bouhia Y, Lyamlouli K (2021) Ulva lactuca extract and fractions as seed priming agents mitigate salinity stress in tomato seedlings. Plants 10(6):1104
Faridvand S, Amirnia R, Tajbakhsh M, Enshasy HE, Sayyed RZ (2021) The effect of foliar application of magnetic water and nano, organic, and chemical fertilizers on phytochemical and yield characteristics of different landraces of fennel (*Foeniculum vulgare* mill). Horticulturae 7:475
Fasciglione G, Casanovas EM, Quillehauquy V, Yommi AK, Goñi MG, Roura SI et al (2015) Azospirillum inoculation effects on growth, product quality and storage life of lettuce plants grown under salt stress. Sci Hortic 195:154–162
Fazeli-Nasab B, Sayyed RZ, Piri R, Rahmani AF (2021) Biopriming and nanopriming: green revolution wings to increase plant yield, growth, and development under stress condition and forward dimensions. In: Antioxidants in plant-microbe interaction. Springer Singapore, Singapore, pp 623–655
Ferioun M, Zouitane I, Bouhraoua S, Elouattassi Y, Belahcen D, Errabbani A, Louahlia S, Sayyed R, El-Ghachtouli N (2025) Applying microbial biostimulants and drought-tolerant genotypes to enhance barley growth and yield under drought stress. Frontiers in Plant Sciences 15:1494987
Franzoni G, Cocetta G, Prinsi B, Ferrante A, Espen L (2022) Biostimulants on crops: their impact under abiotic stress conditions. Horticulturae 8(3):189
Fu L, Wang Z, Dhankher OP, Xing B (2020) Nanotechnology as a new sustainable approach for controlling crop diseases and increasing agricultural production. J Exp Bot 71(2):507–519
Ganugi P, Miras-Moreno B, Garcia-Perez P, Lucini L, Trevisan M (2021) Concealed metabolic reprogramming induced by different herbicides in tomato. Plant Sci 303:110727
Ge Y, Priester JH, Van De Werfhorst LC, Walker SL, Nisbet RM, An YJ et al (2014) Soybean plants modify metal oxide nanoparticle effects on soil bacterial communities. Environ Sci Technol 48(22):13489–13496
Giordano S, Seitanidis I, Ojo M, Adami D, Vignoli F (2018) IoT solutions for crop protection against wild animal attacks. In: In 2018 IEEE international conference on environmental engineering (EE). IEEE, pp 1–5
Gómez-Bellot MJ, Ortuño MF, Nortes PA, Vicente-Sánchez J, Bañón S, Sánchez-Blanco MJ (2015) Mycorrhizal euonymus plants and reclaimed water: biomass, water status and nutritional responses. Sci Hortic 186:61–69
Grasso S (2024) Navigating language maintenance challenges with health professionals: Reflections from Spanish speaking families in Australia. Aust J Linguist 44(1):1–28
Grasso A, Ferrante M, Arena G, Salemi R, Zuccarello P, Fiore M, Copat C (2021) Chemical characterization and quantification of silver nanoparticles (Ag-NPs) and dissolved ag in seafood by single particle icp-ms: Assessment of dietary exposure. Int J Environ Res Public Health 18(8):4076
Greidinger (1997) Symposium on Fertilization and the Environment. Technion, Haifa, Israel
Guleria G, Thakur S, Shandilya M, Sharma S, Thakur S, Kalia S (2023) Nanotechnology for sustainable agro-food systems: the need and role of nanoparticles in protecting plants and improving crop productivity. Plant Physiol Biochem 194:533–549

Hamaoui B, Abbadi J, Burdman S, Rashid A, Sarig S, Okon Y (2001) Effects of inoculation with Azospirillum brasilense on chickpeas (Cicer arietinum) and faba beans (Vicia faba) under different growth conditions. Agrono mie 21(6–7):553–560

Hamid B, Zaman M, Farooq S, Fatima S, Sayyed R, Baba Z, Sheikh T, Reddy M, El Enshasy H, Gafur A (2021) Bacterial plant biostimulants: a sustainable way towards improving growth, productivity, and health of crops. Sustainability 13:2856

Hasan S (2015) A review on nanoparticles: their synthesis and types. Res J Recent Sci 2277:2502

Heil M, Bostock RM (2002) Induced systemic resistance (ISR) against pathogens in the context of induced plant defences. Ann Bot 89(5):503–512

Hijbeek R, Ten Berge H, Whitmore A, Barkusky D, Schroder JJ et al (2018) Nitrogen fertilizer replacement values for organic amendments appear to increase with N application rates. Nutr Cycl Agroecosyst 110(1):105–115

Hong J, Wang C, Wagner DC, Gardea-Torresdey JL, He F, Rico CM (2021) Foliar application of nanoparticles: mechanisms of absorption, transfer, and multiple impacts. Environ Sci Nano 8(5):1196–1210

Huang Z, Zou Z, He C, He Z, Zhang Z, Li J (2011) Physiological and photosynthetic responses of melon (Cucumis melo L.) seedlings to three glomus species under water deficit. Plant Soil 339:391–399

Hussain M, Zahra N, Lang T, Zain M, Raza M, Shakoor N et al (2023) Integrating nanotechnology with plant microbiome for next-generation crop health. Plant Physiol Biochem 196:703–711

Ijaz M, Khan F, Ahmed T, Noman M, Zulfiqar F, Rizwan M et al (2023) Nanobiotechnology to advance stress resilience in plants: current opportunities and challenges. Materials Today Bio 100759

Jakinala P, Lingampally N, Hameeda B, Sayyed RZ, Khan MY, Elsayed EA et al (2021) Silver nanoparticles from insect wing extract: biosynthesis and evaluation for antioxidant and antimicrobial potential. PLoS One 16(3):e0241729

Jiang H, Lv L, Ahmed T, Jin S, Shahid M, Noman M et al (2021) Effect of the nanoparticle exposures on the tomato bacterial wilt disease control by modulating the rhizosphere bacterial community. Int J Mol Sci 23(01):414

Johnson R, Joel JM, Puthur JT (2024) Biostimulants: the futuristic sustainable approach for alleviating crop productivity and abiotic stress tolerance. J Plant Growth Regul 43(3):659–674

Jote CA (2023) The impacts of using inorganic chemical fertilizers on the environment and human health. Organ Med Chem Intl J 13(3):555864

Karnwal A, Martolia S, Dohroo A, Al-Tawaha ARMS, Malik T (2024) Exploring bioremediation strategies for heavy metals and POPs pollution: the role of microbes, plants, and nanotechnology. Front Environ Sci 12:1397850

Keeney DR, Follett RF (1991) Managing nitrogen for ground water quality and farm profitability: overview and introduction. In: Managing nitrogen for ground water quality and farm profitability

Khan MA, Asaf S, Khan AL, Jan R, Kang S-M, Kim K-M et al (2020) Extending thermotolerance to tomato seedlings by inoculation with SA1 isolate of Bacillus cereus and comparison with exogenous humic acid application. PLoS One 15(4):e0232228. https://doi.org/10.1371/journal.pone.0232228

Konappa N, Arakere UC, Krishnamurthy S, Chowdappa S, Jogaiah S (2022) Exploring the potential role of Trichoderma as friends of plants foes for bacterial plant pathogens. In: Biopesticides. Woodhead Publishing, pp 383–399

Koul B, Poonia AK, Yadav D, Jin JO (2021) Microbe-mediated biosynthesis of nanoparticles: applications and future prospects. Biomolecules 11(6):886

Kumar G, Bajpai R, Sarkar A, Mishra RK, Gupta VK, Singh HB et al (2019) Identification, characterization and expression profiles of Fusarium udum stress-responsive WRKY transcription factors in Cajanus cajan under the influence of NaCl stress and Pseudomonas fluorescens OKC. Sci Rep 9:14344

Lephatsi M, Nephali L, Meyer V, Piater LA, Buthelezi N, Dubery IA et al (2022) Molecular mechanisms associated with microbial biostimulant-mediated growth enhancement, priming and drought stress tolerance in maize plants. Sci Rep 12(1):10450

Lin J, Shen Z, Zhang A, Chai Y (2018) Blockchain and IoT based food traceability for smart agriculture. In: Proceedings of the 3rd international conference on crowd science and engineering, pp 1–6

Lindow SE, Brandl MT (2003) Microbiology of the phyllosphere. Appl Environ Microbiol 69:1875–1883

Loureiro RR, Reis RP, Marroig RG (2014) Effect of the commercial extract of the brown alga Ascophyllum nodosum Mont. On Kappaphycus alvarezii (Doty) Doty ex P.C. Silva in situ submitted to lethal temperatures. J Appl Phycol 26(1):629–634

Mahajan Gupta RD, Sharma R (2008) Bio-fertilizers-away to sustainable agriculture. Agrobios Newsletter 6:36–37

Malik N, Cinelli P (2024) Flammability of fiber-based composites. In: Biocomposites for industrial applications. Woodhead Publishing, pp 77–90

Maqsood MF, Shahbaz M, Khalid F, Rasheed Y, Asif K, Naz N et al (2023) Plant-based nanoparticles in ROS mitigation and abiotic stress tolerance: a review. Plant Stress:100281

Martínez-Lorente SE, Martí-Guillén JM, Pedreño MÁ, Almagro L, Sabater-Jara AB (2024) Higher plant-derived biostimulants: mechanisms of action and their role in mitigating plant abiotic stress. Antioxidants 13(3):318

Menon V, Prasad B, Sharma H (2023) Environmental mercury toxicity and its bioremediation. In Mercury toxicity: Challenges and solutions (pp. 325–356). Singapore: Springer Nature Singapore

Mir S, Sirousmehr A, Shirmohammadi E (2015) Effect of nano and biological fertilizers on carbohydrate and chlorophyll content of forage sorghum (Speedfeed hybrid). International Journal of Biosciences (IJB) 6(4):157–164

Moreno-Hernández JM, Benítez-García I, Mazorra-Manzano MA, Ramírez-Suárez JC, Sánchez E (2020) Strategies for production, characterization and application of protein-based biostimulants in agriculture: a review. Chil J Agric Res 80:274–289

Muskan B, Deepanjali B, Ajay K, Riyapi D, Rupesh K, Baljinder S (2024) Biogenic synthesis of nanoparticles using microbes and plants: mechanisms and multifaceted applications. Int J Environ Anal Chem. https://doi.org/10.1080/03067319.2024.2305238

Mutale-Joan C, Sbabou L, Hicham EA (2023) Microalgae and cyanobacteria: how exploiting these microbial resources can address the underlying challenges related to food sources and sustainable agriculture: a review. J Plant Growth Regul 42(1):1–20

Nayeri FD, Mafakheri S, Mirhosseini M, Sayyed R (2021) Phyto-mediated silver nanoparticles via *Melissa officinalis* aqueous and methanolic extracts: synthesis, characterization and biological properties against infectious bacterial strains. Intl J of Adv Biol and Biomed Res 3:270–285

Nazir S, Zhang JM, Junaid M, Saleem S, Ali A, Ullah A, Khan S (2024) Metal-based nanoparticles: basics, types, fabrications and their electronic applications. Z Phys Chem

Niu C, Wang G, Sui J, Liu G, Ma F, Bao Z (2022) Biostimulants alleviate temperature stress in tomato seedlings. Sci Hortic 293:110712

Oluwaseyi M et al (2024) NanoBiofertilizer and its application in sustainable agriculture, crop specific nutrients delivery and environmental sustainability: a review. Appl Sci Eng Progress. https://doi.org/10.14416/j.asep.2024.05.001

Pandiselvi T, Jeyajothiand R, Kandeshwari M (2017) Organic nutrient management a way to improve soil fertility and sustainable agriculture. A review. Int J Adv Life Sci 10(2):175–181

Pereyra MA, Zalazar CA, Barassi CA (2006) Root phospholipids in Azospirillum inoculated wheat seedlings exposed to water stress. Plant Physiol Biochem 44(11–12):873–879

Pereyra MA, García P, Colabelli MN, Barassi CA, Creus CM (2012) A better water status in wheat seedlings induced by Azospirillum under osmotic stress is related to morphological changes in xylem vessels of the coleoptile. Appl Soil Ecol 53(1):94–97

Placidi P, Morbidelli R, Fortunati D, Papini N, Gobbi F, Scorzoni A (2021) Monitoring soil and ambient parameters in the iot precision agriculture scenario: An original modeling approach dedicated to low-cost soil water content sensors. Sensors 21(15):5110

Preininger C, Sauer U, Bejarano A, Berninger T (2018) Concepts and applications of foliar spray for microbial inoculants. Appl Microbiol Biotechnol 102(17):7265–7282

Pukale DD, Adkins-Travis K, Aryal SR, Shriver LP, Patti GJ, Leipzig ND (2024) Investigating post-traumatic syringomyelia and local fluid osmoregulation via a rat model. Fluids and Barriers of the CNS 21(1):19

Rai N, Ashiya P, Singh RD (2014) Comparative study of the effect of chemical fertilizers and organic fertilizers on Eiseniafoetida. Int J Innovat Res Sci Engineer Tech 3(5):12991–12998

Ramírez-Rodríguez S, Preciado-Rangel P, Cabrera-De M, González-Morales S, Ortega-Ortiz H (2024) Chitosan nanoparticles as biostimulant in lettuce. Phyton 93(4):777

Rasouli H, Popović-Djordjević J, Sayyed RZ, Zarayneh S, Jafari M, Fazeli-Nasab B (2020) Nanoparticles: a new threat to crop plants and soil rhizobia? In: Sustainable agriculture reviews 41: nanotechnology for plant growth and development, pp 201–214

Rather MA, Ahmad I, Shah A, Hajam YA, Amin A, Khursheed S et al (2024) Exploring opportunities of artificial intelligence in aquaculture to meet increasing food demand. Food Chem X:101309

Rojas-Tapias D, Moreno-Galván A, Pardo-Díaz S, Obando M, Rivera D, Bonilla R (2012) Effect of inoculation with plant growth-promoting bacteria (PGPB) on amelioration of saline stress in maize (Zea mays). Appl Soil Ecol 61:264–272

Romero AM, Vega D, Correa OS (2014) Azospirillum brasilense mitigates water stress imposed by a vascular disease by increasing xylem vessel area and stem hydraulic conductivity in tomato. Appl Soil Ecol 82:38–43

Rouphael Y, Colla G (2020) Editorial: Biostimulants in Agriculture. Front Plant Sci 11:40

Sangiorgio D (2022) Deciphering the role of plant microbiome on strawberry and raspberry growth, resistance, fruit quality and aroma

Sanjeevi P, Prasanna S, Siva Kumar B, Gunasekaran G, Alagiri I, Vijay Anand R (2020) Precision agriculture and farming using internet of things based on wireless sensor network. Trans Emerg Telecommun Technol 31(12):e3978

Selvakumar G, Kundu S, Joshi P, Nazim S, Gupta AD, Mishra PK et al (2008) Characterization of a cold-tolerant plant growth-promoting bacterium Pantoea dispersa 1A isolated from a sub-alpine soil in the North Western Indian Himalayas. World J Microbiol Biotechnol 24(7):955–960

Sembada AA, Lenggoro IW (2024) Transport of nanoparticles into plants and their detection methods. Nano 14(2):131

Sethi G, Behera KK, Sayyed R, Adarsh V, Sipra BS, Singh L, Alamro AA, Behera M (2025) Enhancing soil health and crop productivity: the role of zinc-solubilizing bacteria in sustainable agriculture. Plant Growth Regul. https://doi.org/10.1007/s10725-025-01294-7

Shahbaz M, Akram A, Mehak A, Haq E, Fatima N, Wareen G, Aamir S, Hassan H, Poczai P, Sayyed RZ, Ilyas N (2023) Evaluation of selenium nanoparticles in inducing disease resistance against spot blotch disease and promoting growth in wheat under biotic stress. Plants 12(4):761. https://doi.org/10.3390/plants12040761

Shaikh A, Jamla M, Joshi S, Patil S, Oak U, Kumar V (2024) Microbial nanotechnology for producing stress smart crops. Plant Nano Biology 100063

Shanware AS, Taiwade LH (2022) Nano-biofertilizers: progressive evolution for sustainable agriculture. Biol Sci 2(2):166–171

Sharma S, Kumari P, Shandilya M, Thakur S, Perveen K, Sheikh I et al (2025) The combination of α-Fe2O3 NP and Trichoderma sp. improves antifungal activity against Fusarium wilt. J Basic Microbiol e2400613

Sillen WM, Thijs S, Abbamondi GR, De La Torre Roche R, Weyens N, White JC, Vangronsveld J (2020) Nanoparticle treatment of maize analyzed through the metatranscriptome: compromised nitrogen cycling, possible phytopathogen selection, and plant hormesis. Microbiome 8:1–17

Singh D, Singh A, Rakhra M, Sarkar T, Cheema GS, Khamparia A (2024) Predictions on the future of agriculture and recent developments in agricultural technology. In: International conference on artificial intelligence of things for smart societies. Springer Nature Switzerland, Cham, pp 297–303

Singhal A, Baweja P, Gupta S, Chopra H, Gandhi PB (2023) Biogenesis of nanoparticles using microorganisms: a review. Plant Science Today 10(4):97–104

Solomon WGO, Ndana RW, Abdulrahim Y (2012) The comparative study of the effect of organic manure cow dung and inorganic fertilizer NPK on the growth rate of maize (Zea mays L.). Int Res J Agri Sci Soil Sci 2(12):516–519

Spann TM, Little HA (2011) Applications of a commercial extract of the brown seaweed Ascophyllum nodosum increases drought tolerance in container-grown "Hamlin" sweet orange nursery trees. Hort Science 46(4):577–582

Suarez C, Cardinale M, Ratering S, Steffens D, Jung S, Montoya AMZ et al (2015) Plant growth-promoting effects of Hartmannibacter diazotrophicus on summer barley (Hordeum vulgare L.) under salt stress. Appl Soil Ecol 95:23–30

Subramanian KS, Santhanakrishnan P, Balasubramanian P (2006) Responses of field grown tomato plants to arbuscular mycorrhizal fungal colonization under varying intensities of drought stress. Sci Hortic 107(3):245–253

Subramanian P, Kim K, Krishnamoorthy R, Mageswari A, Selvakumar G, Sa T (2016) Cold stress tolerance in psychrotolerant soil bacteria and their con ferred chilling resistance in tomato (Solanum lycopersicum mill.) under low temperatures. PLoS One 11(8):e0161592

Tasnim A, Roy A, Akash SR, Ali H, Habib MR, Barasarathi J, Muthukumaran M, Sayyed RZ, Yeasmin T (2024) Hibiscus sabdariffa L. petal biomass: a green source of nanoparticles of multifarious potential. Open Agriculture 2024(9):20220332. https://doi.org/10.1515/opag-2022-0332

Tavares Bruscato L, Heimfarth T, Pignaton de Freitas E (2017) Enhancing time synchronization support in wireless sensor networks. Sensors 17(12):2956

Thakur P, Thakur S, Kumari P, Shandilya M, Sharma S, Poczai P, Alarfaj AA, Sayyed RZ (2022) Nano-insecticide: synthesis, characterization, and evaluation of insecticidal activity of ZnO NPs against *Spodoptera litura* and *Macrosiphum euphorbiae*. Appl Nanosci 12:3835–3850. https://doi.org/10.1007/s13204-022-02530-6

Theocharis A, Bordiec S, Fernandez O, Paquis S, Dhondt-Cordelier S, Baillieul F et al (2012) Burkholderia phytofirmans PsJN primes Vitis vinifera L. and confers a better tolerance to low nonfreezing temperatures. Mol Plant-Microbe Interact 25:241–249

Vijayaram S, Razafindralambo H, Sun YZ, Vasantharaj S, Ghafarifarsani H, Hoseinifar SH, Raeeszadeh M (2024) Applications of green synthesized metal nanoparticles—a review. Biol Trace Elem Res 202(1):360–386

Vitousek PM, Aber JD, Howarth RW, Likens GE, Matson PA et al (1997) Human alterations of the global nitrogen cycle: sources and consequences. Ecol Appl 7(3):737–750

Wei L, Liu J, Jiang G (2024) Nanoparticle-specific transformations dictate nanoparticle effects associated with plants and implications for nanotechnology use in agriculture. Nat Commun 15(1):7389

Wu Q, Zou Y (2009) Mycorrhizal influence on nutrient uptake of citrus exposed to drought stress. Philip Agric Sci 92:33–38

Xu L, Geelen D (2018) Developing biostimulants from agro-food and industrial by-products. Front Plant Sci 9:1567

Yildirim E, Taylor AG, Spittler TD (2006) Ameliorative effects of biological treatments on growth of squash plants under salt stress. Sci Hortic 111:1–6

Zamljen T, Grohar MC, Slatnar A (2024) Effects of pre-and post-transplantation humic acid biostimulant treatment and harvest date on yield quantity and quality parameters of sweet peppers (Capsicum annuum L.). Sci Hortic 338:113747

Zervopoulos A, Tsipis A, Alvanou AG, Bezas K, Papamichail A, Vergis S et al (2020) Wireless sensor network synchronization for precision agriculture applications. Agriculture 10(3):89

Zhang G, Bai J, Zhai Y, Jia J, Zhao Q, Wang W, Hu X (2024) Microbial diversity and functions in saline soils: a review from a biogeochemical perspective. J Adv Res 59:129–140

Controlled Release of Nano-biofertilizers and Nano-biopesticides for Sustainable Agriculture

8

Ahmed M. A. Kenawy, Nehal M. El-Deeb, Mailin Misson, Hadjadj Naima, Ting Ho, Muktiningsih Nurjayadi, Dalia Sukmawati, and Hesham Ali El Enshasy

A. M. A. Kenawy (✉)
Nucleic Acids Research Department, Genetic Engineering and Biotechnology Research Institute (GEBRI), City of Scientific Research and Technological Applications (SRTA-City), New Borg El-Arab, Egypt

N. M. El-Deeb
Pharmaceutical Bioproducts Research Department, Genetic Engineering and Biotechnology Research Institute (GEBRI), City of Scientific Research and Technological Applications SRTA-City, Alexandria, Egypt

M. Misson
Biotechnology Research Institute, University Malaysia Sabah, Sabah, Malaysia

H. Naima
Agri-food Department, Faculty of Natural and Life Sciences, University of Blida1, Ouled Yaïch, Algeria

T. Ho
Back2Nature Organic Farm, Kuala Pilah, Malaysia

M. Nurjayadi · D. Sukmawati
Faculty of Mathematics and Natural Science, Department of Chemistry, Universitas Negeri Jakarta, Jakarta, Indonesia

H. Ali El Enshasy
Innovation Centre in Agritechnology for Advanced Bioprocessing (ICA), Universiti Teknologi Malaysia (UTM), Pagoh, Johor, Malaysia

Faculty of Chemical and Energy Engineering, Universiti Teknologi Malaysia (UTM), Johor, Malaysia

City of Scientific Research and Technological Applications (SRTA-City), New Borg El-Arab, Egypt

R. Sayyed et al. (eds.), *Nanobiostimulants in Innovative Agriculture*,
https://doi.org/10.1007/978-981-95-2477-8_8

Abstract

The rising population causing a significant threat to meet the world food demand results in increased use of chemical fertilizers. Excessive use of chemical fertilizers causes challenges like soil degradation, fertility loss, minerals deficiency, pest resistance, climate change, and chemical pollution. These emerging issues impose a considerable burden on existing agricultural problems. Bio-stimulants are a good initiative to overcome agricultural issues by improving overall production, sustainability, metabolic reactions, and physio-chemical response of plants against various biotic and abiotic stresses. Seaweeds, plant extracts, hydrolyzed proteins, nitrogen-containing compounds, humic substances, and microorganisms are various bio-stimulants. Microbial bio-stimulants are helpful to lower the use of chemical fertilizers but their efficacy is more when combined with nano-stimulants. Nanotechnology is an advanced field that creates nano-sized particles and is used in every field of life. In agriculture, nanoparticles improve crop production and remove agro-waste from water bodies. Nanoparticles (both organic and inorganic) are synthesized by biological, chemical, and mechanical processes. Green synthesized nanoparticles (NP's) are considered environmentally friendly as compared to mechanical and chemical approach due to less production of toxic chemical. Nanomaterials integration with microbial bio-stimulants into agricultural practices boosts crop productivity. Integrating microbial nanotechnology into agricultural practices and biomedicine is promising, but it must overcome synthesis, standardization, and scalability challenges.

Keywords

Hormones · Humic acids · Microorganisms · Plant extracts · Seaweeds

1 Introduction

Fertilizers play a crucial role in boosting crop yields and improving quality in agriculture (Bakker et al. 2024). Currently, agricultural practices cannot satisfy the increasing food demand without relying heavily on fertilizers. Nowadays, the agriculture industry faces notable challenges, such as deteriorating soil health, poor nutrient delivery efficiency, and environmental pollution due to overuse of traditional fertilizers' (Nongbet et al. 2022; Bakker et al. 2024). The use of nanotechnology, especially in the creation of nanofertilizers, presents a promising strategy for improving nutrient use efficiency and addressing nutrient shortages in crops. On the other hand, the use of nanopesticides was also one of the important inventions to reduce the use of toxic chemicals in high concentrations in pest-control formulations.

Nanofertilizers represent a new sustainable technology that revolutionizes the agriculture industry and protects the environment from the drawback impacts of traditional fertilizers. This strategy aims to reduce energy consumption and the rate of fertilization. Also, it increases the efficiency of nutrient uptake via plants.

Altogether, this will reduce the lost amount of fertilizer leaching into the underground water, surface water, air, and other environmental components. Hence, the employment of this technology can reduce the environmental pollution rate and secure sustainable agriculture practices for future generations (Baker et al. 2017). Many methods have been developed to produce nanofertilizers based on the combinations of nanomaterial technology and conventional fertilizers to ensure a higher surface area, more specificity, and highly controlled release (slow release) of nutrients to the plants (Ayenew et al. 2025). Along with all the benefits that nanofertilizers can provide us, several concerns still present challenges against the unconditional applications of nanofertilizers, including cost and safety aspects as the major two factors in this area. On the other hand, utilization of nanotechnology to produce nano-biopesticides showed increased concerns during the last few years. In general, the crop protection market is mainly dominated by chemical pesticides, which are considered toxic, carcinogenic, and responsible for many human and animal diseases. The biopesticides, as a safe alternative for plant protection, are shared by only 5% of the global market (Cappa et al. 2022; Pan et al. 2023). Nano-biopesticides, as one of the new safe agriculture products based on emerging technology, the high cost and public awarenesses are considered as the main two drawbacks which limit their market growth.

In this book chapter, we will shed light on some important points of the production, applications, and development of nanochemicals in the agriculture field. It also covers the essential concepts behind the development of nanofertilizers and highlights the different types of nanofertilizers and nano-biopesticides available along with their potential advantages and challenges. It is imperative that a clear understanding of nanochemicals is essential to harness their potential to transform agriculture and create a more sustainable Good Agriculture Practice (GAP).

2 Nanochemicals in Fertilization and Pest Control

Therefore, application nanochemicals for both fertilization and pest control is a significant innovative breakthrough in the agriculture sector, possessing the ability to minimize the waste associated with conventional fertilizers and enhance nutrient utilization (Nongbet et al. 2022). This approach is a relatively recent development that exploits the distinctive characteristics of nanoparticles to improve nutrient delivery and mitigate negative environmental effects. Nanofertilizers exhibit unique physicochemical traits that provide several advantages over traditional fertilizers. For example, the tiny particle sizes (less than 100 nm) allow nanofertilizers to penetrate plant systems effectively when applied as a foliar spray or through soil application (Seleiman et al. 2021). The creation of these substances has the potential to boost crop yields while offering a sustainable solution to environmental challenges. Although obstacles persist regarding cost, safety, and large-scale application, the outlook for nanofertilizers in contemporary agriculture appears bright. They can facilitate a controlled release of nutrients over time, ensuring a consistent supply of vital elements for plants. Consequently, nutrient-use efficiency is likely to be

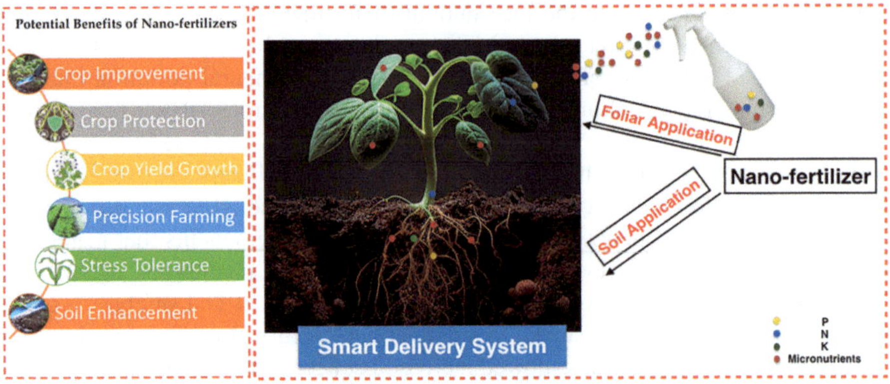

Fig. 8.1 Nanofertilizers and their potential benefits in agriculture

improved, and environmental pollution can be reduced compared to conventional fertilizers (Nongbet et al. 2022). Nanofertilizers are designed to improve the delivery, absorption, and effectiveness of vital plant nutrients, including nitrogen, phosphorus, potassium, and trace elements (Babu et al. 2022). Due to their miniature size and massive surface area, nanoparticles can interact with plant roots, leaves, and soil in innovative manners to improve plant growth and stress tolerance (Fig. 8.1).

3 Nanofertilizers and Nano-biopesticides

The word "nano" comes from the Greek term for "dwarf," indicating the tiny size of particles, usually measuring less than 100 nm in at least one dimension (Saurabh et al. 2024). These nanoparticles exhibit distinctive physicochemical characteristics, such as a high surface-to-volume ratio, crystallinity, and increased reactivity, setting them apart from larger materials (Saurabh et al. 2024). These traits allow nanoparticles to interact more effectively with plant roots and soil microbes, enhancing nutrient absorption and plant metabolism. Nanofertilizers are specially designed fertilizers that use nano-sized materials to encapsulate nutrients, enabling a gradual and controlled release. This leads to more efficient uptake by plant roots and minimizes the need for frequent fertilizer applications (Seleiman et al. 2021). These materials can be engineered to respond to specific plant requirements, environmental situations, and stressors, making them adaptable to various agricultural settings. Furthermore, they help decrease environmental pollution associated with traditional fertilizers by synchronizing nutrient availability with plant growth and enhancing productivity, especially when applied in a sustainable release formulation (Beig et al. 2022). In this essence, nanofertilizers contribute to sustainable agriculture, tackling global issues like food security, environmental contamination, and soil health.

On the other hand, pesticides and biopesticides are widely used for controlling insects, pests, and herbs, which can reduce the yield and quality of plants. They work through either prevention of plant diseases, reduction/prevention of the growth of other unwanted plants, or killing other unwanted insects or small animals. They can be simply divided according to functionality into different categories such as bactericides, fungicides, insecticides, rodenticides, and herbicides. However, chemical pesticides are generally non-specific toxic, and carcinogenic and are able to cause different types of chronic and non-chronic diseases, whereas biopesticides are natural and less toxic with less environmental impact (Abdollahdokhr et al. 2022; Daraban et al. 2023).

4 Nanofertilizers Compared to Traditional Fertilizers

Traditional nitrogen fertilizer substantially enhances plant yield, but overuse of this fertilizer has caused serious damage to the environment and human health. The rate of nitrogen fertilizer application has a close relationship with nitrate accumulation in the environment, groundwater, and leafy and root vegetables (Ahmed et al. 2017). Consumption of diets having high nitrate contents has contributed to endogenous nitronization, which could lead to thyroid disease, various kinds of human cancers, neural tube defects (during fetus development), and diabetes (Iammarino et al. 2013). 50–70% of nitrogen (N) from conventional fertilizers is usually lost to the environment via volatilization, runoff, and leaching, which cause pollution of surface water as well as groundwater and reduce the quality of air (Galloway et al. 2003; Ahmed et al. 2017). Phosphorus (P) fertilizers are also susceptible to being fixed in the soil, making them inaccessible to plants, which leads to the need for frequent applications (Zulfiqar et al. 2019). Such losses decrease the efficiency of fertilizers and also lead to environmental pollution, impacting both soil and water quality. The low efficiency of plant uptake of typical synthetic fertilizers, along with nutrient loss through leaching, volatilization, and gas emissions, are significant contributors to environmental harm (Dimkpa et al. 2015; Kah et al. 2018). Conversely, nanofertilizers possess a larger surface area compared to traditional chemical fertilizers, allowing them to hold more nutrients and release them gradually according to the needs of the crops without negative effects (Siddiqi and Husen 2017).

5 Advantages of Application of Chemicals in Nanoform Over Conventional Forms for Good Agriculture Practice (GAP)

Due to the higher surface area to volume ratio, nanomaterials could effectually hold nutrients, thereby allocating a longer-term and constant nutrient pool to plants to enhance nutrient uptake with the possible design of a targeted delivery system. This will allow the reduction of nutrient loss and increase nutrient use efficiency. Also, it reduces application rates and saves the cost of fertilizers. Nanofertilizers can enhance nutrient availability to plants, minimize loss of nutrients via leaching, and have minimal environmental impact (El-Ghamry et al. 2018). In general, the goal is to minimize energy use and reduce losses of nutrients without losing crop productivity. Due to their diminutive size and expansive surface area, nanoparticles can interact with plant roots and soil innovatively. This distinctive behavior enables nanofertilizers to address certain shortcomings of traditional fertilizers, such as nutrient loss caused by leaching, volatilization, and soil deterioration. The capacity of nanoparticles to supply nutrients more accurately and efficiently presents considerable opportunities for enhancing crop production while reducing environmental repercussions. Reports indicate that the foliar application of nanofertilizers can boost cereal yields by 10–25%, oilseed yields by 20–30%, and pulse yields by 13–15% compared to conventional fertilizers across various agro-climatic and management scenarios (Delfani et al. 2014; Rathore et al. 2019). Likewise, the utilization of P-based nanofertilizers may enhance soybean seed yield by approximately 20% in comparison to traditional synthetic fertilizers (Liu and Lal 2014). Therefore, nanofertilizers present significant potential for achieving sustainable agriculture, especially in underdeveloped and developing countries (Liu and Lal 2014).

6 Types of Nanoparticles and Their Release Control in Fertilization and Pest-Control Applications

Nanofertilizers can be categorized according to their composition and methods of delivery. This includes materials that are nanoporous and covered with polymer films, nanoemulsions, and nanoparticles designed to encapsulate nutrients for precise delivery (Yadva et al. 2022; Yadav et al. 2023). Various coating techniques, such as polymer coatings, clay-based encapsulation, and biopolymer matrices, are utilized to control nutrient release, ensuring that plants receive them at the right time, which minimizes waste and mitigates environmental impact (Galloway et al. 2003). Several basic materials can undergo nanostructuring and be utilized as nanofertilizers, including zeolites, silver (Ag), copper (Cu), aluminum (Al), carbon (C), zinc (Zn), potassium (K), nitrogen (N), silica (Si), iron (Fe), magnesium (Mg), sodium (Na), calcium (Ca), and manganese (Mn). In addition, banana peels (Sharma et al. 2025; Hussein et al. (2022), grape plant residues (Sharma et al. 2025), and a variety of other plant-derived materials are being investigated for the production of nanofertilizers. Recently, natural zeolite, which comprises over 50 minerals, has been transformed into nanofertilizers, primarily due to its widespread availability and economical nature (Eroglu et al. 2017). Nanofertilizers can be customized to address specific nutrient shortages in plants, as the atoms located on the surfaces of nanoparticles can be organized to display distinct properties (Elemike et al. 2019).

7 Plant Absorption Mechanisms

Nanofertilizers can be absorbed by roots when they are introduced into the soil or through leaves when applied as a foliar treatment. The external application of NPs or their aggregates that are of smaller size may facilitate easier penetration through the cuticle using the cuticular pathway compared to larger particles (Sembada and Lenggoro 2024). However, variations in leaf structure and the quantity and dimensions of stomata across different plant species can influence the uptake of various nanofertilizers of differing diameters. Upon their application to the soil, the movement of nanofertilizers can occur via either the apoplastic or symplastic routes (Sattelmacher 2001). In the apoplastic route, NPs first pass through the pores of the cell wall and then move into the intercellular spaces (Farvardin et al. 2020). However, the diameter of the cell wall pores limits their passage if they are larger than 20 nm through the apoplastic pathway. In foliar applications, the sizes of nanoparticles must be smaller than those needed for soil applications (Seleiman et al. 2021).

8 Methods of Nanoparticle Fabrication for Agriculture Applications

In general, three main techniques are available for the fabrication of nanoparticles for agriculture applications:

A. Developing nanoparticles that contain nutrients.
B. Nanoparticle coating in traditional fertilizers.
C. Nanoscale additives in traditional fertilizers.

The predominant technique for creating nanofertilizers using nanomaterials is nutrient encapsulation. Recently, certain nutrients have been encapsulated within nanoporous substances, coated with a thin polymer film, or incorporated into emulsions at the nanoscale. Encapsulating beneficial microorganisms, such as bacteria or fungi, has demonstrated promise for enhancing the availability of nitrogen (N), phosphorus (P), and potassium (K) in the root zone (Bose et al. 2020). Following the extraction and purification stages, which involve various chemical, physical, mechanical, and biological methods, nanofertilizers can be produced from standard bulk fertilizer ingredients. Initial nanomaterials are primarily created using physical (top-down) and chemical (bottom-up) techniques as described in the following lines.

8.1 Physical Top-Down Techniques

This method encompasses photolithography, ball milling, chemical etching, sputtering, and other physical processes. In the physical approaches, mechanical pressure, ultrasonication, and electrical energy are cost-effective, low-toxic, eco-friendly, and efficient methods for creating nanoparticles (Satyanarayana and Sudhakar 2018).

8.2 Chemical and Biological Bottom-Up Techniques

This method utilizes chemical and biological techniques, atomic and molecular assembly, spray, laser pyrolysis (including photochemical, sono-chemical, chemical and electrochemical, sol-gel, microwave, and vapor deposition) approaches (Arole and Munde 2014). The biological method is an emerging green alternative for synthesizing nanoparticles. As biological method can reduce time and energy, utilizes environmentally safe and sustainable materials it become more economically feasible and acceptable approach for field application as safe method (Sharma et al. 2025; Ying et al. 2022).

9 Controlled Release of Nanofertilizers

The conventional fertilizers are not designed to reach the targeted part of plant cells and most of the time are leached out, evaporated, hydrolyzed by soil before reaching the targeted cells (Jakhar et al. 2022). Nanofertilizers with controlled release capabilities are acknowledged for their potential to boost agricultural output (Saurabh et al. 2024). Studies indicate that their implementation can lead to higher crop production, better nutrient use efficiency, and improved overall plant health (Seleiman et al. 2021). These fertilizers can be applied to a range of crops, including rice, wheat, corn, and soybeans, producing favorable results in crop yield and nutrient efficiency. Continuous research and development are crucial for maximizing the advantages of these innovative fertilizers to address the increasing global food production needs. Controlled-release nanofertilizers present numerous advantages over traditional fertilizers, including more efficient and prolonged nutrient release, the simultaneous release of various nutrient types, and an increased water-holding capacity that can mitigate soil erosion. Different materials have been used in the synthesis of slow-release nanofertilizers. For example, chitosan-based nanofertilizers, where chitosan molecules interact with fertilizer molecules to enhance their availability to plants. Chitosan nanoparticles (NPs) can act as carriers for the controlled release of NPK fertilizers in agricultural settings. Chitosan molecules can be readily modified to enhance the uptake and gradual release of plant growth regulators and fertilizers. In addition, it has also been used successfully as a natural herbicide and insecticide. Chitosan can also help in designing a delivery system and a controlled release medium. It can also act as a safeguard for biomolecules from fluctuations in pH, light exposure, and temperature variations, thus prolonging the release of active compounds and averting a rapid discharge that may damage plant cells. Another study also showed the successful production of nanocomposite fertilizer using a combination of NPK, nanochitosan, and ZnO nanoparticles to improve the nanoparticle characteristics and slow-release properties (Madzokere et al. 2021).

Nanoclay is also another example of developing a new fertilizer of significant large surface areas and reactivity at the nanoscale, which makes it ideal for creating controlled-release fertilizer (CRF) formulations. Nanoclay is essential in the production of CRF due to the active surface it provides for a range of physicochemical and biological interactions (Nayak et al. 2025). Organoclays, which are clay minerals with particles smaller than 2 mm, exhibit a high specific surface area, hydrogel capacity, charge, and crystalline structure. These features, combined with their colloidal particle size, are advantageous for the manufacture of CRF (Silva et al. 2020). Also, layered double hydroxides (LDH) are two-dimensional particles characterized by intercalated anionic materials within their layered structure, allowing for the controlled release of anions to balance the positive charge of the LDH. Nanocapsule nanofertilizers are fertilizers contained in tiny capsules made from organic or inorganic materials, designed for the encapsulation and delivery of fertilizers in a slow-release format. Nanogel-based nanofertilizers are porous materials formed by combining a polymer with a liquid, enabling them to absorb and gradually release fertilizers or other substances over time. Additionally, traditional fertilizers could be coated with polyurethane polymer to create slow-release formulations of polyurethane-based nanofertilizers. Moreover, starch-based and zeolite-based nanofertilizers have been formulated to create slow-release products (Channab et al. 2023).

Other forms of nanocomposite hydgrogels, such as PAAm/methyl cellulose/montmorillonite hydrogel, have been successfully used for sustainable release of chemical fertilizers (Bortolin et al. 2013). On the other hand, an efficient sustainable release system for nano-biopesticides was achieved by the incorporation of biopolymers such as chitosan. Microencapsulation of essential oil biopesticides using biopolymers such as chitosan or amphiphilic carboxymethyl chitosan with ricinoleic acid (R-CM chitosan) can also help increase bioactive compound stability (Bakry et al. 2016; Taban et al. 2020).

10 The Promise of Nanofertilizers and Nano-biopesticides in Sustainable Agriculture

Sustainable agriculture is basically the continued crop production without depletion of soil quality and without creating any negative impact on the environment. This principle of the sustainability approach required a careful application of exogenous chemicals to the soil, taking into account different ecosystem elements such as natural macro- and microflora and the whole biological components of the environment. Therefore, nanofertilizers also offer several advantages over conventional fertilizers in terms of sustainable agriculture (Mejias et al. 2021; An et al. 2022). Some of the most significant benefits include:

10.1 Design of Precision Agriculture

Nanofertilizers can be tailored to deliver nutrients more precisely based on the specific needs of crops. By integrating with modern agricultural practices such as precision farming and smart sensors, they could help optimize fertilizer application at a granular level, thus maximizing crop yield with minimal resource wastage.

10.2 Enhanced Soil Health

Some nanofertilizers contain beneficial properties that can improve soil health over time. By delivering nutrients in a controlled manner, this can lead to a proper maintenance of a balanced soil ecosystem, enhancing microbial activity and promoting sustainable agricultural practices. However, sustainable soil health is not easy in the climate change era with the continued threat of abiotic stresses. Addition of nanofertilizers can also mitigate the impact of climate change issues that affect the soil quality, such as soil acidity/alkalinity, which can lead to unfavorable conditions for plant growth, such as droughtiness, salt, and acidic and alkaline stresses (Mejias et al. 2021; Reddy et al. 2024).

10.3 Increased Crop Yield and Quality

Nanofertilizers have the potential to boost crop productivity by providing nutrients in a more readily available form to plants. Studies have shown that nanofertilizers can improve not only the quantity but also the quality of harvested crops, making them more resilient to diseases and environmental stresses (Bastami et al. 2021; Faridvand et al. 2021; Fazeli-Nasab et al. 2021). Therefore, nanofertilizers have a wide array of potential applications in modern agriculture, ranging from large-scale farming to urban gardening. This indeed improves the plant yield, reduces the plant stress, and improves the overall good agriculture practice process. This also decrease the drawbacks of the extensive uses of chemical fertilizers which become one the main threat for sustainability of soil fertility and soil microbiome (Reddy et al. 2024).

On the other hand, application of biopesticides in nanoformulation is as important as nanofertilizers in good agriculture practice for sustainable agriculture. The application of nanotechnology for stabilization of biocontrol agents has been extensively studied by many authors. Nanopesticides are those types of products with molecules in the nanosize range (can be up to 1000 nm) and thus exhibit higher efficacy (Kah et al.2013; Kah and Hofmann 2014; Kah et al. 2018). The production of nanopesticides typically uses biocompatible and natural polymers such as bioclay, cellulose, chitosan, or polyactide, which are usually used for the production of nanomolecules in the form of nanocapsules, nanohydrogels, nanomicelles, and nanospheres to enhance the efficacy, reduce the toxicity, and improve the safety of the product (Walker et al. 2018; Yang et al. 2021; Pan et al. 2023). Nanobotanical insecticides are considered a typical example of a novel nano-biocontrol agent. The main ingredient of the formulation includes plant secondary metabolites such as terpenoids, volatile oils, alkaloids, flavonoids, and other bioactive compounds (Campolo et al. 2020; Wu et al. 2021). However, essential oils are the most studied and applied materials for nano-biopesticide formulations based on their availability, ease of preparation, and cost.

Interestingly, the application of nanotechnology to improve the stability of *Bacillus thuringiensis* toxic proteins has been proven in different ways. It has been reported that adsorption of Bt Cry1Aa toxin on montmorillonite and kaolinite improves stability, adsorption/desorption of the protein in the formulation, efficacy, and prevention of oligomerization/inactivation of the protein (Helassa et al. 2009; Helassa et al. 2011). Other studies also showed that the Bt crystal can be protected from UV irradiation during field application using silicon oxide or silica nanoparticles (Michael et al. 2010; Madliger et al. 2011), a combination of graphene oxide and olive oil (Shahab and Elham 2017), magnesium hydroxide nanoparticles (Pan et al. 2017), and the nanoparticle microencapsulation technique (Zhang et al. 2021). A recent study also reported on the increased pesticidal and antifungal activity of the synthesized silver nanoparticles of Bt-Ag_2O o particle size of 18.24 nm (Ge et al. 2024).

11 Key Applications of Nanofertilizers

Nanofertilizers could revolutionize the cultivation of staple crops such as wheat, rice, and corn by improving nutrient uptake, reducing fertilizer waste, and promoting higher yields. This could be particularly impactful in regions where soil degradation and nutrient deficiencies are prevalent. Horticulture and specialty crops: For crops that require higher precision in nutrient delivery, such as fruits, vegetables, and herbs, nanofertilizers could provide targeted nutrient applications, enhancing both the size and quality of the produce.

11.1 Greenhouse and Vertical Farming

In controlled environments like greenhouses or vertical farms, nanofertilizers could be used to optimize plant growth and minimize nutrient loss. This would allow farmers to achieve higher yields with a fraction of the water and fertilizers typically used.

11.2 Precision Agriculture

The integration of nanofertilizers with smart farming technologies, such as drones, sensors, and automated systems, holds great potential for precision agriculture. These systems could monitor soil and plant conditions in real time, adjusting fertilizer application for maximum efficiency and reduced waste.

11.3 Soil Remediation

Nanofertilizers could also play a role in restoring soils that have been depleted of nutrients or polluted by excessive chemical fertilizers. By delivering nutrients in a more bioavailable form, they may help rejuvenate poor soils and support sustainable land management.

12 Conclusion

Nanofertilizers represent a groundbreaking advancement in agriculture, offering the potential to reduce fertilizer waste, improve nutrient use efficiency, and enhance crop yields in a sustainable way. While challenges remain in terms of cost, safety, and large-scale implementation, the future of nanofertilizers in modern farming looks promising. By integrating these technologies into the current farming practices, we may be able to feed the world's growing population while protecting the planet's precious resources. The industrialization of nanofertilizers needs more efforts to improve the production process and reduce the manufacturing cost to be more comparable to the traditional chemical fertilizers. On the other hand, more research is required, especially in the field trials, to understand the overall impact of nanofertilizers and nanochemicals in general on the environment and life cycle sustainability. In addition, in-depth studies on the risk assessment of nanofertilizers are still needed in different types of soil and applications under different environmental conditions. In addition, the acceptance of nanoparticles in the consumer market is still low due to low public awareness about the safety and efficacy of this technology. Therefore, academia and industries need to collaborate to educate farmers, consumers, policymakers, and other stakeholders to increase their acceptance of this new technology. In general, applications of nanochemicals in the form of nanofertilizers or nano-biocontrol agents are considered one of the important inventions with huge potential applications to replace the traditionally used fertilizers, as they can be used in sustainable release form to minimize the pollution risk and promote sustainable agriculture practices.

References

Abdollahdokhr D, Gao Y, Faramarz S, Poustforoosh A, Abbasi M, Asadikaram G, Nematollahi MH (2022) Conventional agrochemicals towards nano-biopesticides: an overview on recent advances. Chem Biol Technol Agric 9:13

Ahmed M, Rauf M, Saeed NA, Mukhtar Z (2017) Excessive use of nitrogenous fertilizers: an unawareness causing serious threats to environment and human health. Environ Sci Pollut Res 24:26983–26987. https://doi.org/10.1007/s11356-017-0589-7

An C, Sun C, Li N, Huang B, Jiang J, Shen Y, Wang C, Zhao X, Cui B, Wang C, Li X, Zhang S, Gao F, Zeng Z, Cui H, Wang Y (2022) Nanomaterials and nanotechnology for the delivery of agrochemicals: strategies towards sustainable agriculture. J Nanobiotechnol 20:11

Arole VM, & Munde SV (2014) Fabrication of nanomaterials by top-down and bottom-up approaches-an overview. J Mater Sci 1:89–93

Ayenew BM, Satheesh N, Zegeye ZB, Desalegn Adisu Kassie DA (2025) A review on the production of nanofertilizers and its application in agriculture. Heliyon 11(1):e41243. https://doi.org/10.1016/j.heliyon.2024.e41243

Babu S, Singh R, Yadav D, Rathore SS, Raj R, Avas R, Yadav SK, Ds A, Yadav V, Yadav B, Shekhawat K, Upadhyay PK, Yadav DK, Singh VK (2022) Nano-fertilizers for agricultural and environmental sustainability. Chemosphere 292:133451. https://doi.org/10.1016/j.chemosphere.2021.133451

Baker S, Volova T, Prudnikova SV, Satish S, Nagrendra P (2017) Nanoagroparticles emerging trends and future prospect in modern agriculture system. Environ Toxicol Pharmacol 53:10–17

Bakker T, Brauers E, van der Elst G, Wangari E (2024) Nanotechnology in agricultural production nanotechnology: a policy brief was commissioned by the UN policy analysis branch Division for Sustainable Development. https://sustainabledevelopment.un.org/content/documents/12872Policybrief_Agri.pdf. Accessed on 10 Feb 2024

Bakry AM, Abbas S, Ali B, Majeed H, Abouelwafa MY, Mousa A, Liang L (2016) Microencapsulation of oils: a comprehensive review of benefits, techniques, and applications. Compr Rev Food Sci Food Saf 15:143–182

Bastami A, Amirnia R, Sayyed R, Enshasy H (2021) The effect of mycorrhizal fungi and organic fertilizers on quantitative and qualitative traits of two important Satureja species. Agronomy 11:1285. https://doi.org/10.3390/agronomy11071285

Beig B, Niazi MBK, Sher F, Jahan Z, Malik US, Khan MD, Americo-Pinheiro DVN (2022) Nanotechnology -based controlled release of sustainable fertilizers. A review. Environ Chem Lett 20:2709–2726

Bortolin A, Aouda FA, Mattoso LH, Ribeiro C (2013) Nanocomposite PAAm/methyl cellulose/montmorillonite hydrogel: evidence of synergistic effects for the slow release of fertilizers. J Agric Food Chem 61(3):7431–7439

Bose RJC, Tharmalingam N, Choi Y, Madheswaran T, Paulmurugan R, McCarthy JR, … Park H (2020) Combating intracellular pathogens with nanohybrid-facilitated antibiotic delivery. Int J Nanomedicine 15:8437–8449. https://doi.org/10.2147/IJN.S271850

Campolo O, Giunti G, Laigle M, Michel T, Palmeri V (2020) Essential oil-based nano-emulsions: effect of different surfactants, sonication and plant species on physicochemical characteristics. Ind Crop Prod 157:8

Cappa F, Baracchi D, Cervo R (2022) Biopesticides and insect pollinators: detrimental effects, outdated guidelines, and future directions. Sci Total Environ 837:155714

Channab BE, El Idrissi A, Zahouily M, Essamlali Y, White JC (2023) Starch-based controlled release fertilizers: a review. Int J Biol Macromol 31(238):124075. https://doi.org/10.1016/j.ijbiomac.2023.124075

Daraban GM, Hlihor R-M, Suteu D (2023) Pesticides vs. biopesticides: from pest management to toxicity and impacts on the environment and human health. Toxics 11:983. https://doi.org/10.3390/toxics11120983

Delfani M, Baradarn Firouzabadi M, Farrokhi N, & Makarian H (2014) Some Physiological Responses of Black-Eyed Pea to Iron and Magnesium Nanofertilizers. Commun Soil Sci Plant Anal 45(4):530–540. https://doi.org/10.1080/00103624.2013.863911

Dimkpa CO, McLean JE, Britt DW et al (2015) Nano-CuO and interaction with nano-ZnO or soil bacterium provide evidence for the interference of nanoparticles in metal nutrition of plants. Ecotoxicol 24, 119–129. https://doi.org/10.1007/s10646-014-1364-x

Elemike EE, Uzoh IM, Onwudiwe DC, & Babalola OO (2019) The role of nanotechnology in the fortification of plant nutrients and improvement of crop production. Appl Sci 9(3):499

El-Ghamry A, Mosa AA, Alshaal T, & El-Ramady H (2018) Nanofertilizers vs. biofertilizers: new insights. Environ, Biodivers Soil Secur 2(2018):51–72

Eroglu N, Emekci M, & Athanassiou CG (2017) Applications of natural zeolites on agriculture and food production. J Sci Food Agric 97(11):3487–3499. https://doi.org/10.1002/jsfa.8312

Faridvand S, Amirnia R, Tajbakhsh M, Enshasy HE, Sayyed RZ (2021) The effect of foliar application of magnetic water and nano, organic, and chemical fertilizers on phytochemical and yield characteristics of different landraces of fennel (*Foeniculum vulgare* mill). Horticulturae 7:475

Farvardin A, González-Hernández AI, Llorens E, García-Agustín P, Scalschi L, Vicedo B (2020) The Apoplast: a key player in plant survival. Antioxidants 9(7):604. https://doi.org/10.3390/antiox9070604

Fazeli-Nasab B, Sayyed RZ, Piri R, Rahmani AF (2021) Biopriming and nanopriming: green revolution wings to increase plant yield, growth, and development under stress condition and forward dimensions. In: Singh HB et al (eds) Antioxidants in plant-microbe interaction. Springer Singapore, Singapore, pp 623–655

Galloway JN, Aber JD, Erisman JW, Seitzinger SP, Howarth RW, Cowling EB, Cosby BJ (2003) The nitrogen cascade. Bioscience 53(4):341–356

Ge J, Hu J, Cui S, Wang Y, Xu C, Liu W (2024) Biosynthesis of Bt-Ag_2O nanoparticles using *Bacillus thuringiensis* and their pesticidal and antimicrobial activities. Appl Microbiol Biotechnol 108:157

Helassa N, Quiquampoix H, Noinville S, Szponarski W, Staunton S (2009) Adsorption and desorption of monomeric Bt (*Bacillus thuringiensis*) Cry1Aa toxin on motmorillnite and kaolinite. Soil Biol Biochem 41:498–504

Helassa N, Revault M, Quiquampoix H, Dejardin P, Staunton S, Noinville S (2011) Adsorption on montmorillonite prevents oligomerization of Bt Cry1As toxin. J Colloid Interface Sci 356:718–725

Hussein N, Bilal M, & Iqbal HM (2022) Carbon-based nanomaterials with multipurpose attributes for water treatment: Greening the 21st-century nanostructure materials deployment. Biomater PolymHorizon 1(1):48–58

Iammarino M, Di Taranto A, & Cristino M (2013) Endogenous levels of nitrites and nitrates in wide consumption foodstuffs: Results of five years of official controls and monitoring. Food Chem 140(4):763–771. https://doi.org/10.1016/j.foodchem.2012.10.094

Jakhar AM, Aziz I, Kaleri AR, Hasnain M, Haider G, Ma J (2022) Nano-fertilizers: a sustainable technology for improving crop nutrition and food security. NanoImpact 27:100411

Kah M, Hofmann T (2014) Nanopesticide research: current trends and future priorities. Environ Int 63:224–235

Kah M, Beulke S, Tiede K, Hofmann T (2013) Nanopesticides: state of knowledge, environmental fate, and exposure modeling. Crit Rev Environ Sci Tech 43(16):1823–1867. https://doi.org/10.1080/10643389.2012.671750

Kah M, Kookana RS, Gogos A, Bucheli TD (2018) A critical evaluation of nanopesticides and nanofertilizers against their conventional analogues. Nat Nanotechnol 13:677–684

Liu R, Lal R (2014) Synthetic apatite nanoparticles as a phosphorus fertilizer for soybean (Glycine max). Sci Rep 4:5686. https://doi.org/10.1038/srep05686

Madliger M, Gasser CA, Schwarzenbach RP, Sander M (2011) Adsorption of transgenic insecticidal Cry1Ab protein to silica particles. Effect on transport and bioactivity. Environ Sci Technol 45:4377–4384

Madzokere TC, Murombo LT, Chiririwa H (2021) Nano-based slow releasing fertilizers for enahanced agricultural productivity. Mater Today Proc 45(3):3709–3715

Mejias JH, Salazar F, Pérez Amaro L, Hube S, Rodriguez M, Alfaro M (2021) Nanofertilizers: a cutting-edge approach to increase nitrogen use efficiency in grasslands. Front Environ Sci 9:2021

Michael S, Michael M, Schwarzenbach RP (2010) Adsorption of transgenic insecticidal Cry1Ab protein to SiO_2. 1. Forces driving adsorption. Environ Sci Technol 44:8870–8876

Nayak BK, Chatterjee D, Paul R, Das SR, Adak T, Mandal N, Drewer J, Pradhan A, Kundu MC, Pal AK, Santra GH, Sahoo S, Datta SC, Sutton MA, Nayak AK, Pathak H (2025) Review of nano-clay polymer compsites for controlled nitrogen release: prospects and limitations. Agric Res. https://doi.org/10.1007/s40003-025-00848-5

Nongbet A, Mishra AK, Mohanta YK, Mahanta S, Ray MK, Khan M, Baek KH, Chakrabartty I (2022) Nano-fertilizers: a amart and sustainable attribute to modern agriculture. Plants 11:2587. https://doi.org/10.3390/plants11192587

Pan X, Xu Z, Zhang T, Huang T, Li L, Chen Z, Rao W, Chen S, Hong X, Guan X (2017) The adsorption features between insecticidal crystal protein and nano-$Mg(OH)_2$. R Soc Open Sci 4:170883

Pan X, Guo X, Zhai T, Zhang D, Rao W, Cao F, Guan X (2023) Nanobiopesticides in sustainable agriculture: developments, challenges, and perspectives. Environ Sci Nano 10:41–61

Rathore SS, Shekhawat KA, Singh RK, Updhyay PK, Shekhawat R, & Premi OP (2019) Effect of nano-particles on growth, productivity, profitability of Indian mustard (Brassica juncea) under semi-arid conditions. The Indian J Agric Sci 89(7):1145–1150. https://doi.org/10.56093/ijas.v89i7.91669

Reddy MK, Asthana R, Debnath S, Ray P, Mandal N, Arunachalam A, Konduru JR (2024) Nano-fertilizers for sustainable crop production: a comprehensive review. BioNanoSci 14:1918–1939. https://doi.org/10.1007/s12668-024-01413-0

Sattelmacher B (2001) The apoplast and its significance for plant mineral nutrition. New Phytol 149:167–192. https://doi.org/10.1046/j.1469-8137.2001.00034.x

Satyanarayana T, & Sudhakar RS (2018) Methods of nanomaterials. IJRASET, 6(1):2321–8653

Saurabh K, Prakash V, Dubey AK, Ghosh S, Kumari A, Sundaram PK, Jeet P, Sarkar B, Upadhyaya A, Das A, Kumar S, Makarana G, Kumar U, Kumar A, Singh RR (2024) Enhancing sustainability in agriculture with nano-fertilizers. Discov Appl Sci 6:559. https://doi.org/10.1007/s42452-024-06267-5

Seleiman MF, Almutairi KF, Alotaibi M, Shami A, Alhammad BA, Battaglia ML (2021) Nano-fertilization as an emerging fertilization technique: why can modern agriculture benefit from its use? Plants 10(1):2. https://doi.org/10.3390/plants10010002

Sembada AA, Lenggoro IW (2024) Transport of nanoparticles into plants and their detection methods. Nano 14(2):131. https://doi.org/10.3390/nano14020131

Shahab M, Elham J (2017) Noble UV protective agent for *Bacillus thuringiensis* based on a combination of graphene oxide and olive oil. Sci Rep 7:11019

Sharma S, Kumari P, Shandilya M, Thakur S, Perveen K, Sheikh I, Ahmed Z, Sayyed R, Mastinu A (2025) The combination of α-Fe_2O_3NP and *Trichoderma* sp. improves antifungal activity against *Fusarium* wilt. J Basic Microbiol:e2400613. https://doi.org/10.1002/jobm.202400613

Silva IA, Silva DS, Sousa MU, Menezes RR, Ferreira HS, Neves GA (2020) Encapsulation of nitrogen fertilizers in mixtures with organoclays for controlled release. Lean Tech Environm Policy 22:2163–2176

Siddiqi KS, Husen A (2017) Plant Response to Engineered Metal Oxide Nanoparticles. Nanoscale Res Lett (1):92. https://doi.org/10.1186/s11671-017-1861-y

Taban A, Saharkhiz MJ, Khorram M (2020) Formulation and assessment of nano-encapsulated bioherbicides based on biopolymers and essential oil. Ind Crop Prod 149:112348

Walker GW, Kookana RS, Smith NE, Kah M, Doolette CL, Reeves PT, Lovell W, Anderson DJ, Turney TW, Navarro DA (2018) Ecological risk assessment of naon-enabled pesticides: a perspective on problem formulation. J Agric Food Chem 66:6480–6486

Wu Y, Ren D, Gao C, Li J, Du B et al (2021) Recent advances for alkaloids as botanical pesticides for use in organic agriculture. Int J Pest Manag 69(3):288–298. https://doi.org/10.1080/09670874.2021.1917723

Yadav SK, Babu S, Singh R, Yadav D, & Rajanna GA (2023) The role of organic and natural ecosystems in the food industry. In Sustainable development and pathways for food ecosystems 115–128. https://doi.org/10.1016/B978-0-323-90885-6.00007-7

Yang J, Fend J, He K, Chen Z, Chen W, Gao H, Yuan S (2021) Preparation of thermosensitive buprofezin-loaded mesoporous silica nanoparticles by the sol-gel method and their application in pest control. Pest Manag Sci 77:4627–4637

Ying S, Guan Z, Ofoegbu PC, Clubb P, Rico C, He F, Hong J (2022) Green synthesis of nanoparticles: current developments and limitations. Environ Technol Innov 26:102336

Zhang Y, Zhang A, Li M, He K, Guo S (2021) Nanoparticle-loaded microcapsules providing effective UV protection for Cry1Ac. J Microencapsul 38:522–532

Zulfiqar F, Navarro M, Ashraf M, Akram NA, Munné-Bosch S (2019) Nanofertilizer use for sustainable agriculture: advantages and limitations. Plant Sci 289:110270. https://doi.org/10.1016/j.plantsci.2019.11027

Transcriptomic Approaches to Examine the Nanomaterials' Impact on the Physiological and Molecular Responses of Plants

9

L. Jahnavi Reddy, Kavisa Ghosh, and Amesha Panwar

Abstract

Transcriptomics has emerged as one of the powerful tools for investigating the impact of nanomaterials on plants at the physiological and molecular levels. This chapter provides an overview of the various transcriptomic methodologies, integrating transcriptomics with other omics approaches and their applications in elucidating the complex interactions between plants and nanomaterials. RNA sequencing has revealed key effects of nanomaterials on plants, such as the modulation of genes associated with photosynthesis, stress responses, and hormone signaling. Single-cell transcriptomics has provided insights into the cell-specific responses to nanoparticle exposure, identifying changes in gene expression related to oxidative stress, metal detoxification, and defense mechanisms. Integrating transcriptomics with other omics has further enhanced our understanding of various molecular mechanisms underlying plant responses to nanomaterials, including regulating primary and secondary metabolism and the induction of stress tolerance. Transcriptomic studies have also shed light on the differential gene expression patterns and key regulatory pathways modulated by nanoparticle exposure, such as activating transcription factors and altering phy-

The authors L Jahnavi Reddy, Kavisa Ghosh, and Amesha Panwar have equally contributed to this chapter.

L. J. Reddy
Credora Life Sciences, Bangalore, Karnataka, India

K. Ghosh (✉)
Department of Biotechnology, The Oxford College of Science, Bengaluru, Karnataka, India

A. Panwar
Best Agrolife Ltd, New Delhi, Delhi, India

R. Sayyed et al. (eds.), *Nanobiostimulants in Innovative Agriculture*,
https://doi.org/10.1007/978-981-95-2477-8_9

tohormone signaling. Moreover, the chapter discusses the potential applications of nanomaterials in agriculture, such as crop improvement and stress tolerance enhancement, as well as the challenges associated with their use, including toxicity assessment and regulatory concerns. Further, it emphasizes integrating transcriptomics with artificial intelligence and machine learning approaches as promising avenues for predicting nanomaterials' effects on plant health and the environment, facilitating the development of eco-friendly nanomaterials for agricultural applications.

Keywords

Nanomaterials · RNA sequencing · single-cell transcriptomics · sustainable agriculture · transcriptomics

1 Introduction

Nanoparticles are being increasingly studied for sustainable agriculture because they enhance plant growth, nutrient utilization, and stress tolerance (Karnwal et al. 2023; Thabet and Alqudah 2024). They help address food security and environmental sustainability by serving as sources of micronutrients and platforms for bioactive agrochemicals, improving crop yield and quality while minimizing the negative impacts of traditional fertilizers and pesticides (Balusamy et al. 2023; Fincheira et al. 2021). Various nanomaterials, such as carbon-based materials, metal nanoparticles (Ag, Fe, Zn, TiO_2, ZnO, SiO_2 and MgO), functional magnetic nanoparticles, core/shell nanoparticles (CSNs) and bio-based materials like cellulose and chitin, are being used across agricultural applications (Dhiman et al. 2021; Mohanty et al. 2024; Spanos et al. 2021; Saraswat et al. 2023; Zhu et al. 2022). Green synthesis of nanomaterials utilizing plant extracts or microorganisms is favored for their cost-effectiveness and reproducibility (Singh et al. 2021; Sundararajan et al. 2023; Tasnim et al. 2024). Biocompatible materials like gelatin nanoparticles (GNPs) are developed as eco-friendly nanofertilizers and pesticide carriers (Park et al. 2024).

To fully understand the impact of nanomaterials on agricultural applications, it is important to understand and identify their effects on the physiological and molecular levels of the plants. Nanoparticles can enhance biomass, seed germination, grain yield, and stress resistance, but the molecular mechanisms underlying these effects are not fully understood (Ali et al. 2021; Fincheira et al. 2021). Analyzing the nanoparticle's influence on gene expression changes and direct interactions with plant cells is quite crucial in broadening our understanding of nanomaterial effects on plants and helps optimize their application for agricultural use. Nanoparticles can improve nutrient concentrations, photosynthesis, and antioxidant enzyme activities, leading to better shoot length, root development, and fruit yield (Fincheira et al. 2021). This can help create stress-resilient crops and enhance food security (Thabet and Alqudah 2024) while also addressing concerns about absorption, translocation, and toxicity for safer agricultural practices (Ali et al. 2021; Balusamy et al. 2023).

Transcriptomic approaches are essential for understanding how plants respond to nanomaterials (Munir et al. 2023; Xiong et al. 2021). These techniques help provide an in-depth analysis of the changes induced by nanomaterials at physiological and molecular levels in plants (Chen et al. 2021; Sun et al. 2020). For example, lettuce exposed to copper oxide nanoparticles (CuO-NPs) showed altered gene expression that affects growth and photosynthesis (Xiong et al. 2021). Similarly, Arabidopsis treated with single-walled carbon nano-horns (SWCNHs) exhibited altered changes associated with root development and metabolism of the plant (Sun et al. 2020). The impact varies based on nanoparticle type, concentration, and plant species (Chen et al. 2021; Djanaguiraman et al. 2024). Some of the nanomaterials, like gold nanoparticles, can promote the growth of plants at a concentration, and others may induce stress in plants (Ferrari et al. 2021). Transcriptomic studies, often combined with proteomics and metabolomics, are vital for understanding these complex interactions and their long-term agricultural implications (Li et al. 2024a, b; Ruotolo et al. 2018). Thus, by identifying the key gene expression changes in plants, transcriptomics can enhance our understanding of nanotechnology's benefits and risks in agricultural applications.

This chapter provides a constructive overview of the different transcriptomic methodologies, including high-throughput RNA sequencing technologies, single-cell transcriptomics, and integrating transcriptomics with other omics approaches utilized for plant analysis. It highlights how transcriptomics techniques have revealed the intricate interactions between nanomaterials and plants, contributing to advancements in agriculture. Additionally, it addresses emerging trends and future challenges associated with the use of nanomaterials in the agricultural sector.

2 Transcriptomics Approaches in Plant Research

Transcriptomics has revolutionized plant biology research by allowing the detailed analysis of plant responses to environmental stimuli and external treatments. Further, advances in high-throughput sequencing and bioinformatics have enabled researchers to analyze gene expression patterns in previously unachievable ways (Tyagi et al. 2022). In applying nanomaterials for crop improvement, transcriptomic approaches have proved beneficial in evaluating how plants respond to nanomaterial application, particularly concerning potential toxicities, uptake mechanisms, and stress response pathways (Chen et al. 2024a, b, c).

2.1 High Throughput RNA Sequencing Technologies

High-throughput RNA sequencing (RNA-seq) has significantly improved the assessment of gene expression patterns because it yields holistic, unbiased snapshots of the transcriptional dynamics in plant tissues at different developmental stages. RNA-seq helps determine differentially expressed genes (DEGs) in response to nanomaterials, and it can detect signatures of transcriptional reprogramming in stress response, as well as in metabolic responses needed for detoxification processes (Nizamani et al. 2023). RNA-seq can detect coding and non-coding RNAs, thus improving our understanding of regulatory sequence elements activated during stress and homeostatic processes, allowing plant adaptation to nanoparticle treatment (Yadav et al. 2024b). In addition, RNA-seq allows the identification of the stress-response signaling network, the activity of transcription factors, and the epigenetic landscape that controls plant stress tolerance and recovery (Jian et al. 2022; Li et al. 2024c). These properties render RNA-seq a valuable tool for deciphering the mode by which plants adapt to stress induced by nanomaterials and help assess the effectiveness of nanomaterials and nanobiostimulants in stimulating plant growth, stress tolerance, and resistance to diseases (Hayat et al. 2023; Ferioun et al. 2025; Yadav et al. 2024a).

2.1.1 Identifying the Effects of Nanomaterials on Plants Using RNA Sequencing Technologies

RNA sequencing (RNA-seq) technologies have been extensively utilized in plant biology to elucidate gene expression patterns, enhancing the understanding of plant gene regulation and its responses to various stressors. Illumina sequencing is particularly suitable for plant transcriptome analysis due to its high accuracy, depth, and cost-effectiveness. This technology can detect both high- and low-abundance transcripts, thereby facilitating research on gene regulation within complex genomes. The short reads generated by Illumina sequencing enable precise quantification, single-nucleotide polymorphism (SNP) detection, and the investigation of trait variation. Additionally, multiplexing increases the statistical power for differential expression analysis. For instance, it has been employed to detect simple sequence repeat (SSR) loci in *Acacia* coding regions (Ishio et al. 2024). Nanopore sequencing offers real-time, long-read characterization of RNA transcripts, preserving modifications, splicing, and transcript diversity, which are crucial for studies on plant responses to nanomaterials. Its portability allows for on-site evaluations in nano-enabled agriculture (Shearman and Tangphatsornruang 2021; Garg et al. 2024). MP3RNA, a primer-independent RNA-seq method, detects both polyadenylated and non-polyadenylated RNAs, making it well-suited for exploring transcriptional changes in plants treated with nanomaterials. This method reveals mechanisms of action, phytotoxicity, and gene modulation by nanobiostimulants, with multiplexing aiding in studying complex plant responses (Trevisan et al. 2017). Transcriptome sequencing, or Tran Seq, is indispensable for understanding plant reactions to nanoparticles and nanobiostimulants, as it identifies genes associated with stress, growth, and defense that exhibit differential expression. For example, chitosan nanobiostimulants have increased defense gene expression in tomatoes, while ZnO nanoparticles enhance metal transport gene expression in *Sorghum* (Dimkpa et al. 2019). Tran Seq supports the safety evaluation of nano-inputs and the discovery of new genes. Iso-Seq provides full-length transcripts, facilitating the detection of alternative splicing and novel isoforms in plants, which is essential for nanoparticle and nanobiostimulants research. Transcriptome analysis of *Brassica chinensis L.* under foliar application of TiO_2 nanoparticles revealed increased expression of genes involved in photosynthesis, hormone signaling, and stress tolerance, leading to enhanced yield (Wei et al. 2024). Simplified Poly(A) sequencing (SiPAS) is a cost-effective 3' RNA-seq method for examining gene expression changes in plant systems following nanoparticle treatment. It simplifies transcriptome profiling by focusing on transcript expression in the polyadenylated portion, allowing for studies with larger sample sizes. SiPAS has demonstrated differential expression of genes related to stress response and growth regulation in maize treated with ZnO nanoparticles and chitosan nanobiostimulants (Zungu et al. 2023).

RNA-seq has helped identify numerous effects on plants influenced by nanomaterials. Some key findings are briefly elaborated, highlighting the important role RNA-seq plays in identifying and understanding the effects of nanomaterials on plants. RNA-seq helped in demonstrating the role of carbon nanosols (CNS) in enhancing growth in tobacco plants by regulating microRNAs (miRNAs) and transcription factors associated with growth and stress adaptation (Zheng et al. 2024). Gold nanoparticles (AuNPs) were identified to have substantial impacts on gene expression in medicinal plants such as *Nardostachys jatamansi*, wherein positively charged AuNPs performed better than negatively charged ones to enhance growth and modulate stress-related genes (Joshi and Joshi 2024). Metal nanoparticles like CuO can induce phytotoxicity through the release of Cu^{2+} ions. RNA-seq analysis showed that the ions suppressed metal homeostasis and transport genes, indicating the mechanism of nanotoxicity (Landa et al. 2017). Similar results were obtained in nanoplastics-exposed *Orzya*, which influenced changes in gene expression for photosynthesis, defense, and hormone signaling pathways like jasmonic acid and cytokinins (Xu 2024). TiO_2/ZnO nanoparticles treatment on lettuce influenced the genes involved in photosynthetic metabolism, antioxidant activity, and nitrogen metabolism (Wang et al. 2017). Nanobiostimulants like chitosan nanoparticles promoted root growth and plant biomass in cereals, and RNA-seq showed upregulation of development, metabolism, and stress tolerance genes (Mondejar-López et al. 2024). Garlic extract formulations and carbon nanohorns were identified to reprogram Arabidopsis's carbon and nitrogen metabolic pathways. This reprogramming enhanced Arabidopsis's root growth and lateral root development (Sun et al. 2020). RNA-seq also facilitated the identification of spatiotemporal gene expression analysis under nano-plastic stress, which showed organ-specific responses in the roots and leaves of *Oryza sativa* (Xu 2024). Table 9.1 summarizes some of the key effects of nanomaterials identified in plants through RNA-seq.

Although RNA-seq has the potential to revolutionize the use of nanotechnology for crop improvement, it comes with its fair share of challenges. At the top of the list of challenges is transcriptomic data complexity, which needs sophisticated bioinformatics tools and great computational resources. It also necessitates discernment between genuine gene expression changes and possible sequencing artefacts so that they stay relevant biologically (Ungaro et al. 2017). A further limitation is the absence of suitably annotated reference genomes for numerous economically and agronomically relevant plant species. This makes precise alignment and

Table 9.1 Key effects of nanomaterials identified in plants through RNA-seq

Nanomaterial	Effects identified through RNA-seq	References
Chitosan nanoparticles	Enhanced root development, upregulation of metabolism genes	Mondejar-López et al. (2024)
Carbon nanosols	Promoted growth via miRNA regulation	Zheng et al. (2024)
Gold nanoparticles	Modulated stress and growth-related genes	Joshi and Joshi (2024)
CuO nanoparticles	Downregulated metal homeostasis genes	Landa et al. (2017)
Carbon nanohorns	Reprogrammed carbon/nitrogen metabolism	Sun et al. (2020)
TiO_2/ZnO nanoparticles	Regulated photosynthesis and antioxidant genes	Wang et al. (2017)
Nano-plastics	Modulated photosynthesis and defense responses	Xu (2024)

interpretation of RNA-seq reads very challenging. In such situations, scientists must depend on de novo transcriptome assembly, which is a developing field with limitations in accuracy and completeness (Shrestha et al. 2022). In addition, the nanomaterial and nanobiostimulant responses are highly tissue-specific and development stage-dependent. This temporal and spatial heterogeneity introduces a level of complexity, and it becomes challenging to gain a global understanding using standard RNA-seq methodologies. Novel tools such as single-cell RNA-seq provide an attractive solution by allowing cell-type-specific resolution and minimizing the bulk RNA-seq data that masks the subtle effects that are cell or tissue-specific (Cheng et al. 2023). Table 9.2 summarizes the key challenges faced while using RNA-seq technologies and their possible solutions to overcome them.

Other challenges include the potential of some nanomaterials to degrade RNA quality or downstream sequencing reactions, which might create erroneous results. For this reason, optimized RNA extraction processes and sequencing workflows without nanoparticles are critical for preserving RNA integrity and delivering high-quality output (Zhang et al. 2021; Ghamari et al. 2022). Integration of RNA-seq data with other omics platforms, such as proteomics and metabolomics, is yet more demanding. These entail the requirement of robust normalization schemes across platforms and data types. However, multi-omics integration is crucial to building extensive regulatory networks that describe how plants respond to nanomaterial exposure and are stimulated by nanobiostimulants (Chen et al. 2024a, b, c). The important input of RNA-seq to plant nanobiotechnology presents some additional persisting challenges. These encompass the necessity of enhanced computational resources to process big data, the incorporation of multi-omics strategies (e.g., proteomics and metabolomics), and the establishment of standardized protocols for RNA-seq experiments in plants (Munir et al. 2023; Magar et al. 2024). Future studies should utilize these technologies to develop safer and more efficient nanomaterials and nanobiostimulants, extensively identifying the effect of these agents on plants and maximizing their potential benefits for crop improvement.

In summary, although RNA-seq has emerged as an essential tool for elucidating the molecular processes underlying plant responses to nanomaterials and nanobiostimulants, overcoming the technical and analytical challenges will be challenging. Further improvements in genome annotation, single-cell technologies, and

Table 9.2 Summary of the challenges associated with using RNA-seq technologies, along with suggested solutions identified in the literature

Challenge	Description	Suggested solutions
Lack of reference genomes	Poor or absent genomic resources hinder accurate read mapping and annotation.	Development of assembly-free methods and guided transcriptome assembly (Shrestha et al. 2022; Ungaro et al. 2017)
Functional annotation gaps	Limited functional annotations make it difficult to link genes to biological pathways.	Use of ortholog databases and cross-species comparison tools (Li et al. 2023)
High computational demands	De novo assembly is computationally expensive and time-consuming.	Assembly-free approaches and optimized pipelines (Shrestha et al. 2022; Ungaro et al. 2017)
Data accessibility and reusability	Large datasets require specialized skills for reanalysis.	User-friendly platforms like PlantExp for public RNA-Seq data exploration (Liu et al. 2022)
Normalization biases	Traditional normalization methods assume constant total transcript levels.	Use of external RNA spike-ins for accurate normalization (Laosuntisuk et al. 2024)
Lack of standardized tools	Specialized tools are needed for non-model species.	Development of integrated pipelines like PipeOne-NM and ExpressAnalyst (Li et al. 2023; Wang et al. 2023)

multi-omics integration will increasingly make RNA-seq more reliable, developing novel nano-enabled strategies for sustainable agriculture.

2.2 Single-Cell Transcriptomics

Single-cell transcriptomics has been a revolutionary method in plant biology, providing sophisticated insights into cellular heterogeneity and gene expression (Bawa et al. 2022). Studying transcriptomic profiles on a per-cell basis unmasks heterogeneity that is commonly masked in bulk RNA sequencing. In plant biology, single-cell transcriptomics recognizes specialized cell types, follows developmental trajectories, and decodes how plants perceive and adapt to environmental cues (Shaw et al. 2020). For plant-nanoparticle and nanobiostimulants interactions, single-cell transcriptomics provides in-depth cell or tissue-specific information on molecular stress mechanisms and responses. Despite the promise, difficulties in integrating data, tissue complexity, and analysis pipelines persist, hindering the full potential of single-cell transcriptomics in developing safer nanomaterials for sustainable agriculture (Kaur et al. 2023; Kalia and Sharma 2019).

2.2.1 Applications of Single-Cell Transcriptomics for Identifying the Effects of Nanomaterials on Plants

Single-cell transcriptomics has been instrumental in studying how plants react to nanoparticles at the molecular level. This method gives cell-specific gene expression information, making it possible to explain detailed mechanisms by which nanoparticles influence plant development, growth, and stress response (Zheng et al. 2023; Shahbaz et al. 2023). Some key findings are briefed, highlighting the important role of single-cell transcriptomics in identifying and understanding the effects of nanomaterials on plants.

Gold nanoparticles are efficient probes for detecting rare RNA species in plant cells (Patil et al. 2022). Surface-enhanced Raman spectroscopy (SERS) in conjunction with gold nanoprobes has allowed the detection of alternatively spliced variants and microRNAs at single-cell resolution. Such methods, tested in *Arabidopsis thaliana* protoplasts, show promise for real-time monitoring of RNA dynamics and transcript diversity in plant systems (Kadam et al. 2017). Transcriptomic analysis has also identified that various nanoparticles elicit unique gene expression responses in plants. Titanium dioxide (TiO_2) and cerium oxide (CeO_2) nanoparticles, for example, cause changes in the transcriptome of *Arabidopsis thaliana.* Nano-TiO_2 increases photosynthesis-related genes, whereas nano-CeO_2 activates stress-responsive pathways (Tumburu et al. 2017). Similarly, copper oxide (CuO) nanoparticles impact metal homeostasis and oxidative stress response-related genes, similar to ionic copper (Landa et al. 2017). Apart from bulk-level study, plants' spatiotemporal transcriptomic regulation to nanoplastic stress has also been explored. In rice (*Oryza sativa*), nanoplastic treatment has been observed to suppress the expression of chlorophyll and photosynthesis-associated genes while promoting defense-related responses. These studies highlight the relevance of considering both spatial and temporal facets of gene regulation for examining nanoparticle action on plants (Xu 2024). Single-cell transcriptomics has also contributed significantly to investigating the impact of nanobiostimulants used to enhance plant growth and yield. Chitosan nanoparticles, for instance, have been applied as seed nano-priming materials. Transcriptome analysis of nano-primed wheat seeds has shown upregulation of development, metabolic, and stress-related genes, highlighting the potential of such nanobiostimulants to promote crop performance and stress tolerance (Mondejar-López et al. 2024). Besides stimulating growth, nanoparticles have also been identified as elicitors of bioactive compounds in plant systems, such as hairy root cultures. The nanoparticles can activate specific transcription factors responsible for the regulation of biosynthesis of specialized secondary metabolites, with important applications in agriculture and biotechnology (Mmereke et al. 2023).

Integrating single-cell transcriptomics with other omics approaches, such as proteomics and metabolomics, has provided a comprehensive view of plant responses to nanomaterials. For example, transcriptomic meta-analyses in Arabidopsis thaliana exposed to various nanomaterials have identified shared and nanoparticle-specific signaling pathways, including those related to hypoxia and hormone regulation (Wu et al. 2024). Metal-based nanomaterials can improve metal homeostasis for zinc and iron and are beneficial in seed priming to improve abiotic stress (Wu et al. 2024). Carbon-based nanomaterials, instead, trigger hormone-mediated immune systems, reduce water deprivation effects, and serve as signal amplifiers during biotic stress. Moreover, nanoplastics have been shown to regulate processes like chlorophyll biosynthesis, photosynthesis, and defense activation, particularly in rice plants (Xu 2024). Thus, from these omics studies, several key insights have emerged. When combined with cutting-edge omics technologies, single-cell transcriptomics offers profound insights into the molecular interactions between nanoparticles and plants. Such information enhances our knowledge of plant physiology under nano-interventions and presents new opportunities for sustainable agriculture and bioengineering. Table 9.3 summarizes key changes in plants induced by nanoparticles, as identified through single-cell transcriptomics.

Table 9.3 Application of single-cell transcriptomics in identifying key changes in plants induced by nanomaterials

Aspects	Nanoparticle type	Model plant/ system	Key findings	Reference(s)
Detection of rare RNA species	Gold nanoparticles (with SERS)	*Arabidopsis thaliana* protoplasts	Detection of alternatively spliced variants and microRNAs at single-cell level	Kadam et al. (2017)
Transcriptomic changes	TiO_2, CeO_2	*Arabidopsis thaliana*	TiO_2: Upregulates photosynthesis genes; CeO_2: Activates stress pathways	Tumburu et al. (2017)
	CuO	*Arabidopsis thaliana*	Affects genes related to metal homeostasis and stress, like ionic copper	Landa et al. (2017)
Spatiotemporal responses	Nano-plastics	*Oryza sativa* (rice)	Downregulates chlorophyll biosynthesis and photosynthesis; activates defense genes	Xu (2024)
Nanobiostimulant effects	Chitosan nanoparticles	Wheat seeds	Upregulates genes for development, metabolism, and stress response	Mondejar-López et al. (2024)
Bioactive compound elicitation	Various nanoparticles	Hairy root cultures	Activate transcription factors for secondary metabolite biosynthesis	Mmereke et al. (2023)
Integrated omics insights	Metal-based nanomaterials	*Arabidopsis thaliana*	Enhance Zn and Fe homeostasis; useful in abiotic stress relief	Wu et al. (2024)
	Carbon-based nanomaterials	*Arabidopsis thaliana*	Trigger immune hormone signaling; relieve water stress and amplify biotic stress signals	Wu et al. (2024)
	Nano-plastics	*Oryza sativa*	Affect photosynthesis, chlorophyll synthesis, and defense responses	Xu (2024)

In summary, single-cell transcriptomics has transformed the field of plant responses to nanoparticles and nanobiostimulants by offering high-resolution information on gene expression at the single-cell level. Single-cell transcriptomics has dramatically improved our knowledge of the molecular mechanisms of plant–nanoparticle interactions. Nonetheless, several challenges remain, such as the intrinsic complexity of plant tissues, the necessity for robust and standardized data analysis pipelines, and the challenge of integrating multi-omics data (Shaw et al. 2020; Bobrovskikh et al. 2021). Solving these issues is crucial to maximizing the potential of single-cell transcriptomics in plant nanobiotechnology. Future studies must focus on creating novel methods to overcome these limitations, thus propelling plant biotechnology and encouraging sustainable agriculture (Kalia and Sharma 2019; Bian et al. 2024).

2.3 Integrating Transcriptomics with Other Omics Approaches

Combining transcriptomics with other omics strategies has transformed plant research, especially in comprehending the impacts of nanomaterials and nanobiostimulants on plant growth, stress, and metabolic processes. This multi-omics strategy offers an integrated understanding of how plants interact with nanomaterials at the molecular level, allowing the identification of genes, proteins, and metabolites that play significant roles in such interactions (Sanches et al. 2024). Some important studies using transcriptomics, along with other omics methods such as proteomics, metabolomics, and ionomics, in identifying nanomaterials and nanobiostimulants' interaction with plants are discussed further.

Integration with proteomics has the added value of increasing understanding of plant-nanoparticle interactions by detecting protein profile changes. The study of *Arabidopsis thaliana* exposed to cadmium sulfide quantum dots (CdS QDs) showed alteration in protein expression associated with transcriptomic data, though some divergence was noted (Marmiroli et al. 2015). Similarly, proteomic and transcriptomic profiles of TiO_2 nanoparticles in alfalfa revealed that low concentrations upregulated genes and proteins of photosynthesis and carbon fixation and that high concentrations resulted in toxicity and downregulated these processes (Chen et al. 2024a, b, c). Further, metabolomics has been combined with transcriptomics to analyze metabolic modification in nanoparticle- and nanobiostimulant-treated plants. For instance, a research study on *Vallisneria denseserrulata* treated with nanoplastics and arsenic concluded that combining transcriptomic and metabolomic data unveiled the central pathways involved in stress responses, such as photosynthesis, energy conversion, and oxidative regulation (Tang et al. 2024). Also, SiO_2 nanoparticles in *Ehretia macrophylla* seedlings under drought stress changed the metabolite content related to drought tolerance and upregulated genes associated with auxin signal transduction and stress responses (Chen et al. 2024a, b, c). Moreover, the combination of ionomics and transcriptomics has led to the identification of nanoparticle-induced ion transport and homeostasis in plants. For instance, carbon nanosol (CNS) treatment in tobacco plants greatly enhanced shoot potassium levels and changed the expression of genes that play a role in ion transport and homeostasis (Wang et al. 2024a, b). Likewise, TiO_2 nanoparticles in alfalfa influenced the expression of genes that control ion transport, especially at high concentrations, resulting in toxicity (Chen et al. 2024a, b, c). Thus, the integration of omics approaches has revealed key mechanisms underlying plant nanoparticle interactions, which are further summarized in Table 9.4.

2.3.1 Applications of Multi-omics Approaches in Nanoparticle Research for Agriculture

The integration of transcriptomics with other omics approaches has several applications in nanomaterials and nanobiostimulant research for agricultural applications, which has been tabulated in Table 9.5. Transcriptomics and other omics technologies enable the identification of the DEGs and the signaling pathways affected by nanoparticle treatment. This information proved to be helpful in the identification or design of crop varieties with a positive response to nanomaterials to improve nutrient uptake, photosynthetic efficiency, and biomass accumulation, aiding in crop improvement (Daniel et al. 2024; Upadhyay et al. 2024). Integrating transcriptomics with proteomics and metabolomics facilitates understanding the molecular mechanisms involved in nanoparticle-induced abiotic and biotic stress tolerance. This encompasses the induction of antioxidant enzymes, stress-responsive genes, and secondary metabolites, ultimately enhancing resistance to environmental stresses like drought, salinity, or pathogen attack (Chen et al. 2024a, b, c; Wu et al. 2024). Omics techniques allow for the detailed analysis of toxic responses at the molecular, cellular, and body levels. Transcriptomic data can provide insights into how defense mechanisms are triggered, how DNA damage response occurs, and how detoxification occurs. This facilitates a better understanding of nanoparticle toxicity and the development of safer nanomaterials and regulations for their application (Munir et al. 2023; Ruotolo et al. 2018). The use of nanobiostimulants is becoming increasingly popular as a tool to enhance plant growth, nutrient acquisition, and production. Omics tools assist us in understanding the mode of action of these agents on plant processes, hormone regulation, and root design, facilitating environmentally friendly and sustainable agriculture (Mondejar-López et al. 2024; Wang et al. 2024a, b).

Though omics have evolved, nanoparticle and nanobiostimulant research is not without challenges. Advanced bioinformatics resources are necessary to integrate the data, and there is a need for deeper mechanistic understanding (Upadhyay et al. 2024; Munir et al. 2023). Toxicity and the environmental fate of nanoparticles must be determined (Ruotolo et al. 2018), and omics-based frameworks for regulations should be developed (Zemlyanskaya et al. 2021). Combining transcriptomics with other omics has enhanced our knowledge of nanoparticle uptake, transport, and toxicity, but more investigation is needed to optimize their safe and sustainable

Table 9.4 Identification of the molecular mechanisms involved in plant-nanoparticle interactions through multi-omics approaches

Mechanism	Description	Key findings	Molecular components involved	Reference(s)
Uptake and transport	Nanoparticles penetrate via the root epidermis (apoplastic/ symplastic pathways) or leaf stomata, followed by systemic transport via xylem and phloem.	Gene products for membrane transporters, aquaporins, and vesicular trafficking proteins rise; endocytosis is also actively involved.	*Transporters*: ABC transporters, NRAMPs, ZIPs *Aquaporins*: PIP1, PIP2, and TIP1 *Pathways*: Endocytosis, vesicular transport, plasmodesmata-mediated movement	Munir et al. (2023), Wu et al. (2024)
Stress responses	Nanoparticles generate reactive oxygen species (ROS), leading to oxidative stress and cellular damage when not properly regulated.	Antioxidant enzyme genes are activated; ROS-scavenging, and stress signaling pathways are triggered.	*Enzymes*: Catalase (CAT), superoxide dismutase (SOD), ascorbate peroxidase (APX), glutathione S-transferase (GST) *Pathways*: ROS detoxification, MAPK signaling, peroxisomal stress response	Siddique and Aley (2024), Munir et al. (2023)
Hormone signaling	Nanoparticles regulate plant hormone signaling pathways that influence cell elongation, root growth, and stress adaptability.	Enhanced expression of auxin transport and gibberellin perception-related genes; modification of ethylene and abscisic acid pathway recognized.	*Auxin pathway*: AUX1, PIN1, ARFs, TIR1 *Gibberellin Signaling pathway*: GID1, DELLA proteins *Others*: ERFs (ethylene), PP2Cs (ABA), JA-responsive MYCs	Wang et al. (2024a, b), Chen et al. (2024a, b, c)
Photosynthesis and carbon metabolism	Nanoparticles influence chloroplast structure, photosynthetic enzyme contents, and carbohydrate biosynthesis.	Downregulation of RuBisCO and chlorophyll-binding protein genes at high NP levels; PSII activity inhibited.	*Photosynthesis proteins*: RuBisCO (rbcL, rbcS), CABs (chlorophyll a/b-binding proteins), PsbA, PsbD *Pathways*: Calvin cycle, light harvesting complex (LHC), chloroplast redox regulation	Chen et al. (2024a, b, c)

Table 9.5 Summary of multi-omics approaches in nanoparticle research for agriculture applications

Application	Description	Key references
Crop improvement	Identifying genes involved in plant-NP interactions for genetic engineering.	Daniel et al. (2024), Upadhyay et al. (2024)
Stress tolerance	Understanding NP-induced stress tolerance for resilient crop development.	Chen et al. (2024a, b, c), Wu et al. (2024)
Nanotoxicology	Assessing nanoparticle toxicity for risk evaluation and regulatory policies.	Munir et al. (2023), Ruotolo et al. (2018)
Sustainable agriculture	Exploring nanobiostimulants for improved plant growth and metabolism.	Mondejar-López et al. (2024), Wang et al. (2024a, b)

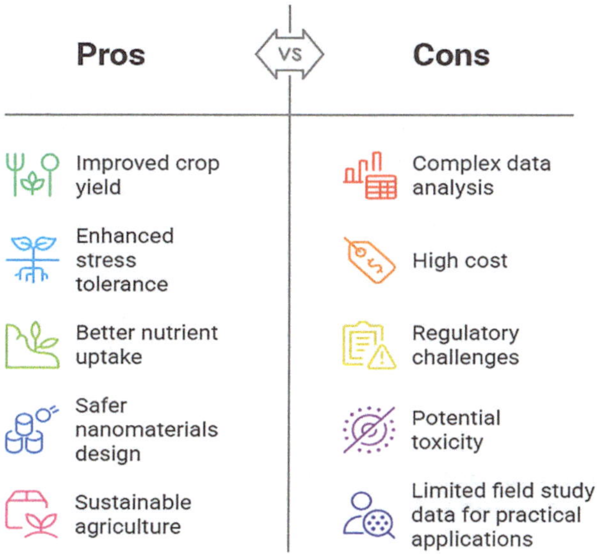

Fig. 9.1 The advantages and challenges of multi-omics approaches for nanomaterial research in agricultural applications

application in agriculture. The advantages and challenges of multi-omics approaches for nanomaterials research in agricultural applications are summarized in Fig. 9.1.

3 Molecular Mechanism Underlying Nanoparticle Action in Plants

Nanomaterials engage with plants at the molecular level, affecting various physiological and biochemical processes. The molecular mechanisms underlying the action of nanomaterials in plants are intricate and are influenced by factors such as the type of nanomaterial, its physicochemical properties, and the plant species involved (Ali et al. 2021; Dev et al. 2017). At the cellular level, nanomaterials can be absorbed and sequestered within plant tissues, positively impacting photosynthetic performance (Liu et al. 2019).

These materials can enhance both the light and dark reactions of photosynthesis, potentially leading to increased crop yields. The uptake and translocation of nanomaterials in plants are influenced by their surface properties, with negatively charged coatings and enhanced biocompatibility facilitating greater uptake (Ali et al. 2021; Liu et al. 2019). Notably, specific nanomaterials, such as gold nanoparticles (AuNPs), can promote plant growth rather than induce stress. Treatment with moderate concentrations of AuNPs has been shown to enhance root growth, increase rosette diameter, and reduce oxidative stress responses. Transcriptomic and proteomic analyses have revealed that oxidative stress responses are downregulated while growth-promoting genes and proteins are upregulated following AuNP treatment (Ferrari et al. 2021). However, the molecular processes of nanomaterial action in plants involve complex interactions with cellular processes, including photosynthesis, oxidative stress responses, and gene expression. Understanding these intricate mechanisms is essential for developing effective nanotechnology applications in agriculture to enhance crop productivity while minimizing environmental impact (Goyal et al. 2023; Liu et al. 2019). However, further research is necessary to fully elucidate the mode of action of nano-enabled agrochemicals and their long-term effects on plant physiology and ecosystem health (Goyal et al. 2023; Sarraf et al. 2022). This section explores the gene expressions and molecular mechanisms associated with nanomaterial-plant interaction elucidated with transcriptomics and multi-omics approaches, which can contribute to developing eco-friendly and beneficial nanomaterials for sustainable agriculture applications.

3.1 Differential Gene Expression in Nanomaterial-Exposed Plants

Plants can absorb nanoparticles (NPs) through their roots or leaves, transporting them either apo-plastically or symplastically to various plant tissues, thereby influencing physiological functions. Polystyrene nanoplastics (PSNPs) have been observed to traverse intercellular air spaces and xylem vessels in dandelions (Li et al. 2024a, b), while ZnO nanoparticles (nano-ZnO) have been shown to modulate the uptake and transport of PSNPs in corn (Guo et al. 2023). Exposure to NPs can induce oxidative stress by generating reactive oxygen species (ROS), activating antioxidant defense systems, as evidenced in Arabidopsis thaliana exposed to silver nanoparticles (Kang et al. 2023; Kaveh et al. 2013). Additionally, NP exposure can impact photosynthesis and carbon metabolism in rice and *Vallisneria denseserrulata* when exposed to nanoplastics and/or arsenic (Tang et al. 2024; Xu 2024). NP exposure also influences hormonal signaling mechanisms and activates defense mechanisms involving WRKY, MYB, and ERF transcription factors (Kaveh et al. 2013; Xu 2024). Metal-based nanoparticles alter the expression of plant genes associated with metal detoxification, oxidative stress, and hormone signaling processes, as observed in lettuce exposed to CuO NPs (Xiong et al. 2021). Graphene nanoplatelets have been found to modify photosynthesis and hormone signaling pathways in Lactuca sativa (Yuan et al. 2024). Polystyrene nanoplastics (PS NPs) affect plant growth and metabolism in wheat, influencing genes involved in carbon metabolism, amino acid biosynthesis, and plant-pathogen interactions (Zhou et al. 2020). Applying nano-fertilizers and phytohormones can enhance plant tolerance to NP toxicity. Nano-ZnO reduces PSNP accumulation in corn and enhances tolerance by regulating photosynthesis and hormone signaling (Guo ct al. 2023). Salicylic acid has been found to alleviate NP stress in Arabidopsis thaliana partially, mitigating transcriptional and phenotypic effects (García-Sánchez et al. 2015). Plants exhibit conserved molecular responses to NP exposure, demonstrating immune-like responses and abscisic acid (ABA) signaling. A meta-analysis revealed that NP exposure generally up-regulates immune-related genes, while down-regulating genes are associated with growth and development (Giudice et al. 2023). ABA-mediated signaling plays a pivotal role in NP-induced responses. In Arabidopsis thaliana, gold NPs have been shown to induce ABA-mediated regulation of genes associated with oxidative stress, metal detoxification, and hormone signaling (Shukla et al. 2014), thereby establishing the role of ABA in plant stress adaptation. Investigating differentially expressed genes can provide insights into the mechanisms underlying the molecular responses of plants to nanomaterial exposure. Understanding NP uptake and transport, along with toxicity in plants, and the role of phytohormones and stress-related genes in mitigating NP exposure is crucial for developing eco-friendly and sustainable nanomaterials for agricultural applications. Table 9.6 summarizes some key differential gene expressions observed in nanomaterial-exposed plants.

3.1.1 Key Regulatory Genes and Pathways Modulated in Nanoparticle-Exposed Plants

Nanoparticles (NPs) improve plant growth, stress tolerance, and metabolism by altering gene expression. Specifically, transcriptomics studies revealed that metal- and carbon-based NPs improved stress mitigation and hormone signaling (Joshi and Joshi 2024; Wu et al. 2024). Transcription Factors (TFs) are essential determinants of gene expression in response to nanoparticle exposure. Several studies identified TF families as important genetic factors differentially expressed upon NP treatment in plants. Myeloblastosis (MYB), tryptophan-arginine-lysine-tyrosine (WRKY), and NAC TF families operate in stress responses, hormone regulation, and secondary metabolite production in plants. MYB and WRKY TFs are found to be upregulated in response to AuNP treatment, which potentially promotes antioxidant potential and stress tolerance (Joshi and Joshi 2024; Zheng et al. 2024). Basic Helix-Loop-Helix (bHLH) and DNA-binding with One Finger (Dof) TF families operate in metal homeostasis and root development in plants. bHLHs are involved in iron and zinc homeostasis and play an important role in plant growth under stress conditions (Tiwari et al. 2016; Wu et al. 2024). Ethylene-Responsive Factor (ERF) and TCP TF families operate in hormone signaling and stress responses. ERF TFs, in particular, are known to mediate ethylene-mediated stress responses to toxic

Table 9.6 Key differential gene expressions observed in nanomaterial-exposed plants

Plant species	Nanoparticle type	Key gene expression changes	Reference(s)
Nardostachys jatamansi	Gold nanoparticles	Upregulation of genes involved in antioxidant potential, hormonal regulation, and stress response.	Joshi and Joshi (2024)
Vallisneria denseserrulata	Nanoplastics	Downregulation of photosynthesis-related genes and upregulation of stress-responsive genes.	Tang et al. (2024)
Oryza sativa	Nanoplastics	Downregulation of chlorophyll biosynthesis and starch metabolism genes.	Xu (2024)
Lactuca sativa	Graphene nanoplatelets	Changes in gene expression related to photosynthesis and plant hormone signal transduction.	Yuan et al. (2024)
Arabidopsis thaliana	Silver nanoparticles	Upregulation of genes involved in oxidative stress and metal detoxification.	Kaveh et al. (2013)
Arabidopsis thaliana	Zinc oxide nanoparticles	Upregulation of stress-related genes and downregulation of cell organization-related genes.	Landa et al. (2015)
Arabidopsis thaliana	Gold nanoparticles	ABA-mediated regulation of genes involved in oxidative stress and hormone signaling.	Shukla et al. (2014)
Arabidopsis thaliana	Cadmium sulfide QDs	Differential expression of genes involved in stress response and detoxification.	Marmiroli et al. (2015)
Arabidopsis thaliana	TiO2 nanoparticles	Repression of transcriptional responses to microbial pathogens and inhibition of root hair development.	García-Sánchez et al. (2015)

conditions that can arise in the presence of nanoparticles (Basak et al. 2023; Zheng et al. 2024).

Genes associated with hormone-related regulation are an important response of plants to nanoparticle-mediated effects. These hormone-related regulations are further elaborated in the consecutive sections and are briefly summarized here. The signal transduction pathways of phytohormones are often altered after exposure to NPs. These pathways control a wide range of physiological activities that include but are not limited to growth, development, and response towards external stressors. Auxins are important to a plant's root architecture and regulate cell elongation. Auxin and ethylene pathways are associated with root development and stress responses. Exposing plants to AuNPs leads to modulated auxin signaling, causing changes in root architecture (Tiwari et al. 2016). Both ethylene and jasmonic acid play a significant role in stress response and defense mechanisms. Carbon nanosols (CNS) regulate the ethylene and jasmonic acid pathways, effectively managing tobacco plants' stress tolerance (Zheng et al. 2024). Moreover, jasmonic acid and salicylic acid pathways are associated with defense responses. This pathway is generally modulated in response to NP exposure and increases the resistance of plants to pathogens and abiotic stresses (Basak et al. 2023; Joshi and Joshi 2024).

NP can enter the plant through roots and/or leaves, then translocate to specific tissues where they can modify gene expression (Munir et al. 2023; Ruotolo et al. 2018). NPs can induce the expression of stress-responsive genes important for plant survival under stress. Antioxidant and detoxification pathways are essential to offset oxidative stress from exposure to NPs. Important pathways include Glutathione S-transferases (GSTs) and ascorbate-glutathione cycle. Genes encoding antioxidant enzymes such as peroxidases (PODs), glutathione S–transferases (GSTs) and superoxide dismutases (SODs) are upregulated in response to NP exposure. These antioxidant–related genes play an important role in scavenging reactive oxygen species (ROS), mitigating oxidative stress, and stabilizing redox homeostasis (Guo et al. 2023; Kang et al. 2023; Joshi and Joshi 2024). The production of ROS is a double-edged sword as ROS can induce oxidative stress when present at high concentrations but also serve as signaling molecules when present at moderate concentrations that trigger the plant's innate antioxidant defenses (Kang et al. 2023; Zheng et al. 2024). Further, the ascorbate-glutathione cycle, which contains enzymes such as ascorbate peroxidase (APX) and monodehydroascorbate reductase (MDHAR), is modulated in NP-exposed plants, resulting in increased antioxidant capacity (Basak et al. 2023; Kang et al. 2023). Moreover, exposure to NPs can promote the expression of heat shock proteins (HSPs) involved in protein folding and stress tolerance, stabilizing cellular homeostasis under stressful conditions (Basak et al. 2023; Kang et al. 2023).

Nanoparticles significantly affect genes related to photosynthesis and carbon metabolism, which are essential for plant growth and productivity. These aspects are further elaborated in the following sections and are briefly summarized here. Photosynthesis-related genes, such as genes involved in chlorophyll biosynthesis (e.g., CHLH, CHLD, CHLM) and photosystem II (e.g., PetH), are modulated during NP exposure, leading to improved photosynthesis productivity and enhancing the plant's ability to tolerate abiotic stress (Guo et al. 2023; Wang et al. 2017). Pathways associated with carbon fixation (e.g., rbcL, rbcS) or sucrose metabolism (e.g., SPS) are reported to be altered with NP exposure plants, resulting in enhanced carbon partitioning and energy production (Siddique and Aley 2024; Wang et al. 2017).

Nanoparticles can also mediate epigenetic modifications, like DNA methylation changes that can lead to heritable responses vis-a-vis stress (Basak et al. 2023; Tripathi et al. 2024). These mechanisms exemplify nanoparticles' potential role in plant adaptation and resilience. Thus, transcriptomic and multi-omics approaches lead to identifying multiple key genes and pathways involved in NP-plant interactions that could improve stress tolerance, growth, and nutrition. Some of the effects of NPs on plants, associated genes and regulated pathways are summarized in Table 9.7.

Table 9.7 Effects of various nanoparticles on plants, associated genes, and regulatory pathways

Nanoparticle type	Key effects on plants	Associated genes	Associated pathways	Reference(s)
Metal nanomaterials	Alleviate biotic/abiotic stresses, boost metal homeostasis	Zinc and iron homeostasis genes, hormone-related immune responses	Zinc and iron homeostasis, hormone-related immune responses	Wu et al. (2024)
Gold nanoparticles (AuNPs)	Regulate growth, metabolism, and defense responses	MYB, WRKY, NAC, bHLH transcription factors	Auxin signaling	Joshi and Joshi (2024), Tiwari et al. (2016)
Carbon nanosols (CNS)	Promote growth, regulate miRNAs and TFs	miR319b, miR160a, NbMYB, NbNAC, NbDof TFs	miRNA regulation, transcription factor modulation	Zheng et al. (2024)
Polystyrene nanoplastics (PSNPs)	Induce genetic changes, affect carbon metabolism, Impairment of photosynthesis and metabolism.		MAPK signaling, plant hormone signal transduction, plant-pathogen interaction	Li et al. (2024a, b), Huang et al. (2023), and Xu (2024)
ZnO nanoparticles	Enhance stress tolerance, modulate antioxidant pathways		Glutathione metabolism, phytohormone signal transduction	Azarin et al. (2022)
TiO2/ZnO nanoparticles	Promote growth, regulate photosynthesis and metabolism		Photosynthetic metabolism, antioxidant enzymes, nitrogen metabolism	Wang et al. (2017)

3.2 Metabolic Pathway Modulation in Nanomaterial-Exposed Plants

Interactions between nanomaterials and plants begin with uptake and translocation influenced by the material's physicochemical properties (Munir et al. 2023; Zaman et al. 2025). Nanomaterial-plant interaction significantly impacts plant's metabolism. Nanomaterials, such as metallic nanoparticles (MNPs), metal oxide nanoparticles (MONPs), and carbon nanomaterials, can regulate both primary and secondary metabolic pathways in plants. These regulations correlate with gene expression changes, enzymatic activity, and hormonal signaling (Chen et al. 2024a, 2024b, 2024c; Li et al. 2023). The molecular mechanisms of nanomaterial-induced metabolic changes and their correlation with the gene expression changes, enzymatic activity, and hormonal signaling changes in plants,focusing on transcriptomic insights, are further discussed below.

3.2.1 Nanomaterial-Induced Modulations in Primary Metabolism in Plants

Nanoparticles influence primary metabolism by altering photosynthesis and nutrient uptake (Munir et al. 2023). Nanomaterials affect photosynthesis by modifying carbon metabolism and chlorophyll biosynthesis. Fluorescent carbon dots (CDs) and silica nanoparticles (SiO_2 NPs) boost photosynthetic efficiency. For instance, NIR-CDs enhance photosynthesis by upregulating porphyrin and chlorophyll metabolism genes (Li and Cardoso 2022: Peng et al. 2022). SiO2 NPs also enhance drought tolerance in *Ehretia macrophylla* by regulating fatty acid metabolism and secondary metabolite accumulation associated with stress tolerance (Chen et al. 2024a, b, c). Further, nanomaterials modulate carbohydrate and nitrogen metabolisms. Polystyrene nanoplastics (PSNPs) affect the carbohydrate metabolism in cucumber leaves. They decrease photosynthetic capability at high concentration levels and increase membrane stability through gene modulation at low concentrations (Huang et al. 2023). Polystyrene nanoplastics (PS-NPs) can alter thylakoid structure and carbohydrate metabolism (Huang et al. 2023; Xu 2024). Likewise, silver nanoparticles (AgNPs) rewire carbon and nitrogen metabolism in *Chlorella pyrenoidosa* by changing the profile of sugar and amino acids (Cao et al. 2021). These reports indicate that nanomaterials can fine-tune carbohydrate and nitrogen metabolism based on concentration and plant type. Moreover, NPs interfere with energy metabolism through oxidative stress induction. For example, in Populus × euramericana, high levels of PSNPs induced oxidative damage, impaired photosynthesis, and remodeled energy conversion processes (Xu 2024). The interference with growth and development can be affected, showing the need to consider the assessment of nanoplastic influence on plant energy homeostasis.

3.2.2 Nanomaterial-Induced Elicitation of Secondary Metabolism in Plants

Nanomaterials play a significant role in secondary metabolism by modulating the biosynthesis of phenolic compounds and flavonoids. Carbon dots (CDs) induce phenolic acid accumulation but inhibit flavonoid biosynthesis in *Tetrastigma hemsleyanum* (Peng et al. 2022). Similarly, gold nanoparticles (AuNPs) induce genes responsible for secondary metabolite biosynthesis, especially for phenolic compound biosynthesis in *Nardostachys jatamansi* (Joshi and Joshi 2024). Nanomaterials also control the biosynthesis of terpenoids and alkaloids by controlling TFs. Cationic CTAB-AuNPs produce a significant upregulation effect on Myb, WRKY, and NAC TFs compared to anionic citrate-AuNPs, increasing metabolite yield in *Nardostachys jatamansi* (Joshi and Joshi 2024). MONPs have also been shown to have a positive influence on secondary metabolite biosynthesis to further aid plant-derived product production (Inam et al. 2023). These observations suggest that surface charge and composition are critical for nanomaterial-induced secondary metabolite production. Further, nanomaterials also regulate phytohormone pathways, affecting secondary metabolism and growth. Carbon nanosol (CNS) increases auxin buildup in tobacco plants, enhancing growth and regulating metabolite profiles (Wang et al. 2024a, b). This indicates that nanomaterials may serve as bio-stimulants that enhance hormone-regulated metabolic responses in plants, influencing their growth and metabolite profiles. Moreover, transcriptomics provides evidence of nanomaterial-induced metabolic changes at the genetic level. SiO2 NPs upregulate genes related to auxin signaling and MAPK pathways in *Ehretia macrophylla* under drought stress (Chen et al. 2024a, b, c). Similarly, AgNPs modulate carbohydrate and amino acid metabolism-related gene expression in *Chlorella pyrenoidesa* (Cao et al. 2021). These insights demonstrate how nanomaterials influence plant metabolism through differential gene regulation. Nanomaterials also influence enzyme activity, impacting metabolite profiles in plants. AgNP exposure enhances catalase and peroxidase activity in *Chlorella pyrenoidesa*, improving stress resistance (Cao et al. 2021). These observations emphasize the connection between enzymatic modulation and metabolic reprogramming. Metabolomic studies show that nanomaterials alter the accumulation of metabolites, including amino acids, sugars, and fatty acids (Huang et al. 2023). SiO_2 nanoparticles mitigate oxidative stress encountered during drought stress by promoting antioxidant enzyme activity and inducing the accumulation of secondary metabolites (Chen et al. 2024a, b, c). These findings highlight the potential of nanomaterials as agents for the alleviation of stress-imposed metabolic disturbance in plants. Table 9.8 summarizes some key modulations of primary and secondary metabolic pathways observed in nanomaterials-exposed plants.

3.2.3 Nonmaterial-Induced Hormonal Regulations in Plants

Plant hormones oversee plant metabolic functions, controlling plant growth, development, and stress response. Phytohormones such as auxins, gibberellins, cytokinins, abscisic acid (ABA), and ethylene control primary and secondary metabolism. Nanomaterials provide new strategies for improving the resilience of plants by interacting with these processes. Molecular processes and transcriptomic basis of

Table 9.8 Modulation of primary and secondary metabolic pathways by nanomaterials

Nanomaterials	Effect on primary metabolism	Effect on secondary metabolism	Reference(s)
Carbon nanosol	Increased potassium uptake and auxin accumulation	Increased production of secondary metabolites	Li et al. (2023), Wang et al. (2024a, b)
Silica NPs	Enhanced photosynthetic rates and nutrient assimilation	Increased production of flavonoids and phenolic acids	Chen et al. (2024a, b, c)
Fluorescent CDs	Increased expression of photosynthetic genes	Increased production of phenolic acids	Li et al. (2023), Peng et al. (2022)
Gold NPs	Increased expression of genes involved in respiration	Increased production of alkaloids	Joshi and Joshi (2024)
Carbon Nano-horns	Increased expression of photosynthetic genes	Increased production of terpenoids	Sun et al. (2020)

nanomaterial-hormone interaction for high plant productivity are further discussed below.

Phytohormones are master regulators of plant metabolism, regulating gene expression and enzyme activity to modulate growth and stress responses. Gibberellins regulate starch metabolism and photosynthesis (Manoharlal et al. 2018), and strigolactones regulate secondary metabolism, such as catechin and caffeine biosynthesis in tea leaves (Yu et al. 2024). These hormones can interact synergistically or antagonistically, as observed in ABA and other phytohormones under drought stress (Yoshida and Fernie 2023). At the molecular level, hormonal regulation involves complex signaling networks in which TFs such as MYB, WRKY, and NAC regulate gene expression (Joshi and Joshi 2024). Gibberellins regulate saccharide metabolism and terpenoid biosynthesis, and strigolactones regulate flavonoid biosynthesis in a dose-dependent manner (Yu et al. 2024).

Nanoparticles affect metabolism by inducing oxidative stress, which initiates signaling from antioxidant defense mechanisms in plants and alters phytohormone signaling for stress adaptation (Zia-Ur-Rehman et al. 2022). Thus, nanomaterials are effective modulators of these processes, altering phytohormone content and metabolic responses. Gold nanoparticles regulate genes involved in hormonal metabolism (Joshi and Joshi 2024), and silica nanoparticles enhance drought tolerance through the modulation of auxin signaling and fatty acid metabolism (Chen et al. 2024a, b, c). These are dose-dependent, with low nanomaterial concentrations enhancing stress resilience and higher doses potentially causing toxicity (Kumar et al. 2023). Nano-ZnO inhibits nanoplastic phytotoxicity by regulating photosynthesis, glutathione metabolism, and phytohormone signaling genes (Guo et al. 2023). RNA sequencing of *Arabidopsis thaliana* exposed to metal and carbon-based nanomaterials validated that metallic nanomaterials significantly impact metal homeostasis, and carbon-based nanomaterials trigger hormone-related immunity (Wu et al. 2024). Transcriptome analysis of drought stress-treated plants with SiO2 nanoparticles showed differently regulated genes related to auxin signal transduction and mitogen-activated protein kinase signaling pathways (Chen et al. 2024a, b, c). Functional enrichment analysis, such as Gene Ontology (GO) and Kyoto Encyclopedia of Genes and Genomes (KEGG) pathways, can also illustrate the biological processes influenced by exposure to nanomaterials. For example, a GO analysis of plants treated with PS-NPs showed significant changes in photosynthesis, antioxidant activity, and hormone metabolism (Joshi and Joshi 2024; Xu 2024). Table 9.9 summarizes the effect of some nanomaterials on plant metabolism and hormonal pathways.

Nanomaterials also serve as a good vehicle for the controlled release of plant hormones, which solves the problems of dose control and targeted delivery. For instance, nanoparticles can encapsulate plant hormones like auxins and cytokinin, allowing for sustained release and delivery to target tissues (Liu et al. 2023). This method increases the efficiency of plant hormones and minimizes the application dose and its effect on non-targeted organisms. Further, nanomaterials are also applied in plant tissue culture and metabolic engineering. Nanoparticles can be precisely designed to act as bioelicitors to enhance the production of valuable

Table 9.9 Nanomaterials and their effects on plant metabolism and hormonal pathways

Nanomaterial types	Key effects on plant metabolism	Reference(s)
Gold nanoparticles (AuNPs)	Regulate genes involved in hormonal metabolism and stress responses.	Joshi and Joshi (2024)
Silica nanoparticles (SiO_2 NPs)	Enhance drought tolerance by modulating auxin signaling and fatty acid metabolism	Chen et al. (2024a, b, c)
Zinc oxide nanoparticles (nano-ZnO)	Alleviate nanoplastic-induced phytotoxicity by regulating photosynthesis and glutathione metabolism	Guo et al. (2023)
Carbon nanomaterials	Induce hormone-related immune responses under stress conditions	Wu et al. (2024)
Designer nanoparticles	Stimulate specialized metabolite production in plant cell cultures	Arya et al. (2021)
Graphene-based nanomaterials (GBNs)	Induction of oxidative stress and activation of antioxidant defense systems.	Li et al. (2023)

compounds in plant cell cultures (Arya et al. 2021). Nanoparticles can also be utilized to transfer genes, assisting in transforming plants to enable them to produce the compound of interest (Pathak et al. 2023; Yan et al. 2023).

Combining plant biology and nanotechnology will improve hormone regulation and metabolic functions. The current research focus areas in plant biology are mechanistic understanding, nano-delivery of phytohormones, omics-based biomarkers, and sustainability. These will facilitate precision agriculture, improve stress tolerance, and promote high crop yields.

4 Future Prospects and Challenges

The advancements in the field of nanotechnology have led to the development of numerous novel nanomaterials, like nano-fertilizers (Akhtar et al. 2022), nano-insecticides (Thakur et al. 2022) and nano-pesticides, that can be utilized for various agriculture applications such as enhancing plant growth efficiency, which would directly help promote food security (Xin et al. 2020; Javaid et al. 2024). However, nanomaterials can also be hazardous to the ecosystem and human health. Extensive toxicity studies before broader applications are necessary to avoid these adverse effects, and eco-friendly nanomaterials must be developed for agriculture applications (Tripathi et al. 2023; Rasouli et al. 2020). Future research should focus on optimizing biodegradable and safe nanomaterials and performing life cycle analyses that will help assess environmental impacts (Basu et al. 2021; Kumar et al. 2018).

While nanotechnology in agriculture presents opportunities such as improved crop protection and resource efficiency, it also raises challenges like the development of delivery systems, regulatory and ethical challenges, public acceptance and safety concerns (Fig. 9.2) (Hofmann et al. 2020; Rahim et al. 2021).

These issues can be primarily tackled by comprehensive risk governance and environmental assessments (Chávez-Hernández et al. 2024; Hegde et al. 2016). Conducting field trials, educating farmers on nanotechnology use, and developing

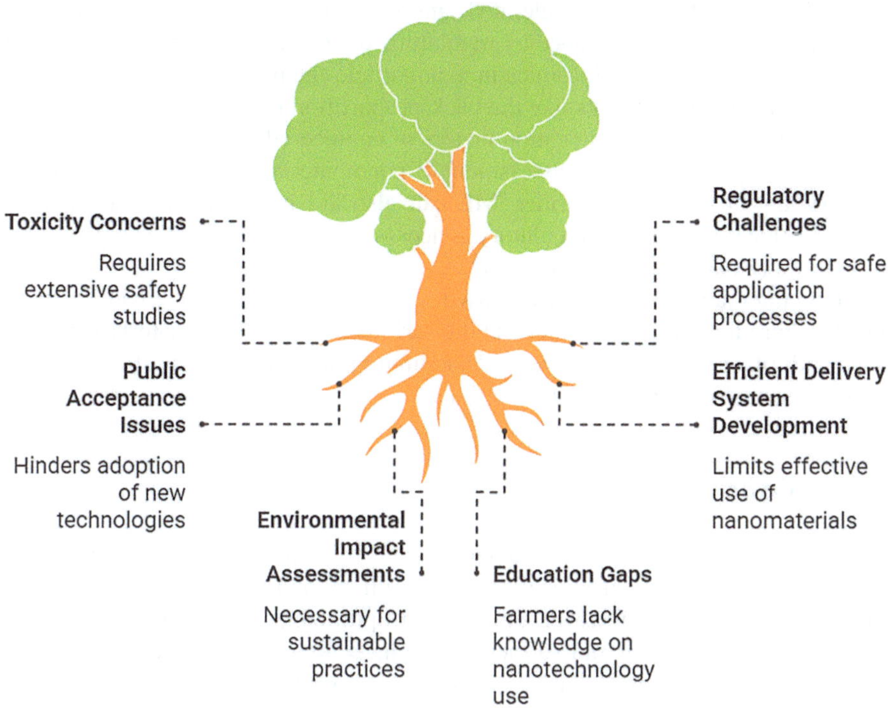

Fig. 9.2 Challenges faced in utilizing nanomaterials for agricultural applications

effective regulatory frameworks are essential for facilitating responsible development and agricultural use of nanomaterials (Kaphle et al. 2017; Miguel-Rojas and Pérez-De-Luque 2023). Addressing these concerns is vital for leveraging nanomaterials in sustainable agriculture.

CRISPR nanoparticles have opened the gateway for accurate plant trait modification, where the nanoparticles are used as effective delivery vehicles, transforming breeding while maintaining ecological balance (Afzal et al. 2023). Moreover, combining artificial intelligence (AI) and machine learning (ML) along with omics data would further facilitate enhancing the efficiency to model and predict the effects of nanomaterials on plant health and the environment (Ahmad et al. 2021). These tools enable sophisticated processing of multi-omics data, leading to better risk assessments and safer nano-enabled agricultural products (Ahmad et al. 2021; Nguyen and Falagan-Lotsch 2023). AI and ML can identify subtle plant responses that traditional methods might miss, improving our understanding of nanomaterial impacts (Ji et al. 2022; Gupta et al. 2024). Key insights into gene expression from transcriptomic data and the utilization of AI and ML can also help researchers develop predictive models incorporating environmental factors like soil composition and temperature (Deng et al. 2023). Further, incorporating machine learning models (MLMs), such as random forests, neural networks, etc., can help analyze data for predicting plant growth, stress resistance, and toxic effects of nanomaterials

(Ahmadi et al. 2024). ML-accelerated zinc oxide (ZnO) nanopriming facilitated enhanced maize stress resistance by regulating key metabolic pathways, highlighting the role of ML and multi-omics in nanoparticle-mediated stress mitigation (Yu et al. 2023). Neural networks like the back-propagation neural network can predict root concentration and translocation factors of engineered nanoparticles, highlighting the ML's potential in risk assessment of nanomaterials for agriculture applications (Wang et al. 2021). Further, ML can also aid in optimizing nanomaterial synthesis that could address regulatory concerns and ensure their safe application in agriculture (Konstantopoulos et al. 2022).

Thus, integrating multi-omics data with AI and ML can offer a wide range of advanced prediction tools to elucidate the various modes of nanomaterials' interactions with plants and the environment, facilitating the development of eco-friendly nanomaterials for agricultural applications.

5 Conclusion

Nanomaterials have proven to be valuable agents for enhancing agricultural productivity by modulating plant physiology, metabolism, and tolerance to stress. Their successful application, however, depends on optimizing physicochemical properties, such as size, shape, and surface functionalization, to realize their maximum effectiveness as metabolic elicitors with minimal toxicity. For their safe and sustainable use in agriculture, it is crucial to thoroughly comprehend nanomaterial-plant interactions. Transcriptomic approaches provide significant information regarding the molecular mechanisms underlying the influence of nanomaterial response in plants. Next-generation sequencing platforms, including Illumina sequencing, Nanopore sequencing, and Iso-Seq, have revealed significant changes in gene expression involving photosynthesis, nutrient acquisition, stress response, and hormone signaling. The integration of single-cell transcriptomics provides further insight by providing cell-specific gene expression information, which shows fine-scale regulatory mechanisms regulating plant adaptation to nanoparticle exposure. Integrating transcriptomics with proteomics and metabolomics provides a wider understanding of the physiological and biochemical changes in plants induced by nanomaterials. Nanomaterials profoundly impact plant metabolism by regulating primary and secondary metabolic processes, enhancing photosynthesis efficiency, and changing carbohydrate and nitrogen metabolism. They also interact with phytohormones to regulate growth and stress acclimation.

However, the ecotoxicological implications of nanomaterials, such as their uptake, translocation, and toxicity, must be thoroughly assessed to ensure ecological safety. While nanotechnology is full of promising prospects in agricultural sciences, some barriers exist, such as data complexity, non-standard procedures, and demands for integral regulatory procedures. Integrating multi-omics data and ML, coupled with AI, is a feasible approach to unravelling complex interactions between nanomaterials and plants, forming a basis for the rational and sustainable design of nanomaterials. Further research should focus on highlighting the significance of biodegradable nanomaterials in bringing about the convergence of agricultural benefits and environmental sustainability. With the advancement of transcriptomic methods and integration of multi-omics strategies, AI and ML scientists can create safer and more efficient nanomaterials aligned to support sustainable and precision-based agriculture.

References

Afzal A, Syed S, Khizar M, Iqbal J, Ashraf S, Altaf A, Mehmood B, Khan MR (2023) Advancement of crop productivity via crispr-nanoparticle interface. Pakistan J Biotechnol. https://doi.org/10.34016/pjbt.2023.20.02.826

Ahmad F, Muhmood T, Mahmood A (2021) Machine learning-integrated omics for the risk and safety assessment of nanomaterials. Biomater Sci 9(5):1598–1608. https://doi.org/10.1039/d0bm01672a

Ahmadi M, Ayyoubzadeh SM, Ghorbani-Bidkorpeh F (2024) Toxicity prediction of nanoparticles using machine learning approaches. Toxicology 501:153697. https://doi.org/10.1016/j.tox.2023.153697

Akhtar N, Ilyas N, Meraj TA, Pour-Aboughadareh A, Sayyed RZ, Mashwani Z, Poczai P (2022) Improvement of plant responses by Nanobiofertilizer: a step towards sustainable agriculture. Nano 12(6):965. https://doi.org/10.3390/nano12060965

Ali S, Mehmood A, Khan N (2021) Uptake, translocation, and consequences of nanomaterials on plant growth and stress adaptation. J Nanomater 2021:1–17. https://doi.org/10.1155/2021/6677616

Arya SS, Arya SS, Lenka SK, Cahill DM, Rookes JE (2021) Designer nanoparticles for plant cell culture systems: mechanisms of elicitation and harnessing of specialized metabolites. BioEssays 43(11):2100081. https://doi.org/10.1002/BIES.202100081

Azarin K, Usatov AV, Minkina T, Duplii N, Kasyanova AS, Fedorenko A, Khachumov VA, Mandzhieva S, Rajput VD (2022) Effects of bulk and nano-ZnO particles on functioning of photosynthetic apparatus in barley (Hordeum vulgare L.). Environ Res 216:114748. https://doi.org/10.1016/j.envres.2022.114748

Balusamy SR, Joshi AS, Perumalsamy H, Mijakovic I, Singh P (2023) Advancing sustainable agriculture: a critical review of smart and eco-friendly nanomaterial applications. J Nanobiotechnol 21(1). https://doi.org/10.1186/s12951-023-02135-3

Basak A, Amarshettiwar S, Bordoloi N, Shahni YS, Mubeen S, Beese S, Kushwaha TN, Bhushan S (2023) Nanomaterial-induced changes in plant physiology and genetics: implications for crop improvement strategies. Int J Enviorn Climate Change 13(11):2043–2048. https://doi.org/10.9734/ijecc/2023/v13i113364

Basu A, Prasad P, Das SN, Kalam S, Sayyed RZ, Reddy MS, Enshasy HE (2021) Plant growth promoting Rhizobacteria (PGPR) as green bioinoculants: recent developments, constraints, and prospects. Sustainability 13(3):1140. https://doi.org/10.3390/su13031140

Bawa G, Liu Z, Yu X, Qin A, Sun X (2022) Single-cell RNA sequencing for plant research: insights and possible benefits. Int J Mol Sci 23(9):4497. https://doi.org/10.3390/ijms23094497

Bian J, Zhuang Z, Ji X, Tang R, Li J, Chen J, Li Z, Peng Y (2024) Research Progress of single-cell transcriptome sequencing Technology in Plants. Agronomy 14(11):2530. https://doi.org/10.3390/agronomy14112530

Bobrovskikh A, Doroshkov A, Mazzoleni S, Cartenì F, Giannino F, Zubairova U (2021) A sight on single-cell transcriptomics in plants through the prism of cell-based computational modeling approaches: benefits and challenges for data analysis. Front Genet 12. https://doi.org/10.3389/fgene.2021.652974

Cao M, Huang X, Wang F, Zhang Y, Zhou B, Chen H, Yuan R, Ma S, Geng H, Xu D, Yan C, & Xing B (2021) Transcriptomics and Metabolomics Revealed the Biological Response of Chlorella pyrenoidesa to Single and Repeated Exposures of AgNPs at Different Concentrations. Environ Sci & Technol 55(23):15776–15787.https://doi.org/10.1021/acs.est.1c04059

Chávez-Hernández JA, Velarde-Salcedo AJ, Gonzalez C, Navarro-Tovar G (2024) Safe nanomaterials: from their use, application, and disposal to regulations. Nanoscale Adv 6(6):1583–1610. https://doi.org/10.1039/d3na01097j

Chen Z, Guo Z, Niu J, Xu N, Sui X, Kareem HA, Hassan MU, Yan M, Zhang Q, Wang Z, Mi F, Kang J, Cui J, Wang Q (2021) Phytotoxic effect and molecular mechanism induced by graphene towards alfalfa (Medicago sativa L.) by integrating transcriptomic and metabolomics analysis. Chemosphere 290:133368. https://doi.org/10.1016/j.chemosphere.2021.133368

Chen Z, Han M, Guo Z, Feng Y, Guo Y, Yan X (2024a) An integration of physiology, transcriptomics, and proteomics reveals carbon and nitrogen metabolism responses in alfalfa (Medicago sativa L.) exposed to titanium dioxide nanoparticles. J Hazard Mater 474:134851. https://doi.org/10.1016/j.jhazmat.2024.134851

Chen M, Jiao S, Xie L, Geng X, Qi S, Fan J, Cheng S, Shi J, Cao X (2024b) Integrated physiological, transcriptomic, and metabolomic analyses of drought stress alleviation in Ehretia macrophylla wall. Seedlings by SiO2 NPs (silica nanoparticles). Front Plant Sci 15. https://doi.org/10.3389/fpls.2024.126

Chen L, Zhu L, Liu X, Chen L, Zhou H, Ma H, Sun G, Nyande A, Li Z, Wu H (2024c) Recent omics progress in nanobiotechnology for plant abiotic stress tolerance improvement. Crop J 12(5):1274–1279. https://doi.org/10.1016/j.cj.2024.05.017

Cheng C, Chen W, Jin H, Chen X (2023) A review of single-cell RNA-Seq annotation, integration, and cell–cell communication. Cells 12(15):1970. https://doi.org/10.3390/cells12151970

Daniel AI, Hüsselmann L, Shittu OK, Gokul A, Keyster M, Klein A (2024) Application of nanotechnology and proteomic tools in crop development towards sustainable agriculture. J Crop Sci Biotechnol. https://doi.org/10.1007/s12892-024-00235-6

Deng P, Gao Y, Mu L, Hu X, Yu F, Jia Y, Wang Z, Xing B (2023) Development potential of nanoenabled agriculture projected using machine learning. Proc Natl Acad Sci USA 120(25):e2301885120. https://doi.org/10.1073/pnas.2301885120

Dev A, Karmakar S, Srivastava AK (2017) Nanomaterial toxicity for plants. Environ Chem Lett 16(1):85–100. https://doi.org/10.1007/s10311-017-0667-6

Dhiman S, Yadav A, Das S, Debnath N (2021) Application of Core/Shell nanoparticles in smart farming: a paradigm shift for making the agriculture sector more sustainable. J Agric Food Chem 69(11):3267–3283. https://doi.org/10.1021/acs.jafc.0c05403

Dimkpa CO, Singh U, Bindraban PS, Elmer WH, Gardea-Torresdey JL, White JC (2019) Zinc oxide nanoparticles alleviate drought-induced alterations in sorghum performance, nutrient acquisition, and grain fortification. Sci Total Environ 688:926–934. https://doi.org/10.1016/j.scitotenv.2019.06.392

Djanaguiraman M, Anbazhagan V, Dhankher OP, Prasad PVV (2024) Uptake, translocation, toxicity, and impact of nanoparticles on plant physiological processes. Plants (Basel, Switz) 13(22):3137. https://doi.org/10.3390/plants13223137

Ferioun M, Zouitane I, Bouhraoua S, Elouattassi Y, Belahcen D, Errabbani A, Louahlia S, Sayyed R, Ghachtouli NE (2025) Applying microbial biostimulants and drought-tolerant genotypes to enhance barley growth and yield under drought stress. Front Plant Sci 15. https://doi.org/10.3389/fpls.2024.1494987

Ferrari E, Kemmerling B, Barbero F, Franz-Wachtel M, Busquets-Fité M, Köhler H-R, Puntes V (2021) Growth-promoting gold nanoparticles decrease stress responses in Arabidopsis seedlings. Nano 11(12):3161. https://doi.org/10.3390/nano11123161

Fincheira P, Quiroz A, Diez MC, Seabra AB, Rubilar O, Tortella G (2021) Nanotechnology advances for sustainable agriculture: current knowledge and prospects in plant growth modulation and nutrition. Planta 254(4). https://doi.org/10.1007/s00425-021-03714-0

García-Sánchez S, Bernales I, Cristobal S, Cristobal S (2015) Early response to nanoparticles in the Arabidopsis transcriptome compromises plant defence and root-hair development through salicylic acid signalling. BMC Genomics 16(1):341. https://doi.org/10.1186/S12864-015-1530-4

Garg S, Nain P, Kumar A, Joshi S, Punetha H, Sharma PK, Siddiqui S, Alshaharni MO, Algopishi UB, Mittal A (2024) Next generation plant biostimulants & genome sequencing strategies for sustainable agriculture development. Front Microbiol 15. https://doi.org/10.3389/fmicb.2024.1439561

Ghamari R, Ahmadikhah A, Tohidfar M, Bakhtiarizadeh MR (2022) RNA-Seq analysis of Magnaporthe grisea transcriptome reveals the high potential of ZnO nanoparticles as a Nanofungicide. Front Plant Sci 13. https://doi.org/10.3389/fpls.2022.896283

Giudice G, Serra A, Saarimäki LA, Kotsis K, Rouse IP, Jagiello K, Mikolajczyk A, Fratello M, Papadiamantis AG, Lazzari N, Sanabria JA, Annala M, Morikka J, Kinaret PAS, Melagraki G, Afantitis A, Tämm K, Puzyn T, Gulumian M et al (2023) An ancestral molecular response to nanomaterial particulates. Nat Nanotechnol. https://doi.org/10.1038/s41565-023-01393-4

Goyal V, Deng C, Rani D, Ritika R, Mehrotra S, Wang Y (2023) Unlocking the potential of nano-enabled precision agriculture for efficient and sustainable farming. Plants (Basel, Switz) 12(21):3744. https://doi.org/10.3390/plants12213744

Guo S, Zhang X, Sun H (2023) Transcriptomic mechanism for foliar applied nano-ZnO alleviating phytotoxicity of nanoplastics in corn (Zea mays L.) plants. Sci Total Environ 166818. https://doi.org/10.1016/j.scitotenv.2023.166818

Gupta DK, Pagani A, Zamboni P, Singh AK (2024) AI-powered revolution in plant sciences: advancements, applications, and challenges for sustainable agriculture and food security. Explor Foods Foodomics 2(5):443–459. https://doi.org/10.37349/eff.2024.00045

Hayat F, Khanum F, Li J, Iqbal S, Khan U, Javed HU, Razzaq MK, Altaf MA, Peng Y, Ma X, Li C, Tu P, Chen J (2023) Nanoparticles and their potential role in plant adaptation to abiotic stress in horticultural crops: a review. Sci Hortic 321:112285. https://doi.org/10.1016/j.scienta.2023.112285

Huang D, Shi Z, Shan X, Yang S, Zhang Y, & Guo X (2023) Insights into growth-affecting effect of nanomaterials: Using metabolomics and transcriptomics to reveal the molecular mechanisms of cucumber leaves upon exposure to polystyrene nanoplastics (PSNPs). The Sci Total Environ 866:161247. https://doi.org/10.1016/j.scitotenv.2022.161247

Hegde K, Verma M, Surampalli RY, Brar SK (2016) Current understandings of toxicity, risks and regulations of engineered nanoparticles with respect to environmental microorganisms. Nanotechnol Environ Eng 1(1). https://doi.org/10.1007/s41204-016-0005-4

Hofmann T, Unrine JM, Tufenkji N, Ghoshal S, Gilbertson LM, Dutcher JR, Giraldo JP, Brambilla D, White JC, Wilkinson KJ, Lowry GV, Kinsella JM, Paret M, Landry MP, Pedersen JA, Lovell W, Naccache R (2020) Technology readiness and overcoming barriers to sustainably implement nanotechnology-enabled plant agriculture. Nature Food 1(7):416–425. https://doi.org/10.1038/s43016-020-0110-1

Inam M, Attique I, Zahra M, Khan AK, Hahim M, Hano C, Anjum S (2023) Metal oxide nanoparticles and plant secondary metabolism: unraveling the game-changer nano-elicitors. Plant Cell Tissue Organ Culture (PCTOC) 155(2):327–344. https://doi.org/10.1007/s11240-023-02587-3

Ishio S, Kusunoki K, Nemoto M, Kanao T, Tamura T (2024) Illumina-based transcriptomic analysis of the fast-growing leguminous tree acacia crassicarpa: functional gene annotation and identification of novel SSR-markers. Front Plant Sci 15. https://doi.org/10.3389/fpls.2024.1339958

Javaid A, Hameed S, Li L, Zhang Z, Zhang B, Rahman M-U (2024) Can nanotechnology and genomics innovations trigger agricultural revolution and sustainable development? Funct Integr Genomics 24(6). https://doi.org/10.1007/s10142-024-01485-x

Ji Z, Sakkiah S, Liu J, Patterson TA, Hong H, Wood EL, Xu X, Guo W (2022) Machine learning models for predicting cytotoxicity of nanomaterials. Chem Res Toxicol 35(2):125–139. https://doi.org/10.1021/acs.chemrestox.1c00310

Jian H, Sun H, Liu R, Zhang W, Shang L, Wang J, Khassanov V, Lyu D (2022) Construction of drought stress regulation networks in potato based on SMRT and RNA sequencing data. BMC Plant Biol 22(1). https://doi.org/10.1186/s12870-022-03758-8

Joshi S, Joshi R (2024) Comparative transcriptome profiling unveils the role of differentially charged gold nanoparticle in growth, metabolic regulation and defence response of micropropagated Nardostachys jatamansi wall. Ex DC. S Afr J Bot. https://doi.org/10.1016/j.sajb.2024.02.052

Kadam US, Schulz B, Irudayaraj J (2017) Multiplex single-cell quantification of rare RNA transcripts from protoplasts in a model plant system. Plant J 90(6):1187–1195. https://doi.org/10.1111/TPJ.13537

Kalia A, Sharma SP (2019) Single-cell omics in crop plants: opportunities and challenges. Academic Press, pp 341–355. https://doi.org/10.1016/B978-0-12-817532-3.00020-7

Kang M, Weng Y-J, Liu Y, Wang H, Ye L, Gu Y, Bai X (2023) A review on the toxicity mechanisms and potential risks of engineered nanoparticles to plants. Rev Environ Contam Toxicol 261(1):1–33. https://doi.org/10.1007/s44169-023-00029-x

Kaphle A, Daima HK, Umapathi A, Navya PN (2017) Nanomaterials for agriculture, food and environment: applications, toxicity and regulation. Environ Chem Lett 16(1):43–58. https://doi.org/10.1007/s10311-017-0662-y

Karnwal A, Malik T, Dohroo A (2023) Unveiling the potential of bioinoculants and nanoparticles in sustainable agriculture for enhanced plant growth and food security. Biomed Res Int 2023(1):1–23. https://doi.org/10.1155/2023/6911851

Kaur H, Jha P, Ochatt S, Kumar V (2023) Single-cell transcriptomics is revolutionizing the improvement of plant biotechnology research: recent advances and future opportunities. Crit Rev Biotechnol 1–16. https://doi.org/10.1080/07388551.2023.2165900

Kaveh R, Li Y-S, Ranjbar S, Tehrani R, Brueck CL, Van Aken B (2013) Changes in Arabidopsis thaliana gene expression in response to silver nanoparticles and silver ions. Environ Sci Technol 47(18):10637–10644. https://doi.org/10.1021/ES402209W

Konstantopoulos G, Koumoulos EP, Charitidis CA (2022) Digital innovation enabled nanomaterial manufacturing; machine learning strategies and green perspectives. Nano 12(15):2646. https://doi.org/10.3390/nano12152646

Kumar A, Gupta K, Srivastava S, Mishra K, Dixit S (2018) A review on positive and negative impacts of nanotechnology in agriculture. Int J Environ Sci Technol 16(4):2175–2184. https://doi.org/10.1007/s13762-018-2119-7

Kumar DS, Singh R, Upadhyay SK, Verma KK, Tripathi RM, Liu H, Dhankher OP, Tripathi RD, Sahi SV, Seth CS (2023) Review on interactions between nanomaterials and phytohormones: novel perspectives and opportunities for mitigating environmental challenges. Plant Sci 111964. https://doi.org/10.1016/j.plantsci.2023.111964

Landa P, Prerostova S, Petrová Š, Knirsch V, Vankova R, Vanek T (2015) The transcriptomic response of Arabidopsis thaliana to zinc oxide: a comparison of the impact of nanoparticle, bulk, and ionic zinc. Environ Sci Technol 49(24):14537–14545. https://doi.org/10.1021/ACS.EST.5B03330

Landa P, Dytrych P, Prerostova S, Petrová Š, Vankova R, Vanek T (2017) Transcriptomic response of Arabidopsis thaliana exposed to CuO nanoparticles, bulk material, and ionic copper. Environ Sci Technol 51(18):10814–10824. https://doi.org/10.1021/ACS.EST.7B02265

Laosuntisuk K, Vennapusa AR, Somayanda IM, Leman AR, Jagadish SVK, Doherty CJ (2024) A normalization method that controls for total RNA abundance affects the identification of differentially expressed genes, revealing bias toward morning-expressed responses. Plant J. https://doi.org/10.1111/tpj.16654

Li M, Cardoso JS (2022) Multi-omics analyses revealed key factors involved in fluorescent carbon-dots-regulated secondary metabolism in Tetrastigma hemsleyanum. J Nanobiotechnol 20(1). https://doi.org/10.1186/s12951-022-01271-6

Li P, Ewald J, Pang Z, Legrand E, Jeon YS, Hacariz O, Zhou G, Head JA, Basu N, Xia J (2023) Express analyst: a unified platform for RNA-sequencing analysis in non-model species. Nat Commun 14(1). https://doi.org/10.1038/s41467-023-38785-y

Li X, Du X, He Z, Zhou R, Guo Y, Liu H, Lian J, Guo X, Tang Z (2024a) Transcriptome mechanisms of dandelion under stress of polystyrene and dibutyl phthalate and quantitative tracing of nanoplastics. Sci Total Environ 948:174894. https://doi.org/10.1016/j.scitotenv.2024.174894

Li Y, Vulpe C, Lammers T, Pallares RM (2024b) Assessing inorganic nanoparticle toxicity through omics approaches. Nanoscale 16(34). https://doi.org/10.1039/d4nr02328e

Li Z, Yang J, Cai X, Zeng X, Zou J, & Xing W (2024c) A systematic review on the role of miRNAs in plant response to stresses under the changing climatic conditions. Plant Stress 14:100674. https://doi.org/10.1016/j.stress.2024.100674

Liu Y, Xing B, Yue L, Wang Z (2019) Processes and mechanisms of photosynthesis augmented by engineered nanomaterials. Environ Chem 16(6):430. https://doi.org/10.1071/en19046

Liu J, Zhang Y, Zheng Y, Zhu Y, Shi Y, Guan Z, Lang K, Shen D, Huang W, Dou D (2022) PlantExp: a platform for exploration of gene expression and alternative splicing based on public plant RNA-seq samples. Nucleic Acids Res 51(D1):D1483–D1491. https://doi.org/10.1093/nar/gkac917

Liu Z, Wen F, Cheng X, Wu Z (2023) Nano-controlled release of phytohormones will broaden its application on plant protection. Adv Agrochem. https://doi.org/10.1016/j.aac.2023.11.004

Magar S, Jadhav P, Shinde U, Vaidya ER (2024) Transcriptomics in crop improvement (pp 1–35). https://doi.org/10.58532/v3bdbt7p1ch1

Manoharlal R, Saiprasad GVS, Thambrahalli A, Madhavakrishna K (2018) Dissecting the transcriptional networks underlying the gibberellin response in Nicotiana tabacum. Biol Plant 62(4):647–662. https://doi.org/10.1007/S10535-018-0809-0

Marmiroli M, Imperiale D, Pagano L, Villani M, Zappettini A, Marmiroli N (2015) The proteomic response of Arabidopsis thaliana to cadmium Sulfide quantum dots, and its correlation with the transcriptomic response. Front Plant Sci 6:1104. https://doi.org/10.3389/FPLS.2015.01104

Miguel-Rojas C, Pérez-De-Luque A (2023) Nanobiosensors and nanoformulations in agriculture: new advances and challenges for sustainable agriculture. Emerging Topics Life Sci 7(2):229–238. https://doi.org/10.1042/etls20230070

Mmereke KM, Venkataraman S, Moiketsi BN, Khan MR, Hassan SH, Rantong G, Masisi K, Kwape TE, Gaobotse G, Zulfiqar F, Sharma SK, Malik S, Makhzoum A (2023) Nanoparticle elicitation: a promising strategy to modulate the production of bioactive compounds in hairy roots. Food Res Int. https://doi.org/10.1016/j.foodres.2023.113910

Mohanty P, Singh PK, Lenka B, Adhya TK, Verma SK, Ayreen Z, Patro S, Sarkar B, Mohapatra RK, Mishra S (2024) Biofabricated nanomaterials in sustainable agriculture: insights, challenges and prospects. Biofabrication 16(4):042003. https://doi.org/10.1088/1758-5090/ad60f7

Mondéjar-López M, García-Simarro MP, Navarro-Simarro P, Gómez-Gómez L, Ahrazem O, Niza E (2024) A review on the encapsulation of "eco-friendly" compounds in natural polymer-based nanoparticles as next generation nano-agrochemicals for sustainable agriculture and crop management. Int J Biol Macromol 280(Pt 3):136030. https://doi.org/10.1016/j.ijbiomac.2024.136030

Munir N, Gulzar W, Abideen Z, Hasanuzzaman M, El-Keblawy A, Zhao F (2023) Plant–nanoparticle interactions: transcriptomic and proteomic insights. Agronomy. https://doi.org/10.3390/agronomy13082112

Nguyen NHA, Falagan-Lotsch P (2023) Mechanistic insights into the biological effects of engineered nanomaterials: a focus on gold nanoparticles. Int J Mol Sci 24(4):4109. https://doi.org/10.3390/ijms24044109

Nizamani MM, Zhang Q, Muhae-Ud-Din G et al (2023) High-throughput sequencing in plant disease management: a comprehensive review of benefits, challenges, and future perspectives. Phytopathol Res 5:44. https://doi.org/10.1186/s42483-023-00199-5

Park S, Park S, Park S, Park S, Son HI, Kim D, Seol J, Kim J, Safdar M, Kim D, Kim J, Kim W, Kim W, Lee K, Safdar M, Lee K, Safdar M, Seol J, Lee K et al (2024) Gelatin nanoparticles can improve pesticide delivery performance to plants. Small (Weinheim an Der Bergstrasse, Germany) 20(42). https://doi.org/10.1002/smll.202402899

Pathak A, Ul Haq S, Meena N, Dwivedi P, Kothari SL, Kachhwaha S (2023) Multifaceted role of nanomaterials in modulating in vitro seed germination, plant morphogenesis, metabolism and genetic engineering. Plants 12(17). https://doi.org/10.3390/plants12173126

Patil T, Gambhir R, Vibhute A, Tiwari AP (2022) Gold nanoparticles: synthesis methods, functionalization and biological applications. J Clust Sci 34(2):705–725. https://doi.org/10.1007/s10876-022-02287-6

Peng X, Xie Z, Wang X, Zhao Y, Yang C, Zhang Z, Li M, Zheng J, Wang Y (2022) Multi-omics analyses revealed key factors involved in fluorescent carbon-dots-regulated secondary

metabolism in Tetrastigma hemsleyanum. J Nanobiotechnol 20(1). https://doi.org/10.1186/s12951-022-01271-6
Rahim HU, Qaswar M, Uddin M, Giannini C, Herrera ML, Rea G (2021) Nano-enable materials promoting sustainability and resilience in modern agriculture. Nano 11(8):2068. https://doi.org/10.3390/nano11082068
Rasouli H, Popović-Djordjević J, Sayyed RZ, Zarayneh S, Jafari M, Fazeli-Nasab B (2020) Nanoparticles: a new threat to crop plants and soil rhizobia? In: Sustainable agriculture reviews, pp 201–214. https://doi.org/10.1007/978-3-030-33996-8_11
Ruotolo R, Marmiroli M, White JC, Pagano L, Maestri E, Marmiroli N (2018) Plant response to metal-containing engineered nanomaterials: An omics-based perspective. Environ Sci Technol 52(5):2451–2467. https://doi.org/10.1021/acs.est.7b04121
Sanches PHG, De Melo NC, Porcari AM, De Carvalho LM (2024) Integrating molecular perspectives: strategies for comprehensive multi-omics integrative data analysis and machine learning applications in transcriptomics, proteomics, and metabolomics. Biology 13(11):848. https://doi.org/10.3390/biology13110848
Saraswat P, Singh S, Prasad M, Misra R, Rajput VD, Ranjan R (2023) Applications of bio-based nanomaterials in environment and agriculture: a review on recent progresses. Hybrid Adv 4:100097. https://doi.org/10.1016/j.hybadv.2023.100097
Sarraf M, Fujita M, Vishwakarma K, Janeeshma E, Das S, Kumar V, Arif N, Hasanuzzaman M, Puthur JT, Aliniaeifard S, Chauhan DK, Johnson R (2022) Metal/metalloid-based nanomaterials for plant abiotic stress tolerance: an overview of the mechanisms. Plants 11(3):316. https://doi.org/10.3390/plants11030316
Shahbaz M, Akram A, Mehak A, Haq EU, Fatima N, Wareen G, Fitriatin BN, Sayyed RZ, Ilyas N, Sabullah MK (2023) Evaluation of selenium nanoparticles in inducing disease resistance against spot blotch disease and promoting growth in wheat under biotic stress. Plants 12(4):761. https://doi.org/10.3390/plants12040761
Shaw R, Tian X, Xu J (2020) Single-cell transcriptome analysis in plants: advances and challenges. Mol Plant 14(1):115–126. https://doi.org/10.1016/j.molp.2020.10.012
Shearman JR, Tangphatsornruang S (2021) Nanopore sequencing in agricultural and food applications. In: Handbook of nanotechnology applications. Elsevier, pp 443–459
Shrestha AMS, Guiao JEB, Santiago KCR (2022) Assembly-free rapid differential gene expression analysis in non-model organisms using DNA-protein alignment. BMC Genomics 23(1). https://doi.org/10.1186/s12864-021-08278-7
Shukla D, Krishnamurthy S, Sahi SV (2014) Genome wide transcriptome analysis reveals ABA mediated response in Arabidopsis during gold (AuCl−4) treatment. Front Plant Sci 5:652. https://doi.org/10.3389/FPLS.2014.00652
Siddique A, Aley P (2024) Nano-omics approaches for studying plant growth, yield, and nutritional value. IGI Global, pp 19–41. https://doi.org/10.4018/979-8-3693-1890-4.ch002
Singh NB, De A, Tomar R, Jain P (2021) Green synthesis and applications of nanomaterials. Curr Pharm Biotechnol 22(13):1705–1747. https://doi.org/10.2174/1389201022666210412142734
Spanos A, Athanasiou K, Krasia-Christoforou T, Fotopoulos V, Ioannou A (2021) Functionalized magnetic nanomaterials in agricultural applications. Nano 11(11):3106. https://doi.org/10.3390/nano11113106
Sun L, Ju Q, Xu J, Wang R (2020) Physiological, metabolic, and transcriptomic analyses reveal the responses of Arabidopsis seedlings to carbon Nanohorns. Environ Sci Technol 54(7):4409–4420. https://doi.org/10.1021/acs.est.9b07133
Sundararajan N, Habeebsheriff HS, Dhanabalan K, Cong VH, Wong LS, Rajamani R, Dhar BK (2023) Mitigating global challenges: harnessing green synthesized nanomaterials for sustainable crop production systems. Global Challenges (Hoboken, NJ) 8(1). https://doi.org/10.1002/gch2.202300187
Tang N, Li X, Gao X, Liu X, Xing W (2024) Drilling into the physiology, transcriptomics, and metabolomics to enhance insight on Vallisneria denseserrulata responses to nanoplastics and metalloid co-stress. J Clean Prod. https://doi.org/10.1016/j.jclepro.2024.141653

Tasnim A, Roy A, Akash SR, Ali H, Habib MR, Barasarathi J, Muthukumaran M, Sayyed RZ, Yeasmin T (2024) Hibiscus sabdariffa L. petal biomass: a green source of nanoparticles of multifarious potential. Open Agric 9(1). https://doi.org/10.1515/opag-2022-0332

Thabet SG, Alqudah AM (2024) Unraveling the role of nanoparticles in improving plant resilience under environmental stress condition. Plant Soil 503(1–2):313–330. https://doi.org/10.1007/s11104-024-06581-2

Thakur P, Thakur S, Kumari P, Shandilya M, Sharma S, Poczai P, Alarfaj AA, Sayyed RZ (2022) Nano-insecticide: synthesis, characterization, and evaluation of insecticidal activity of ZnO NPs against Spodoptera litura and Macrosiphum euphorbiae. Appl Nanosci 12(12):3835–3850. https://doi.org/10.1007/s13204-022-02530-6

Tiwari M, Krishnamurthy S, Shukla D, Kiiskila JD, Jain A, Datta R, Sharma NC, Sahi SV (2016) Comparative transcriptome and proteome analysis to reveal the biosynthesis of gold nanoparticles in Arabidopsis. Sci Rep 6(1):21733. https://doi.org/10.1038/SREP21733

Trevisan S, Manoli A, Ravazzolo L, Franceschi C, Quaggiotti S (2017) mRNA-sequencing analysis reveals transcriptional changes in root of maize seedlings treated with two increasing concentrations of a new biostimulant. J Agric Food Chem 65(46):9956–9969. https://doi.org/10.1021/acs.jafc.7b03069

Tripathi S, Mahra S, Tripathi DK et al (2023) Recent advances and perspectives of nanomaterials in agricultural management and associated environmental risk: a review. Nano 13(10):1604. https://doi.org/10.3390/nano13101604

Tripathi S, Sharma S, Suri S, Tiwari K, Tripathi DK, Sharma S (2024) Insights into physiological and molecular responses of plants under metal-nanoparticle stresses. BENTHAM SCIENCE PUBLISHERS, pp 147–173. https://doi.org/10.2174/9789815179699124010009

Tumburu L, Andersen CP, Rygiewicz PT, Reichman JR (2017) Molecular and physiological responses to titanium dioxide and cerium oxide nanoparticles in Arabidopsis. Environ Toxicol Chem 36(1):71–82. https://doi.org/10.1002/ETC.3500

Tyagi P, Singh D, Mathur S, Singh A, Ranjan R (2022) Upcoming progress of transcriptomics studies on plants: an overview. Front Plant Sci 13. https://doi.org/10.3389/fpls.2022.1030890

Ungaro A, Pech N, Martin J-F, McCairns RJS, Mévy J-P, Chappaz R, Gilles A (2017) Challenges and advances for transcriptome assembly in non-model species. *bioRxiv*, 084145. https://doi.org/10.1101/084145

Upadhyay P, Navvuru S, Yadav P, Lalotra S, Singh A, Rajput VD, Minkina T, Ghazaryan K (2024) Nano-omics-based abiotic and biotic stresses management. IGI Global, pp 348–371. https://doi.org/10.4018/979-8-3693-1890-4.ch017

Wang Y, Chen R, Hao Y, Liu H, Song S, Sun G (2017) Transcriptome analysis reveals differentially expressed genes (DEGs) related to lettuce (Lactuca sativa) treated by TiO2/ZnO nanoparticles. Plant Growth Regul 83(1):13–25. https://doi.org/10.1007/s10725-017-0280-5

Wang X, Liu L, Zhang W, Ma X (2021) Prediction of plant uptake and translocation of engineered metallic nanoparticles by machine learning. Environ Sci Technol 55(11):7491–7500. https://doi.org/10.1021/acs.est.1c01603

Wang Y, Nong B, Zhou W, Xu S, Songyang Z, Xiong Y (2023) Comprehensive RNA-Seq analysis pipeline for non-model organisms and its application in Schmidtea mediterranea. Genes 14(5):989. https://doi.org/10.3390/genes14050989

Wang Z, Adusei-Fosu K, Wang Y, Wang H, Li X (2024a) Integrative chemical, physiological, and metabolomics analyses reveal nanospecific phytotoxicity of metal nanoparticles. J Environ Manag 354:120338. https://doi.org/10.1016/j.jenvman.2024.120338

Wang C, Hua Y, Liang T, Guo Y, Wang L, Zheng X, Liu P, Zheng Q, Kang Z, Xu Y, Cao P, Chen Q (2024b) Integrated analyses of ionomics, phytohormone profiles, transcriptomics, and metabolomics reveal a pivotal role of carbon-nano sol in promoting the growth of tobacco plants. BMC Plant Biol 24(1):473. https://doi.org/10.1186/s12870-024-05195-1

Wei L, Ji L, Rico CM, He C, Shakoor I, Fakunle M, Xia Y, Hou Y, Hong J (2024) Transcriptomics reveals the pathway for increasing Brassica chinensis L. yield under foliar application of titanium oxide nanoparticles. J Agric Food Chem. https://doi.org/10.1021/acs.jafc.4c04075

Wu Y, Wang Y, Liu X, Zhang C (2024) Unveiling key mechanisms: transcriptomic meta-analysis of diverse nanomaterial applications addressing biotic and abiotic stresses in Arabidopsis Thaliana. Sci Total Environ:172476. https://doi.org/10.1016/j.scitotenv.2024.172476

Xin X, He Z, Judy JD, Sumerlin BB (2020) Nano-enabled agriculture: from nanoparticles to smart nanodelivery systems. Environ Chem 17(6):413. https://doi.org/10.1071/en19254

Xiong T, Li S, Zhang S, Zhang T, Kang Z (2021) Dose-dependent physiological and transcriptomic responses of lettuce (Lactuca sativa L.) to copper oxide nanoparticles-insights into the Phytotoxicity mechanisms. Int J Mol Sci 22(7):3688. https://doi.org/10.3390/ijms22073688

Xu C (2024) The Oryza sativa transcriptome responds spatiotemporally to nanoplastic stress. Sci Total Environ:172449. https://doi.org/10.1016/j.scitotenv.2024.172449

Yadav A, Alhan S, Dhania G (2024a) Nanobiostimulants for enhancing plant growth and yield. In: Nanobiostimulants: emerging strategies for agricultural sustainability, pp 111–134. https://doi.org/10.1007/978-3-031-68138-7_6

Yadav A, Mathan J, Dubey AK, Singh A (2024b) The emerging role of non-coding RNAs (NCRNAs) in plant growth, development, and stress response signaling. Non-Coding RNA 10(1):13. https://doi.org/10.3390/ncrna10010013

Yan T, Hou Q-G, Wei X, Qi Y, Pu A, Wu S, An X, Wan X (2023) Promoting genotype-independent plant transformation by manipulating developmental regulatory genes and/or using nanoparticles. Plant Cell Rep. https://doi.org/10.1007/s00299-023-03037-2

Yoshida T, Fernie AR (2023) Hormonal regulation of plant primary metabolism under drought. J Exp Bot. https://doi.org/10.1093/jxb/erad358

Yu H, Luo D-L, Li SFY, Qu M, Liu D, He Y, Cheng F (2023) Interpretable machine learning-accelerated seed treatment by nanomaterials for environmental stress alleviation. *arXiv.Org, abs/2304.03928*. https://doi.org/10.48550/arXiv.2304.03928

Yu S, Zuo H, Li P, Lu L, Li J, Zhou Z, Zhao S, Huang J, Liu Z, Zhu M, Zhao J (2024) Strigolactones regulate secondary metabolism and nitrogen/phosphate Signaling in tea plants via transcriptional reprogramming and hormonal interactions. J Agric Food Chem. https://doi.org/10.1021/acs.jafc.4c05100

Yuan XD, Wang Z, Peijnenburg WJGM (2024) Molecular characterization and transcriptional response of Lactuca sativa seedlings to co-exposure to graphene nanoplatelets and titanium dioxide nanoparticles. Ecotoxicol Environ Saf 282:116761. https://doi.org/10.1016/j.ecoenv.2024.116761

Zaman W, Ayaz A, Park S (2025) Nanomaterials in agriculture: a pathway to enhanced plant growth and abiotic stress resistance. Plants 14(5):716. https://doi.org/10.3390/plants14050716

Zemlyanskaya EV, Dolgikh VA, Levitsky VG, Mironova V (2021) Transcriptional regulation in plants: using omics data to crack the cis-regulatory code. Curr Opin Plant Biol 63:102058. https://doi.org/10.1016/j.pbi.2021.102058

Zhang H, Goh NS, Wang JW, Pinals RL, González-Grandío E, Demirer GS, Butrus S, Fakra SC, Del Rio Flores A, Zhai R, Zhao B, Park S, Landry MP (2021) Nanoparticle cellular internalization is not required for RNA delivery to mature plant leaves. Nat Nanotechnol 17(2):197–205. https://doi.org/10.1038/s41565-021-01018-8

Zheng D, Xu J, Lu Y, Chen H, Chu Q, Fan L (2023) Recent progresses in plant single-cell transcriptomics. Crop Design 2(2):100041. https://doi.org/10.1016/j.cropd.2023.100041

Zheng X, Wang C, Zhang K-L, Xu Y, Qu X, Cao P, Zhou T, Chen Q (2024) Revealing critical mechanisms involved in carbon nanosol-mediated tobacco growth using small RNA and mRNA sequencing in silico approach. BMC Plant Biol 24(1). https://doi.org/10.1186/s12870-024-05992-8

Zhou Q et al (2020) Transcriptome mechanisms underlying interaction of polystyrene nanoplastics and wheat Triticum aestivum L. https://doi.org/10.21203/RS.3.RS-98748/V1

Zhu L, Ma H, Wu H, Gu J, Chen L (2022) Carbon-based nanomaterials for sustainable agriculture: their application as light converters, nanosensors, and delivery tools. Plants (Basel, Switz) 11(4):511. https://doi.org/10.3390/plants11040511

Zia-Ur-Rehman M, Anayatullah S, Irfan E, Hussain SM, Rizwan M, Sohail MI, Jafir M, Ahmad T, Usman M, Alharby HF (2022) Nanoparticles assisted regulation of oxidative stress and antioxidant enzyme system in plants under salt stress: a review. Chemosphere 314:137649. https://doi.org/10.1016/j.chemosphere.2022.137649

Zungu B, Paumo HK, Gaorongwe JL, Tsuene GN, Ruzvidzo O, Katata-Seru L (2023) Zn nutrients-loaded chitosan nanocomposites and their efficacy as nanopriming agents for maize (Zea mays) seeds. Front Chem 11. https://doi.org/10.3389/fchem.2023.1243884

Proteomics Approaches to Examine the Impact of Nanomaterials on Plant Biochemical, Physiological, and Molecular Responses

10

Abdel Rahman Mohammad Al-Tawaha, Arun Karnwal, Abdel Razzaq Al-Tawaha, Natalia Nesterova, and Riyaz Sayyed

Abstract

Nanomaterials (NMs) have emerged as transformative tools in agriculture, offering novel solutions to enhance soil health, improve crop productivity, and manage plant diseases with precision. This review explores the biochemical, physiological, and molecular responses of plants to NMs, emphasizing their role in nutrient uptake, stress tolerance, and overall plant health. The integration of nano-fertilizers, nano-pesticides, and nano-sensors has demonstrated significant improvements in plant growth and yield, primarily due to enhanced nutrient efficiency and targeted pest control mechanisms. However, concerns remain regarding the long-term ecological and environmental impacts of NM applications, necessitating further research and regulatory frameworks to ensure sustainable usage. Biotechnological advancements, particularly in proteomics, have provided insights into NM–plant interactions at the molecular level. Proteomic stud-

A. R. M. Al-Tawaha (✉)
Department of Biological Sciences, Al-Hussein Bin Talal University, Maan, Jordan
e-mail: abdel-al-tawaha@ahu.edu.jo

A. Karnwal
Department of Microbiology, Graphic Era (Deemed to be University), Dehradun, Uttarakhand, India

A. R. Al-Tawaha
National Agricultural Research Center (NARC), Baq'a, Jordan

N. Nesterova
National University of Life and Environmental Sciences of Ukraine, Kyiv, Ukraine

R. Sayyed
Department of Biological Science and Chemistry, College of Arts and Science, University of Nizwa, Nizwa, Oman

R. Sayyed et al. (eds.), *Nanobiostimulants in Innovative Agriculture*,
https://doi.org/10.1007/978-981-95-2477-8_10

ies reveal substantial alterations in protein expression and post-translational modifications that influence plant metabolism, photosynthesis, and stress responses. The potential of NMs to mitigate abiotic stresses such as drought and salinity through modulation of antioxidant enzyme activity highlights their promise for improving agricultural resilience. Despite these benefits, knowledge gaps persist regarding the dose-dependent effects of NMs, their bioaccumulation in plant tissues, and their impact on microbial ecosystems. To address these concerns, this review underscores the necessity of developing environmentally friendly nanotechnologies that optimize plant performance while minimizing risks. Future research should focus on integrating multi-omics approaches, refining application methods, and establishing comprehensive risk assessments to balance productivity with ecological safety. By leveraging advanced proteomics and nanotechnology, sustainable agricultural practices can be enhanced, contributing to food security and environmental conservation.

Keywords

Crop growth · Drought · Nanomaterial · Stress physiology,

1 Introduction

Agriculture has been revolutionized as a result of the growing interest in nanomaterials (NMs), which have provided new solutions to problems such as the deterioration of soil, the inefficiency of nutrient availability, and the control of pests (Sethi et al. 2025; Ferioun et al. 2025; Sharma et al. 2025; Akhtar et al. 2022; Praveen et al. 2024). Materials that display unique physicochemical qualities such as large surface area, enhanced reactivity, and controlled release capabilities are referred to as nanomaterials (NMs). These materials are classified as having at least one dimension in the nanometer scale, which ranges from 1 to 100 nanometers (Khan et al. 2020). Due to these features, they have proven to be very well-suited for a wide range of agricultural applications, such as nano-pesticides, nano-fertilizers, and soil additives (Tasnim et al. 2024; Bastami et al. 2021).

Over the past few years, there has been a substantial amount of attention paid to the investigation of the function that nano-fertilizers play in the enhancement of nutrient usage efficiency (NUE) (Faridvand et al. 2021). Nano-fertilizers are shown to lessen environmental losses (Rai et al. 2021). This is accomplished through the provision of targeted delivery and regulated nutrient release. Ghormade et al. (2011) investigated the potential of nanobiotechnology to revolutionize plant protection and nutrition, with a particular focus on the significance of this technology in sustainable agriculture. They highlight the ways in which nanotechnology might successfully address nutrient deficiencies, improve insect resistance, and reduce environmental issues by boosting the effectiveness of agrochemicals while simultaneously minimizing the negative effects that they have. According to Ghormade et al. (2011), nano-biotechnology presents great prospects for improving agricultural yield, plant defense systems, and soil health. These improvements can be achieved while simultaneously reducing dependency on chemical inputs and minimizing environmental concerns. However, in order to assure both the sustainability and safety of agroecosystems, it is necessary to do extensive research, establish regulatory frameworks, and conduct risk assessments. This is necessary for the successful and safe deployment of this technology in agriculture. For example, nano-encapsulated fertilizers release nutrients in accordance with the requirements of the plant, hence reducing the amount of leaching and volatilization that occurs (Ghormade et al. 2011). In a similar vein, the role of nano-pesticides has grown increasingly significant due to their capacity to improve the control of diseases and pests by enhancing the stability and effectiveness of the active components. As an illustration, silver nanoparticles, also known as AgNPs, have been shown to possess antibacterial effects specifically against plant diseases (Kumar et al. 2021).

Nanomaterials have also been the subject of extensive research due to the fact that they are relevant to the process of amending and restoring soil. Nanocomposites that are based on biochar, for instance, increase the structure of the soil and its ability to retain water (Lal 2020). Additionally, nano-clays are utilized for the immobilization of heavy metals, which results in increased productivity in soils that have been contaminated (Mukhopadhyay et al. 2021). These applications demonstrate the potential of NMs to solve soil health issues that are of crucial importance in agricultural settings. There has been a significant amount of research published on the topic of the incorporation of natural resources (NMs) in agricultural production. This research has highlighted the possibility of NMs lessening dependency on conventional chemical inputs and minimizing the environmental imprint. As a result of the effectiveness of nano-pesticides, misuse is reduced, which in turn minimizes the amount of contamination that occurs in water and soil resources (Parisi et al. 2015).

In addition, nanotechnologies provide economic benefits by lowering the costs of inputs and increasing crop yields, which makes a substantial contribution to the implementation of sustainable agriculture practices (Servin and White 2015). A significant amount of study has been conducted, but very little is known about the long-term effects that NMs have on the environment and the ecology. There are still issues that need to be tackled, including those concerning safety, regulatory systems, and environmental dangers. Despite the fact that a great number of studies have been conducted to investigate the possible advantages of NMs, there is an increasing demand for complete risk assessments to guarantee the ecological safety of their use in agricultural environments (Kah et al. 2018). In order to maximize the sustainable potential of nanomaterials, future research should concentrate on the development of environmentally friendly nanomaterials and the integration of multidisciplinary techniques.

2 Biotechnological Advancements in Plant Protection and Nutrition

There has been a growing interest in biotechnology solutions for plant protection and nutrition, which has led to the development of new ways for boosting the resilience and productivity of farmers' crops (Bright et al. 2025) (Fig. 10.1). According to Kumar et al. (2021), in recent years, there has been a substantial focus on understanding how genetic and microbial technologies might replace or minimize the reliance on chemical agro-inputs while also maintaining sustainable agricultural practices.

Because of the growing desire for plants that are resistant to disease and intolerant of stress, the role of agricultural improvement through genetic engineering has become increasingly essential. Among the most noteworthy examples are transgenic insect-resistant maize and cotton, which contain insecticidal genes from *Bacillus thuringiensis* (Bt) in order to improve yields and increase resistance to pests (James 2018). By reducing the need for synthetic insecticides, these genetically modified crops hope to minimize the amount of pollution that is released into the environment, as well as the expenses associated with production. Although there has been a significant amount of research conducted, there is still a lack of knowledge on the long-term ecological impacts of transgenic crops. Furthermore, the acceptance of these crops is still restricted in many countries due to ethical and regulatory issues.

The importance of biopesticides and biofertilizers to organic and sustainable agricultural techniques has led to a significant amount of research being conducted on these approaches, despite the fact that various studies have investigated them as alternative strategies. As natural antagonists against pests and pathogens,

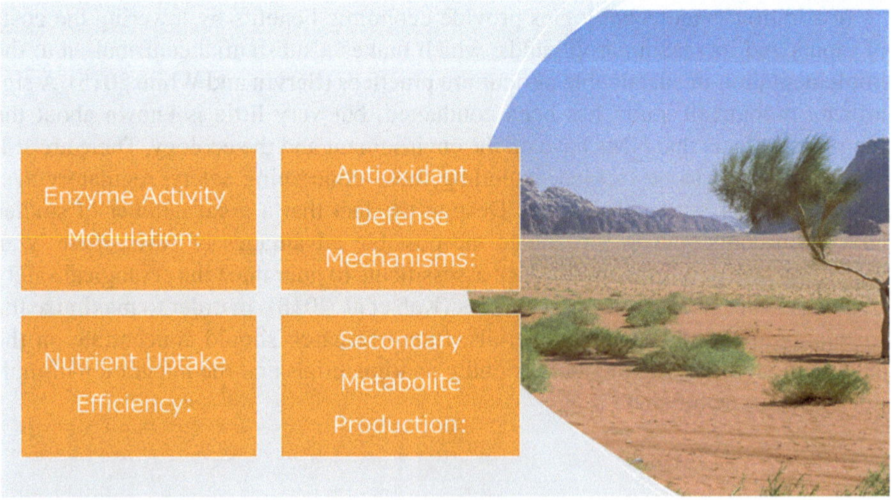

Fig. 10.1 Biochemical responses of plants to nanomaterials

biopesticides, which are comprised of living organisms or the bioactive substances that arise from them, are extremely effective. In integrated pest management (IPM), their compatibility with chemical pesticides, target specificity, environmental safety, and acceptance in organic farming are among the most important advantages that they offer (Glare et al. 2012). Biofertilizers are beneficial to soil fertility and plant growth because they improve nutrient availability and uptake. A significant amount of research has been conducted and published on the subject of the advantages of biofertilizers. These biological inputs contribute to the long-term health of the soil by facilitating microbial symbiosis, nitrogen fixation, and phosphate solubilization. Additionally, they lessen the impact that synthetic fertilizers have on the ecosystem (Bashan et al. 2014). This research has revealed major gaps that require additional investigation. These deficiencies include the scalability, efficiency, and formulation stability of biopesticides and biofertilizers when they are used in field circumstances. Overall, the present body of literature offers a solid basis for investigating biotechnological advancements in plant protection and nutrition; nevertheless, additional study that draws from a variety of disciplines is required to guarantee the broad acceptance of these breakthroughs and their efficiency in a variety of agroecosystems (Kumar et al. 2021). The authors Bashan et al. (2014) investigate the developments that have occurred in plant growth-promoting bacterial (PGPB) inoculants between the years 1998 and 2013, with a particular emphasis on their formulations and practical uses. A growing number of people are becoming interested in using PGPB as biofertilizers because of its capacity to promote the intake of nutrients, to stimulate root development (Bright et al. 2025; Kalam et al. 2020; Basu et al. 2021; Hamid et al. 2021), increase the stress tolerance of plants (Ferioun et al. 2025), and inhibit the growth of pathogens (Vinothini et al. 2024; Anbalagan et al. 2024; Praveen et al. 2024). Over the past few years, there has been a substantial emphasis placed on gaining a knowledge of the ways in which bacterial strains like *Azospirillum* and *Rhizobium* contribute to the growth of plants by fixing nitrogen, solubilizing phosphates, and creating phytohormones to promote plant growth. Because of the growing demand for inoculants that are not only stable but also effective and simple to administer, the role that PGPB formulations play has become increasingly significant (Bashan et al. 2014). Polymer-based and biofilm-forming bacterial formulations have been extensively researched due to their significance in boosting bacterial survival under difficult environmental circumstances (Bashan et al. 2014). As a result, a significant quantity of literature has been written on the subject of their application. There has been a significant amount of research conducted, but there is still a lack of knowledge on the long-term performance of these formulations in a variety of agricultural systems and soil types. This lack of knowledge highlights the necessity for additional study into their scalability and commercial viability. In general, the existing body of literature offers a solid basis for investigating the incorporation of PGPB inoculants into contemporary agricultural systems. For the purpose of facilitating adoption, the review highlights the significance of standardized production methods, regulatory frameworks, and large-scale field trials. During the course of this research, several major holes have been highlighted that require additional investigation. These gaps include the compatibility of

PGPB inoculants with conventional fertilization tactics and the impact that these inoculants have on the populations of soil microbes. For the purpose of promoting global food security and sustainable agriculture, future research should concentrate on the development of PGPB formulations that are both environmentally friendly and cost-effective.

3 Importance of Understanding Plant Responses to Nanomaterials

There is a growing interest in nanomaterials (NMs) for agricultural applications, which has shown that there is tremendous potential to improve crop yield and address important environmental concerns (Fig. 10.2).

In recent years, there has been a substantial focus on understanding how plants react to NMs. This is because it is essential to optimize the use of NMs while also ensuring the safety of the environment and food. According to Ghormade et al. (2011), NMs have the ability to interact with plants on cellular, biochemical, and molecular levels, which in turn influences things like growth, stress responses, and overall plant health. Due to the fact that nano-fertilizers have the potential to increase the availability of nutrients and the amount that plants are able to absorb, their role has become increasingly crucial. According to Kah et al. (2018), nano-encapsulated fertilizers, for example, release nutrients in a controlled manner, which aligns with the needs demanded by plants and improves the efficiency with which nutrients are used. It is essential to have a solid understanding of these nutrient delivery responses in order to make the most of the benefits that nano-fertilizers offer without overwhelming the soil with an excessive amount of nutrients, which could have

Fig. 10.2 Nanoparticles play a key role in enhancing drought stress tolerance in plants

hazardous effects. As a result of their usefulness in reducing abiotic stresses, including drought and salinity by modifying plant stress responses, NMs have been the subject of a significant amount of investigative research. According to Dimkpa and Bindraban (2018), zinc oxide nanoparticles (ZnO NPs) have the ability to boost the activity of antioxidant enzymes, hence lowering the level of oxidative stress that occurs in saline environments. For the purpose of developing tailored methods to successfully manage abiotic stresses, it is vital to have a solid understanding of these stress responses. The effect that NMs have on photosynthesis and growth has been the subject of a significant amount of research that has been produced and published. Certain natural products (NMs) enhance photosynthesis by either boosting the amount of chlorophyll present or shielding the machinery involved in photosynthesis from the effects of stress. For instance, it has been demonstrated that silver nanoparticles, also known as AgNPs, can improve the efficiency of photosynthesis and stimulate the growth of plants when subjected to particular conditions (Raliya et al. 2015). For the purpose of avoiding the possibility of phytotoxicity, however, it is vital to have a grasp of dose-response correlations. There is a lack of knowledge regarding the long-term effects of excessive or incorrect applications of NM on plants, despite the fact that extensive studies have been conducted. For example, it has been demonstrated that high quantities of titanium dioxide nanoparticles (TiO_2 NPs) can accumulate in plant tissues, causing disruptions to cellular functioning and metabolic processes (Servin and White 2015). In order to reduce the likelihood of these dangers occurring, it is vital to establish safe application thresholds through ongoing study. Despite the fact that a great number of studies have been conducted to investigate the molecular-level interactions of NMs with plants, the influence that NMs have on the expression of genes and the synthesis of proteins continues to be an important topic of research. (Kumar et al. 2021) Research that was conducted using proteomic and transcriptome techniques revealed significant alterations in stress-related genes and pathways that were triggered by NMs. The findings presented here shed light on the fact that NMs have the ability to produce advantageous characteristics, such as increased stress tolerance and resistance to pathogens. Root exudation and absorption pathways are two examples of plant responses that can have an effect on the behavior of NMs in the soil-plant system (Parisi et al. 2015). These interactions have the potential to influence the mobility of NM, its bioavailability, and its long-term impact on the health of the soil and the dynamics of the ecosystem. In general, the existing body of literature offers a solid basis for investigating the possibilities of NMs in environmentally responsible agriculture. Researchers have the ability to optimize the uses of NMs, boost agricultural productivity, and address concerns regarding safety when they have a better understanding of how plants react to NMs. During the course of this assessment, important gaps have been found that require additional investigation. One of these gaps is the development of NM-based solutions for agriculture that are both effective and environmentally friendly. It is essential to continue studying in order to fully harness the potential of NMs while also ensuring that their use is both sustainable and safe.

4 Role of Proteomics in Elucidating Nanomaterial (NM)–Plant Interactions

The increasing interest in nanomaterials (NMs) for agricultural applications has shown considerable potential to improve crop yield and tackle urgent environmental issues. Recent years have seen considerable emphasis on comprehending plant responses to NMs, which is essential for optimizing their applications while safeguarding environmental and food safety. NMs engage with plants at cellular, biochemical, and molecular levels, affecting growth, stress responses, and overall plant health (Ghormade et al. 2011). The significance of nano-fertilizers has grown due to their capacity to improve nutrient accessibility and absorption by plants. For example, nano-encapsulated fertilizers deliver nutrients in a regulated fashion, according to plant requirements and enhancing nutrient utilization efficiency (Kah et al. 2018). Comprehending these nutrient delivery responses is essential for optimizing the advantages of nano-fertilizers while preventing the soil from becoming overloaded with excessive nutrients, which may lead to toxicity. Nanomaterials have been extensively researched for their significance in alleviating abiotic stresses, including drought and salinity, via influencing plant stress responses. Zinc oxide nanoparticles (ZnO NPs) augment antioxidant enzyme activity, mitigating oxidative stress in salt-water environments (Dimkpa and Bindraban 2018). Comprehending these stress responses is crucial for formulating tailored methods to control abiotic pressures efficiently. A significant body of literature has been produced regarding the effects of nanomaterials on photosynthesis and growth. Specific nanomaterials enhance photosynthesis by augmenting chlorophyll levels or safeguarding photosynthetic apparatus under stress conditions. Silver nanoparticles (AgNPs) have been documented to improve photosynthetic efficiency and stimulate plant development under certain conditions (Raliya et al. 2015). Nonetheless, comprehending dose-response connections is essential to prevent possible phytotoxicity. Despite a comprehensive study, little knowledge exists regarding the long-term effects of excessive or incorrect NM treatments on plants. Elevated levels of titanium dioxide nanoparticles (TiO_2 NPs) have been demonstrated to accumulate in plant tissues, interfering with cellular functioning and metabolic processes (Servin and White 2015). It is crucial to establish safe application thresholds through ongoing study to mitigate these dangers. Although many studies have investigated the molecular interactions of nanomaterials with plants, their effects on gene expression and protein synthesis remain a significant focus of research. Proteomic and transcriptome analyses indicate substantial alterations in stress-related genes and pathways induced by NMs (Kumar et al. 2021). These observations underscore the capacity of NMs to elicit advantageous characteristics, including improved stress resilience and pathogen resistance. The conduct of nanoparticles in the soil-plant system is affected by plant reactions, such as root exudation and absorption mechanisms. These interactions can influence NM mobility, bioavailability, and the long-term effects on soil health and ecosystem dynamics (Parisi et al. 2015). The current research establishes a robust basis for investigating the possibilities of nanomaterials in sustainable agriculture. Comprehending plant reactions to nanomaterials enables researchers to enhance their utilization, augment agricultural yield, and mitigate safety issues. This research has pinpointed critical deficiencies that necessitate additional inquiry, particularly the creation of sustainable and efficient NM-based agricultural solutions. Ongoing research is essential to completely realize the promise of N. The increasing interest in nanomaterials (NMs) has resulted in a proliferation of research aimed at comprehending their interactions with plants. Proteomics, the extensive examination of proteins, has become an essential instrument for clarifying the biochemical, physiological, and molecular reactions of plants to nanomaterials. Recent years have witnessed considerable attention on comprehending the effects of nanomaterials on plant proteomes. Research indicates that nanoparticles produce significant alterations in protein expression, post-translational modifications, and pathways, demonstrating their impact on biological processes. Proteomics plays a crucial role in examining NM–plant interactions by elucidating the molecular processes that govern these interactions. Proteomics enhances the optimization of nanomaterial uses in agriculture by analyzing protein expression and changes, while also addressing safety and environmental issues. The impact of nanomaterials on plant stress responses has been extensively researched because of their significance to agricultural output. Proteomics has demonstrated that NMs influence plant defense mechanisms by modifying the expression of stress-related proteins, including heat shock proteins and antioxidant enzymes. A substantial body of literature has been produced regarding the application of proteomics in the investigation of NM–plant interactions. Proteomic investigations on zinc oxide nanoparticles have demonstrated elevated expression of stress-related proteins, underscoring the importance of nanomaterials in influencing plant defense mechanisms. Despite comprehensive research, limited knowledge exists regarding the specific biochemical pathways influenced by NMs in plants. Proteomics has revealed proteins linked to abiotic and biotic stress responses induced by nanoparticles, such as silver nanoparticles. Although much research has investigated the impact of NMs on plant proteomes, the function of post-translational modifications in modulating protein activity and signaling cascades is still inadequately comprehended. Proteomics has demonstrated that nanoparticles affect phosphorylation events in proteins associated with photosynthesis and energy metabolism. This analysis has pinpointed critical gaps necessitating future examination, particularly the molecular mechanisms driving NM-induced alterations in plant nutrient absorption and transport pathways. Proteomics offers a robust framework for investigating these pathways and creating sustainable nanotechnologies for agriculture. The current literature establishes a robust basis for investigating the application of proteomics in examining NM–plant interactions. By merging proteomics with other omics methodologies, researchers can attain a more profound comprehension of the molecular pathways that regulate plant responses to nanomaterials, while guaranteeing their sustainable and safe application.

5 Types of Nanomaterials Used in Agriculture

As interest in nanomaterials (NMs) has grown, they have become revolutionary instruments in agriculture, providing creative ways to boost soil health, control pests, and increase crop yield (Karnwal et al. 2024; Rajput et al. 2024; Singh et al. 2024). NMs are very useful in agricultural applications because of their special qualities, which include high surface area, adjustable porosity, and controlled release capabilities. Understanding how NMs can enhance crop nutrition has received a lot of attention in recent years. For example, nano-fertilizers have grown in significance because of their capacity to release nutrients in a regulated manner, minimizing losses from volatilization and leaching (Kah et al. 2018). Because of its importance to sustainable agriculture, the effect of nano-fertilizers in increasing nutrient usage efficiency has been extensively researched. Nitrogen and phosphorus fertilizers that are nanoencapsulated and improve nutrient use efficiency are two examples (Dimkpa and Bindraban 2018). In order to improve growth and productivity in crops with zinc deficiencies, zinc oxide nanoparticles, or ZnO NPs, are commonly utilized (Raliya et al. 2015). There has been a significant quantity of published research on the application of nanopesticides in agriculture. By improving the active components' solubility, stability, and effectiveness, these materials provide targeted disease and insect management while reducing environmental contamination (Thakur et al. 2022). For example, bacterial and fungal infections can be efficiently controlled by silver nanoparticles (AgNPs), which have broad-spectrum antibacterial qualities (Kumar et al. 2021). Little is known about the long-term impacts of nanopesticides on microbial communities and soil health, despite a great deal of research. Even though the effectiveness of nanopesticides has been the subject of numerous studies, more investigation is required to fully grasp the advantages and disadvantages of these treatments (Rasouli et al. 2020). Because nano-sensors can provide real-time monitoring of environmental variables, plant nutrition, and soil health, their usage in precision agriculture has grown in importance. For instance, precise fertilization techniques are made possible by the use of carbon nanotubes to measure soil pH and nutrient levels (Servin and White 2015). Key gaps that require more research have been found by this review, including the possible advantages and disadvantages of using nanomaterials in agriculture. All things considered, the body of current research offers a solid starting point for investigating the application of nanomaterials in sustainable agriculture. Because of its potential to lessen environmental contamination, the function of nano-carriers in the targeted distribution of agrochemicals has been extensively researched. For example, lipid-based nanocarriers guarantee effective absorption and prolonged release of active ingredients, lowering the frequency of applications and their negative effects on the environment (Parisi et al. 2015). The application of nanomaterials in soil remediation has been the subject of a sizable body of published literature. For instance, stabilizing soil particles with nano-clays has improved water infiltration and decreased erosion (Mukhopadhyay et al. 2021).

6 Uptake and Translocation of Nanomaterials (NMS) in Plants

The increasing interest in nanomaterials (NMs) has resulted in an increase in the amount of research that is centered on gaining knowledge of how these materials interact with plant systems and how they might be utilized in agricultural settings. The uptake and translocation of NMs in plants is an important area of research that provides insights into the influence that NMs have on the development, metabolism, and stress responses of plants (Kah et al. 2018). Over the past few years, there has been a substantial emphasis placed on gaining better knowledge of the function of NMs in plant nutrition and the potential of these substances to enhance agricultural output. It is possible for NMs to enter plant systems by a variety of channels and then be transported to various tissues, where they have the ability to influence growth, metabolism, and stress responses (Kah et al. 2018). This is because of their small size and the unique physicochemical features that they possess. Because of the significance it has in terms of comprehending the interactions between NM and plants, the role of root absorption in NM translocation has evolved into an increasingly crucial one. The root uptake pathway is the one that has received the most attention from researchers. According to Dimkpa and Bindraban (2018), nanoparticles reach the root system through the apoplastic and symplastic channels. By way of illustration, it has been demonstrated that zinc oxide nanoparticles, also known as ZnO NPs, are capable of penetrating root epidermal cells and travelling along both channels, hence increasing the availability of zinc in crops (Raliya et al. 2015).Due to the fact that it is relevant to the optimization of NM applications in agriculture, the efficiency of NM uptake has been the subject of extensive research. Their size, shape, surface charge, and coating all play a role in determining how effectively they are able to absorb NM. It has been demonstrated by Ghormade et al. (2011) that nanoparticles with a size of less than 50 nanometers are more easily absorbed because of their capacity to pass through cell membranes and cell walls. The translocation of NMs in plants has been the subject of a significant quantity of research writing that has been published. Once they have entered the root system, NMs are transported to different regions of the plant by way of the vascular tissues. It is primarily through the process of transpiration pull that the xylem is responsible for transporting NMs from the roots to the shoots, whereas the phloem is responsible for facilitating movement to sinks such as fruits, seeds, and young leaves (Fazeli-Nasab et al. 2021; Kumar et al. 2021). The long-term behavior of NMs in plants and the ecological implications they have are not well understood, despite the substantial studies that have been conducted on the subject. There have been great advancements achieved, but there are still many obstacles to overcome in order to comprehend the destiny and movement of NMs inside plant systems. Despite the fact that a great number of research studies have been conducted to investigate the uptake and translocation of NMs in plants, the function that NM physicochemical qualities play in determining the effectiveness of their translocation is still not well known. The physicochemical features of NMs, such as their size, shape, surface charge, and coating, have a substantial impact on the efficiency with which they translocate. As a result of this analysis, key gaps have been highlighted that require additional investigation. One of these gaps is the creation of biodegradable and environmentally safe NMs that maximize plant absorption and translocation while simultaneously minimizing the risks of toxicity. For the most part, the existing body of literature offers a solid basis for investigating the application of NMs in agricultural settings. On the other hand, additional research is required in order to completely comprehend the interactions that occur between NMs and plant systems and to develop environmentally friendly technologies that are based on NMs for agricultural purposes.

7 General Impacts of NMS on Plant Biochemical, Physiological, and Molecular Processes

Agricultural applications of nanomaterials (NMs) are on the rise due to their remarkable physicochemical features, such as their large surface area, increased reactivity, and controlled release capacities. Exciting opportunities and possible obstacles have prompted renewed interest in studying the effects of NMs on plants' biochemical, physiological, and molecular processes (Kah et al. 2018). If we want to maximize the benefits of NM applications while minimizing the hazards, we must understand these implications. Because of their capacity to regulate biochemical pathways and enzymatic activity, NMs have assumed a more central position in plant metabolism. Zinc oxide nanoparticles (ZnO NPs) are one example; they boost the effectiveness of antioxidant enzymes like catalase and superoxide dismutase (SOD), which aids plants in dealing with oxidative stress in adverse environments (Dimkpa and Bindraban 2018). Furthermore, AgNPs have antibacterial characteristics that indirectly benefit plant health by destroying microbial cell walls (Kumar et al. 2021). The overproduction of reactive oxygen species (ROS), which can cause oxidative stress and cellular damage, can occur at high concentrations of NM, though (Raliya et al. 2015). Because of their importance in plant development, photosynthesis, and nutrient absorption, NMs have been the subject of extensive research. Studies have demonstrated that nano-fertilizers increase the uptake of nutrients and stimulate the development of new shoots and roots. For instance, while cerium oxide nanoparticles (CeO_2 NPs) improve photosynthetic performance by increasing chlorophyll content and water use efficiency, high concentrations of titanium dioxide nanoparticles (TiO_2 NPs) have been found to disrupt stomatal conductance, which in turn affects gas exchange and water balance negatively (Ghormade et al. 2011). Researchers have found that NMs affect gene expression and protein synthesis at the molecular level, according to a large body of published literature. Proteomic research has demonstrated that silica nanoparticle (SiO2 NP) treatment increases levels of stress-responsive proteins that are implicated in pathways for stress signaling and detoxification (Servin and White 2015). Plants treated with ZnO NPs showed changes in genes involved in photosynthesis, energy metabolism, and hormone biosynthesis, according to transcriptomic analysis (Raliya et al. 2015). The long-term consequences of NMs on plant resistance to abiotic conditions like heat, salinity, and drought remain largely unknown, despite substantial research in the field. Excessive exposure to NMs can activate stress pathways, resulting in metabolic imbalances and growth suppression, although silicon nanoparticles (Si NPs) increase cell wall strength and decrease osmotic stress in salty settings (Dimkpa and Bindraban 2018; Kah et al. 2018). Although several studies have looked at how NMs affect plant hormone regulation, researchers are still trying to figure out how they affect development and growth. Example: a study by Mukhopadhyay et al. (2021) indicated that nano-copper particles (Cu NPs) increased auxin levels, which in turn stimulated root extension and the development of lateral roots. But critical developmental processes can be hindered when NM concentrations are not appropriately adjusted.

Notably, the possible toxicity of NMs when applied excessively is one of the important knowledge gaps that this analysis has uncovered. By interfering with cellular homeostasis, researchers have demonstrated that high concentrations of AgNPs can hinder seed germination and root elongation (Kumar et al. 2021). To guarantee the safe and sustainable deployment of NMs in agriculture, it is necessary to understand these biochemical, physiological, and molecular interactions. All things considered, there is a solid basis in the current research to investigate how to best use NMs in crop management. Researchers can enhance plant production with NMs while reducing environmental and ecological concerns by adjusting dosages and formulations (Parisi et al. 2015).

8 Proteomics: A Tool for Understanding Nanomaterial (NM)–Plant Interactions

There have been considerable breakthroughs in agricultural research as a result of the increased interest in proteomics as a technique for examining the molecular and biochemical interactions that occur between nanomaterials (NMs) and plants. In the past few years, there has been a substantial emphasis placed on the knowledge of how proteomics can provide insights into protein expression, function, and post-translational modifications (PTMs). This has enabled researchers to unravel the intricate mechanisms that are responsible for plant responses to NMs. It is essential to take this strategy in order to maximize the utilization of NMs in agricultural settings while simultaneously maintaining their safety and effectiveness (Kumar et al. 2021).

In the field of natural medicine (NM) research, the function of proteomics has grown increasingly significant as a result of its capacity to assist protein analysis on a large scale. Proteins are the primary molecules responsible for cellular processes; hence, the study of proteins is crucial in order to comprehend the connections between NM and plants. Due to the fact that proteomics is relevant to the investigation of the effects that NMs have on plant physiology, it has been extensively researched since the introduction of high-throughput techniques such as mass spectrometry (MS) and two-dimensional gel electrophoresis (2-DE) (Kah et al. 2018).

The way in which NMs change the expression of proteins that are involved in essential plant activities like photosynthesis, stress responses, and metabolism has been the subject of a significant amount of research that has been published. For instance, the presence of zinc oxide nanoparticles (ZnO NPs) in wheat led to an increase in the quantity of photosynthetic proteins and antioxidant enzymes. This finding exemplifies the dual role that ZnO NPs play in promoting growth and reducing the effects of oxidative stress (Raliya et al. 2015).

The field of proteomics makes it easier to investigate protein-transfer mechanisms (PTMs), which are essential for the regulation of proteins and include phosphorylation, glycosylation, and ubiquitination. According to research conducted by (Servin and White 2015), titanium dioxide nanoparticles (TiO2 NPs) have the ability to stimulate the phosphorylation of proteins that are involved in energy metabolism and stress signaling. This results in an increase in the plant's resistance to abiotic stress conditions.

It is not well understood how NMs regulate stress-responsive proteins at the molecular level, despite the fact that substantial studies have been conducted on the subject. There has been a great amount of research that has investigated the effect that NMs have on the stress responses of plants; nevertheless, proteomic analyses have shown that plants that have been treated with NMs have an increased expression of stress-related proteins. For example, the presence of silver nanoparticles (AgNPs) was found to boost the production of heat shock proteins (HSPs) and superoxide dismutase (SOD), two enzymes that are extremely important in the process of protecting plants from oxidative damage (Jakinala et al. 2021; Dimkpa and Bindraban 2018). During the course of this research, major gaps that require future investigation have been highlighted. These gaps include the molecular pathways that are impacted by NMs. In rice that was subjected to silica nanoparticles (SiO_2 NPs), advanced proteomic analyses have revealed significant changes in proteins related to nutrient transport, hormone metabolism, and the detoxification of reactive oxygen species (ROS). These findings demonstrate the complex effects that silica nanoparticles have on plant physiology (Kumar et al. 2021). Using proteomics, researchers were able to discover that tomato plants that were treated with cerium oxide nanoparticles (CeO2 NPs) had their lignin biosynthesis proteins upregulated, which resulted in an increase in structural integrity and resistance to disease (Ghormade et al. 2011). According to Parisi et al. (2015), the use of nano-iron fertilizers in maize led to an increase in the levels of chlorophyll-related proteins, which in turn led to an increase in the efficiency of photosynthesis and the generation of biomass. In general, the existing body of literature offers a solid basis for investigating the possibility of combining proteomics with other omics methodologies, including transcriptomics and metabolomics, in order to achieve a more comprehensive comprehension of the effects that NM has on plants.

The use of proteomics in NM research, on the other hand, is fraught with difficulties, including as the difficulty of data interpretation and the requirement for more sophisticated bioinformatics tools. According to Servin and White (2015), the focus of future research should be on overcoming these constraints through the development of standardized methodology and the integration of multi-omics approaches in order to improve our understanding of the interactions between nutrients and plants. There have been considerable breakthroughs in agricultural research as a result of the increased interest in proteomics as a technique for examining the molecular and biochemical interactions that occur between nanomaterials (NMs) and plants. In the past few years, there has been a substantial emphasis placed on the knowledge of how proteomics can provide insights into protein expression, function, and post-translational modifications (PTMs). This has enabled researchers to unravel the intricate mechanisms that are responsible for plant responses to NMs. It is essential to take this strategy in order to maximize the utilization of NMs in agricultural settings while simultaneously maintaining their safety and effectiveness (Kumar et al. 2021). In the field of natural medicine (NM) research, the function of proteomics has grown increasingly significant as a result of its capacity to assist protein analysis on a large scale. Proteins are the primary molecules responsible for cellular processes; hence, the study of proteins is crucial in order to comprehend the connections between NM and plants. Due to the fact that proteomics is relevant to the investigation of the effects that NMs have on plant physiology, it has been extensively researched since the introduction of high-throughput techniques such as mass spectrometry (MS) and two-dimensional gel electrophoresis (2-DE) (Kah et al. 2018).

The way in which NMs change the expression of proteins that are involved in essential plant activities like photosynthesis, stress responses, and metabolism has been the subject of a significant amount of research that has been published. For instance, the presence of zinc oxide nanoparticles (ZnO NPs) in wheat led to an increase in the quantity of photosynthetic proteins and antioxidant enzymes. This finding exemplifies the dual role that ZnO NPs play in promoting growth and reducing the effects of oxidative stress (Raliya et al. 2015). The field of proteomics makes it easier to investigate protein-transfer mechanisms (PTMs), which are essential for the regulation of proteins and include phosphorylation, glycosylation, and ubiquitination. According to research conducted by Servin and White (2015), titanium dioxide nanoparticles (TiO_2 NPs) have the ability to stimulate the phosphorylation of proteins that are involved in energy metabolism and stress signaling. This results in an increase in the plant's resistance to abiotic stress conditions. It is not well understood how NMs regulate stress-responsive proteins at the molecular level, despite the fact that substantial studies have been conducted on the subject. A great number of studies have investigated the effect that NMs have on the stress responses of plants; nevertheless, proteomic analyses have shown that plants that have been treated with NMs have an increased expression of stress-related proteins. For example, the presence of silver nanoparticles (AgNPs) was found to boost the production of heat shock proteins (HSPs) and superoxide dismutase (SOD), two enzymes that are extremely important in the process of protecting plants from oxidative damage (Dimkpa and Bindraban 2018). During the course of this research, major gaps that require future investigation have been highlighted. These gaps include the molecular pathways that are impacted by NMs. In rice that was subjected to silica nanoparticles (SiO_2 NPs), advanced proteomic analyses have revealed significant changes in proteins related to nutrient transport, hormone metabolism, and the detoxification of reactive oxygen species (ROS). These findings demonstrate the complex effects that silica nanoparticles have on plant physiology (Kumar et al. 2021). Using proteomics, researchers were able to discover that tomato plants that were treated with cerium oxide nanoparticles (CeO_2 NPs) had their lignin biosynthesis proteins upregulated, which resulted in an increase in structural integrity and resistance to disease (Ghormade et al. 2011).

According to Parisi et al. (2015), the use of nano-iron fertilizers in maize led to an increase in the levels of chlorophyll-related proteins, which in turn led to an increase in the efficiency of photosynthesis and the generation of biomass. In general, the existing body of literature offers a solid basis for investigating the possibility of combining proteomics with other omics methodologies, including transcriptomics and metabolomics, in order to achieve a more comprehensive comprehension of the effects that NM has on plants. The use of proteomics in NM research, on the other hand, is fraught with difficulties, including as the difficulty of data interpretation and the requirement for more sophisticated bioinformatics tools. According to Servin and White (2015), the focus of future research should be on overcoming these constraints through the development of standardized methodology and the integration of multi-omics approaches in order to improve our understanding of the interactions between nutrients and plants.

9 Biochemical Responses of Plants to Nanomaterials

Nanomaterials (NMs) have become increasingly popular in recent years, which has led to their emergence as potent instruments in the agricultural sector. These nanomaterials offer novel techniques to improve plant growth and resistance. In recent years, there has been a large focus on understanding the biochemical interactions that occur between natural resources (NMs) and plant systems. This is due to the fact that these interactions ultimately result in a variety of metabolic responses, the regulation of oxidative stress, and changes in enzyme activity. For the purpose of optimizing NM applications while simultaneously minimizing potential hazards, it is essential to have a solid understanding of these biochemical responses (Kah et al. 2018). The ability of nanoparticles (NMs), such as zinc oxide nanoparticles (ZnO NPs) and silver nanoparticles (AgNPs), to generate reactive oxygen species (ROS) during their contact with plant cells has led to an increase in the significance of the function that oxidative stress control plays (Dimkpa and Bindraban 2018). A lack of control over these reactive oxygen species (ROS) can result in damage to essential cellular components such as DNA, proteins, and lipids. These antioxidant defense systems are activated in plants so that they can combat the effects of oxidative stress. There is an increase in the expression of enzymes including superoxide dismutase (SOD), catalase (CAT), and peroxidase (POD) when NMs are present. This phenomenon has been extensively researched because of its significance to the responses of plants to stress. Raliya et al. (2015) conducted research on wheat that was subjected to ZnO nanoparticles and found that the activity of superoxide dismutase (SOD) and catalase (CAT) increased, resulting in improved stress tolerance and enhanced ROS detoxification. NMs have been shown to have a significant impact on primary metabolic activities, such as photosynthesis and respiration, according to a substantial body of research that has been published. The use of titanium dioxide nanoparticles, also known as TiO2 NPs, has been demonstrated to improve the chlorophyll content and photosynthetic efficiency of plants. This is accomplished by optimizing the light absorption and electron transport processes (Ghormade et al. 2011). Cerium oxide nanoparticles, also known as CeO2 NPs, have been shown to enhance the efficiency with which plants use water and assimilate carbon, hence enhancing plant development even when under stressful conditions (Kumar et al. 2021). The specific role that NMs play in plant defense is still a topic of investigation, despite the fact that various studies have been conducted to investigate the effects that NMs have on secondary metabolites. Secondary metabolites, which include flavonoids, phenolics, and alkaloids, are extremely important in the process of defending plants from the damaging effects of both biotic and abiotic stress. For example, it has been observed that silicon nanoparticles (Si NPs) can boost the amount of phenolic compounds found in rice, which in turn increases the plant's tolerance to oxidative stress and infections (Mukhopadhyay et al. 2021).

The long-term effects of NMs on major enzyme activities linked with nitrogen metabolism are not well understood, despite the fact that an exhaustive study has been conducted on the subject. According to Kah et al. (2018), nano-fertilizers, such as nano-encapsulated nitrogen, have been shown to promote nitrate reductase (NR) activity, which in turn leads to enhanced nitrogen assimilation, which is ultimately beneficial to plant growth. In addition, phosphorus-based NMs have the ability to control phosphatase activity, which in turn optimizes phosphate absorption and utilization (Parisi et al. 2015). Within the scope of this analysis, major gaps that require future exploration have been identified. These gaps include the influence of NMs on the regulation of hormones and the transmission of signals in plants. According to research conducted by Raliya et al. (2015), copper nanoparticles (Cu NPs) have been demonstrated to increase auxin levels, which in turn promotes root extension and the production of lateral roots. Furthermore, calcium-dependent protein kinases (CDPKs), which are involved in the process of stress signal transduction, are influenced by NMs (Dimkpa and Bindraban 2018). In general, the bodies of literature that are now available offer a solid basis for investigating the ways in which proteome investigations show the influence of NMs on the synthesis of proteins and the metabolism of amino acids in plants. For instance, it was discovered that silver nanoparticles increased the expression of stress-related proteins in tomato plants while simultaneously decreasing the expression of growth-inhibitory proteins (Servin and White 2015). According to Kumar et al.'s research from 2021, the levels of amino acids like proline, which is an important osmoprotectant, are increased in response to exposure to NM, which helps the body better tolerate stress. The use of NMs can have considerable benefits, but when they are present in high amounts, they can be hazardous, causing disruptions in metabolic processes and restricting growth. In the future, research should concentrate on optimizing the formulations and dosages of NM in order to maximize the benefits to agriculture while simultaneously minimizing the dangers that may be involved (Parisi et al. 2015).

10 Advanced Proteomics Techniques in Nanomaterial (NM) Studies

An increasing number of people are using proteomics as a method for investigating how plants react to nanomaterials (NMs), which has resulted in tremendous progress being made in the knowledge of the molecular interactions between these substances. In recent years, there has been a growing focus on understanding how modern proteomics techniques can provide precise insights into NM-induced changes in protein expression, post-translational modifications (PTMs), and stress signaling pathways. This has been the case because these techniques involve the study of proteomics. Within the context of agricultural and environmental applications, these methodologies make it possible for researchers to evaluate the safety, effectiveness, and potential dangers associated with NMs (Kumar et al. 2021). The ability of mass spectrometry (MS) to enable high-resolution identification and quantification of proteins has led to an increase in the significance of the function that MS plays in proteomics. Because of their applicability to the investigation of NM-induced alterations in the plant proteome, techniques such as liquid chromatography-tandem mass spectrometry (LC-MS/MS) have been the subject of a significant amount of research interest. In rice that was subjected to zinc oxide nanoparticles (ZnO NPs), for instance, LC-MS/MS was able to disclose the overexpression of stress-related proteins, which offered light on the processes that are responsible for mitigating the effects of stress (Raliya et al. 2015).

The field of quantitative proteomics has been the subject of a significant amount of published research, which includes label-free quantification as well as isotope-based labelling techniques as SILAC and iTRAQ. These techniques make it possible to obtain accurate measurements of the amount of protein present in NM treatments. Label-free methods have been utilized to evaluate the impacts of AgNPs on tomato plants. These methods have resulted in the identification of significant alterations in the proteins responsible for photosynthesis and antioxidants (Dimkpa and Bindraban 2018). Despite the fact that a great amount of research has investigated the use of two-dimensional gel electrophoresis (2-DE) in proteomics, it continues to be an effective method for separating proteins according to their molecular weights and isoelectric points. Through the use of MS in conjunction with 2-DE, differential protein expression analysis can be made easier. For example, a two-dimensional differential equation analysis was performed on wheat that had been treated with cerium oxide nanoparticles (CeO2 NPs). The results showed that the proteins related to energy metabolism and oxidative stress responses had changed (Parisi et al. 2015). Few people are aware of the full potential of targeted proteomics, which includes techniques such as multiple reaction monitoring (MRM) and parallel reaction monitoring (PRM), when it comes to validating important proteins that are involved in NM–plant interactions. This is despite the fact that substantial research has been conducted. These techniques are especially helpful for establishing the function of particular antioxidant enzymes in plants that have been subjected to titanium dioxide nanoparticles (TiO2 NPs) (Kah et al. 2018). During the course of this assessment, major gaps have been highlighted that call for additional research.

One of these gaps is the investigation of post-translational modifications (PTMs) in plants that have been exposed to NM. Nano-silica, also known as SiO_2 nanoparticles, has been shown to modify the phosphorylation states of stress-responsive proteins in maize, hence influencing the signaling pathways that regulate plant responses to environmental stress (Servin and White 2015). This discovery was made according to advanced proteomics techniques. In general, the present body of literature offers a solid basis for investigating the incorporation of multi-omics methodologies, including proteomics, transcriptomics, and metabolomics, in order to achieve a comprehensive comprehension of the interactions between NM and plants.

Mukhopadhyay et al. (2021) conducted proteo-transcriptomic investigations on rice that had been treated with copper nanoparticles (CuNPs). The results of this research revealed the presence of co-regulated genes and proteins that are linked with stress signaling and nutrient absorption. Research on proteomics has been further improved by the increasing reliance on bioinformatics tools, which have made it possible to conduct data analysis on a massive scale. MaxQuant, Proteome Discoverer, and Perseus are examples of several types of software that can be used to facilitate the identification, quantification, and functional annotation of proteins. It is possible to discover biological processes that are affected by NMs through the use of functional enrichment analyses that make use of Gene Ontology (GO) and KEGG pathways (Kumar et al. 2021). Improved proteomics applications will continue to be made possible by future developments in mass spectrometry apparatus that have increased sensitivity and resolution. The accuracy of NM effect evaluations on plant systems will be improved through the combination of proteomics, machine learning, and predictive modelling, which will provide significant insights for the implementation of sustainable agriculture practices (Kah et al. 2018).

11 Case Studies and Insights from Proteomics

There have been substantial breakthroughs in the understanding of stress-responsive pathways, photosynthetic processes, and molecular adaptations as a result of the increased interest in proteomics as a method for researching plant responses to nanomaterials (NMs). In recent years, there has been a growing focus on understanding how proteome analyses might provide insights into these interactions. These insights have valuable implications for enhancing agricultural production and sustainability; hence, it is important to understand how these analyses can be used. Proteomic studies have become more relevant as a result of their capacity to evaluate the impacts of nanoparticles (NPs) on plants. These nanoparticles include zinc oxide nanoparticles (ZnO NPs), silver nanoparticles (AgNPs), and titanium dioxide nanoparticles (TiO2 NPs). These materials have been the subject of extensive research due to the fact that they have the potential to improve both stress tolerance and nutritional efficiency. Proteomics research conducted on rice (*Oryza sativa*) that had been treated with zinc oxide nanoparticles (ZnO NPs) found an increase in the expression of proteins that are involved in antioxidant defense and nutrient uptake, indicating that these proteins play a role in strengthening plant resilience (Raliya et al. 15). According to Dimkpa and Bindraban (2018), it was discovered that AgNPs have the ability to change the protein profiles of tomato plants (*Solanum lycopersicum*), hence increasing the levels of proteins that are related with disease resistance and the control of oxidative stress. There is a substantial body of research that has been published on the topic of antioxidant enzyme regulation in plants that have been treated with NM (Shahbaz et al. 2023). The generation of reactive oxygen species (ROS) is frequently triggered by exposure to NMs, which in turn necessitates the overexpression of antioxidant enzymes such as superoxide dismutase (SOD), catalase (CAT), and peroxidase (POD). Increasing the activity of these enzymes was demonstrated by proteomics research conducted on wheat that had been treated with cerium oxide nanoparticles (CeO2 NPs) (Parisi et al. 2015). This elevated activity significantly mitigated the effects of oxidative stress. Despite the fact that a great number of studies have investigated heat shock proteins (HSPs) in response to NM exposure, there is still a need for additional research to determine their precise function in stress mitigation.

According to Kumar et al.'s research from 2021, AgNPs have been shown to stimulate the development of HSP, which protects plants from the denaturation of proteins under detrimental conditions. In a similar vein, it has been demonstrated that ROS scavenging proteins, such as those that are impacted by nano-silica (SiO2 NPs), can improve maize's stress tolerance through the use of detoxifying mechanisms (Servin and White 2015). In spite of the significant study that has been conducted, very little is known about the influence that NMs have on photosynthetic proteins in various plant species. Based on the findings of Ghormade et al. (2011), it has been discovered that TiO_2 nanoparticles have the ability to improve the efficiency of photosynthetic processes in soybean (Glycine max) by enhancing chlorophyll production and upregulating light-harvesting complex proteins. Further, the enzyme known as ribulose-1,5-bisphosphate carboxylase/oxygenase (Rubisco), which plays a crucial role in the process of carbon fixation, was shown to be upregulated in rice that was treated with ZnO nanoparticles, which resulted in enhanced rates of photosynthetic activity (Kah et al. 2018). Additionally, it was demonstrated that CeO2 nanoparticles would improve the efficiency of energy conversion by enhancing proteins that are involved in the electron transport chain of the photosynthetic pathway (Raliya et al. 2015). During the course of this research, several major gaps have been highlighted that require future exploration. These gaps include the long-term effects of NMs on plant metabolism as well as their role in decreasing stress. According to Dimkpa and Bindraban (2018), NMs boost plant tolerance to abiotic stimuli such as drought, salinity, and heat by upregulating stress-responsive proteins. This results in increased yield stability under demanding conditions. In general, the present body of literature offers a solid basis for investigating the ways in which proteomics-based insights about the interactions between NM and plants might be utilized in precision agriculture.

According to Parisi et al. (2015), enhanced rates of photosynthetic protein expression result in increased levels of biomass production and improved nutrient utilization, both of which contribute to increased levels of agricultural output. By gaining an understanding of the molecular pathways that are influenced by nanomaterials, it is possible to produce nano-pesticides and nano-fertilizers that are more targeted, thereby lowering the negative impact on the environment while simultaneously boosting crop performance (Kumar et al. 2021). Integrating NMs, which is led by proteomics, helps to support sustainable agriculture by reducing the amount of chemical inputs, lowering the amount of pollution in the environment, and increasing the efficiency with which resources are used. According to the findings of proteomics research, for instance, nano-fertilizers maximize the intake of nutrients while simultaneously minimizing waste, which is in line with the objectives of global sustainability (Ghormade et al. 2011). A continuation of the investigation of the ways in which proteomics might improve NM applications for increased agricultural sustainability should be continued in future studies.

12 Conclusion

Nanomaterials present a promising frontier in modern agriculture, offering innovative approaches to plant nutrition, protection, and stress management. Their ability to enhance nutrient uptake, regulate biochemical pathways, and improve stress resilience positions them as valuable tools for sustainable crop production. However, while research has demonstrated significant benefits, challenges related to NM toxicity, regulatory compliance, and long-term ecological implications remain unresolved. Proteomics has proven instrumental in elucidating NM–plant interactions, shedding light on molecular adaptations that govern plant responses to nanomaterials. The identification of stress-related proteins, antioxidant mechanisms, and metabolic shifts provides a foundation for optimizing NM applications in agriculture. Nevertheless, discrepancies in research findings necessitate standardized methodologies and large-scale field trials to validate laboratory-based observations.

Moving forward, a balanced approach that incorporates scientific innovation with environmental stewardship is crucial. The development of biodegradable and non-toxic NMs, coupled with stringent regulatory guidelines, will ensure their responsible integration into agricultural systems. Collaborative efforts between researchers, policymakers, and industry stakeholders are essential to harness the full potential of nanotechnology while mitigating associated risks. By advancing our understanding of NM interactions with plants, the agricultural sector can achieve enhanced productivity without compromising environmental integrity.

References

Akhtar N, Ilyas N, Meraj TA, Aboughadareh AP, Sayyed RZ, Mashwani Z, Poczai P (2022) Improvement of plant responses by nanobiofertilizer: a step towards sustainable agriculture. Nano 12:965

Anbalagan SA, Appusamy S, Kumeresan PV, Chellappan GCS et al (2024) Deciphering the biocontrol potential of *Trichoderma asperellum* (Tv1) against *Fusarium*-nematode wilt complex in tomato. J Basic Microbiol e2400595:1–13. https://doi.org/10.1002/jobm.202400595

Bashan Y et al (2014) Advances in plant growth-promoting bacterial inoculant technology: formulations and practical perspectives (1998–2013). Plant Soil 378(1–2):1–33

Bastami A, Amirnia R, Sayyed R, Enshasy H (2021) The effect of mycorrhizal fungi and organic fertilizers on quantitative and qualitative traits of two important Satureja species. Agronomy 11:1285

Basu A, Prasad P, Das SN, Kalam S, Sayyed RZ, Reddy MS, El Enshasy H (2021) Plant growth promoting rhizobacteria (PGPR) as green bioinoculants: recent developments, constraints, and prospects. Sustainability 13:1140

Bright JP, Maheshwari HS, Thangappan S, Perveeen K, Bukhari NA, Mitra D, Sayyed RZ, Mantinu A (2025) Biofilmed-PGPR: a next-generation bioinoculant for plant growth promotion in rice (*Oryza sativa* L.) under changing climate. Rice Sci 32:10.1016/j.rsci.2024.08.008

Dimkpa CO, Bindraban PS (2018) Nano-fertilizers: new products for the industry? J Agric Food Chem 66(26):6462–6473

Faridvand S, Amirnia R, Tajbakhsh M, Enshasy HE, Sayyed RZ (2021) The effect of foliar application of magnetic water and nano, organic, and chemical fertilizers on phytochemical and yield characteristics of different landraces of fennel (*Foeniculum vulgare* mill). Horticulturae 7:475

Fazeli-Nasab B, Sayyed RZ, Piri R, Rahmani AF (2021) Biopriming and nanopriming: green revolution wings to increase plant yield, growth, and development under stress condition and forward dimensions. In: Antioxidants in plant-microbe interaction. Springer Singapore, Singapore, pp 623–655

Ferioun M, Zouitane I, Bouhraoua S, Elouattassi Y, Belahcen D, Errabbani A, Louahlia S, Sayyed R, El-Ghachtouli N (2025) Applying microbial biostimulants and drought-tolerant genotypes to enhance barley growth and yield under drought stress. Front Plant Sci 15:1494987. https://doi.org/10.3389/fpls.2024.1494987

Ghormade V, Deshpande MV, Paknikar KM (2011) Perspectives for nano-biotechnology enabled protection and nutrition of plants. Biotechnol Adv 29(6):792–803

Glare TR, Caradus JR, Gelernter WD, Jackson TA, Keyhani NO, Köhl J, Stewart A (2012) Have biopesticides come of age? Trends Biotechnol 30(5):250–258

Hamid B, Zaman M, Farooq S, Fatima S, Sayyed R, Baba Z, Sheikh T, Reddy M, El Enshasy H, Gafur A (2021) Bacterial plant biostimulants: a sustainable way towards improving growth, productivity, and health of crops. Sustainability 13:2856

Jakinala P, Lingampally N, Hameeda B, Sayyed RZ, Khan MY, Elsayed EA et al (2021) Silver nanoparticles from insect wing extract: biosynthesis and evaluation for antioxidant and antimicrobial potential. PLoS One 16(3):e0241729

James C (2018) Global status of commercialized biotech/GM crops: 2018. ISAAA Brief No. 54

Kah M, Kookana RS, Gogos A, Bucheli TD (2018) A critical evaluation of nano-pesticides and nano-fertilizers against their conventional analogues. Nat Nanotechnol 13(8):677–684

Kalam S, Basu A, Ahmad I, Sayyed RZ, El-Enshasy HA, Dailin DJ, Suriani NL (2020) Recent understanding of soil acidobacteria and their ecological significance: a critical review. Front Microbiol 11:580024

Karnwal A, Jassim AY, Mohammed AA, Sharma V, Al-Tawaha ARMS, Sivanesan I (2024) Nanotechnology for healthcare: plant-derived nanoparticles in disease treatment and regenerative medicine. Pharmaceuticals 17(12):1711

Khan S, Anjum MA, Nazir S, Arshad M (2020) Nano-fertilizers and their role in sustainable agriculture: a review. J Agric Sci 55(2):97–112

Kumar S, Singh RK, Tiwari KN, Pandey P (2021) Advances in nano-materials for combating plant diseases. Plant Dis 105(1):26–34

Lal R (2020) Managing soils for advancing the sustainable development goals. Soil Sci Soc Am J 84(1):1–21

Mukhopadhyay R, Ghosh M, Mukherjee A (2021) Role of nano-materials in soil health management: a comprehensive review. Environ Nanotechnol Monit Manag 15:100431

Parisi C, Vigani M, Rodríguez-Cerezo E (2015) Agricultural nano-technologies: what are the current possibilities? Nano Today 10(2):124–127

Praveen V, Sudha A, Amala SA, Ramjegathesh R, Kavitha S, Perveen K, Bukhari NA, Sayyed RZ (2024) Harnessing *Trichoderma asperellum*: tri-trophic interactions for enhanced black gram growth and root rot resilience. J Basic Microbiol:e2400569. https://doi.org/10.1002/jobm.202400569

Rai M, Ribeiro C, Mattoso L, Duran N (2021) Nanotechnology for plant protection. Springer International Publishing

Rajput VD, Singh A, Ghazaryan K, Movsesyan HS, Minkina TM, Al-Tawaha ARM et al (2024) Nanotechnology and biotechnology: transforming agriculture for food security and sustainability. In: Sustainable agriculture: nanotechnology, biotechnology, management and food security. Walter de Gruyter GmbH & Co KG, p 213

Raliya R, Nair R, Chavalmane S, Wang W-N, Biswas P (2015) Mechanistic evaluation of translocation and physiological impact of titanium dioxide and zinc oxide nanoparticles on the tomato (*Solanum lycopersicum* L.) plant. Metallomics 7(12):1584–1594

Rasouli H, Popović-Djordjević J, Sayyed RZ, Zarayneh S, Jafari M, Fazeli-Nasab B (2020) Nanoparticles: a new threat to crop plants and soil rhizobia? In: Sustainable agriculture reviews 41: nanotechnology for plant growth and development. Springer International Publishing, Cham, pp 201–214

Servin A, White JC (2015) Nanotechnology for sustainable agriculture: improvements in plant growth and protection. Front Plant Sci 6:249

Sethi G, Behera KK, Sayyed R, Adarsh V, Sipra BS, Singh L, Alamro AA, Behera M (2025) Enhancing soil health and crop productivity: the role of zinc-solubilizing bacteria in sustainable agriculture. Plant Growth Regul. https://doi.org/10.1007/s10725-025-01294-7

Shahbaz M, Akram A, Mehak A, Haq E, Fatima N, Wareen G, Aamir S, Hassan H, Poczai P, Sayyed RZ, Ilyas N (2023) Evaluation of selenium nanoparticles in inducing disease resistance against spot blotch disease and promoting growth in wheat under biotic stress. Plants 12(4):761. https://doi.org/10.3390/plants12040761

Sharma S, Kumari P, Shandilya M, Thakur S, Perveen K, Sheikh I, Ahmed Z, Sayyed R, Mastinu A (2025) The combination of α-Fe_2O_3NP and *Trichoderma* sp. improves antifungal activity against *Fusarium* wilt. J Basic Microbiol:e2400613. https://doi.org/10.1002/jobm.202400613

Singh A, Rajput VD, Agrawal S, Ghazaryan K, Minkina T, Al Tawaha ARM et al (2024) Nanoparticles mediated salt stress resilience: a holistic exploration of physiological, biochemical, and Nano-omics approaches. Rev Environ Contam Toxicol 262(1):19

Tasnim A, Roy A, Akash SR, Ali H, Habib MR, Barasarathi J, Muthukumaran M, Sayyed R, Yeasmin T (2024) Hibiscus sabdariffa L. petal biomass: a green source of nanoparticles of multifarious potential. Open Agric 9:20220332. https://doi.org/10.1515/opag-2022-0332

Thakur P, Thakur S, Kumari P, Shandilya M, Sharma S, Poczai P, Alarfaj AA, Sayyed RZ (2022) Nano-insecticide: synthesis, characterization, and evaluation of insecticidal activity of ZnO NPs against *Spodoptera litura* and *Macrosiphum euphorbiae*. Appl Nanosci 12:3835–3850. https://doi.org/10.1007/s13204-022-02530-6

Vinothini K, Nakkeeran S, Saranya N, Jothi P, Richard JI, Perveen K, Bukhari NA, Glick BR, Sayyed RZ, Mastinu A (2024) Rhizosphere engineering of biocontrol agents enriches soil microbial diversity and effectively controls root-knot nematode. Microb Ecol 87:120. https://doi.org/10.1007/s00248-024-02435-7

Employing Metabolomics to Systematically Investigate How Nanoparticles Influence the Physio-Biochemical Responses of Plants Under Stress

11

Virgilio Gavicho Uarrota

Abstract

Metabolomics serves as a crucial instrument for scientists employing advanced post-genomics methodologies to uncover metabolite networks and their interrelationships. It pertains to the examination of the metabolome, which refers to the array of small molecules synthesized by cells that govern the metabolic functions of an organism. This concept has emerged and progressed significantly over the past decade. It encompasses the thorough analysis of metabolites within a biological specimen, or the systematic identification and quantification of all metabolites present within a specific organism or biological sample. As direct indicators of metabolic processes, metabolomics aligns more closely with phenotypic expressions when compared to genomics, transcriptomics, and proteomics. Researchers can explore the metabolome through both targeted and untargeted methodologies. Untargeted metabolomics, often referred to as "global metabolomics," employs an unbiased screening approach to detect thousands of metabolites in a single analytical run, facilitating exploratory investigations of unidentified metabolites. In contrast, targeted metabolomics concentrates on quantifying a specific set of metabolites relevant to the pathway of interest. The domain of small molecule biochemistry in plant research has advanced significantly, attributed to the enhanced resolution afforded by techniques such as nuclear magnetic resonance (NMR) spectroscopy and mass spectrometry (MS), along with LC-MS, GC-MS, and robust chemometric analytical software. These methodologies enable the concurrent analysis and comparison of thousands of compounds. Nanomaterials are attracting global interest due to their ability to protect plant growth from environmental stressors, including drought, salinity,

V. G. Uarrota (✉)
Instituto de Ciencias Agroalimentarias, Animales y Ambientales (ICA3),Campus Colchagua, Universidad de O'Higgins, San Fernando, Chile
e-mail: virgilio.gavicho@uoh.cl

R. Sayyed et al. (eds.), *Nanobiostimulants in Innovative Agriculture*,
https://doi.org/10.1007/978-981-95-2477-8_11

heavy metal toxicity, extreme temperatures, and flooding. In light of current and forecasted challenges in sustainable agricultural production, nanoparticles (NPs) are viewed as viable and efficient approaches to improving crop yields by increasing a plant's resilience to abiotic stress conditions. Nanomaterials have been connected to plant biochemical, physiological, and molecular responses to drought stress, and they have been used as a strategy to boost plant temperature resilience. This chapter covers and updates the use of metabolomic technologies to investigate the physio-biochemical reactions of stressed plants.

Keywords

Metabolomic tools · Nanoparticles · Plant responses · Stress conditions

1 Introduction

1.1 Metabolomics and Nanoparticles' Use in Agriculture

Metabolomics is characterized as a comprehensive analysis of metabolites present in a biological sample (Clish 2015; Doroghazi et al. 2014; Barkal et al. 2016; Wen et al. 2014). The primary methodologies employed encompass nuclear magnetic resonance (NMR) and gas chromatography coupled with mass spectrometry (GC-MS). Additional techniques utilized include high-performance liquid chromatography (HPLC), ultraviolet (UV) spectroscopy, gas chromatography-time-of-flight/time-of-flight (GC-TOF/TOF), liquid chromatography-time-of-flight (LC-TOF), ultra-performance liquid chromatography coupled with mass spectrometry (UPLC-FMS), liquid chromatography-tandem mass spectrometry (LC-MS/MS), ultra-high-performance liquid chromatography-mass spectrometry (UHPLC-MS/MS), and gas chromatography-time-of-flight mass spectrometry (GC-TOF-MS). Recent advancements in mass spectrometry (MS) have provided valuable insights into the regulation of cellular metabolic processes. While NMR quantifies metabolites in the micromolar range, mass spectrometry is capable of identifying concentrations as minute as pico- and nanomolar. The predominant technique for the analysis of polar and non-polar metabolites is liquid chromatography-mass spectrometry (LC-MS), while gas chromatography-mass spectrometry (GC-MS) is utilized for volatile and primary metabolites following derivatization, alongside capillary electrophoresis-mass spectrometry (CE-MS) for polar and charged metabolites (Cajka and Fiehn 2015). Though metabolomics can be executed through a range of technological platforms, liquid chromatography/mass spectrometry (LC/MS) remains the preferred choice for untargeted discovery experiments, as it frequently identifies thousands of unique signals from the metabolic extracts of biological samples (Zamboni et al. 2015). Fig. 11.1 provides a concise overview of the applications of metabolomics in agriculture (CP 2025; Zhou et al. 2022; Yadav et al. 2023).

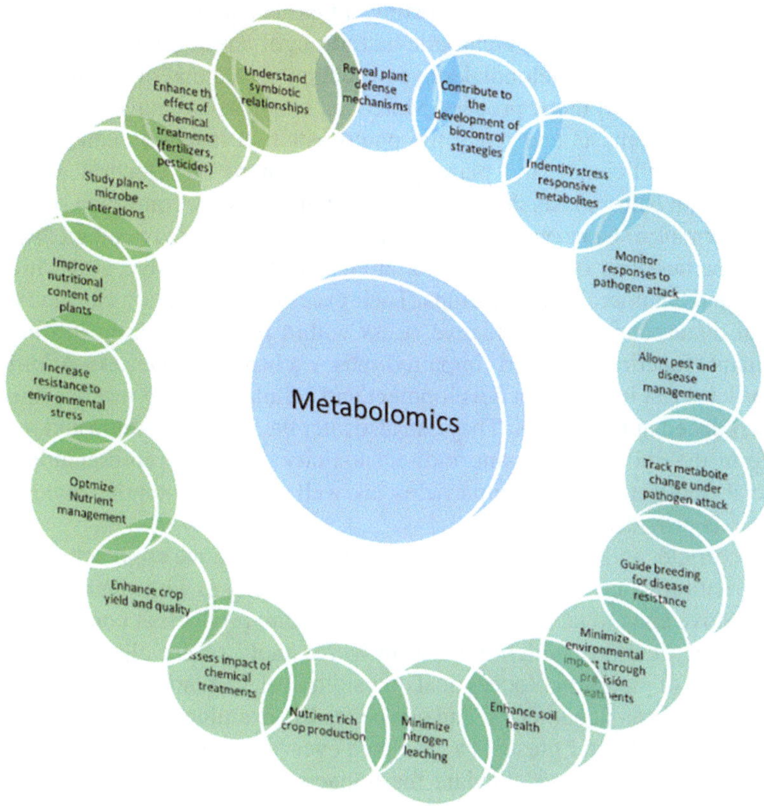

Fig. 11.1 Summarized applications of metabolomics in agriculture

Nanotechnology, which has emerged as a significant area of focus, finds extensive applications in the food, healthcare, and agriculture industries. Various nanoparticles, such as titanium dioxide, iron oxide, zinc oxide, silicon oxide, copper nanoparticles, and selenium nanoparticles, have garnered considerable interest lately due to their beneficial uses in agricultural practices. While nanoparticles can facilitate the growth and development of plants, their impacts are influenced by factors including the type of crop, nanoparticle size, concentration, and the timing of application. Recently, nanomaterials have gained considerable attention because of their distinctive physical and chemical characteristics, which enable their deployment in various applications. It has become increasingly clear that the integration of nanomaterials in agriculture is essential for enhancing crop yields while also preserving environmental health. There are several approaches to utilizing nanoparticles for the enhancement of crop growth, such as seed enhancement, leaf spraying, and root treatment. It would be intriguing to investigate the effective application of nanoparticles and their potential ability to penetrate various plant tissues, from roots to shoots. Recently, nanoparticles have been shown to enhance plant tolerance to both biotic and abiotic stresses by boosting antioxidant activity and shielding plants from oxidative damage. They can alleviate drought stress effects by diminishing hydrogen peroxide and malondialdehyde production, thereby maintaining the efficiency of the photosynthetic process. Moreover, nanoparticles have the capability to enter plant chloroplasts and reach the photosystem II reaction center to enhance photosynthetic efficiency and plant growth. By improving electron transfer and light capture in chloroplasts during drought conditions, they contribute positively to plant performance. Concerns have been raised regarding the toxicological and ecological repercussions of nanoparticles, alongside their extensive usage in diverse products and commercial applications. Their excessive application can lead to physiological disturbances and oxidative stress within plants, resulting in diminished antioxidant functionality and impaired gas exchange processes. Additionally, nanoparticles can impact root development and cellular division mechanisms, leading to a reduced mitotic index. They can also alter the growth medium and microbial activity in the soil, thus causing indirect toxicity. A visual representation of the beneficial applications of nanoparticles, as well as their potential toxicity when overused, is provided in Fig. 11.2.

Nanoparticles (NPs) are considered viable and innovative approaches for modifying agricultural output by enhancing a plant's capacity to withstand abiotic stress. This objective aims to tackle both existing and future production constraints within sustainable agriculture, which arise from increasingly severe climatic conditions. The reactions of plants on biochemical, physiological, and molecular levels to nanomaterials have been documented and can be readily examined through metabolomic techniques. This chapter presents a refreshed overview of studies utilizing nanomaterials as effective tools to alleviate the impacts of various abiotic stress factors. Figure 11.3 illustrates the applications of nanoparticles in reducing different forms of stress.

2 Nanotechnology as a Promising Tool for Boosting Plants' Capacity to Withstand Stress

Adverse changes in the environment have significantly impacted the global agricultural industry. The abiotic surroundings of plants have been altered due to climate change, influencing their physiological processes, development, and overall productivity. The abiotic stress faced by plants poses a substantial challenge to both global agricultural output and food security. To tackle these issues and ensure sustainability, the development of new approaches is imperative (Al-Khayri et al. 2023). Nanotechnology represents a groundbreaking method for boosting agricultural output, utilizing nanoscale materials like nanopesticides, nanofungicides, nanoherbicides, and nanofertilizers. The ability of nanoparticles to penetrate cellular barriers makes them suitable for agricultural applications. Because of their reduced size, enhanced solubility, and efficiency in plant absorption, nanoparticles are gaining traction as a contemporary agricultural innovation. Research has shown that the

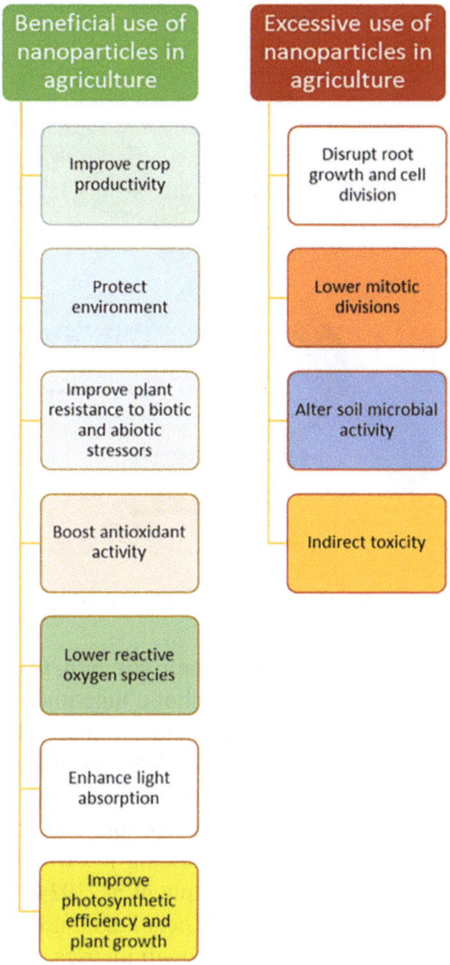

Fig. 11.2 Brief summary of the beneficial uses of nanoparticles in agriculture and the consequences of excessive use

application of nanoparticles in crop cultivation under various biotic and abiotic stress conditions improves both the quality and quantity of crop yields (Al-Khayri et al. 2023).

2.1 The Role of Nanoparticles on Salt-Stress Tolerance in Plants

Water scarcity, exacerbated by climate change, compels agricultural sectors globally to adopt saline water for irrigation, which in turn elevates soil salinity levels. Salinity, defined as the excessive build-up of salts in the soil, poses a critical challenge to contemporary agriculture, as it ultimately obstructs plant growth and development, potentially leading to plant mortality (Al-Khayri et al. 2023). The presence

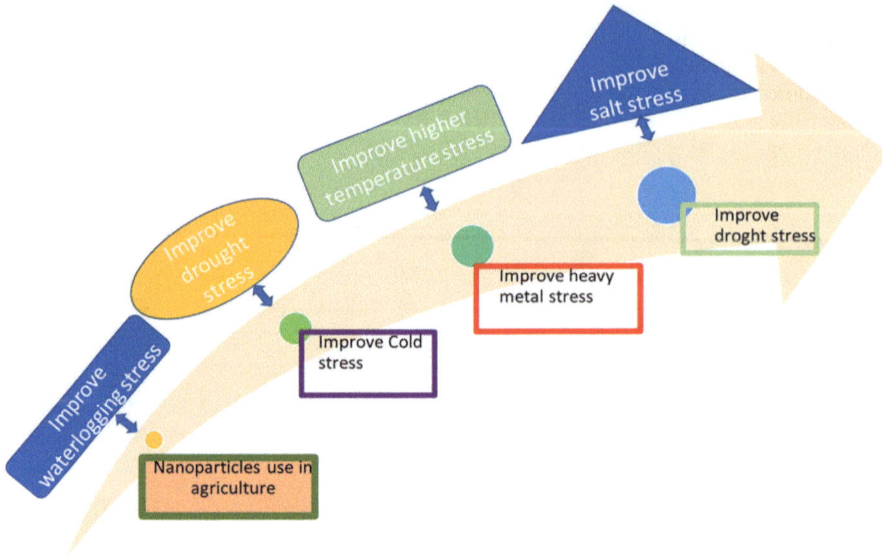

Fig. 11.3 The effect of applying nanoparticles on different types of abiotic stress

of salt stress modifies ionic strength, thereby impacting a multitude of biochemical, physiological, and metabolic functions. It also incites osmotic pressure, water deficit, oxidative challenges, nutritional imbalance, and retarded cellular division. The application of Zinc Oxide Nanoparticles (ZnO NPs) has been shown to mitigate salt stress by enhancing indicators of physiological development. Through the regulation of plant nutrient equilibrium and chlorophyll fluorescence, these nanoparticles foster growth while reducing the formation of reactive oxygen species (ROS) in Phaseolus vulgaris (Gupta et al. 2024). In Pisum sativum, treatment with 50 ppm ZnO NPs yielded both beneficial and detrimental effects, including extended root length and improved growth metrics, alongside reduced levels of malondialdehyde (MDA) and hydrogen peroxide. The combination of 50 mM NaCl with 50 ppm ZnO NPs positively influenced pea physiology, markedly increasing root length and improving other physiological parameters. Additionally, this treatment resulted in decreased concentrations of MDA, glycine betaine, and hydrogen peroxide. Research conducted by Pirzada et al. (2024) examined the synergistic effects of ascorbic acid (AsA), silver nanoparticles (NPs), and aqueous leaf extract of Salvadora oleoides on the salinity tolerance of wheat. Under conditions free from stress, the pairing of 40 NPs and 5 AsA proved most effective in enhancing chlorophyll levels and 1,1-diphenyl-2-picrylhydrazyl (DPPH) activity, while the 10% aqueous leaf extract of Salvadora oleoides significantly improved superoxide dismutase (SOD) activity and alleviated lipid peroxidation during saline stress (Heikal et al. 2022). Nanoparticles were noted to lower the activity of antioxidant enzymes while increasing levels of peroxidase (POD), proline, and sugars in Triticum aestivum (Mohamed et al. 2017). Alabdallah and Alzahrani (2020) reported enhancements in catalase (CAT) and SOD activity, alongside increased photosynthetic pigments and reduced amounts of total soluble sugars and proline in Abelmoschus esculentus. Furthermore, decreases in proline and MDA levels, diminished antioxidant enzyme activity, expanded leaf area, and elevated contents of anthocyanins, flavonoids, and phenolic compounds, alongside enhanced activities of catalase, glutathione reductase, ascorbate peroxidase, and guaiacol peroxidase enzymes have also been documented. Nanoparticles can alleviate salinity stress by lowering Na + and Cl- concentrations while enhancing levels of nitrogen, phosphorus, potassium, calcium, magnesium, iron, zinc, total chlorophyll, carotenoids, proline, carbohydrates, crude proteins, and both enzymatic and non-enzymatic antioxidants. They also increase non-enzymatic substances, bolster antioxidant activities, and potassium content, while decreasing production of reactive oxygen species, Na + concentration, and lipid peroxidation (Heikal et al. 2022; Askary et al. 2017). In Medicago species, potassium nanoparticles were observed to decrease electrolyte leakage and elevate catalase activity (El-Sharkawy et al. 2017). Silver nanoparticles (AgNPs) were found to reduce MDA levels, enhance antioxidant capabilities, and minimize Na + uptake in the leaves of Pearl Millet (Khan et al. 2020), while cerium oxide facilitated photosynthetic processes and nutrient absorption in Brassica napus (Rossi et al. 2017; Al-Khayri et al. 2023). A comprehensive examination of the role of nanoparticles in mitigating salinity stress can be found in the study by Sarkar and Kalita (2023).

2.2 The Role of Nanoparticles on Drought-Stress Tolerance in Plants

Nanoparticles (NPs) enhance the hydraulic conductivity of plant roots and facilitate water absorption, thereby improving plants' resistance to drought conditions. They induce a range of morphological, physiological, and biochemical changes in vegetation, along with variations in protein levels associated with oxidation-reduction, reactive oxygen species detoxification, stress response signaling, and hormonal pathways (Du et al. 2011). Research has indicated that the topical application of metal oxide nanoparticles, including iron oxide (Fe3O4), zinc oxide (ZnO), and titanium dioxide (TiO2), positively influences the physiological and metabolic functions of plants experiencing drought stress (Alabdallah et al. 2021). In experiments with drought-stressed pomegranate plants treated with Si NPs, improvements were observed in photosynthetic pigments, nutrient levels, physical and chemical characteristics (especially those related to fruit cracking), phenolic content, and concentrations of osmolytes, antioxidant enzymes, and abscisic acid (Zahedi et al. 2021). El-Zohri et al. (2021) reported that topical application of green ZnO NPs at lower concentrations effectively enhances the drought resistance of tomato plants. Furthermore, green-synthesized Fe3O4 NPs, along with nanofertilizers, have been shown to alleviate drought stress impacts on fenugreek plants (Bishta et al. 2022). Potter et al. (2021) discovered that the advantageous effects of nanoparticles in enhancing plant drought tolerance are only evident under particular environmental circumstances.

2.3 The Role of Nanoparticles on Cold-Stress Tolerance in Plants

Nanoparticles (NPs) can enhance the ability of plants to endure cold stress by fortifying cellular structures, boosting photosynthetic efficiency, and reducing oxidative damage (Al-Khayri et al. 2023). For instance, nanoparticles like graphene oxide can reinforce cellular walls and improve membrane flexibility (Zaman et al. 2025), elevate chlorophyll levels, and enhance the photosynthesis rate (Khalid et al. 2022), as well as eliminate reactive oxygen species (ROS) while stimulating antioxidant defense systems. They also encourage the accumulation of osmolytes, which effectively reduces the freezing point of fluids within cells (Hayat et al. 2023) and maintains metabolic stability in cooler environments (Zaman et al. 2025).

2.4 The Role of Nanoparticles on Heavy Metal-Stress Tolerance in Plants

The usage of nanoparticles (NPs) on plant surfaces or within the soil can mitigate the adverse consequences associated with heavy metal (HM) stress, enhance vegetation growth and photosynthetic processes, as well as diminish toxicity due to oxidative stress (El-Saadony et al. 2022). As stated by Asgher et al. (2024), the application of silicon nanoparticles during heavy metal exposure can strengthen the plant's antioxidant systems and promote the biosilification of cell walls through the formation of phytoliths. This includes increasing the concentration of osmolytes such as proline and glycine-betaine, reducing the transfer of metals from the roots to the shoots, activating the OVPI transporter that facilitates vacuolar sequestration of heavy metals, and suppressing the expression of transporter-related genes (NRAMP5, HMA2/3, LCT1). As noted by Faizan et al. (2023), the responses of plants to heavy metals mediated by nanoparticles encompass the induction of systemic resistance, enhanced root development, improved nutrient absorption, diminished heavy metal toxicity, promotion of phytohormone production, and the triggering of antioxidant enzyme generation.

Elevated concentrations of heavy metals (HMs) within the soil can detrimentally affect agricultural yield by diminishing crop outputs and modifying the nutritional composition of harvested grains. Additionally, HMs may adversely influence reproductive processes, leading to the production of smaller seeds and a reduction in overall grain yield. Research has demonstrated that cadmium can impair pollen viability as well as seed output. Moreover, contamination from heavy metals may alter the nutritional quality of grains, potentially causing deficiencies in micronutrients for individuals whose diets predominantly consist of grains. Consequently, there is a depletion of essential metals such as iron, manganese, and zinc (Faizan et al. 2023). Nanoparticles possess the capability to decrease the mobility and bioaccumulation of HMs through their capacity to absorb and modify them within soil environments. The incorporation of Fe3O4 NPs has led to a reduction of cadmium availability in soil (Wang et al. 2020). Hydroxyapatite nanoparticles (NPs) can help in managing soil pH and minimizing detrimental effects from metals by releasing phosphate ions (Cui et al. 2018). In addition, NPs can promote the formation of apoplastic barriers, which in turn lowers the concentration of heavy metals in roots. Certain nanoparticles are effective in modulating the expression of plant metal transporter genes, thus preventing the translocation of heavy metals by forming stable complexes with them (Wang et al. 2021). According to the findings of Cui et al. (2017) and Wu et al. (2017), nanoparticles such as SiNPs facilitate the production of organic acids that alleviate HM stress effects. NPs enhance the antioxidant mechanisms, leading to a reduction in stress caused by reactive oxygen species (ROS). The ramifications of HMs on plant growth and metabolic processes encompass diminished biomass, weakened cellular growth, and metabolism, increased chlorophyll levels, enhanced photosynthetic efficiency, reduced accumulation of ROS, inhibited growth, and nutritional imbalances. They also disrupt the electron transport chain from PSII, alter the ultrastructure of mesophyll cells, affect starch grain quantity within chloroplasts, and escalate levels of malondialdehyde, reactive carbonyl species, and H2O2. Furthermore, seed germination is inhibited alongside a decrease in rubisco activity. Chromium supplementation has been shown to adversely affect plant height, leaf number, and overall biomass. It leads to elevated intracellular ROS, negatively impacting membrane potential and resulting in solute leakage (Al-Khayri et al. 2023).

2.5 The Role of Nanoparticles on Waterlogging-Stress Tolerance in Plants

Nanoparticles have demonstrated efficacy in aiding plants experiencing flooding conditions. Research conducted by Mustafa and colleagues in 2015 and 2016 revealed that silver nanoparticles enhanced the growth of soybean plants under flood stress by influencing proteins, amino acids, glycolysis, and wax synthesis. Another investigation focused on soybean plants facing flooding stress utilized Al2O3 nanoparticles to mitigate growth restrictions caused by this stressor. The Al2O3 nanoparticles inhibited proteins associated with glycolysis, facilitated the scavenging of reactive oxygen species by boosting the ascorbate/glutathione pathway, promoted the production of ribosomal proteins, and contributed to the elongation of roots and hypocotyls (Al-Khayri et al. 2023).

2.6 The Role of Nanoparticles on Higher Temperature-Stress Tolerance in Plants

Nanoparticle treatment significantly improves the anatomical modifications of plants, enabling them to endure challenging environmental conditions. Prior to the onset of heat stress, alfalfa seedlings were treated with ZnO nanoparticles to mitigate heat-related injury and facilitate plant growth. The administration of 90 mg/L of ZnO nanoparticles before the initiation of heat stress proved to be more effective in addressing the ultrastructural changes in chloroplasts, mitochondria, and cell walls induced by high temperatures. Additionally, this treatment substantially reduced damage to mesophyll cells and chloroplasts (Sidhu et al. 2024). The distinctive physicochemical characteristics of nanoparticles allow them to influence plant reactions to thermal stress. Temperature stress, resulting from either heat or cold, can considerably affect the growth, development, and overall performance of plants. Research indicates that when nanoparticles are applied to plants experiencing temperature stress, they can trigger both beneficial and detrimental responses. On the positive side, certain nanoparticles have demonstrated the ability to enhance plants' resilience to extreme temperatures by acting as stress mitigators. These nanoparticles may improve the absorption of water and nutrients, enhance photosynthetic efficiency, and strengthen antioxidant defenses, thus enabling plants to cope with temperature-related challenges. Conversely, some nanoparticles might exacerbate the detrimental effects of temperature stress on plants. They could disrupt cellular processes, compromise membrane integrity, or induce oxidative stress, leading to stunted growth and reduced resistance to extreme temperatures (Sidhu et al. 2024). Physiological changes induced by nanoparticles under elevated temperature stress encompass the regulation of PSII, an increase in water conductance and transpiration rates, elevated levels of carotenoids and chlorophyll, enhanced stomatal conductance, improved net photosynthesis, better plant growth and flower quality, increased pollen germination, seed set percentage, seed yield, as well as greater fresh and dry weight of plants, along with extended root and shoot lengths, increased leaf area, and a higher number of leaves.

3 Metabolomics as a Powerful Tool to Analyze How Nanoparticles Influence the Plant Response to Stress

Metabolomics serves as a valuable approach for examining the influence of nanoparticles on the stress responses of plants. As noted by Silva et al. (2023), this method facilitates the identification of affected metabolites and alterations in metabolic pathways, analyzes the chemical composition of botanical samples, identifies metabolites that experience elevation during stress, uncovers biomarkers associated with diseases or stressors, and enhances understanding of the biochemical reactions of plants under stress (Dawid and Hille 2018). Furthermore, it reveals the specific metabolic pathways targeted by nanoparticles and elucidates the implications of these particles on plant biochemistry and physiology (Manickam et al. 2023). Additionally, it sheds light on how plants respond to abiotic stress factors such as salinity, drought, and elevated temperatures. The presence of nanoparticles influences the levels of secondary metabolites in plants by altering the production of reactive nitrogen and oxygen species, modifying gene expression, and affecting signaling pathways (Selvakesavan et al. 2023). A comprehensive study involving floating seedling cultures of Hypericum perforatum L. explored whether the introduction of nanoparticles to the root system impacts secondary metabolism within the plants. Physiological and metabolomic analyses indicated that Fe3O4 and ZnO nanoparticles mitigate Cd toxicity in tobacco plants, as observed through a UHPLC system (Zou et al. 2022). The influence of silver nanoparticles on the metabolic profiles of two transgenic soybean varieties was examined through an NMR-based metabolomics strategy. To evaluate the impact of abiotic stress on soybean plants, physiological assessments, anatomical investigations utilizing microscopy, and studies of metabolic profiles were conducted using 1H NMR spectroscopy (Quintela et al. 2024). Yu et al. (2023) executed an integrated transcriptomics and metabolomics analysis to assess the overall effect of nano-titanium dioxide on mulberry seedlings using ICP-5000 and Focused Photonics technology. Additionally, mass spectrometry-based metabolomics was employed to study the absorption of silver nanoparticles in Arabidopsis thaliana (Soria et al. 2017). Rico et al. (2022) also explored alterations in plant metabolite compositions following exposure to synthetic nanoparticles. HPLC and UV-visible spectroscopy techniques were utilized to confirm the induction of secondary metabolites in callus cultures of Artemisia annua L. in response to nanoparticle-induced stress (Fatima et al. 2021). The findings presented highlight a growing trend in the integration of nanoparticles with metabolomic methodologies to enhance the understanding of both biotic and abiotic stressors across various plant species.

4 Conclusion and Future Perspectives

Metabolomics provides direct indicators of metabolic processes, facilitating the connection with different symptoms. Numerous nanoparticles can be employed to enhance the resilience of plants due to their safe applications in agricultural settings. While nanoparticles have beneficial effects on the growth and development of plants, these outcomes differ based on factors such as the origin, size, concentration, and timing of application of the crops; nonetheless, they have proven advantageous in farming for tackling various abiotic stress factors. Subsequent research needs to concentrate on investigating the potential of nanoparticles to alleviate ongoing climate change impacts, especially concerning certain types of fruits and vegetables.

Acknowledgments The author thanks Universidad de O'Higgins, Instituto de Ciencias Agroalimentarias, Animales y Ambientales (ICA3), for research facilities.

References

Alabdallah NM, Alzahrani HS (2020) The potential mitigation effect of ZnO nanoparticles on [*Abelmoschus esculentus* L. Moench] metabolism under salt stress conditions. Saudi J Biol Sci 27:3132–3137

Alabdallah NM, Hasan M, Hammami I, Alghamdi AI, Alshehri D, Alatawi HA (2021) Green synthesized metal oxide nanoparticles mediate growth regulation and physiology of crop plants under drought stress. Plants 10:1730

Al-Khayri JM, Rashmi R, Surya Ulhas R, Sudheer WN, Banadka A, Nagella P, Aldaej MI, Rezk AA, Shehata WF, Almaghasla MI (2023) The role of nanoparticles in response of plants to abiotic stress at physiological, biochemical, and molecular levels. Plants (Basel) 12(2):292

Asgher M, Rehaman A et al (2024) Multifaceted roles of silicon nano particles in heavy metals-stressed plants. Environ Pollut 341:122886

Askary M, Talebi SM, Amini F, Bangan AD (2017) Effects of iron nanoparticles on Mentha piperita L. under salinity stress. Biologija 63:65–75
Barkal LJ, Theberge AB, Guo CJ, Spraker J, Rappert L, Berthier J, Brakke KA, Wang CC, Beebe DJ, Keller NP, Berthier E (2016) Microbial Metabolomics in open microscale platforms. Nat Commun 7(1). https://doi.org/10.1038/ncomms10610
Bishta S, Sharma V, Kumari N (2022) Biosynthesized magnetite nanoparticles from Polyalthia longifolia leaves improve photosynthetic performance and yield of Trigonella foenum-graecum under drought stress. Plant Stress 5:100090
Cajka T, Fiehn O (2015) Toward merging untargeted and targeted methods in mass spectrometry-based metabolomics and Lipidomics. Anal Chem 88(1):524–545. https://doi.org/10.1021/acs.analchem.5b04491
Clish CB (2015) Metabolomics: An emerging but powerful tool for precision medicine. Mole Case Stud 1(1). https://doi.org/10.1101/mcs.a000588
Creative Proteomics (CP) (2025) Metabolomics in agriculture: transforming sustainability and crop quality. https://www.creative-proteomics.com/resource/metabolomics-agriculture-sustainability-crops.htm. Accessed June 2025
Cui J, Liu T, Li F, Yi J, Liu C, Yu H (2017) Silica nanoparticles alleviate cadmium toxicity in rice cells: mechanisms and size effects. Environ Pollut 228:363–369
Cui H, Shi Y, Zhou J, Chu H, Cang L, Zhou D (2018) Effect of different grain sizes of hydroxyapatite on soil heavy metal bioavailability and microbial community composition. Agric Ecosyst Environ 267:165–173
Dawid C, Hille K (2018) Functional metabolomics – a useful tool to characterize stress-induced metabolome alterations opening new avenues towards tailoring food crop quality. Agronomy 8(8):138
Doroghazi JR, Albright JC, Goering AW, Ju KS, Haines RR, Tchalukov KA, Labeda DP, Kelleher NL, Metcalf WW (2014) A roadmap for natural product discovery based on large-scale genomics and Metabolomics. Nat Chem Biol 10(11):963–968. https://doi.org/10.1038/nchembio.1659
Du W, Sun Y, Ji R, Zhu J, Wu J, Guo H (2011) TiO2 and ZnO nanoparticles negatively affect wheat growth and soil enzyme activities in agricultural soil. J Environ Monit 13:822–828
El-Saadony MT, Saad AM, Soliman SM, Salem HM, Desoky ESM, Babalghith AO, El-Tahan AM, Ibrahim OM, Ebrahim AA, Abd El-Mageed TA, Elrys AS, Elbadawi AA, El-Tarabily KA, AbuQamar SF (2022) Role of nanoparticles in enhancing crop tolerance to abiotic stress: A comprehensive review. Front Plant Sci 13. https://doi.org/10.3389/fpls.2022.946717
El-Sharkawy MS, El-Beshsbeshy TR, Mahmoud EK, Abdelkader NI, Al-Shal RM, Missaoui AM (2017) Response of alfalfa under salt stress to the application of potassium Sulfate nanoparticles. Am J Plant Sci 8:1751–1773
El-Zohri M, Al-Wadaani NA, Bafeel SO (2021) Foliar sprayed green zinc oxide nanoparticles mitigate drought-induced oxidative stress in tomato. Plants 10:2400
Faizan M, Alam P, Rajput VD, Faraz A, Afzal S, Ahmed SM, Yu F-Y, Minkina T, Hayat S (2023) Nanoparticle mediated plant tolerance to heavy metal stress: ≌what we know? Sustainability 15(2):1446
Fatima K, Abbas SR, Zia M, Sabir SM, Khan RT, Khan AA, Hassan Z, Zaman R (2021) Induction of secondary metabolites on nanoparticles stress in callus culture of artemisia annua L. Braz J Biol 81(2):474–483
Gupta A, Bharati R, Kubes J, Popelkova D, Praus L, Yang X, Severova L, Skalicky M, Brestic M (2024) Zinc oxide nanoparticles application alleviates salinity stress by modulating plant growth, biochemical attributes and nutrient homeostasis in phaseolus vulgaris L. Front Plant Sci 15. https://doi.org/10.3389/fpls.2024.1432258
Hayat F, Khanum F, Li J, Iqbal S, Khan U, Javed HU, Razzaq MK, Altaf MA, Peng Y, Ma X, Li C, Panfeng T, Chen J (2023) Nanoparticles and their potential role in plant adaptation to abiotic stress in horticultural crops: a review. Sci Hortic 321:112285
Heikal YM, El-Esawi MA, El-Ballat EM, Abdel-Aziz HMM (2022) Applications of nanoparticles for mitigating salinity and drought stress in plants: an overview on the physiological, biochemical and molecular genetic aspects. N Z J Crop Hortic Sci 51(3):297–327

Khalid MF, Iqbal Khan R, Jawaid MZ, Shafqat W, Hussain S, Ahmed T, Rizwan M, Ercisli S, Pop OL, Alina MR (2022) Nanoparticles: the plant saviour under abiotic stresses. Nanomaterials (Basel) 12(21):3915

Khan I, Raza MA, Awan SA, Shah GA, Rizwan M, Ali B, Tariq R, Hassan MJ, Alyemeni MN, Brestic M et al (2020) Amelioration of salt induced toxicity in pearl millet by seed priming with silver nanoparticles (AgNPs): the oxidative damage, antioxidant enzymes and ions uptake are major determinants of salt tolerant capacity. Plant Physiol Biochem 156:221–232

Manickam S, Rajagopalan VR, Kambale R, Rajasekaran R, Kanagarajan S, Muthurajan R (2023) Plant metabolomics: current initiatives and future prospects. Curr Issues Mol Biol 45(11):8894–8906

Mohamed AKSH, Qayyum MF, Abdel-Hadi AM, Rehman RA, Ali S, Rizwan M (2017) Interactive effect of salinity and silver nanoparticles on photosynthetic and biochemical parameters of wheat. Arch Agron Soil Sci 63(12):1736–1747

Pirzada AM, Anwar T, Qureshi WA, Qureshi H, Siddiqi EH, Zaman W, Soufan W (2024) Salinity stress mitigation in wheat through synergistic application of ascorbic acid, nanoparticles and Salvadora oleoides extract. Sci Rep 14(1). https://doi.org/10.1038/s41598-024-76194-3

Potter M, Deakin J, Cartwright A, Hortin J, Sparks D, Anderson AJ, McLean JE, Jacobson A, Britt DW (2021) Absence of nanoparticle-induced drought tolerance in nutrient sufficient wheat seedlings. Environ Sci Technol 55:13541–13550

Quintela AL, Santos MF, de Lima RF, Mayer JL, Marcheafave GG, Arruda MA, Tormena CF (2024) Influence of silver nanoparticles on the metabolites of two transgenic soybean varieties: an NMR-based metabolomics approach. J Agric Food Chem 72(21):12281–12294

Rico CM, Ofoegbu PC, Kirwa NJ, Wagner DC et al (2022) Chapter 4 – changes in metabolite profile of plants exposed to engineered nanomaterials. In: Rico CM (ed) Nanomaterial-plant interactions, plant exposure to engineered nanoparticles, vol 3. Academic Press, pp 119–149

Rossi L, Zhang W, Ma X (2017) Cerium oxide nanoparticles alter the salt stress tolerance of Brassica napus L. by modifying the formation of root apoplastic barriers. Environ Pollut 229:132–138

Sarkar RD, Kalita MC (2023) Alleviation of salt stress complications in plants by nanoparticles and the associated mechanisms: an overview. Plant Stress 7:100134

Selvakesavan RK, Kruszka D, Shakya P, Mondal D, Franklin G (2023) Impact of nanomaterials on plant secondary metabolism. In: Al-Khayri JM, Alnaddaf LM, Jain SM (eds) Nanomaterial interactions with plant cellular mechanisms and macromolecules and agricultural implications. Springer, Cham

Sidhu AK, Sharma M, Bhickchand Agrawal S et al (2024) Nanomaterial strategies for enhancing plant resilience in the face of temperature stress. CABI Agric Biosci 5:60

Silva S, Dias MC, Pinto DCGA, Silva AMS (2023) Metabolomics as a tool to understand nano-plant interactions: the case study of metal-based nanoparticles. Plants 12(3):491

Soria CNG, Montes A, Bisson MA, Atilla-Gokcumen GE, Aga DS (2017) Mass spectrometry-based metabolomics to assess uptake of silver nanoparticles by Arabidopsis thaliana. Environ Sci Nano 4(10):1944–1953. https://doi.org/10.1039/c7en00555e

Wang Y, Liu Y, Zhan W, Zheng K, Lian M, Zhang C, Ruan X, Li T (2020) Long-term stabilization of cd in agricultural soil using mercapto-functionalized nano-silica (MPTS/nano-silica): a three-year field study. Ecotoxicol Environ Saf 197:110600

Wang K, Wang Y, Wan Y, Mi Z, Wang Q, Wang Q, Li H (2021) The fate of arsenic in rice plants (*Oryza sativa* L.): influence of different forms of selenium. Chemosphere 264:128417

Wen W, Li D, Li X, Gao Y, Li W, Li H, Liu J, Liu H, Chen W, Luo J, Yan J (2014) Metabolome-based genome-wide association study of maize kernel leads to novel biochemical insights. Nat Commun 5(1). https://doi.org/10.1038/ncomms4438

Wu H, Tito N, Giraldo JP (2017) Anionic cerium oxide nanoparticles protect plant photosynthesis from abiotic stress by scavenging reactive oxygen species. ACS Nano 11:11283–11297

Yadav G, Prajapati P, Singh D, Hora S, Singh S, Vishwakarma K, Ahmad IZ (2023) Emerging trends in plant metabolomics and hormonomics to study abiotic stress tolerance associated with rhizospheric probiotics. In: Plant hormones in crop improvement, pp 283–306. https://doi.org/10.1016/b978-0-323-91886-2.00012-4

Yu D, Lu Q, Wei Y, Hou D, Yin X, Cai K, Qiu C, Xu K (2023) Combined análisis of transcriptomics and metabolomics on the cumulative effect of nano-titanium dioxide on mulberry seedlings. Front Plant Sci 14:1175012

Zahedi SM, Hosseini MS, Meybodi NDH, Peijnenburg W (2021) Mitigation of the effect of drought on growth and yield of pomegranates by foliar spraying of different sizes of selenium nanoparticles. J Sci Food Agric 101:5202–5213

Zaman W, Ayaz A, Park S (2025) Nanomaterials in agriculture: a pathway to enhanced plant growth and abiotic stress resistance. Plants 14(5):716

Zamboni N, Saghatelian A, Patti GJ (2015) Defining the metabolome: Size, Flux, and regulation. Mol Cell 58(4):699–706. https://doi.org/10.1016/j.molcel.2015.04.021

Zhou R, Jiang F, Niu L, Song X, Yu L, Yang Y, Wu Z (2022) Increase crop resilience to heat stress using OMIC strategies. Front Plant Sci 13. https://doi.org/10.3389/fpls.2022.891861

Zou C, Lu T, Wang R et al (2022) Comparative physiological and metabolomic analyses reveal that Fe_3O_4 and ZnO nanoparticles alleviate Cd toxicity in tobacco. J Nanobiotechnol 20:302

Nanomaterials' Impact in Phytohormone Signalling of Plants

12

Beatrice Farda, Rihab Djebaili, Ramila Fares, Gökhan Boyno, Younes Rezaee Danesh, and Marika Pellegrini

Abstract

Nanotechnology can improve agriculture by enabling plant growth and crop productivity. Nanomaterials (NMs) constitute a pivotal technology for modulating phytohormone signalling, which is essential for regulating plant development and environmental responses. Through interactions with the signalling molecules at cellular levels, nanoparticles (NPs) can alter phytohormone pathways like auxins, gibberellins, cytokinins, and abscisic acid. Metallic NPs such as silver and zinc oxide have also been shown to regulate auxin transport and biosynthesis, stimulating roots and shoots. In addition, nanocarriers optimize the administration of specific hormones in a given placement, enabling precise regulation of plant responses to abiotic stresses (Akhtar et al. 2022). Mechanically, nanoparticles affect phytohormone signaling systems via receptors or by regulating oxidative stress levels. However, there are concerns about the toxicity of nanomaterials. Plant tissues treated with NPs have the potential to lead to oxida-

B. Farda · R. Djebaili (✉) · M. Pellegrini
Department of Life, Health and Environmental Sciences, University of L'Aquila, L'Aquila, Italy
e-mail: rihab.djebaili@univaq.it

R. Fares
Department of Life, Health and Environmental Sciences, University of L'Aquila, L'Aquila, Italy

Laboratory Biotechnology, Water, Environment, and Health, Abbes Laghrour University of Khenchela, Khenchela, Algeria

Laboratory of Mycology, Biotechnology, and Microbial Activity, Brothers Mentouri University of Constantine, Constantine, Algeria

G. Boyno · Y. R. Danesh
Faculty of Agriculture, Department of Plant Protection, Van Yuzuncu Yil University, Van, Turkey

R. Sayyed et al. (eds.), *Nanobiostimulants in Innovative Agriculture*,
https://doi.org/10.1007/978-981-95-2477-8_12

tive processes, which can disrupt cellular processes, hormonal balance, and homeostasis. In addition, nanoparticles interacting with soil and water systems could provide long-term environmental effects. A thorough risk assessment is needed to balance the effectiveness of using nanotechnology in agriculture and the risks to plant and ecosystem health. Nanotechnology modulation of particles should be focused on biocompatible and biodegradable nanomaterials that will be less toxic and more efficient. This integrated approach will enable nanotechnology to be an environmentally friendly tool in improving crop productivity and food safety in the changing global scene.

Keywords

Abiotic stress · Environmentally friendly · Nanoparticles · Phytohormone · Signaling molecules

1 Introduction

The integration of nanotechnology in plant biology has revolutionized the understanding of molecular interactions within plant materials, particularly with regard to phytohormone signaling pathways (Tripathi et al. 2019). This convergence represents a crucial point in botanical studies, given that manufactured nanomaterials are increasingly penetrating agricultural environments and natural ecosystems, impacting plant growth, development and stress responses through interactions with hormonal networks. These specific interactions between cellular elements and signalling molecules are favoured by the particular physicochemical characteristics of nanomaterials, which are their quantum properties and their exceptional surface/volume ratios (Francis et al. 2024; Wang et al. 2022). These properties vary according to the different categories of nanomaterials, such as carbon-based structures and metallic nanoparticles, each with its requirements. Carbon nanotubes (CNTs) have demonstrated a high potential in penetrating the cellular membranes of plants, thereby interfering with auxin transport mechanisms and inducing remarkable changes in root morphology and developmental processes (Ulhassan et al. 2022). Similarly, metallic nanoparticles, especially silver nanoparticles (AgNPs) and zinc oxide nanoparticles (ZnONPs), strongly affect ethylene perception and related signalling mechanisms, which may revolutionise current strategies for controlling fruit ripening and senescence (Kusiak et al. 2022). Understanding how nanoparticle features, particularly size, surface chemistry, and aggregation state, influence their interactions with hormone signalling pathways has significantly evolved (Mirakhorli et al. 2021).

Nanoparticles with smaller sizes (<50 nm) have shown an increased potential for cellular uptake and distribution, which might allow access to cellular compartments previously thought to be inaccessible (Sun et al. 2022). The other primary component is surface modification. The precise interactions between specialised nanoparticles, hormone receptors, and transport proteins present novel strategies for hormone regulation. The interaction between nanoparticles and hormones is significantly more intricate (Ali et al. 2021; Mathur et al. 2023). Graphene oxide affects auxin, cytokinin, and abscisic acid signaling systems simultaneously, resulting in complex plant growth responses (Abideen et al. 2022). Since several hormones interact with nanomaterials, a systems biology approach is necessary to understand how they impact plant development and response to stress. Recently, analytical techniques such as high-resolution imaging and omics technologies have shown previously unknown interaction mechanisms between nanomaterials and hormones (Abideen et al. 2022; Khan et al. 2021). Transcriptomic research disclosed new genes and regulatory elements responsive to nanomaterials exposure, while proteomic studies showed changes in hormone receptor abundance and post-translational modifications (Kumar et al. 2024). Ecologically, these effects may extend beyond individual plant organisms, since they might modify ecosystem functioning through altered plant–microbe interactions and changing stress events (Ji et al. 2023). Alterations in hormone signaling induced by nanomaterials can modify the composition of root exudates. Thus, this influence impacts the fundamental resource of the native population about processes linked to nutrient cycling (Jiao et al. 2023). Other potential consequences include changes in flowering and the reproductive cycle. These alterations are linked to disturbances in the synchronisation of hormonal signals, significantly influencing plant species populations and evolutionary trajectories (Li et al. 2021). Specialised nanoscale materials are being developed as advanced delivery systems for phytohormones, ensuring greater precision in agriculture. On the other hand, the same advance brings enormous concerns about the long-term consequences of nanomaterial accumulation in agricultural soils and non-target organisms, specifically under the effects of altered hormone signalling pathways (Francis et al. 2024; Roy 2023). Despite advancements in the interactions between nanomaterials and phytohormones, difficulties and possibilities continue to emerge. Advanced analytical methods and methodologies are required to investigate the intricate molecular mechanisms governing their interaction (Anand et al. 2023; Choudhary et al. 2024). While difficulties are associated with the more sophisticated engineered nanomaterials that have been generated thus far, such as hybrid and multifunctional ones, there are also new prospects for the specific regulation of plant hormone responses (Wu and Li 2022). This discipline is currently at a crucial intersection of fundamental research and applied science, extending its relevance into the immediate future. The research about the interaction between nanomaterials and hormone function will further increase the current understanding of plant biology and encourage innovative approaches toward agriculture (Etesami et al. 2021). This advancement raises significant concerns over the long-term impacts of nanomaterial build-up in agricultural soils and non-target organisms, particularly about altered hormone signalling pathways. More creative, complicated analytical tools and approaches will be needed to explore complicated molecular mechanisms underlying their interaction. The increased complexity of the developed engineered nanomaterials, including hybrid and multifunctional ones, poses challenges and opens new opportunities for the precise regulation of plant hormone responses (Zhang et al. 2022). This field is an essential crossroads between fundamental research and applied science, reinforcing its short and long-term importance (Raval et al. 2023). This chapter provides an outline of current advancements in nanotechnology in plants, the role of nanoparticles in phytohormone regulation, the mechanisms underlying this interaction, and the phytotoxicity of these materials, along with risk assessment.

2 Plants and Nanotechnology

2.1 Applications of Nanotechnology in Plant Sciences

The term "nanotechnology" describes the exact manipulation of matter at the atomic and molecular levels ($\leq$100 nm) to create objects with specific properties, like size and shape. Due to their unique properties, nanoparticles (NPs) have been a focal point in advanced interdisciplinary research in the fields related to chemistry, biology, physics, medicine, material science, and engineering. One of the high-profile focus areas is using nanotechnology in plants (Jangid et al. 2024). NPs are compatible with plant-related systems. Their small sizes and large surface areas lead to quantifiable modification possibilities. The multiple functions of these versatile NPs include delivery mechanisms as reservoirs, coating materials, electrical systems, and imaging agents. They have been shown to interact with plant cells in intra- and extracellular locations, influencing root- and foliar-based processes, including transport across the cell membrane. Nanotechnology in plant sciences impacts plant processes such as growth, absorption and transport of nutrients, pests, and pesticides (Thakur et al. 2022; Sharma et al. 2025). The emergence of nanotechnologies significantly impacts plant fertilisation more efficiently without the need for run-off chemicals and enhances pest control by delivering active substances directly to affected pests and plants (El-Kady et al. 2023). Increased crop survival and improved crops through direct and specific up-regulation of particular (high-yield) proteins will massively reduce the agricultural area needed for producing food. This will increase the efficiency and use of biologically based NPs, redefining a sustainable agricultural economy with a much-reduced environmental burden (Zandalinas et al. 2024; Fazeli-Nasab et al. 2021). In recent decades, the growing field of plant nanotechnology has addressed the interest in the effects of nanomaterials on plant production and the rhizosphere (Francis et al. 2024). There is significant interest in the design of nanoparticles used in plant systems; however, it is crucial to comprehend the plant response. Nowadays, nanoparticle–plant system interactions have been the subject of intense study. Many critical reviews, scientific investigations, and meta-analysis studies have become available so far, investigating the impact of nanoparticles on various plant species, soil dynamics, and ecosystem health in general. Reports of these studies have elucidated the uptake mechanism of nanoparticles by plants and their consequent impact on plant growth, development, and physiological processes (Rajput et al. 2020) (Fig. 12.1).

Further, several researchers have started exploring the possibility of using nanoparticles in agriculture by increasing nutrient use efficiency and enhancing

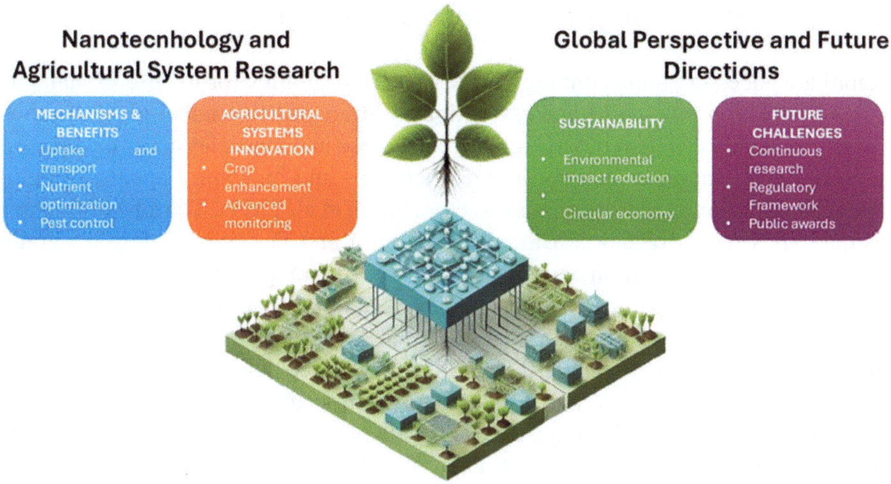

Fig. 12.1 Nanotechnology and plant sciences

abiotic stress tolerance for higher yields of crops (Faridvand et al. 2021). However, it is vital to ensure that the integration of nanoparticles into plant systems does not harm the environment and human health (Alabdallah et al. 2021; Rasouli et al. 2020). More research is needed to fully understand the long-term implications of using nanoparticles in plant nanotechnology and to develop appropriate guidelines for their safe and sustainable use in agriculture (Martínez et al. 2020). A complex interplay of biological, chemical, and physical interactions affects primary nanoparticle uptake and dynamics in plants. Recently, great attention has been directed toward the factors affecting the accumulation of nanoparticles in plants due to their size, charge, and surface potential (Ali et al. 2021). There are two primary uptake pathways: passive uptake through plant roots and active uptake through foliar pathways. The properties of nanoparticles affect the mode of transport through plants but also depend on plant species-specific characteristics. Water and vascular flow are the two primary transport mechanisms that plants can use to translocate nanoparticles (Fares et al. 2024). The introduction of nanoparticles can influence various aspects of plant physiology. Within the plant organism, nanoparticles can influence stress response mechanisms, growth, and development while inducing modifications at the genetic and transcriptional levels (Zhou et al. 2020). The trend in the literature has also concentrated on the bioavailability and fate of nanoparticles in the environment since the plants exposed to pesticides in the treated soil or groundwater that can also contain nanomaterials will most definitely be affected by these multiway interactions. These studies have shown that the interactions between nanoparticles and plants can have far-reaching consequences, influencing the plants and the surrounding ecosystem (Basak et al. 2023). By comprehending and manipulating these interactions, scientists aim to create plans to minimise possible hazards and optimise the advantages of using nanoparticles in environmental clean-up and agriculture (Khan et al. 2021).

2.2 Plant Pathology and Crop Improvement

Rapid advances in nanotechnology offer vast opportunities for the use of nanoparticles to develop innovative and revolutionary strategies to combat plant diseases. Acting as highly efficient naturalising agents, these nanoparticles act as exceptional vectors for the targeted delivery of agrochemicals and biocontrol agents, bringing about a revolution in the agricultural sector (Singh et al. 2021). The advent of nanotechnologies has attracted considerable attention and has been widely integrated into various crop protection approaches. The use of nanoparticles in agriculture has proven highly effective, offering promising results in controlling plant pathogens and optimising crop yields (Zehra et al. 2021). An exciting aspect of disease management using nanoparticles is their ability to deliver antifungal agents progressively, providing continuous and prolonged protection against pathogenic infections. The specific properties of nanoparticles provide plants with an effective and reliable defence system that prevents the intrusion of harmful pathogens. In addition, it has been shown that nanoparticles can encapsulate various natural compounds or pesticides, thereby limiting the emergence of pesticide resistance in pathogens. This multidimensional approach makes it possible to ensure adequate disease management and promote sustainable agricultural practices (Fu et al. 2020). Numerous formulations containing nanoparticles have already demonstrated their effectiveness in disease control strategies, underlining the vast potential of nanotechnology in agriculture. For example, using nisin nanoparticles to combat soft rot has produced remarkable results in effectively reducing the occurrence and severity of this devastating pathology (Wang et al. 2024). In addition, using CuO nanoparticles effectively controls bacterial spot disease on tomato plants, providing farmers with a sustainable, high-performance alternative (Yadav et al. 2022). Integrating nanotechnology in agriculture offers superior disease management capabilities and promises to improve crop growth and overall agricultural sustainability. It may change the face of agriculture by helping farmers effectively fight plant diseases and improving the productivity of crops, thereby improving food security (Usman et al. 2020). Another promising sphere of integration between plant breeding and nanotechnology applications is genetic engineering. Indeed, a specific type of nanomaterials was shown to act as nanocarriers for alien nucleic acids to intervene in the gene expression of plants and crops directly (Torres-Vanegas et al. 2021). This new 'target' approach will be used more widely to enhance the ability of crops or plants to resist pests and diseases. There is further scope to develop new concepts extending nanoscale interventions in agriculture into economic and environmental issues (Bigini et al. 2021). Integrated pest management is a valid strategy, and it can be extended to reduce the use of pesticides for managing diseases in a synergistic manner to enhance the resistance of crops against disease attacks (Ahmar et al. 2021).

3 Nanoparticles-Mediated Phytohormone Regulation

In recent years, exploring nanoparticle interactions in plants has garnered significant interest (Khan and Hossain 2022). Nanoparticles can move across the plasmodesmata in meristematic tissues after physical entry into the apoplast or are internalised inside the cell. The nanoparticle features and the co-cultured plant species determine the transport of nanoparticles. Interactions between nanoparticles and plants include alteration in root architecture, influencing seed germination, photoprotection, and protection from oxidative stress (Pociecha et al. 2021; Verma et al. 2020). Numerous studies have shown that nanoparticles exert accelerative or inhibitory effects on plant growth and development. To gain a clear insight into the biological effects of nanoparticles, it is necessary to investigate how nanoparticles and plant cells interact to activate phytochemicals. Phytohormones are vital in regulating various processes of growth, development and immunity in plants (Santás-Miguel et al. 2023). In addition, artificially adding natural phytohormones is proving to be an effective way of stimulating plant development and improving crop quality in agriculture (Tripathi et al. 2022). Three types of newly developed nanoparticles are commonly utilised in agricultural applications. Metallic nanoparticles include silver, gold, and zinc oxide, the regulation of which plant phytohormones has been experimentally confirmed. Carbon-based nanoparticles are usually combined with other materials as horticultural remedies to a more significant extent, and, to date, graphene oxide has been used as a unique carbon-based nanoparticle for plant phytohormone regulation (Cruz-Luna et al. 2021). Graphene oxide within the plant cells retained a favourable reactive oxygen species accumulation that promoted 1-aminocyclopropane-1-carboxylate synthetase activity and ethylene generation. As the third type, polymeric nanoparticles from synthetic or natural polymers are initially available as nanoparticle-based agricultural products. They are widely used, such as delivering pesticides, fertilisers, or biocontrol agents (Hakim et al. 2021). Plant hormone-responsive transcriptome analyses have indicated that polystyrene nanoparticles in *Arabidopsis* leaves can change the genes involved in the auxin signalling pathway. Recently, it has been reported that polystyrene nanoparticles are involved in cytokinin generation, asphyxiating maize interveinal mesophyll cells. In addition, nanoparticles and phytohormones can be applied to different plant species, such as lupins, rice, and tomato hybrids (Huang et al. 2023). These differences are mainly explained by fluctuations in nanoparticle concentration and surface charge and by experimental and environmental conditions (Weiss et al. 2021).

The modulation of endogenous hormone levels and amplification/inhibition activities of different signalling components might be serious steps toward control of the diseased state and augmentation of resistance or tolerance (Tripathi et al. 2022). Several studies have reported partially comparative or analogous observations, which showed the potential of nanoparticles as substitutes or opponents for specific applications of endogenous hormones concerning managing disease conditions. The nanoparticles reportedly participated in various hormone-related pathways that resulted in different levels of modulations, plant physiological responses, and abiotic stress regulation in some instances (Rana et al. 2021; Ranjan et al.

2024). Applications of nanoparticles in nanotechnology have shown huge potential in the manipulation of endogenous hormone levels and controlling several component activities, which, in turn, leads to further disease control, increased resistance, and augmented tolerance (Shahbaz et al. 2023). Several reports have indicated how nanoparticles replace or act in opposition to indigenous hormones in disease management. Notably, it was found that nanoparticles affect different hormone-related pathways, showing a different level of modulation, plant physiological response, and abiotic stress regulation (Dutta et al. 2022; Mukarram et al. 2022). The Application of nanotechnology-based methods offers excellent promise for plant disease control and resistance. Such nanoparticles modulate the endogenous levels of hormones, allowing a revolutionary solution in managing diseases and stress tolerance. Particles have previously been proven to replace the activity of endogenous hormones. More interestingly, the modulation caused by nanoparticles occurred at several pathways concerning hormones and different modulation levels that control plants' physiological responses and the regulation of abiotic stresses. The role of nanoparticles mediating the interaction between phytohormone, secondary metabolite, and endogenous defence proteins has been established as indispensable through extensive experimental evidence. These interactions begin at the level of signalling, growth, and development, demonstrating a variety of effects and responses to environmental stresses (Karabulut 2024). Nanoparticles induce various hormone-related signalling pathways that influence different modulations of plant physiological responses, and the regulation of abiotic stress (Ahmed et al. 2022).

4 Mechanism of Nanomaterial–Phytohormone Interactions

Using nanomaterials in agriculture can be important in advancing so-called 'intelligent' or 'precision' agriculture. Nanomaterials are characterised by the presence of particles or fibres of micro or nanometric dimensions, which exhibit distinctive properties such as ionic radiation, confined electrons, increased ionic discharge capacity, a significant surface/volume ratio, high surface free energy, and surface charge. Unlike bulk materials, nanoparticles exhibit unique physicochemical and biological properties because of their increased surface area (Mekuye and Abera 2023). All these properties, described in many contexts of their interactions with phytohormones, make nanomaterials necessary compounds for plant biology, opening new applications of nanotechnology in agriculture. Although phytohormones are produced naturally by plants at low concentrations, they play a number of essential roles in regulating plant development and growth, and their response to biotic and abiotic stimuli. Phytohormones regulate physiological and biochemical processes according to their characteristics, which classify them into various groups. Interactions between nanomaterials and phytohormones present an interdisciplinary research area combining nanotechnology, biology, biochemistry, and physiology (Adeleke et al. 2022; Raval et al. 2023; Sabagh et al. 2022).

Various methods have been used to examine the interactions between nanomaterials and phytohormones, including light and electron microscopy, scanning and transmission electron microscopy, and confocal laser scanning microscopy. These methods offer the possibility of observing the penetration of the nanomaterial into the plant tissue and analysing its interaction with phytohormones (Kumar et al. 2024). Recently, more precise methods have been developed to analyse the metabolites present in plants and microbes, focusing on metabolomics and, in particular, on specific phytohormones or phytohormonal pathways (Wang et al. 2020). The majority of research into the evaluation of the interaction between nanomaterials and phytohormones is carried out under rigorously controlled laboratory conditions. Field experimentation is uncommon due to the methodological challenges involved, such as soil pH fluctuations and the impact of environmental conditions, as highlighted by Alhaithloul and his collaborators (Alhaithloul et al. 2024). During in-depth field research, it is essential to assess the dissemination and impact of nanomaterials beyond the confined area to prevent any unwanted exposure or potential risk to human health and the environment (Driesen et al. 2020). To investigate the interaction between nanomaterials and phytohormones, several simple methodologies are used, which focus on a single aspect and are part of different research areas: material engineering, physicochemical characteristics, or biological effects. However, a comprehensive analysis of the outcome and detail of this interaction requires a multidisciplinary approach and may help to obtain a compound and reliable picture of the process. Moreover, the different nanomaterial compositions and concentrations, experimental timing, methodologies, and plant growth conditions in the different studies contribute to the heterogeneous outcomes (Yang et al. 2023). These comparisons show that when the same type of nanomaterials is adopted, their impact can differ according to their concentration in the study system. A similar concept is found if different nanomaterials are used; the biological outcome could be significantly different depending on the nanomaterial's type. In horticulture, abscisic acid, auxin, cytokinins, ethylene, gibberellins, jasmonic acid, and salicylic acid are the main phytohormones investigated in the interactions with nanomaterials, and their level change was studied to estimate the nanomaterial effect (Kandhol et al. 2023).

Metal nanomaterials and metal oxides interact with phytohormones, promoting their transport and exerting a significant influence on various physiological processes such as seed germination, root development, and biomass accumulation. The interaction leads to their accumulation in the roots and aerial parts. It enhances the uptake and assimilation of essential nutrients, demonstrating the dual role of metal and metal oxides in plant nutrient defences (Karabulut 2024). The synergistic endogenous phytohormone levels and moderate activity of phytohormones can accelerate plant development, whereas their overexpression is linked to higher stress levels that hamper seed germination (Bhatt et al. 2020). Nanoparticles can indirectly affect the germination potential to trigger endocrine hormone synthesis by altering a possible upsurge in activity during seed germination. The direct or indirect regulation of nanoparticles on phytohormones has also been considered using different case studies to improve plant growth and development, thereby enhancing the natural defence response against adverse conditions (Chandrasekaran et al. 2020).

Various signalling pathways may mediate nanomaterial-phytohormone interactions. Carbonaceous nanomaterials act as auxin/melatonin molecules, building crosstalk between the auxin and melatonin pathways (Kido et al. 2023). They may act with the blockage of brassinosteroid inhibitory proteins. Since brassinosteroids interact with gibberellin signals, they could also affect gibberellin homeostasis and perception. Although nanomaterials have not yet been identified as triggers of jasmonic acid signals, they have been observed to inhibit wound-induced resetting of the jasmonic acid pathway (Cai et al. 2020; Shanmugaiah et al. 2023). Nanomaterials interact with phytohormones at different levels, such as genetic, biochemical, cellular, and overall physiological responses to the plant. At the genetic and biochemical level, phytohormone and nanomaterial interaction is reflected in i) changes in the expression of the phytohormone metabolism and signalling pathway genes in nanomaterial-exposed plants and ii) the presence or altered composition of phytohormones in plant extracts where nanomaterials are also present (Ali and Mosa 2022). Temporal variations in the response of the plant transcriptome to exposure to nanomaterials indicate that, at various times after exposure, a more or less efficient transcriptional state for a specific biological process will be established, distinctly under the influence of nanomaterials (Mascarenhas et al. 2023).

In contemporary agriculture, emerging nanotechnology, as a promising technology, can be exploited to increase the yield and quality of crops and enhance their resistance to pests and pathogens. Since phytohormones are commonly used as stimulants, applying external phytohormones to plants induces specific responses that can potentially improve agronomic techniques (Khan et al. 2021). Several studies have used nanomaterials as nanocarriers for enhancing the efficacy, stability, and targeted delivery of various exogenous phytohormones. This suggests that comparative regulation of the molecular response and interaction of different nanomaterials and phytohormones can help apply crop management practices. The process of administering these phytohormones helps to limit the excessive use of chemical fertilisers and pesticides, thus meeting the needs of sustainable and precision agriculture (Butova et al. 2023). Several limiting factors, such as moisture stress, light intensity, temperature effects, or even biotic and abiotic stresses, may provide substantial new directions in nano agronomy (Ansari et al. 2020; Iavicoli et al. 2017). Since most of the experiments took place in the laboratory under controlled conditions or in open fields at the micro-plot level, applying these new methods to industrial-scale agriculture is impeded. In nano agriculture, focusing on such limitations may be crucial. Regulatory systems and guidelines vary from nation to nation (Danai-Varsou et al. 2023). For commercial applications in agriculture, safety assessments for products or materials must guarantee no harmful side effects. The plant system comprises numerous positive and negative aspects to consider when creating a risk assessment and assessment protocol. Based on this information, additional interdisciplinary research is required to explain and understand not only the behaviour of plants when interacting with nanomaterials and phytohormones but also human health and the environment, such as soil and water (Chanu et al. 2021).

5 Nanophytotoxicity and Risk Assessment

Considering the potential benefits of nano-based products in the agricultural sector, numerous governments globally are making significant efforts to assess the feasibility and compatibility of integrating nanotechnology with agricultural practices (Mishra et al. 2017). Agricultural scientists are addressing knowledge gaps regarding the toxicity of nanoparticles on plants, soil, and soil biota to decrease risk assessment factors and control their application for safe agricultural use. Table 12.1 summarises the toxic effects of the mainly used nanoparticle groups on plants and soils.

The interaction between plants, soil, and nanoparticles points out some critical features of agricultural production, mainly regarding toxicity and nanoparticle impact on both plant-associated microbiota and plants. Soil serves as the major sink for nanoparticles and, as such, has a controlling effect on their fate, transport, and behaviour. Several physicochemical properties of soil, including pH, organic matter content, and cation exchange capacity, determine the mobility, bioavailability, and toxicity of nanoparticles (Anjum et al. 2013; Huang et al. 2022). AgNPs show significant antimicrobial activity, changing soil properties and affecting microbial composition. Lower pH, organic matter, and cation exchange capacity decrease silver sorption; this increases the mobility and, therefore, the toxicity of AgNPs, while high values of all these factors enhance sorption, reducing toxicity accordingly (Klitzke et al. 2015; Kyziol-Komosinska et al. 2024). AgNP toxicity in plant-associated microbiota is dose-dependent. Although AgNPs at low concentrations

Table 12.1 Effects on plants of the main nanoparticle classes

NMs	Target plants	NM effects	References
Silver nanoparticles (AgNPs)	*Phaseolus radiatus*, *Sorghum bicolor*	High toxicity for denitrifying bacterial soil community; 5–10 nm particle uptake	Lee et al. (2012)
Zinc oxide (ZnO-NPs)	*Triticum aestivum* *Zea mays*	Low toxicity in acid soil; high toxicity neutral soil	Du et al. (2011), Zhao et al. (2013)
Titanium dioxide (TiO_2-NPs)	*Spinacia oleracea*, *Vitex*	High negative impact on soil microbial community; inhibited hydraulic conductivity and transpiration; at low concentration, stimulation of growth, chlorophyll, protein, and nitrogen content	Ge et al. (2012), Farahi et al. (2023), Yang et al. (2007)
Copper nanoparticles (cu-NPs)	*Cucumis sativus* *Allium cepa*	Higher toxicity in sandy loam soil; genotoxicity and oxidative stress; positive impact on onion growth at lower concentrations	Mosa et al. (2018), Mottaleb et al. (2021)
Multi-walled carbon nanotubes (CNTs)	*Lolium perenne*, *Solanum lycopersicum*,	Enhanced seed germination, root elongation, and yield; low or neutral effect on others	Khodakovskaya et al. (2013), Lin and Xing (2007)

(0.2 ppm) increase the nitrate reductase activity in *Azotobacter*, at higher dosages, it is harmful to denitrifying bacteria (Chen et al. 2023; Shahrokh et al. 2014). Plants like *Phaseolus radiatus* are highly susceptible to the toxicity of AgNPs in the 5–10 nm size range. In contrast, the plants are less affected in the soil system compared to the agar medium in *Sorghum bicolor* (Lee et al. 2012). These findings emphasise the complex interaction of AgNPs with soil and biological systems, so a careful evaluation of environmental risks is required. ZnO-NPs exhibit variable toxicity regarding soil characteristics, and organic matter in soil significantly modifies their transport and toxicity. In loamy sand soil with a pH of 5.5, no phytotoxicity toward *Triticum aestivum* was noticed at a concentration of 2000 mg kg^{-1}. Instead, in loamy clay soil (pH 7.36), toxicity was significant at a concentration of 45.45 mg kg^{-1} (Du et al. 2011). In addition, the safety issues of the NPs are significantly influenced by the soil organic matter, as evidenced by the reduced toxicity of the ZnO NPs in *Zea mays* when alginate is administered at a concentration of 400–800 mg kg^{-1}. ZnO NPs substantially decrease plant biomass (Zhao et al. 2013). Moreover, it was demonstrated that zinc oxide nanoparticles showed apparent toxicity to microbial communities, as reflected by affecting ammonification, respiration, and enzymatic activities under acidic and neutral soils. In contrast, the toxicity was much reduced in alkaline soils (Strekalovskaya et al. 2024). Plant-associated microbiota and plant growth are affected by titanium-based nanoparticles. Several investigations have demonstrated the impact on microbial community diversity, resulting in a decrease in the abundance of taxa linked to nitrogen fixation, including *Rhizobiales* and *Bradyrhizobium*, while simultaneously increasing the prevalence of taxa such as Sphingomonadaceae and Streptomycetaceae, which are involved in the decomposition of organic pollutants (Farahi et al. 2023; Ge et al. 2012). Instead, under particular conditions, TiO2-NPs may promote plant growth. For instance, TiO2-NPs promoted the increase of fresh and dry weight of *Spinacia oleracea* by promoting nitrogen photoreduction and photosynthetic pigments and biochemical and antioxidant properties of *Vitex* plant (Jakinala et al. 2021; Farahi et al. 2023; Yang et al. 2007). Nevertheless, the toxicity of TiO2-NPs to the health of plants and the equilibrium of soil microorganisms is significantly influenced by the concentration of nanoparticles and the environmental conditions. More comprehension is still needed regarding the long-term environmental effects and the necessity of additional research (Mittal et al. 2020). Copper and Iron nanoparticle toxicity is affected by soil types. For instance, sandy loam soils exhibited more negative impacts than sandy clay loam soils, where higher clay content can mitigate nanoparticle toxicity (Frenk et al. 2013). Although these nanoparticles impact microbial activity and soil health, their effects on plants are less documented than those of AgNPs and ZnO-NPs. It is imperative to comprehend their conduct in a variety of soil matrices in order to create a secure agricultural application. Nanoparticles directly impact plants through their ability to translocate within plant tissues, which is facilitated by their high reactivity and nanoscale size. They can enter plants by adsorption on the surface or through openings such as stomata, impacting their physiological processes. For example, the study conducted by Mosa and colleagues examined the toxic effects of copper nanoparticles on *Cucumis sativus*. Adverse phenotypic alterations were observed, including a reduction in biomass, photosynthetic pigments, and lesions in the root plasma membrane (Mosa et al. 2018). Phytotoxicity also depends on the size and concentration of nanoparticles; nanoparticles less than 5 nm in size can efficiently translocate and induce toxicity even at a low concentration (Ma et al. 2010). Conversely, under certain conditions, nanoparticles can stimulate plant growth. Mottaleb and his collaborators revealed positive effects regarding growth parameters (plant height, fresh and dry weight, yield, and bulb quality) and chemical constituents in *Allium cepa* (Mottaleb et al. 2021). Other nanomaterials, such as multi-walled carbon nanotubes (CNT), have been found to improve seed germination and root elongation of ryegrass (*Lolium perenne*) at a concentration of 2 g L^{-1}. However, no significant effects were observed on radish (*Raphanus sativus*), lettuce (*Lactuca sativa*), and cucumber (*Cucumis sativus*) seeds (Lin and Xing 2007). Studies have also shown that applying 50 mg L^{-1} increased the yield of tomato (*Solanum lycopersicum*) plants by optimising the efficiency of irrigation by plants (Khodakovskaya et al. 2013). These findings underscore that the phytotoxicity of nanomaterials is predominantly influenced by the concentration and the plant growth system. The toxicity of nanoparticles to plants and associated microbiota points out the importance of soil, acting as a moderating factor. Soil-based studies offer realistic insights into the effects of nanoparticles, hence the need for better soil management to reduce risks. Measures such as increasing the level of soil organic matter, adjusting the optimum pH, and preventing the excessive release of nanoparticles into the ecosystem can help to mitigate negative impacts. Future studies should focus on the interrelationship between soil attributes, nanoparticle characteristics and agroecosystems to promote the development of sustainable agriculture.

6 Conclusion and Future Perspectives

Nanotechnology has emerged as a crucial catalyst for change in the agricultural sector. It proposes innovative approaches to optimise plant growth, control disease, and increase crop yield by combining nanoparticles and phytohormones. Nanoparticles regulate plant growth processes at molecular and physiological levels. Despite all these developments, plant nanotechnology has many challenges to overcome. One involves the toxicity related to nanoparticles: several studies have shown that NPs can disrupt the hormonal balance in plants, producing oxidative stress that changes growth patterns and negatively impacts soil and water systems. The size, concentration, and surface properties of nanoparticles significantly influence the bioavailability, accumulation, and interaction of nanoparticles with plant tissues and soil microbes. In addition, the environmental disposition of these materials raises concerns over long-term ecological impacts and hence necessitates a careful risk assessment. Future investigations are necessary for developing new biocompatible and biodegradable nanomaterials. There is also a distinct requirement for engineered nanoparticles to optimise efficacy while minimising environmental toxicity. Green nanotechnology approaches utilise plant-derived or eco-friendly nanoparticles for sustainable applications. In this context, it is essential that regulatory frameworks evolve to establish appropriate protocols promoting a universal approach to the safe integration of nanotechnology in the agricultural sector while considering crop safety and environmental preservation. In addition, it is strongly recommended that interdisciplinary research be promoted to deepen understanding of the interactions between nanoparticles and phytohormones. Approaches including high-resolution imaging and omics technologies allow accurate tracking of nanoparticles within plant systems, showing their mechanism of action and off-target effects. Integrating these insights with artificial intelligence and machine learning will optimise nanoparticles for design and predict their behaviour in complex ecosystems. It is imperative to carry out an in-depth field study to confirm experimental findings and take into account the diversity induced by environmental parameters such as soil type, humidity, and pH. Researchers, policy-makers, and agricultural stakeholders must work together to facilitate the transfer of nanotechnological advances from the laboratory to the field. Public awareness and acceptance are also critical; hence, there is a need for transparent communication of the benefits and risks associated with nanotechnology. Incorporating nanotechnology into agriculture is one of the critical milestones toward ensuring the next generation of food supplies under the threats of global warming and ever-growing population pressures. Overcoming the present limitations and with the help of advanced technological features, nanotechnology could foster sustainable agricultural development that would balance productivity with environmental care. Plant Nanotechnology has a bright future in balancing this equation for the emergence of greener and more resilient agriculture.

References

Abideen Z, Hanif M, Munir N, Nielsen BL (2022) Impact of nanomaterials on the regulation of gene expression and metabolomics of plants under salt stress. Plants 11:691. https://doi.org/10.3390/plants11050691

Adeleke BS, Akinola SA, Adedayo AA, Glick BR, Babalola OO (2022) Synergistic relationship of endophyte-nanomaterials to alleviate abiotic stress in plants. Front Environ Sci 10. https://doi.org/10.3389/fenvs.2022.1015897

Ahmar S, Mahmood T, Fiaz S, Mora-Poblete F, Shafique MS, Chattha MS, Jung K-H (2021) Advantage of nanotechnology-based genome editing system and its application in crop improvement. Front Plant Sci 12. https://doi.org/10.3389/fpls.2021.663849

Ahmed M, Decsi K, Tóth Z (2022) Different tactics of synthesized zinc oxide nanoparticles, homeostasis ions, and phytohormones as regulators and adaptatively parameters to alleviate the adverse effects of salinity stress on plants. Life 13:73. https://doi.org/10.3390/life13010073

Akhtar N, Ilyas N, Meraj TA, Aboughadareh AP, Sayyed RZ, Mashwani Z, Poczai P (2022) Improvement of plant responses by nanobiofertilizer: a step towards sustainable agriculture. Nano 12:965

Alabdallah NM, Hasan MM, Hammami I, Alghamdi AI, Alshehri D, Alatawi HA (2021) Green synthesized metal oxide nanoparticles mediate growth regulation and physiology of crop plants under drought stress. Plants 10(8):1730

Alhaithloul HAS, Awad NS, Shoala T, Alqahtani MM, Alzuaibr FM, Alasmari A, Abdein MA (2024) Comparison of various hormone combinations and the potential of chitosan nanoparticles for growth stimulation in Astragalus spp. Plant Cell Tissue Organ Culture (PCTOC) 157:74. https://doi.org/10.1007/s11240-024-02790-w

Ali MA, Mosa KA (2022) Metal nanoparticles applications and their release into surrounding: perspectives of plant uptake and effects on phytohormones. In: Agricultural biocatalysis: biological and chemical applications, pp 103–130

Ali S, Mehmood A, Khan N (2021) Uptake, translocation, and consequences of nanomaterials on plant growth and stress adaptation. J Nanomater:1–17. https://doi.org/10.1155/2021/6677616

Anand KJ, Nagre SP, Shrivastava MK, Amrate PK, Patel T, Katara VK (2023) Enhancing crop improvement through synergistic integration of advanced plant breeding and proximal remote sensing techniques: a review. Int J Plant Soil Sci 35:121–138. https://doi.org/10.9734/ijpss/2023/v35i193533

Anjum NA, Gill SS, Duarte AC, Pereira E, Ahmad I (2013) Silver nanoparticles in soil–plant systems. J Nanopart Res 15:1896. https://doi.org/10.1007/s11051-013-1896-7

Ansari M, Shahzadi K, Ahmed S (2020) Nanotechnology: a breakthrough in agronomy. In: Nanoagronomy. Springer International Publishing, Cham, pp 1–21. https://doi.org/10.1007/978-3-030-41275-3_1

Basak A, Amarshettiwar SB, Bordoloi N, Shahni YS et al (2023) Nanomaterial-induced changes in plant physiology and genetics: implications for crop improvement strategies. Int J Environ Climate Change 13:2043–2048. https://doi.org/10.9734/ijecc/2023/v13i113364

Bhatt D, Nath M, Sharma M, Bhatt MD, Bisht DS, Butani NV (2020) Role of growth regulators and phytohormones in overcoming environmental stress. In: Protective chemical agents in the amelioration of plant abiotic stress. Wiley, pp 254–279. https://doi.org/10.1002/9781119552154.ch11

Bigini V, Camerlengo F, Botticella E, Sestili F, Savatin DV (2021) Biotechnological resources to increase disease-resistance by improving plant immunity: a sustainable approach to save cereal crop production. Plants 10:1146. https://doi.org/10.3390/plants10061146

Butova VV, Bauer TV, Polyakov VA, Minkina TM (2023) Advances in nanoparticle and organic formulations for prolonged controlled release of auxins. Plant Physiol Biochem 201:107808. https://doi.org/10.1016/j.plaphy.2023.107808

Cai L, Cai L, Jia H, Liu C, Wang D, Sun X (2020) Foliar exposure of Fe3O4 nanoparticles on Nicotiana benthamiana: evidence for nanoparticles uptake, plant growth promoter and defense response elicitor against plant virus. J Hazard Mater 393:122415. https://doi.org/10.1016/j.jhazmat.2020.122415

Chandrasekaran U, Luo X, Wang Q, Shu K (2020) Are there unidentified factors involved in the germination of nanoprimed seeds? Front Plant Sci 11. https://doi.org/10.3389/fpls.2020.00832

Chanu NB, Alice AK, Thokchom A, Singh MC, Chanu NT, Singh YD (2021) Engineered nanomaterial and their interactions with plant–soil system: a developmental journey and opposing facts. Nanotechnol Environ Eng 6:36. https://doi.org/10.1007/s41204-021-00130-3

Chen S, Yan X, Peralta-Videa JR, Su Z, Hong J, Zhao L (2023) Biological effects of AgNPs on crop plants: environmental implications and agricultural applications. Environ Sci Nano 10:62–71. https://doi.org/10.1039/D2EN00801G

Choudhary A, Sharma S, Kaur H, Sharma N, Gadewar MM, Mehta S, Husen A (2024) Plant system, abiotic stress resilience, reactive oxygen species, and coordination of engineered nanomaterials: a review. S Afr J Bot 171:45–59. https://doi.org/10.1016/J.SAJB.2024.05.053

Cruz-Luna AR, Cruz-Martínez H, Vásquez-López A, Medina DI (2021) Metal nanoparticles as novel antifungal agents for sustainable agriculture: current advances and future directions. J Fungi 7:1033. https://doi.org/10.3390/jof7121033

Danai-Varsou D, Zhang P, Afantitis A, Guo Z, Lynch I, Melagraki G (2023) Nanoinformatics and artificial intelligence for nano-enabled sustainable agriculture. In: Nano-enabled sustainable and precision agriculture, pp 503–531. https://doi.org/10.1016/B978-0-323-91233-4.00015-6

Driesen E, den Ende W, Van P, De M, Saeys W (2020) Influence of environmental factors light, CO2, temperature, and relative humidity on stomatal opening and development: a review. Agronomy 10:1975. https://doi.org/10.3390/agronomy10121975

Du W, Sun Y, Ji R, Zhu J, Wu J, Guo H (2011) TiO2 and ZnO nanoparticles negatively affect wheat growth and soil enzyme activities in agricultural soil. J Environ Monit 13:822. https://doi.org/10.1039/c0em00611d

Dutta A, Banerjee S, Dinda S, Chowdhury I, Haldar S, Bandyopadhyay S (2022) A critical analysis on the roles of exopolysaccharides and ACC deaminase in salinity stress tolerance in crop plants. Biocatal Agric Biotechnol 42:102372. https://doi.org/10.1016/j.bcab.2022.102372

El-Kady MM, Ansari I, Arora C, Rai N, Soni S, Verma DK, Singh P, Mahmoud AED (2023) Nanomaterials: a comprehensive review of applications, toxicity, impact, and fate to environment. J Mol Liq 370:121046. https://doi.org/10.1016/j.molliq.2022.121046

Etesami H, Fatemi H, Rizwan M (2021) Interactions of nanoparticles and salinity stress at physiological, biochemical and molecular levels in plants: a review. Ecotoxicol Environ Saf 225:112769. https://doi.org/10.1016/J.ECOENV.2021.112769

Farahi SMM, Yazdi MET, Einafshar E, Akhondi M, Ebadi M, Azimipour S, Mahmoodzadeh H, Iranbakhsh A (2023) The effects of titanium dioxide (TiO2) nanoparticles on physiological, biochemical, and antioxidant properties of Vitex plant (Vitex agnus – Castus L). Heliyon 9:e22144. https://doi.org/10.1016/J.HELIYON.2023.E22144

Fares A, Mahdy A, Ahmed G (2024) Unraveling the mysteries of silver nanoparticles: synthesis, characterization, antimicrobial effects and uptake translocation in plant – a review. Planta 260:7. https://doi.org/10.1007/s00425-024-04439-6

Faridvand S, Amirnia R, Tajbakhsh M, Enshasy HE, Sayyed RZ (2021) The effect of foliar application of magnetic water and nano, organic, and chemical fertilizers on phytochemical and yield characteristics of different landraces of fennel (*Foeniculum vulgare* mill). Horticulturae 7:475

Fazeli-Nasab B, Sayyed RZ, Piri R, Rahmani AF (2021) Biopriming and nanopriming: green revolution wings to increase plant yield, growth, and development under stress condition and forward dimensions. In: Antioxidants in plant-microbe interaction. Singapore, Springer Singapore, pp 623–655

Francis DV, Abdalla AK, Mahakham W, Sarmah AK, Ahmed ZFR (2024) Interaction of plants and metal nanoparticles: exploring its molecular mechanisms for sustainable agriculture and crop improvement. Environ Int 190:108859. https://doi.org/10.1016/j.envint.2024.108859

Frenk S, Ben-Moshe T, Dror I, Berkowitz B, Minz D (2013) Effect of metal oxide nanoparticles on microbial community structure and function in two different soil types. PLoS One 8:e84441. https://doi.org/10.1371/JOURNAL.PONE.0084441

Fu L, Wang Z, Dhankher OP, Xing B (2020) Nanotechnology as a new sustainable approach for controlling crop diseases and increasing agricultural production. J Exp Bot 71:507–519. https://doi.org/10.1093/jxb/erz314

Ge Y, Schimel JP, Holdena PA (2012) Identification of soil bacteria susceptible to TiO2 and ZnO nanoparticles. Appl Environ Microbiol 78:6749–6758

Hakim S, Naqqash T, Nawaz MS, Laraib I, Siddique MJ, Zia R, Mirza MS, Imran A (2021) Rhizosphere engineering with plant growth-promoting microorganisms for agriculture and ecological sustainability. Front Sustain Food Syst 5. https://doi.org/10.3389/fsufs.2021.617157

Huang D, Dang F, Huang Y, Chen N, Zhou D (2022) Uptake, translocation, and transformation of silver nanoparticles in plants. Environ Sci Nano 9:12–39. https://doi.org/10.1039/D1EN00870F

Huang D, Shi Z, Shan X, Yang S, Zhang Y, Guo X (2023) Insights into growth-affecting effect of nanomaterials: using metabolomics and transcriptomics to reveal the molecular mechanisms of cucumber leaves upon exposure to polystyrene nanoplastics (PSNPs). Sci Total Environ 866:161247. https://doi.org/10.1016/j.scitotenv.2022.161247

Iavicoli I, Leso V, Beezhold DH, Shvedova AA (2017) Nanotechnology in agriculture: opportunities, toxicological implications, and occupational risks. Toxicol Appl Pharmacol 329:96–111. https://doi.org/10.1016/j.taap.2017.05.025

Jakinala P, Lingampally N, Hameeda B, Sayyed RZ, Khan MY, Elsayed EA et al (2021) Silver nanoparticles from insect wing extract: biosynthesis and evaluation for antioxidant and antimicrobial potential. PLoS One 16(3):e0241729

Jangid H, Singh S, Kashyap P, Singh A, Kumar G (2024) Advancing biomedical applications: an in-depth analysis of silver nanoparticles in antimicrobial, anticancer, and wound healing roles. Front Pharmacol 15. https://doi.org/10.3389/fphar.2024.1438227

Ji B, Xuan L, Zhang Y, Mu W, Paek K-Y, Park S-Y, Wang J, Gao W (2023) Application of data modeling, instrument engineering and nanomaterials in selected Medid the scientific Recinal plant tissue culture. Plants 12:1505. https://doi.org/10.3390/plants12071505

Jiao L, Cao X, Wang C, Chen F, Zou H, Yue L, Wang Z (2023) Crosstalk between in situ root exudates and rhizobacteria to promote rice growth by selenium nanomaterials. Sci Total Environ 878:163175. https://doi.org/10.1016/j.scitotenv.2023.163175

Kandhol N, Singh VP, White JC, Tran L-SP, Tripathi DK (2023) Plant growth hormones and nanomaterial Interface: exploring the connection from development to defense. Plant Cell Physiol 63:1840–1847. https://doi.org/10.1093/pcp/pcac147

Karabulut F (2024) The impact of nanoparticles on plant growth, development, and stress tolerance through regulating phytohormones. In: Nanoparticles in plant biotic stress management. Springer Nature Singapore, Singapore, pp 87–105. https://doi.org/10.1007/978-981-97-0851-2_4

Khan S, Hossain MK (2022) Classification and properties of nanoparticles. In: Nanoparticle-based polymer composites. Elsevier, pp 15–54. https://doi.org/10.1016/B978-0-12-824272-8.00009-9

Khan M, Khan AU, Hasan MA, Yadav KK, Pinto MMC, Malik N, Yadav VK, Khan AH, Islam S, Sharma GK (2021) Agro-nanotechnology as an emerging field: a novel sustainable approach for improving plant growth by reducing biotic stress. Appl Sci 11:2282. https://doi.org/10.3390/app11052282

Khodakovskaya MV, Kim BS, Kim JN, Alimohammadi M, Dervishi E, Mustafa T, Cernigla CE (2013) Carbon nanotubes as plant growth regulators: effects on tomato growth, reproductive system, and soil microbial community. Small 9:115–123. https://doi.org/10.1002/SMLL.201201225

Kido ÉA, de Andrade Luz G, da Silva V, da Costa Gomes MF, Neto JRCF (2023) Hormonal regulation of heavy metal toxicity and tolerance in crop plants. In: Heavy metal toxicity and tolerance in plants. Wiley, pp 425–460. https://doi.org/10.1002/9781119906506.ch20

Klitzke S, Metreveli G, Peters A, Schaumann GE, Lang F (2015) The fate of silver nanoparticles in soil solution – sorption of solutes and aggregation. Sci Total Environ 535:54–60. https://doi.org/10.1016/j.scitotenv.2014.10.108

Kumar P, Thakur N, Kumar K, Kumar S, Dutt A, Thakur VK, Gutiérrez-Rodelo C, Thakur P, Navarrete A, Thakur N (2024) Catalyzing innovation: exploring iron oxide nanoparticles – origins, advancements, and future application horizons. Coord Chem Rev 507:215750. https://doi.org/10.1016/j.ccr.2024.215750

Kusiak M, Oleszczuk P, Jośko I (2022) Cross-examination of engineered nanomaterials in crop production: application and related implications. J Hazard Mater 424:127374. https://doi.org/10.1016/J.JHAZMAT.2021.127374

Kyziol-Komosinska J, Dzieniszewska A, Czupioł J (2024) Behavior of silver species in soil: Ag nanoparticles vs ionic Ag. Molecules 29:5531. https://doi.org/10.3390/molecules29235531

Lee W-M, Kwak JI, An Y-J (2012) Effect of silver nanoparticles in crop plants Phaseolus radiatus and Sorghum bicolor: media effect on phytotoxicity. Chemosphere 86:491–499. https://doi.org/10.1016/j.chemosphere.2011.10.013

Li Y, Wang J, Li B, Geng M, Wang Y, Zhao J, Jin B, Li Y (2021) Response of extracellular polymeric substances and microbial community structures on resistance genes expression in wastewater treatment containing copper oxide nanoparticles and humic acid. Bioresour Technol 340:125741. https://doi.org/10.1016/j.biortech.2021.125741

Lin D, Xing B (2007) Phytotoxicity of nanoparticles: inhibition of seed germination and root growth. Environ Pollut 150:243–250. https://doi.org/10.1016/J.ENVPOL.2007.01.016

Ma X, Geiser-Lee J, Deng Y, Kolmakov A (2010) Interactions between engineered nanoparticles (ENPs) and plants: phytotoxicity, uptake and accumulation. Sci Total Environ 408:3053–3061. https://doi.org/10.1016/J.SCITOTENV.2010.03.031

Martínez G, Merinero M, Pérez-Aranda M, Pérez-Soriano E, Ortiz T, Villamor E, Begines B, Alcudia A (2020) Environmental impact of nanoparticles' application as an emerging technology: a review. Materials 14:166. https://doi.org/10.3390/ma14010166

Mascarenhas R, Mathur T, Maheshwari J, Nagella P (2023) Interaction of nanomaterials with plant macromolecules: nucleic acid, proteins and hormones. In: Nanomaterial interactions with plant cellular mechanisms and macromolecules and agricultural implications, pp 231–271. https://doi.org/10.1007/978-3-031-20878-2_9

Mathur P, Chakraborty R, Aftab T, Roy S (2023) Engineered nanoparticles in plant growth: phytotoxicity concerns and the strategies for their attenuation. Plant Physiol Biochem 199:107721. https://doi.org/10.1016/J.PLAPHY.2023.107721

Mekuye B, Abera B (2023) Nanomaterials: an overview of synthesis, classification, characterization, and applications. Nano Select 4:486–501. https://doi.org/10.1002/nano.202300038

Mirakhorli T, Ardebili ZO, Ladan-Moghadam A, Danaee E (2021) Bulk and nanoparticles of zinc oxide exerted their beneficial effects by conferring modifications in transcription factors, histone deacetylase, carbon and nitrogen assimilation, antioxidant biomarkers, and secondary metabolism in soybean. PLoS One 16:e0256905. https://doi.org/10.1371/journal.pone.0256905

Mishra S, Keswani C, Abhilash PC, Fraceto LF, Singh HB (2017) Integrated approach of Agri-nanotechnology: challenges and future trends. Front Plant Sci 8:254477. https://doi.org/10.3389/fpls.2017.00471

Mittal D, Kaur G, Singh P, Yadav K, Ali SA (2020) Nanoparticle-based sustainable agriculture and food science: recent advances and future outlook. Front Nanotechnol 2:10. https://doi.org/10.3389/fnano.2020.579954

Mosa KA, El-Naggar M, Ramamoorthy K, Alawadhi H, Elnaggar A, Wartanian S, Ibrahim E, Hani H (2018) Copper nanoparticles induced genotoxicty, oxidative stress, and changes in superoxide dismutase (SOD) gene expression in cucumber (cucumis sativus) plants. Front Plant Sci 9:365054. https://doi.org/10.3389/FPLS.2018.00872/BIBTEX

Mottaleb SA, Hassan AZA, El-Bahbohy R, Mahmoud AWM (2021) Are copper nanoparticles toxic to all plants? A case study on onion (allium cepa l.). Agronomy 11:1006. https://doi.org/10.3390/AGRONOMY11051006/S1

Mukarram M, Petrik P, Mushtaq Z, Khan MMA, Gulfishan M, Lux A (2022) Silicon nanoparticles in higher plants: uptake, action, stress tolerance, and crosstalk with phytohormones, antioxidants, and other signalling molecules. Environ Pollut 310:119855. https://doi.org/10.1016/j.envpol.2022.119855

Pociecha E, Gorczyca A, Dziurka M, Matras E, Oćwieja M (2021) Silver nanoparticles and silver ions differentially affect the phytohormone balance and yield in wheat. Agriculture 11:729. https://doi.org/10.3390/agriculture11080729

Rajput V, Minkina T, Mazarji M, Shende S, Sushkova S, Mandzhieva S, Burachevskaya M, Chaplygin V, Singh A, Jatav H (2020) Accumulation of nanoparticles in the soil-plant systems and their effects on human health. Ann Agric Sci 65:137–143. https://doi.org/10.1016/j.aoas.2020.08.001

Rana R, Siddiqui M, Skalicky M, Brestic M, Hossain A, Kayesh E, Popov M, Hejnak V, Gupta D, Mahmud N, Islam T (2021) Prospects of nanotechnology in improving the productivity and quality of horticultural crops. Horticulturae 7:332. https://doi.org/10.3390/horticulturae7100332

Ranjan A, Rajput VD, Prazdnova EV, Gurnani M, Sharma S, Bhardwaj P, Shende SS, Mandzhieva SS, Sushkova S, Minkina T, Chauhan A, Jindal T, Prasad R, Wong MH (2024) Augmenting abiotic stress tolerance and root architecture: the function of phytohormone-producing PGPR and their interaction with nanoparticles. S Afr J Bot 167:612–629. https://doi.org/10.1016/j.sajb.2024.02.041

Rasouli H, Popović-Djordjević J, Sayyed RZ, Zarayneh S, Jafari M, Fazeli-Nasab B (2020) Nanoparticles: a new threat to crop plants and soil rhizobia? In: Sustainable agriculture reviews 41: nanotechnology for plant growth and development. Springer International Publishing, Cham, pp 201–214

Raval JB, Mehta VN, Singhal RK, Basu H, Jha S, Kailasa SK (2023) Insights of nanomaterials integrated analytical approaches for detection of plant hormones in agricultural and environmental samples. Trends Environ Anal Chem 39:e00205. https://doi.org/10.1016/j.teac.2023.e00205

Roy I (2023) Interactions of nanoparticles with plants. In: Bioinspired and green synthesis of nanostructures. Wiley, pp 157–188. https://doi.org/10.1002/9781394174928.ch7

Sabagh AEL, Islam MS, Hossain A, Iqbal MA, Mubeen M, Waleed M, Reginato M, Battaglia M, Ahmed S, Rehman A, Arif M, Athar H-U-R, Ratnasekera D, Danish S, Raza MA, Rajendran K, Mushtaq M, Skalicky M, Brestic M, Soufan W, Fahad S, Pandey S, Kamran M, Datta R, Abdelhamid MT (2022) Phytohormones as growth regulators during abiotic stress tolerance in plants. Front Agron 4. https://doi.org/10.3389/fagro.2022.765068

Santás-Miguel V, Arias-Estévez M, Rodríguez-Seijo A, Arenas-Lago D (2023) Use of metal nanoparticles in agriculture. A review on the effects on plant germination. Environ Pollut 334:122222. https://doi.org/10.1016/j.envpol.2023.122222

Shahbaz M, Akram A, Mehak A, Haq E, Fatima N, Wareen G, Aamir S, Hassan H, Poczai P, Sayyed RZ, Ilyas N (2023) Evaluation of selenium nanoparticles in inducing disease resistance against spot blotch disease and promoting growth in wheat under biotic stress. Plants 12(4):761. https://doi.org/10.3390/plants12040761

Shahrokh S, Hosseinkhani B, Emtiazi G (2014) The impact of nano-silver on bacterial aerobic nitrate reductase. J Bioprocess Biotech 4

Shanmugaiah V, Gauba A, Hari SK, Prasad R, Ramamoorthy V, Sharma MP (2023) Effect of silicon micronutrient on plant's cellular signaling cascades in stimulating plant growth by mitigating the environmental stressors. Plant Growth Regul 100:391–408. https://doi.org/10.1007/s10725-023-00982-6

Sharma S et al (2025) The combination of α-Fe_2O_3NP and *Trichoderma* sp. improves antifungal activity against *Fusarium* wilt. J Basic Microbiol:e2400613

Singh RP, Handa R, Manchanda G (2021) Nanoparticles in sustainable agriculture: an emerging opportunity. J Control Release 329:1234–1248. https://doi.org/10.1016/j.jconrel.2020.10.051

Strekalovskaya EI, Perfileva AI, Krutovsky KV (2024) Zinc oxide nanoparticles in the "soil–bacterial community–plant" system: impact on the stability of soil ecosystems. Agronomy 14:1588. https://doi.org/10.3390/agronomy14071588

Sun Y, Yang Y, Tou F-Y, Niu Z-S, Guo X-P, Liu C, Yan J, Wu J-Y, Xu M, Hou L-J, Liu M (2022) Extraction and quantification of metal-containing nanoparticles in marine shellfish based on single particle inductively coupled plasma-mass spectrometry technique. J Hazard Mater 424:127383. https://doi.org/10.1016/j.jhazmat.2021.127383

Thakur P, Thakur S, Kumari P, Shandilya M, Sharma S, Poczai P, Alarfaj AA, Sayyed RZ (2022) Nano-insecticide: synthesis, characterization, and evaluation of insecticidal activity of ZnO NPs against *Spodoptera litura* and *Macrosiphum euphorbiae*. Appl Nanosci 12:3835–3850

Torres-Vanegas JD, Cruz JC, Reyes LH (2021) Delivery systems for nucleic acids and proteins: barriers, cell capture pathways and nanocarriers. Pharmaceutics 13:428. https://doi.org/10.3390/pharmaceutics13030428

Tripathi D, Raikhy G, Kumar D (2019) Chemical elicitors of systemic acquired resistance – salicylic acid and its functional analogs. Curr Plant Biol 17:48–59

Tripathi D, Singh M, Pandey-Rai S (2022) Crosstalk of nanoparticles and phytohormones regulate plant growth and metabolism under abiotic and biotic stress. Plant Stress 6:100107. https://doi.org/10.1016/j.stress.2022.100107

Ulhassan Z, Bhat JA, Zhou W, Senan AM, Alam P, Ahmad P (2022) Attenuation mechanisms of arsenic induced toxicity and its accumulation in plants by engineered nanoparticles: a review. Environ Pollut 302:119038. https://doi.org/10.1016/J.ENVPOL.2022.119038

Usman M, Farooq M, Wakeel A, Nawaz A, Cheema SA et al (2020) Nanotechnology in agriculture: current status, challenges and future opportunities. Sci Total Environ 721:137778. https://doi.org/10.1016/j.scitotenv.2020.137778

Verma DK, Patel S, Kushwah KS (2020) Effects of nanoparticles on seed germination, growth, phytotoxicity and crop improvement. Agric Rev

Wang Z, Yue L, Dhankher OP, Xing B (2020) Nano-enabled improvements of growth and nutritional quality in food plants driven by rhizosphere processes. Environ Int 142:105831. https://doi.org/10.1016/J.ENVINT.2020.105831

Wang Z, Tang J, Zhu L, Feng Y, Yue L, Wang C, Xiao Z, Chen F (2022) Nanomaterial-induced modulation of hormonal pathways enhances plant cell growth. Environ Sci Nano 9:1578–1590. https://doi.org/10.1039/D2EN00251E

Wang J, Liu R, Huang X, Bao Y, Wang X, Yi H, Lu Y (2024) The effect of nanoscale modification of Nisin by different milk-derived proteins on its physicochemical properties and antibacterial activity. Foods 13:1606. https://doi.org/10.3390/FOODS13111606/S1

Weiss M, Fan J, Claudel M, Sonntag T, Didier P, Ronzani C, Lebeau L, Pons F (2021) Density of surface charge is a more predictive factor of the toxicity of cationic carbon nanoparticles than zeta potential. J Nanobiotechnol 19:5. https://doi.org/10.1186/s12951-020-00747-7

Wu H, Li Z (2022) Recent advances in nano-enabled agriculture for improving plant performance. Crop J 10:1–12. https://doi.org/10.1016/J.CJ.2021.06.002

Yadav K, Pradhan M, Singh D, Singh MR (2022) Targeting autoimmune disorders through metal nanoformulation in overcoming the fences of conventional treatment approaches. In: Translational autoimmunity. Elsevier, pp 361–393. https://doi.org/10.1016/B978-0-12-824390-9.00017-7

Yang F, Liu C, Gao F, Su M, Wu X, Zheng L, Hong F, Yang P (2007) The improvement of spinach growth by nano-anatase TiO2 treatment is related to nitrogen photoreduction. Biol Trace Elem Res 119:77–88. https://doi.org/10.1007/S12011-007-0046-4/FIGURES/9

Yang M, Wang L, Lu H, Dong Q, Li H, Liu S (2023) Graphene and graphene-like carbon nanomaterials-based electrochemical biosensors for phytohormone detection. Carbon Lett 33:1343–1358. https://doi.org/10.1007/s42823-022-00419-6

Zandalinas SI, Peláez-Vico MÁ, Sinha R, Pascual LS, Mittler R (2024) The impact of multifactorial stress combination on plants, crops, and ecosystems: how should we prepare for what comes next? Plant J 117:1800–1814. https://doi.org/10.1111/TPJ.16557

Zehra A, Rai A, Singh SK, Aamir M, Ansari WA, Upadhyay RS (2021) An overview of nanotechnology in plant disease management, food safety, and sustainable agriculture. In: Food security and plant disease management. Elsevier, pp 193–219. https://doi.org/10.1016/B978-0-12-821843-3.00009-X

Zhang X, Cao H, Wang H, Zhao J, Gao K, Qiao J, Li J, Ge S (2022) The effects of graphene-family nanomaterials on plant growth: a review. Nano 12:936. https://doi.org/10.3390/nano12060936

Zhao L, Hernandez-Viezcas JA, Peralta-Videa JR, Bandyopadhyay S, Peng B, Munoz B, Keller AA, Gardea-Torresdey JL (2013) ZnO nanoparticle fate in soil and zinc bioaccumulation in corn plants (Zea mays) influenced by alginate. Environ Sci: Processes Impacts 15:260–266. https://doi.org/10.1039/C2EM30610G

Zhou P, Adeel M, Shakoor N, Guo M, Hao Y, Azeem I, Li M, Liu M, Rui Y (2020) Application of nanoparticles alleviates heavy metals stress and promotes plant growth: An overview. Nano 11:26. https://doi.org/10.3390/nano11010026

Regulatory and Safety Concerns Related to the Use of Nanomaterials

13

Rihab Djebaili, Beatrice Farda, Ramila Fares, and Marika Pellegrini

Abstract

Nanomaterials (NMs) have gained attention and increased interest from many scientific disciplines due to their diversified fields of application and specific physicochemical properties. The remarkable advancement of nanotechnology has revolutionised various sectors, including agriculture, where nanomaterials are progressively being examined to enhance productivity, sustainability, and environmental health. NMs' application in agriculture raises concerns for their unique properties and benefits related to increasing yields and plant stress resistance. One of the significant challenges facing agriculture today is the efficient and sustainable use of resources such as water, fertilisers and pesticides. Traditional methods often result in considerable waste and environmental pollution. Nanomaterials offer solutions through targeted delivery systems, controlled release mechanisms and improved efficacy of agrochemicals. This chapter intends to discuss the classification of nanomaterials, production procedures, problems related to toxicity, and regulatory restrictions on their use. The NMs classification, focusing on the organic and inorganic classes of NMs, is presented. Efficiency, scalability, and environmental impact have been summarised

R. Djebaili · B. Farda (✉) · M. Pellegrini
Department of Life, Health and Environmental Sciences, University of L'Aquila, L'Aquila, Italy
e-mail: beatrice.farda@graduate.univaq.it

R. Fares
Department of Life, Health and Environmental Sciences, University of L'Aquila, L'Aquila, Italy

Laboratory Biotechnology, Water, Environment, and Health, Abbes Laghrour University of Khenchela, Khenchela, Algeria

Laboratory of Mycology, Biotechnology, and Microbial Activity, Brothers Mentouri University of Constantine, Constantine, Algeria

R. Sayyed et al. (eds.), *Nanobiostimulants in Innovative Agriculture*,
https://doi.org/10.1007/978-981-95-2477-8_13

with respect to NMs' chemical, physical, and biological processes synthesis. The toxicity of NMs has been considered thoroughly, underlining the health and environmental hazards possibly linked with these particles and hence suggesting a deep understanding of their interaction with a biological system. The chapter further describes the present regulatory frameworks and improvements made for uniform regulations concerning safety in NMs creation, manufacturing process, and application.

Keywords

Agriculture · Manufacturing · NMs production · Regularity-framework · Toxicity

1 Introduction

Nanomaterials (NMs) are nanostructures with sizes of up to 100 nanometres. Because of their small size, these particles have configurable functions with surface molecules, which changes their molecular interactions, resulting in novel applications (Saleh 2020; Devi et al. 2024; Eker et al. 2024). The particles' attributes, in combination with their structural elements, outline their principal functions, activity profiles, and utility traits. The size and shape of NMs can influence the optical responses and cells' intake. These are of crucial importance in medicinal applications in treating cancer and in studies of imaging (Siddique and Chow 2020; Devi et al. 2024). Another critical factor taken into consideration is related to the surface charge of the NPs, which influences cellular interaction and the toxicity potential, together with the characteristics of the material (Augustine et al. 2020; Eker et al. 2024). Different physical properties can be dependent on the NMs' size and shape. The importance of NMs was realised when researchers found that size influences the physicochemical properties of a substance, including its chemical, electrical, mechanical, and optical properties. Thus, NMs have attracted considerable interest because of their unique properties (Saleh 2020; Osman et al. 2024; Truong et al. 2024). Such NPs have opened up possibilities for applications in water treatment plants, oil refineries, the petrochemical industry, catalytic processes, diagnostics of buildings and building materials, and drug delivery. Several techniques are involved in the production of NPs, including condensation, attrition, chemical precipitation, ion implantation, pyrolysis, and hydrothermal synthesis (Saleh 2020; Osman et al. 2024). In recent years, the utilisation of nanotechnology in agriculture has gained considerable interest. Nanotechnology is a new approach to combining productivity, sustainability, and efficiency (Mukhopadhyay 2014; Osman et al. 2024).

Miniaturisation of the delivery system increases efficacy in the creation of high-capacity storage and controlled release of potent fertilisers, insecticides, herbicides, and plant growth regulators (Thakur et al. 2022; Usman et al. 2020). In this regard, nanotechnology enhances the capacity of farmers to address climate change and food security by using fewer natural resources and increasing agricultural sustainability. Application of NMs in the manufacturing of highly improved sensor monitors for real-time soil moisture, nutrient levels, and the plant's health. This improves water purification, enhances irrigation techniques, and reduces water stress (Adetutu et al. 2024). Thus far, different world governments have shied away from efficiently regulating NMs. So far, the most significant barriers to the establishment of solid regulatory regimes have been the lack of a standard definition and the heterogeneity of NMs in themselves (El-Kady et al. 2023; Chávez-Hernández et al. 2024). Notable regulatory advancements encompass: (I) The European Union defines and governs NMs through the REACH (Registration, Evaluation, Authorisation and Restriction of Chemicals) legislation. Manufacturers and importers must inform the European Chemicals Agency about using NMs (Bleeker et al. 2017). (II) In the United States, the Environmental Protection Agency (EPA) can regulate NMs under the Toxic Substances Control Act (TSCA). The EPA has promulgated rules on reporting and record-keeping requirements for NMs (Klabunde et al. 2010). (III) Other nations, including Canada, Japan, and China either have or are developing some sort of regulatory schemes for NMs (Assessment 2009; Chowdhury et al. 2023).

Apart from the property problems associated with small size/high-surface-area materials, special concerns arise with regard to their toxicity and environmental impact. The health adverse effects associated with exposure to NMs are in the respiratory, cardiovascular, and nervous systems, besides genotoxicity and carcinogenicity (Tang et al. 2024). Other environmental risks include bioaccumulation in biological organisms and changes in ecosystem dynamics (Keller et al. 2024). More intensive research is required to enhance the understanding of NMs' safety and possible toxicity (Budamagunta et al. 2023; Petersen et al. 2015; Corsi et al. 2023). In this context, the present chapter gives an overview of recent progress made in classifying NMs, methods of synthesis, and related toxicity problems. It gives an overview of the current legislative framework and developments concerning the establishment of harmonised legislation in the field of safety in NMs production, manufacturing processes, and applications, as well as the legal limits concerning their uses (Fig. 13.1).

2 Classification

2.1 Based on Their Origin NMs Can Be Classified as

Natural NMs Natural NMs exist in diverse forms, including viruses, protein molecules, minerals such as clay, natural colloids like milk and blood (liquid colloids), fog (aerosol type), gelatin (gel type), mineralised organic materials like shells, corals, and bones, insect wings, opals, spider silk, lotus leaves, gecko feet, volcanic ash, and ocean spray (Khan and Hossain 2022).

Artificial NMs Examples of artificial NMs are carbon nanotubes and semiconductor NMs like quantum dots (QDs). NMs are classified as metal-based materials, dendrimers, or composites based on their structural composition (Khan and Hossain 2022).

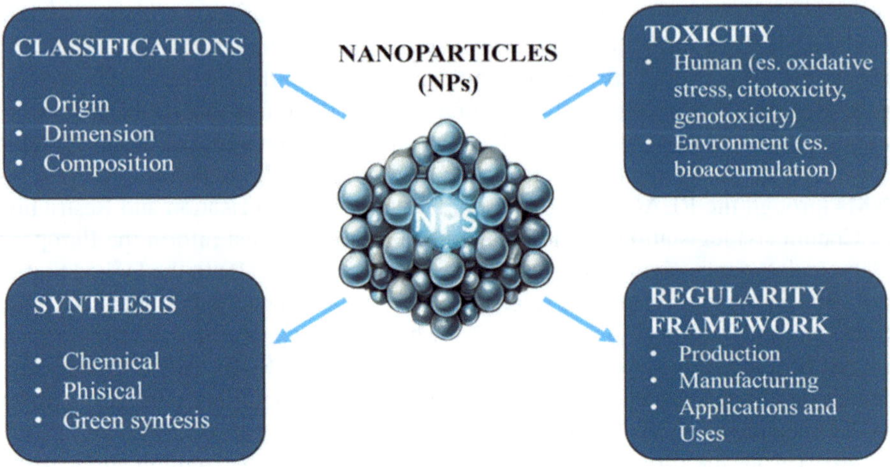

Fig. 13.1 Summary of the main topics addressed in the chapter

2.2 Based on Their Dimensionality We Can Find

Zero-Dimensional NMs Nanomaterial 0D has exterior dimensions between 1 and 100 nm. It includes quantum dots, semiconductor nanocrystals with diameters less than 10 nm and can be considered potential wells. The 0D NMs also include many forms of NMs, including whole spheres, highly symmetrical dendrimers, branching macromolecules, and hollow spheres (Dolez 2015).

One-Dimensional NMs One-dimensional (1D) NMs possess two exterior dimensions at the nanoscale, while the third dimension often exists at the microscale. This encompasses nanofibers, nanotubes, nanowires, and nanorods (Dolez 2015). NMs in this category possess two dimensions (*x*, *y*) within the nanoscale range, whereas one dimension of the nanostructure exceeds the non-metric threshold (>10 nm). One-dimensional NMs, including nanofibers, nanotubes, nanorods, thin films, and nanowires, are like needle-shaped NMs. They may be amorphous or crystalline, single or polycrystalline, chemically pure or impure, independent materials, or integrated inside another medium, such as metallic, ceramic, or polymeric. One-dimensional NMs may be classified as metallic, ceramic, or polymeric (Mekuye and Abera 2023).

Two-Dimensional NMs NMs 2D, characterised by a singular exterior dimension at the nanoscale, include thin films, nanocoatings, and nanoplates. Thin films comprise ceramic or metallic coatings (Dolez 2015). They are mostly utilised in physics and electronics, for instance, to produce electronic components with insulating or conductive surface characteristics or to modify the optical reflectance of surfaces. Different materials, such as polymers and composites, can fabricate nanocoatings (Mekuye and Abera 2023).

Three-Dimensional NMs NMs 3D exhibit internal nanoscale characteristics lacking outward dimensions at the nanoscale. This encompasses nanocomposites and nanostructured materials. Nanocomposites are multiphase solid materials characterised by at least one phase with a nanoscale exterior dimension. The word nanocomposite often refers to nanofillers distributed inside a bulk matrix. Nanofillers can be chosen from 0D, 1D, or 2D NMs. Matrices can be polymers, ceramics, or metals. The ultimate material may be one-dimensional, two-dimensional, or three-dimensional, namely a fibre, a film, or a solid volume (Camargo et al. 2009).

2.3 Based on Their Composition We Can Group NMs in

Organic NMs In the nanoscale, organic molecules are transformed into organic NMs. Examples of organic NMs or polymers include liposomes, dendrimers, micelles, and ferritin. Biodegradable, non-toxic NMs with hollow interiors, called nano capsule micelles and liposomes, are sensitive to thermal energy, electromagnetic radiation, and light (Khan and Hossain 2022).

Carbon-Based NMs Carbon nanotubes, graphene, fullerenes, carbon nanofibers, and carbon black are the five most commonly used carbon-based NMs. Most applications of carbon-based NMs have been used as reinforcement. The material is an insulator, although there is thermal conductivity along the axis of the carbon-based nanomaterial (Gu et al. 2022; Eker et al. 2024).

Composites NMs Composite NMs include aggregated forms that interface with larger materials, and NMs combined with bulk materials. Current applications involve the improvement of mechanical and thermal properties of a range of products, from automotive components to packaging materials (Gu et al. 2022).

Inorganic NMs Within this group we can find:

Metal-Based NMs Metal-based NMs can be produced using destructive or constructive methods. Aluminium (Al), cadmium (Cd), cobalt (Co), copper (Cu), gold (Au), iron (Fe), lead (Pb), silver (Ag), and zinc (Zn) are metallic substances commonly employed in the creation of NMs. Due to their quantum effects and substantial surface-to-volume ratio, metal NMs exhibit remarkable ultraviolet-visible sensitivity and exceptional electrical, catalytic, thermal, and antibacterial capabilities. Metal NMs are utilised throughout several study domains due to their exceptional optical characteristics (Eker et al. 2024).

Metal Oxide NMs Metal oxide nanomaterial contain positive metallic ions and negative oxygen ions. Most common syntheses involve the metal oxides like SiO_2, TiO_2, ZnO, and Al_2O_3. All these NMs illustrate exceptional properties when differentiated from their metallic counterparts (Khan and Hossain 2022).

Ceramic NMs These are inorganic solids of carbides, carbonates, oxides, and phosphates prepared by heat treatment followed by cooling. Ceramic NMs can be utilised in medication delivery systems, particularly for targeting tumours, glaucoma, and certain bacterial infections. NMs garner significant attention from researchers for their applications in catalysis, photocatalysis, photodegradation of dyes, and imaging technologies (Khan et al. 2019).

Lipid-Based NMs The general shape of the lipid-based nanoparticle is spherical, ranging from 10 to 100 nm. It has a solid lipid core with a matrix containing soluble lipophilic compounds. In the biomedical sector, the vectors of drugs and providing the release of RNA in cancer treatment are lipid-based NMs (Khan et al. 2019; Eker et al. 2024).

2.4 Based on Their Potential Toxicity We Can Organize NMs in

Fiber-Like Although fibre-like NMs lack asbestos-like features, they are similar to carbon nanotubes, fibrer-like metal oxides, and permanent carbon nanotubes. Workplace exposure limits for persistent biocarbon nanotubes and stiff NMs range from 10^4 to 10^5 fibres m^{-3} (El-Kady et al. 2023; Mekuye and Abera 2023).

Biopersistent Granular NMs The suggested exposure limits for persistent granular NMs are 2×10^7 particles m^{-3}, comparable to gold, silver, cobalt, lanthanum, lead, iron, iron oxide, cerium oxide, antimony oxide, and tin oxide. There is also a recommendation for insoluble NMs without a workplace exposure limit of a value of 0.3 mg m^{-3} (El-Kady et al. 2023; Mekuye and Abera 2023).

CMAR (Carcinogenic, Mutagenic, Asthma Genic, Reproductive Toxin) NMs The toxic CMAR NMs are nickel, cadmium-containing QDs, chromium VI, beryllium, arsenic, and zinc chromate. The recommended occupational exposure limits are 2×10^7 to 4×10^7 particles m^{-3}. Whenever there is a lack of any work exposure limit for insoluble NMs, the recommended threshold is 0.003 mg m^{-3}. The recommended threshold for soluble NMs lacking any work exposure limit is 1.5 mg m^{-3} (El-Kady et al. 2023; Mekuye and Abera 2023).

3 Synthesis and Production of NMs

3.1 Chemical Methods

3.1.1 Polyol and Alcohol Reduction Processes

Organic and inorganic molecules act as reducing agents when chemically synthesising nanometals. Chemical disposal of NMs occurs through two methods, namely, the alcohol reduction method and the polyol method (Zhang and Erkey 2006; Wang et al. 2009; Rasool et al. 2024). Silver NMs are synthesised using various organic solvents, such as n-butyl alcohol and ethylene glycol (Corrêa et al. 2016). Copper NMs were synthesised using polyvinyl pyrrolidone in nitrogen (Tam and Ng 2015; Din and Rehan 2017). Ethanol and water have been employed to synthesise high-velocity gold NMs (AuNPs) by a plasma dissolution method (Rasool et al. 2024). Circular gold NMs, ranging in size from 23 to 79 nm, were synthesised in a one-step procedure within a polymer matrix grafted with PEG-g-PVA. The reduction of silver nitrate facilitates nanosynthesis utilising ethylene glycol in polyvinylpyrrolidone (PVP) (Alkilany et al. 2015; Dheyab et al. 2022).

3.1.2 Micro-emulsion Method

Various materials have been successfully prepared using the oil-in-water or water-in-oil microemulsion approach. ZnO nanoparticle was synthesised by the technique of microemulsion (Wang et al. 2014). Silver nano dendrites were synthesised by the electrochemical method using BaTiO with the addition of surfactants such as polyoxyethylene ether, alcohol, nonylphenol, and hexadecyl dimethyl ammonium bromide. Zinc oxide-Au NMs were synthesised using an electrochemical method (Rasool et al. 2024). Silver nanostructures were created using an electrochemical process that included potentiostatic pyrrole polymerisation and galvanic deposition (Latif-ur-Rahman et al. 2015). The potentiostat has been utilised to synthesise TiO_2 NMs (Hu et al. 2010).

3.1.3 Thermal Breakdown of Metallic Salts

The electrochemical method involving the potentiostatic polymerisation of pyrrole and galvanic deposition resulted in silver nanostructures. The potentiostat has also synthesised TiO_2 NMs (Feng et al. 2016). Polymeric NMs are synthesised with regulated properties from iron-based precursors, including iron pentacarbonyl, iron oleate, and iron oxide (Nodera and Kanai 2004). The thorough burning of boron oxide, urea, copper nitrate, and h-BNN produces Cu NMs. Copper NMs have been synthesised through the thermal decomposition of copper oleate molecules (Patil et al. 1968).

3.1.4 Electrochemical Synthesis

Synthesis of titanium dioxide NMs has been carried out using the electrochemical technique with propyl ammonium bromide salt as a stabilising agent (Anandgaonker et al. 2019). Meso-porous gold (Au) has been synthesised via soft templates from stable electrochemical deposition of block copolymers (Nugraha et al. 2020). The electrochemical technique has also been applied to synthesise silver nanodendrites (Rasool et al. 2024). The synthesis of zinc oxide-Au NMs and an epidermal carbon electrode modification have been carried out using the electrochemical method (Starowicz and Stypuła 2008). Silver nanostructures were synthesised by an electrochemical method involving potentiostatic polymerisation of pyrrole and galvanic deposition (Latif-ur-Rahman et al. 2015). Titanium dioxide NMs have been synthesised using a chemical method using potentiostat. Cobalt NMs have been synthesised via the reduction of $COCl_4^2$ anthracene assisted by potentiostatic electrolysis, which induces ion generation in the dissolution of the cobalt anode (Hu et al. 2010; Rasool et al. 2024).

3.2 Physical Method

This process involves the synthesis of metallic NMs from fragmented metallic cables (Wang et al. 2018). Metallic NMs originating from copper and its alloys are synthesised by burning metal wires in aqueous solutions, producing extremely conductive hues (Nawab et al. 2023). The aluminium wire detonated, forming aluminium nanosilicon, which subsequently built microporous membranes (Liu et al. 2013). Silver and copper NMs were synthesised by applying high voltage current, resulting in the evaporation of the wire, a technique known as explosive wire technology. The diameter of the NMs was controlled by changing the voltage conditions (Zhang et al. 2022). The blasting process was used in order to produce nanoaluminium in nitrogen and argon atmospheres (Antony et al. 2010) (Fig. 13.2).

3.3 Green Synthesis

3.3.1 Microbial Synthesis

The synthesis of green NMs from microbes has attracted heightened interest in green synthesis owing to its environmental sustainability and superior manufacturing simplicity compared to alternative approaches (Park et al. 2016; Dahoumane et al. 2017; Sundararajan et al. 2024). Diverse microorganisms, such as bacteria, algae, and fungi, are presently employed for nanoparticle synthesis in green nanotechnology. Methods include sonication for cell wall lysis, centrifugation for biomass separation, and washing procedures for nanoparticle purification, which obviates the need for additional cellular recovery of NMs (Sundararajan et al. 2024). Bacteria that can be used in the green synthesis of NMs belong to a vast category of unicellular organisms with cell walls without organelles or an organised nucleus. Various strains, such as *Escherichia coli* and *Bacillus subtilis*, are easily cultivable, and their genetic sequences can be quickly altered. The features of bacteria render nanoparticle manufacturing a feasible method. Bacteria are first cultured aerobically to reach a predetermined optical density, after which the growth medium containing the cells is combined with the nanoparticle precursor for nanomaterial synthesis. Following an incubation period and a discernible change in media colour, the media is subjected to high-speed centrifugation. The supernatant from this centrifugation contains a suspension of NMs (Huston et al. 2021). Silver, gold, and cadmium NMs are formed in various forms due to interactions with various biomolecules. Gurunathan et al. showed that the proteins on the surface of the *E. coli* cell wall interacted with silver nitrate and chloroauric solution, resulting in the formation of irregular and triangular morphologies of silver and gold NMs (Gurunathan et al. 2009). Gold NMs, ranging from 10 to 20 nm, were synthesised extracellularly using photosynthetic bacteria such as *Rhodopseudomonas capsulata*. The bacterial enzyme Nicotinamide Adenine Dinucleotide Hydride (NADH)-dependent reductase is crucial in reducing gold ions to gold NMs (Prabhu Balaraman et al. 2021). Schluter reported the extracellular synthesis of palladium NMs utilising *Pseudomonas* (Schlüter et al. 2014). Singaravelu et al. proposed the synthesis of extracellular gold NMs from *Sargassum wightii* algae (Singaravelu et al. 2007). Research on the creation of NMs utilising algae is little investigated. The drawbacks of this procedure include the presence of hazardous bacteria, fungi, and algae, necessitating the implementation of safety measures (Kolahalam et al. 2019). The fungus *Fusarium oxysporum* was used to synthesise extracellular silver NMs. These NMs exhibit long-term stability attributable to the enzymatic activity of NADH-reductase. Compared to bacterial cells, fungal cells produce more protein (Ahmad et al. 2003).

3.3.2 Plant Extracts

The application of plant extracts for the reduction and stabilisation of agents has generated scientific interest in the creation of NMs, particularly concerning food safety. The utilisation of plant extracts is simple, cost-effective, environmentally

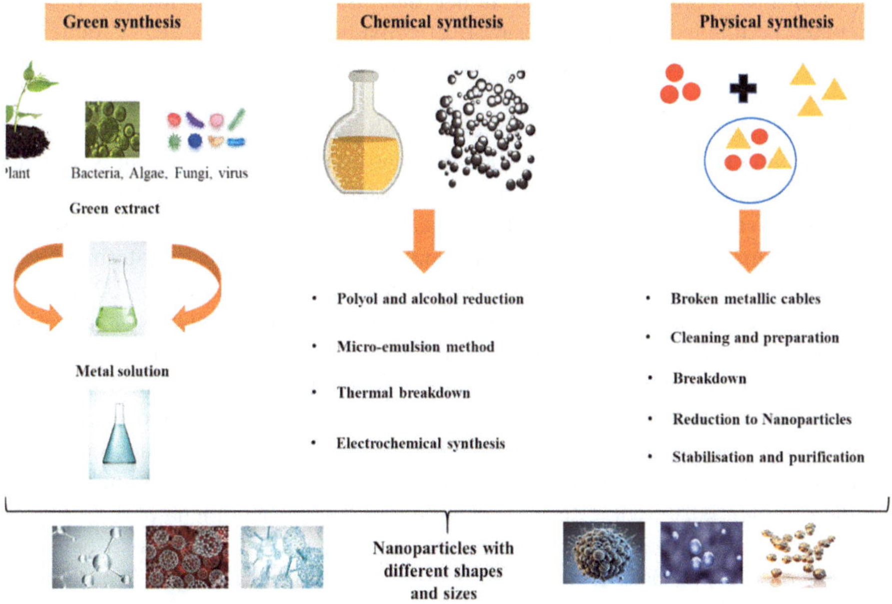

Fig. 13.2 Synthesis methods of NMs

friendly, and reduces the risk of contamination. Various plant components, such as stems, roots, leaves, flowers, fruits, seeds, bark, or entire plants, can synthesise NMs eco-friendly (Nayeri et al. 2021; Daneshmand et al. 2024; Alsulami et al. 2023; Tasnim et al. 2024; Shah et al. 2022; Jafarzadeh et al. 2024; Sharma et al. 2025). Plants' phytochemicals reduce metal NMs. Phytochemicals such as flavones, organic acids, and quinones naturally serve as effective reducing agents for nanoparticle synthesis. Gold NMs of various shapes are synthesised from the biomass of *Medicago sativa* and *Pelargonium graveolens* leaves (Kolahalam et al. 2019). Bimetallic Au, Ag, and bimetallic Au core-Ag shell NMs are synthesised from *Azadirachta indica* leaves. This plant's sugars and terpenoids function as reducing agents (Shankar et al. 2004). Gold nanotriangles are synthesised using *Aloe vera* leaf extract. NMs of silver, nickel, cobalt, zinc, and copper are synthesised using diverse plants such as *Helianthus annuus* (Banerjee et al. 2014).

4 Toxicity and Risk Assessment

There is a recognised requirement for standardised risk assessment methods in the agricultural sector that consider the toxicity of NMs, their behaviour in ecosystems, and their potential accumulation in soils and food crops. Regulatory authorities consent to a risk evaluation before approving any nanotechnology-based agricultural product. This encompasses assessing the toxicity of the utilised NMs, the likelihood of environmental dispersion, and the effects on human health. Research shows the necessity for preliminary assessment exposure studies and dose-dependent safety evaluations in natural environments, rather than just relying on in vitro experiments, to address ecological consequences more effectively (Mishra et al. 2017). Table 13.1 summarises the principal human and ecological toxicological consequences and the risk assessment of the primary types of NMs.

Titanium dioxide NMs (TiO_2) are extensively utilised in multiple fields, potentially harming human health and the environment. The primary way that NMs enter the body is through inhalation, causing transportation to several lung tissues, capillaries and alveoli and translocated to the heart, liver, and nervous systems. Studies show titanium NPs do not cause severe pulmonary overload or long-term adverse effects at inhalation concentrations of up to 35 mg m^{-3} (Rashid et al. 2021). Other exposure modes are recognised, like ingestion, dermal contact, or injection. Although TiO_2 NMs typically exhibit minimal acute toxicity, prolonged exposure may induce oxidative stress, inflammation, and genotoxicity, particularly in working environments or via ingestion (Warheit and Donner 2015; Baranowska-Wójcik et al. 2020). Environmental impacts include bioaccumulation in aquatic organisms and disruptions in soil ecosystems. The size and shape of NMs determine the magnitude of potentially toxic effects, causing aggregation phenomena derived by pH, ionic strength, and ionic identity (inorganic and organic) of aqueous suspensions. Ecotoxicological hazards of nano-TiO_2 were found in bacteria, algae, invertebrates, nematodes, and rainbow trout (Sharma 2009). Risk assessments emphasise regulated usage, safety measures, and the development of green nanotechnologies to mitigate potential health and ecological hazards. NMs from zinc oxides are applied in industries, agriculture (Faridvand et al. 2021; Akhtar et al. 2022) and biomedical treatments (Jakinala et al. 2021; Kapadia et al. 2021). Toxicological and ecotoxicological studies underlying their high-risk assessment, mainly due to their cytotoxicity and genotoxicity in humans. ZnO-NPs, thanks to their ability to generate ROS and release zinc ions (Zn^{2+}), interact with biological targets, causing cytotoxicity, genotoxicity, and oxidative stress (Fernández-Bertólez et al. 2024). Many reports are on the adverse effects on the soil microbiota, marine ecosystems, and human health. For instance, it has been stated that ZnO-NPs interfere with enzyme activities of soil microbes, affecting marine crustacean survival, and impair cellular DNA and viability in human glial cells. Knowing these risks is important to

Table 13.1 Toxicity and risk assessment of the most relevant NM classes

NM class	Human toxicity	Ecological toxicity	Risk assessment	References
Titanium dioxide (TiO_2)	Irritation of respiratory tracks, oxidative stress, and cellular damage	Accumulation in sediments and toxicity for bacteria and algae	*Moderate*—Low toxicity if controlled, risky with high exposure.	Warheit and Donner (2015), Baranowska-Wójcik et al. (2020), Rashid et al. (2021)
Zinc oxide (ZnO)	Respiratory track toxicity, cytotoxic and genotoxic effects, and skin and eye irritation	Highly toxic to aquatic organisms due to Zn^{2+} ion release	*High*—Significant risk due to cytotoxicity, genotoxicity, and aquatic toxicity	Shen et al. (2015), Vimercati et al. (2020), Fernández-Bertólez et al. (2024), Murali et al. (2023)
Silver (AgNPs)	Cytotoxic, oxidative stress and systemic effects	Toxic to aquatic life, bioaccumulation, and ion release toxic	*High*—Widespread use increases exposure risk, significant environmental and human toxicity	Jaswal and Gupta (2023), Noga et al. (2023)
Iron oxide (Fe_2O_3, Fe_3O_4)	Low toxicity, high doses, toxic (cellular stress)	Low ecotoxicity, risks at high concentrations	*Low to Moderate*—Biocompatible, localised risks at high doses	Caixeta et al. (2021), Chen and Hou (2023), Tao et al. (2023)
Gold (AuNPs)	Biocompatible but may cause cellular damage	Low ecological toxicity, but may affect aquatic feeding behaviour	*Low*—Minimal health and ecological risks	Hadrup et al. (2015), Sani et al. (2021)
Copper (CuNPs)s	Cytotoxicity, oxidative stress, DNA damage	Toxic to aquatic life, plants, and soil microbe's bioaccumulation.	*High*—High reactivity, cytotoxicity, ecological toxicity, and increased exposure from antimicrobial use	Ameh and Sayes (2019), Naz et al. (2020)

ensure sustainable and safe applications of NMs (Shen et al. 2015; Vimercati et al. 2020). It is evidently known that AgO NPs used in various industries might be potentially toxic to humans and the environment. Toxicity studies in human cells indicate that exposure to AgO NPs results in the generation of ROS through their penetration via inhalation, ingestion, or skin contact. Such oxidative stress gives rise to cellular injuries at higher levels, such as in the liver, lungs, kidneys, and nervous system, and then develops into DNA damage, apoptosis, or inflammatory responses (Jaswal and Gupta 2023). In the environment, AgO NPs could interact with several organisms, from procaryotic to higher organisms, through interference with cellular processes, finally leading to their accumulation in ecosystems. Such interactions could alter metabolic functions in aquatic and soil organisms, thus modulating ecological balance. Moreover, lower soil pH, reduced organic matter content, and diminished cation exchange capacity impede silver (Ag) sorption to soil, hence increasing the risks of mobility, toxicity, and bioavailability of silver NMs (AgNPs). There is an emphasis on extending risk assessment to include extensive evaluation for dose-dependent responses and developing less toxic green synthesis methods for mitigation without compromising the antimicrobial benefits of AgO NPs (Noga et al. 2023).

Iron oxide NMs (IONPs) have shown rather low toxicity in humans; therefore, they are found in many biomedical applications, including drug delivery and imaging. They are biocompatible and biodegradable, and the iron from IONPs is finally metabolised and stored safely in proteins such as ferritin. IONPs can be metabolised safely by cells through lysosomal pathways, where they regulate any released iron to minimise possible cytotoxic effects. Only at high concentrations, above 50 $\mu g\ mL^{-1}$, IONPs possibly increase ROS levels, which might result in oxidative stress; however, in the standard dosing of biomedicine, they are generally safe (Caixeta et al. 2021; Chen and Hou 2023).

For environmental risk, it was recorded that IONPs tend to bioaccumulate within aquatic organisms. However, the controlled usages and applications of the low-risk concentration generally showed good tolerance in most ecosystems (Tao et al. 2023). Overall, risk assessments underline the demand for further data on long-term environmental exposure and confirm low human toxicity under controlled conditions. Gold NMs (AuNPs) hold significant promise in medical and industrial fields. In general, toxic influence develops according to their concentration, size, shape, and modifications of their surface. For instance, it was stated that DNA damage and induction of oxidative stress occurred upon treatment with 10–20 $\mu g\ mL^{-1}$ of AuNPs in lung and liver cells. In animal models, 5–10 mg kg^{-1} administered has caused accumulation in the liver and kidneys, sometimes crossing the blood-brain barrier when the particles are small (under 10 nm) (Hadrup et al. 2015; Sani et al. 2021).

Copper NMs have huge applications; however, they present serious issues in human and environmental health. This is mainly due to the generation of ROS and the release of copper ions, which in turn cause oxidative stress and result in cellular component damage. Humans are exposed via inhalation, ingestion, and dermal exposure, leading to inflammation, cellular damage, or accumulation in specific organs like the liver and kidneys (Naz et al. 2020). Environmental exposure enhances transformation in CuNPs such as sedimentation and dissolution. These processes eventually modify the bioavailability and toxicity of CuNPs to various aquatic organisms, vegetation, and microorganisms. For example, CuNPs may suppress the photosynthetic activity and uptake of nutrients in plants or provoke oxidative stress and damage to organs in fish (Ameh and Sayes 2019). The assessment of CuNP risk has to be founded on a deeper understanding of its physicochemical properties, coupled with long-lasting effects, to ensure environmental safety as well as proper management.

5 Regularity Framework

Globally, the regulations concerning the application of NMs in agriculture are constantly evolving. Although using NMs in agriculture can improve sustainability and efficiency, many regulatory gaps must be addressed for their safe and practical integration. Nowadays, each country maintains a distinct regulatory framework regarding the application of nanotechnology in agriculture. Among them, Europe includes specific regulations, guidelines, and recommendations for NMs in agriculture, governed by the European Chemicals Agency (ECHA) and the European Food Safety Authority (EFSA) (Kumari et al. 2023). The EU Regulation 2019/1009 specifies marketing regulations for fertilizing products and guarantees safety for human health and the environment. The Regulation (EC) 1107/2009 provides a framework for authorising and placing nanopesticide products on the market (Amenta et al. 2015). Other regulations, such as Regulation (EC) 767/2009-(EC) 396/2005-(EC) 1935/2004, establish the authorising and marketing rules of feed additives, the maximum residue levels for nanopesticides, and the general safety standards for products in contact with food, respectively. In the United States, Pesticides are governed by the Environmental Protection Agency (EPA), while agricultural biotechnology products are governed by the U.S. Department of Agriculture (USDA). Currently, the EU is recognised as the sole global region that has integrated nano-specific provisions into legislation concerning agriculture, feed, and food. This includes distinct information requirements for the risk assessment of NMs, legally binding definitions of the term "nanomaterial," and mandates for labelling or reporting the presence of NMs in products (Subramanian and Rajkishore 2018).

In the US, the use of NMs in the agricultural sector, specifically for pesticide regulation, is regulated by the Food and Drug Administration (FDA), the Environmental Protection Agency (EPA), and the United States Department of Agriculture (USDA). Some US Agencies, like the US Environmental Protection Agency (US-EPA), regulate NMs' safety, authorisation, and market regulation. For example, the Toxic Substances Control Act (TSCA) and the Federal Insecticide, Fungicide, and Rodenticide Act (FIFRA) allow EPA to regulate the production, importation, use, and disposal of chemical substances, and the registration and use of pesticides, including nano pesticides, respectively. Although the presence of these regulatory laws, no specific requirements for NMs are available under this regulation. US-EPA is studying several nano-compounds, such as nano-silver, carbon nanotubes, cerium dioxide, titanium dioxide, iron, and micronized copper, to develop efficient testing strategies for engineered NMs. For this reason, the Nanotechnology Research and Development Act (NRDA) is legislation that may instruct federal agencies to synchronise research and development initiatives about nanotechnology. The objective is to guarantee a comprehensive understanding of the hazards and benefits of nanotechnology with the establishment of suitable legislation.

In Asia, some countries are active in producing and regulating NMs, establishing standards and certification systems for nano-enabled products. Several government organisations in China, including the Ministry of Agriculture and Rural Affairs (MARA), the State Administration for Market Regulation (SAMR), the National Health Commission (NHC), and the Ministry of Ecology and Environment (MEE), are in charge of regulating agricultural products based on nanotechnology. In China, food safety is regulated under the Food Safety Law, which does not include any specifications for NMs. The National Centre for Nanoscience and Technology (NCNST) and the Commission on Nanotechnology Standardisation are responsible for developing national standards in the nanotechnology area. Regulation on the Safety Assessment of Agricultural Genetically Modified Organisms (MARA Order No. 7)-2001 and Administrative Measures for Safety Evaluation of New Varieties of Agricultural Genetically Modified Organisms (SAMR Order No. 8)-2020 regulate the safety procedures for the new varieties' approval of genetically modified agricultural products, including those that utilise nanotechnology. Furthermore, Technical Guidelines for Safety Assessment of Nano-Scale Agricultural Products (MARA No. 198)-2014 underlines the guidelines for safety assessment of production, processing, and application of nanotechnology-based agricultural products.

The regulation of nanotechnology-based agricultural products in India is overseen by multiple agencies and laws, including the Department of Biotechnology (DBT), the Ministry of Environment, Forest and Climate Change (MoEFCC), the Food Safety and Standards Authority of India (FSSAI), and the Indian Council of Agricultural Research (ICAR). Several national acts have been enacted since the 1960s: the Insecticides (1968) and Seeds Acts (1966), which include nano-based materials for regulating insecticides and genetically modified seeds. In 2001, the Indian government initiated the Nano Science and Technology Initiative (NSTI) and, in 2007, the Nano Mission, emphasising research, risk management, and standardisation. In 2010, the government provided some guidelines on Nanotechnology Research in Agriculture (ICAR) for the safe use of NMs in agricultural practices. Studies emphasise the need for comprehensive and coordinated regulatory frameworks to ensure the safety and efficacy of nanotechnology-based agri-products in India, as the current regulatory framework is still evolving due to a lack of resources and expertise (Barpujari 2011).

In recent years, there has been a global shift towards frameworks incorporating "One Health" concepts as regulatory organisations strive to update standards that concurrently address dangers to humans, animals, and ecosystems (Lombi et al. 2019). A globally consistent regulatory framework that includes nano-specific assessments and international collaboration is crucial for the safe adoption of nanotechnologies in agriculture. Ongoing investigation into nanoparticle toxicity, bioaccumulation, and ecological effects will enhance regulatory requirements that guarantee production and safety in agricultural techniques.

6 Conclusion and Future Perspectives

Nanotechnology serves as the novelty enabler for nearly all areas in general, specifically agriculture, medicine, environment and manufacturing. NMs are unique. They possess nano properties, which make them suitable for new applications such as targeted drug delivery, agriculture in which nutrition gets enhanced, as well as remediation of environmental pollution, and manufacturing for reinforcement of the materials and the conductivity issues. The positive effects of NMs are linked to the high degree of surface area and reactivity, and the ability to interact with biological tissues specifically. However, these features of the NMs, possessing all the qualities that one may need, are also fraught with the issue of their regulation and risk management. Because of their tiny size, potentially harmful effects can make nanotechnology useful for agriculture. For crop protection, instead of the conventional broadcasting of fertilisers on crops, a new methodology of precision farming that takes all factors of production, including water, fertilisers, and pesticides, has emerged near the root zone of crops. On the broader scale of the environment, the various types of NMs, such as membranes, nano-powders and so on, are used for water treatment, namely filtration and photocatalytic degradation, help in the recovery processes of useful metals from liquid waste, assist the production of petrochemicals and other industries by catalysis respectively. The medical and pharmaceutical fields also embraced nanotechnology in similar ways. Targeted cancer treatments have NMs at their core, thereby increasing the accuracy of drug delivery while reducing side effects. Yet despite these promising developments, serious questions persist about NMs' safety, toxicity, and environmental impact. Their small size enables NMs to easily pass through biological barriers into the human body, which can become toxic to the respiratory, cardiovascular, and nervous systems. More specifically, the environmental interactions also create risks for the bioaccumulation process, whereby nano-particulates may result in unforeseen ecological outcomes, such as changes in soil composition or distortion in aquatic ecosystems. Some studies indicate that different types of NMs can induce oxidative stress, inflammation, and DNA damage, threatening human health and the environment.

These risks make it urgently necessary to establish a regulatory framework to keep pace with the fast growth of nanotechnology. The existing regulation structures often lack uniformity. The EU has proactively created nano-specific regulations through REACH legislation, mandating the need for safety assessments of NMs used in various products, especially by agri-food and chemical industries. On the other hand, the United States and other countries have yet to adopt such a rigorous process. Several factors complicate NM regulation: there is no uniform definition of NMs, and NMs differ in their composition, size, and application. Moreover, the toxicity profiles and applications are so different that it is hard to determine standard risk assessment practices. Current practices still need to be completed, as they often fail to adequately consider long-term exposure or cumulative environmental exposure. These limitations point to the development of improved risk assessment methodologies and uniform and comprehensive testing standards that can capture the unique interactions of NMs with biological and environmental systems. The future of nanotechnology needs progress in innovation and safety. Research should prioritise exploring the mechanisms of NM toxicity, bioaccumulation, and long-term ecological effects to understand and mitigate potential hazards. Additionally, green synthesis techniques for NMs may reduce environmental risks since these employ eco-friendly solvents, biological agents, and low-energy processes, leading to safer and more sustainable NMs. Green nanotechnology has the potential to make NM production safer and more economical, hence extending its applications to very environmentally sensitive industries such as agriculture and food safety. The broad and diverging applications of nanotechnology require more coordinated regulatory frameworks at the global level. One possible underpinning for international regulation cooperation is "One Health," which recognises an interconnection among humans, animals, and ecosystems. A balance of innovation with stringent measures for safety and sustainable practices will be crucial as nanotechnology reshapes industries and impacts global ecosystems. Coordinated research, enhanced risk assessment, and robust international regulation will unleash nanotechnology's potential while safeguarding public health and the environment.

References

Adetutu BO, Buba AB, Akinhanmi F et al (2024) Relevance of nanotechnology in agriculture. In: Handbook of agricultural biotechnology. Wiley, pp 413–462

Ahmad A, Mukherjee P, Senapati S et al (2003) Extracellular biosynthesis of silver NMs using the fungus Fusarium oxysporum. Colloids Surf B Biointerfaces 28:313–318. https://doi.org/10.1016/S0927-7765(02)00174-1

Akhtar N, Ilyas N, Meraj TA, Aboughadareh AP, Sayyed RZ, Mashwani Z, Poczai P (2022) Improvement of plant responses by nanobiofertilizer: a step towards sustainable agriculture. Nano 12:965. https://www.mdpi.com/2079-4991/12/6/965

Alkilany AM, Bani Yaseen AI, Kailani MH (2015) Synthesis of monodispersed gold NMs with exceptional colloidal stability with grafted polyethylene glycol- *g* -polyvinyl alcohol. J Nanomater 2015:712359. https://doi.org/10.1155/2015/712359

Alsulami JA, Perveen K, Alothman MR, Al-Humaid LA, Munshi FM, Adawiyah Ahmad R, Sayyed R, Khan S (2023) Microwave-assisted green synthesis of silver nanoparticles by extracts of fig fruits and myrrh oleogum resin and their role in antibacterial activity. J King Saud Univ Sci 35(10):102959. https://doi.org/10.1016/j.jksus.2023.102959

Ameh T, Sayes CM (2019) The potential exposure and hazards of copper NMs: a review. Environ Toxicol Pharmacol 71:103220. https://doi.org/10.1016/j.etap.2019.103220

Amenta V, Aschberger K, Arena M et al (2015) Regulatory aspects of nanotechnology in the agri/feed/food sector in EU and non-EU countries. Regul Toxicol Pharmacol 73:463–476. https://doi.org/10.1016/j.yrtph.2015.06.016

Anandgaonker P, Kulkarni G, Gaikwad S, Rajbhoj A (2019) Synthesis of TiO2 NMs by electrochemical method and their antibacterial application. Arab J Chem 12:1815–1822. https://doi.org/10.1016/j.arabjc.2014.12.015

Antony JK, Vasa NJ, Chakravarthy SR, Sarathi R (2010) Understanding the mechanism of nanoaluminum particle formation by wire explosion process using optical emission technique. J Quant Spectrosc Radiat Transf 111:2509–2516. https://doi.org/10.1016/j.jqsrt.2010.07.011

Assessment UENC for E (2009) EMERGNANO: a review of completed and near completed environment, health and safety research on nanomaterials and nanotechnology

Augustine R, Hasan A, Primavera R et al (2020) Cellular uptake and retention of NMs: insights on particle properties and interaction with cellular components. Mater Today Commun 25:101692. https://doi.org/10.1016/j.mtcomm.2020.101692

Banerjee M, Dey B, Talukdar J, Chandra Kalita M (2014) Production of biodiesel from sunflower oil using highly catalytic bimetallic gold–silver core–shell nanoparticle. Energy 69:695–699. https://doi.org/10.1016/J.ENERGY.2014.03.065

Baranowska-Wójcik E, Szwajgier D, Oleszczuk P, Winiarska-Mieczan A (2020) Effects of titanium dioxide NMs exposure on human health—a review. Biol Trace Elem Res 193:118–129. https://doi.org/10.1007/s12011-019-01706-6

Barpujari I (2011) Attenuating risks through regulation: issues for nanotechnology in India. J Biomed Nanotechnol 7:85–86. https://doi.org/10.1166/jbn.2011.1214

Bleeker EAJ, Cassee FR, Geertsma RE et al (2017) Interpretation and implications of the European Commission's definition on nanomaterials

Budamagunta V, Shameem N, Irusappan S, Parray JA, Thomas M, Marimuthu S, Kirubakaran R, Jothi KNA, Sayyed RZ, Lim HR (2023) Microbial nanovesicle and extracellular polymeric substances mediate nickel tolerance and remediate heavy metal ions from soil. Environ Res 219(15):114997. https://doi.org/10.1016/j.envres.2022.114997

Caixeta MB, Araújo PS, Rodrigues CC et al (2021) Risk assessment of iron oxide NMs in an aquatic ecosystem: a case study on Biomphalaria glabrata. J Hazard Mater 401:123398. https://doi.org/10.1016/j.jhazmat.2020.123398

Camargo PHC, Satyanarayana KG, Wypych F (2009) Nanocomposites: synthesis, structure, properties and new application opportunities. Mater Res 12:1–39. https://doi.org/10.1590/S1516-14392009000100002

Chávez-Hernández JA, Velarde-Salcedo AJ, Navarro-Tovar G, Gonzalez C (2024) Safe nanomaterials: from their use, application, and disposal to regulations. Nanoscale Adv 6:1583–1610. https://doi.org/10.1039/D3NA01097J

Chen Y, Hou S (2023) Recent progress in the effect of magnetic iron oxide NMs on cells and extracellular vesicles. Cell Death Discov 9(19):1–7. https://doi.org/10.1038/s41420-023-01490-2

Chowdhury M, Kushwah A, Satpute AN et al (2023) A comprehensive review on potential application of nanomaterials in the field of agricultural engineering. J Biosyst Eng 48:457–477. https://doi.org/10.1007/s42853-023-00204-x

Corrêa CM, Bizeto MA, Camilo FF (2016) Direct synthesis of silver NMs in ionic liquid. J Nanopart Res 18:132. https://doi.org/10.1007/s11051-016-3436-8

Corsi I, Venditti I, Trotta F, Punta C (2023) Environmental safety of nanotechnologies: the eco-design of manufactured nanomaterials for environmental remediation. Sci Total Environ 864:161181. https://doi.org/10.1016/j.scitotenv.2022.161181

Dahoumane S, Jeffryes C, Mechouet M, Agathos S (2017) Biosynthesis of inorganic NMs: a fresh look at the control of shape, size and composition. Bioengineering 4:14. https://doi.org/10.3390/bioengineering4010014

Daneshmand S, Niazi M, Fazeli-Nasab B, Asili J, Golmohammadzadeh S, Sayyed RZ (2024) Solid lipid nanoparticles of Platycladus orientalis L. possessing 5-alpha reductase inhibiting activity for treating hair loss and hirsutism. J Med Plants By-prod 1:233–246. https://doi.org/10.22034/jmpb.2023.364389.1634

Devi L, Kushwaha P, Ansari TM et al (2024) Recent trends in biologically synthesized metal NMs and their biomedical applications: a review. Biol Trace Elem Res 202:3383–3399. https://doi.org/10.1007/s12011-023-03920-9

Dheyab MA, Aziz AA, Moradi Khaniabadi P et al (2022) Monodisperse gold NMs: a review on synthesis and their application in modern medicine. Int J Mol Sci 23:7400. https://doi.org/10.3390/ijms23137400

Din MI, Rehan R (2017) Synthesis, characterization, and applications of copper NMs. Anal Lett 50:50–62. https://doi.org/10.1080/00032719.2016.1172081

Dolez PI (2015) Nanomaterials definitions, classifications, and applications. In: Nanoengineering. Elsevier, pp 3–40

Eker F, Duman H, Akdaşçi E et al (2024) A comprehensive review of NMs: from classification to application and toxicity. Molecules 29:3482. https://doi.org/10.3390/molecules29153482

El-Kady MM, Ansari I, Arora C et al (2023) Nanomaterials: a comprehensive review of applications, toxicity, impact, and fate to environment. J Mol Liq 370:121046. https://doi.org/10.1016/j.molliq.2022.121046

Faridvand S, Amirnia R, Tajbakhsh M, Enshasy HE, Sayyed RZ (2021) The effect of foliar application of magnetic water and nano, organic, and chemical fertilizers on phytochemical and yield characteristics of different landraces of fennel (*Foeniculum vulgare* Mill). Horticulturae 7:475. https://www.mdpi.com/2311-7524/7/11/475

Feng J, Carpanese C, Fina A (2016) Thermal decomposition investigation of ABS containing Lewis-acid type metal salts. Polym Degrad Stab 129:319–327. https://doi.org/10.1016/j.polymdegradstab.2016.05.013

Fernández-Bertólez N, Alba-González A, Touzani A et al (2024) Toxicity of zinc oxide NMs: cellular and behavioural effects. Chemosphere 363:142993. https://doi.org/10.1016/j.chemosphere.2024.142993

Gu Q, Lv J, Mo X, Jiang X (2022) High aspect ratio metamaterials and their applications. Sens Actuators A Phys 335:113376. https://doi.org/10.1016/j.sna.2022.113376

Gurunathan S, Kalishwaralal K, Vaidyanathan R et al (2009) Biosynthesis, purification and characterization of silver NMs using Escherichia coli. Colloids Surf B Biointerfaces 74:328–335. https://doi.org/10.1016/j.colsurfb.2009.07.048

Hadrup N, Sharma AK, Poulsen M, Nielsen E (2015) Toxicological risk assessment of elemental gold following oral exposure to sheets and NMs—a review. Regul Toxicol Pharmacol 72:216–221. https://doi.org/10.1016/j.yrtph.2015.04.017

Hu Y, Yang T, Wang X, Jiao K (2010) Highly sensitive indicator-free impedance sensing of DNA hybridization based on poly(*m* -aminobenzenesulfonic acid)/TiO_2 nanosheet membranes with pulse potentiostatic method preparation. Chem Eur J 16:1992–1999. https://doi.org/10.1002/chem.200901870

Huston M, DeBella M, DiBella M, Gupta A (2021) Green synthesis of nanomaterials. Nano 11:2130. https://doi.org/10.3390/nano11082130

Jafarzadeh S, Nooshkam M, Zargar M et al (2024) Green synthesis of nanomaterials for smart biopolymer packaging: challenges and outlooks. J Nanostructure Chem 14:113–136. https://doi.org/10.1007/s40097-023-00527-3

Jakinala P, Lingampally N, Hameeda B, Sayyed RZ, Khan MY, Elsayed EA et al (2021) Silver nanoparticles from insect wing extract: biosynthesis and evaluation for antioxidant and antimicrobial potential. PLoS One 16(3):e0241729. https://journals.plos.org/plosone/article?id=10.1371/journal.pone.0241729

Jaswal T, Gupta J (2023) A review on the toxicity of silver NMs on human health. Mater Today Proc 81:859–863. https://doi.org/10.1016/j.matpr.2021.04.266

Kapadia C, Lokhandwala F, Patel N, Elesawy BH, Sayyed RZ, Alhazmi A, Haque S, Datta R (2021) Nanoparticles combined with cefixime as an effective synergistic strategy against *Salmonella enteric typhi*. Saudi J Biol Sci 28:4164–4172. https://doi.org/10.1016/j.sjbs.2021.05.032

Keller AA, Zheng Y, Praetorius A et al (2024) Predicting environmental concentrations of nanomaterials for exposure assessment - a review. NanoImpact 33:100496. https://doi.org/10.1016/j.impact.2024.100496

Khan S, Hossain MK (2022) Classification and properties of NMs. In: Nanoparticle-based polymer composites. Elsevier, pp 15–54

Khan I, Saeed K, Khan I (2019) NMs: properties, applications and toxicities. Arab J Chem 12:908–931. https://doi.org/10.1016/j.arabjc.2017.05.011

Klabunde KJ, Erickson L, Koper O, Richards R (2010) Review of nanoscale materials in chemistry: environmental applications, ACS symposium series. American Chemical Society, pp 1–13

Kolahalam LA, Kasi Viswanath IV, Diwakar BS et al (2019) Review on nanomaterials: synthesis and applications. Mater Today Proc 18:2182–2190. https://doi.org/10.1016/j.matpr.2019.07.371

Kumari R, Suman K, Karmakar S et al (2023) Regulation and safety measures for nanotechnology-based agri-products. Front Genome Ed 5:1200987. https://doi.org/10.3389/fgeed.2023.1200987

Latif-ur-Rahman SA, Khan SB et al (2015) Synthesis, characterization, and application of Au–Ag alloy NMs for the sensing of an environmental toxin, pyrene. J Appl Electrochem 45:463–472. https://doi.org/10.1007/s10800-015-0807-2

Liu L, Zhang Q, Zhao J et al (2013) Study on characteristics of nanopowders synthesized by nanosecond electrical explosion of thin aluminum wire in the argon gas. IEEE Trans Plasma Sci 41:2221–2226. https://doi.org/10.1109/TPS.2013.2240318

Lombi E, Donner E, Dusinska M, Wickson F (2019) A One Health approach to managing the applications and implications of nanotechnologies in agriculture. Nat Nanotechnol 14:523–531. https://doi.org/10.1038/s41565-019-0460-8

Mekuye B, Abera B (2023) Nanomaterials: an overview of synthesis, classification, characterization, and applications. Nano Select 4:486–501. https://doi.org/10.1002/nano.202300038

Mishra S, Keswani C, Abhilash PC et al (2017) Integrated approach of agri-nanotechnology: challenges and future trends. Front Plant Sci 8:254477. https://doi.org/10.3389/fpls.2017.00471

Mukhopadhyay SS (2014) Nanotechnology in agriculture: prospects and constraints. Nanotechnol Sci Appl 7:63. https://doi.org/10.2147/NSA.S39409

Murali M, Gowtham HG, Shilpa N, Singh BS, Mohammed A, Sayyed RZ, Shivamallu C, Achar RR, Silina E, Victor S, Natalia M, Shati AA, Alfaifi MY, Elbehairi SEI, Kollur SP (2023) Zinc oxide nanoparticles prepared through microbial mediated synthesis for therapeutic applications: a possible alternative for plants. Front Microbiol 14:1227951. https://doi.org/10.3389/fmicb.2023.1227951

Nawab R, Iqbal A, Niazi F et al (2023) Review featuring the use of inorganic nano-structured material for anti-microbial properties in textile. Polym Bull 80:7221–7245. https://doi.org/10.1007/s00289-022-04418-5

Nayeri FD, Mafakheri S, Mirhosseini M, Sayyed R (2021) Phyto-mediated silver nanoparticles via *Melissa officinalis* aqueous and methanolic extracts: synthesis, characterization and biological properties against infectious bacterial strains. Int J Adv Biol Biomed Res (3):270–285. http://www.ijabbr.com/article_245059.html

Naz S, Gul A, Zia M (2020) Toxicity of copper oxide NMs: a review study. IET Nanobiotechnol 14:1–13. https://doi.org/10.1049/iet-nbt.2019.0176

Nodera A, Kanai T (2004) Thermal decomposition behavior and flame retardancy of polycarbonate containing organic metal salts: effect of salt composition. J Appl Polym Sci 94:2131–2139. https://doi.org/10.1002/app.21091

Noga M, Milan J, Frydrych A, Jurowski K (2023) Toxicological aspects, safety assessment, and green toxicology of silver NMs (AgNPs)—critical review: state of the art. Int J Mol Sci 24:5133. https://doi.org/10.3390/ijms24065133

Nugraha AS, Na J, Hossain MSA et al (2020) Block copolymer-templated electrodeposition of mesoporous Au-Ni alloy films with tunable composition. Appl Mater Today 18:100526. https://doi.org/10.1016/j.apmt.2019.100526

Osman AI, Zhang Y, Farghali M et al (2024) Synthesis of green NMs for energy, biomedical, environmental, agricultural, and food applications: a review. Environ Chem Lett 22:841–887. https://doi.org/10.1007/s10311-023-01682-3

Park TJ, Lee KG, Lee SY (2016) Advances in microbial biosynthesis of metal NMs. Appl Microbiol Biotechnol 100:521–534. https://doi.org/10.1007/s00253-015-6904-7

Patil KC, Chandrashekhar GV, George MV, Rao CNR (1968) Infrared spectra and thermal decompositions of metal acetates and dicarboxylates. Can J Chem 46:257–265. https://doi.org/10.1139/v68-040

Petersen EJ, Diamond SA, Kennedy AJ et al (2015) Adapting OECD aquatic toxicity tests for use with manufactured nanomaterials: key issues and consensus recommendations. Environ Sci Technol 49:9532–9547. https://doi.org/10.1021/acs.est.5b00997

Prabhu Balaraman R, Mendel J, Flores L, Choudhary M (2021) Nanoparticle biosynthesis and interaction with the microbial cell, antimicrobial and antibiofilm effects, and environmental impact. In: Nanotechnology in the life sciences. Springer, Cham, pp 371–405

Rashid MM, Tavčer PF, Tomšič B (2021) Influence of titanium dioxide NMs on human health and the environment. Nano 11:2354. https://doi.org/10.3390/NANO11092354

Rasool A, Sri S, Zulfajri M, Sri Herwahyu Krismastuti F (2024) Nature inspired nanomaterials, advancements in green synthesis for biological sustainability. Inorg Chem Commun 169:112954. https://doi.org/10.1016/j.inoche.2024.112954

Saleh TA (2020) Nanomaterials: classification, properties, and environmental toxicities. Environ Technol Innov 20:101067. https://doi.org/10.1016/j.eti.2020.101067

Sani A, Cao C, Cui D (2021) Toxicity of gold NMs (AuNPs): a review. Biochem Biophys Rep 26:100991. https://doi.org/10.1016/j.bbrep.2021.100991

Schlüter M, Hentzel T, Suarez C et al (2014) Synthesis of novel palladium(0) nanocatalysts by microorganisms from heavy-metal-influenced high-alpine sites for dehalogenation of polychlorinated dioxins. Chemosphere 117:462–470. https://doi.org/10.1016/j.chemosphere.2014.07.030

Shah S, Shah SA, Faisal S et al (2022) Engineering novel gold NMs using Sageretia thea leaf extract and evaluation of their biological activities. J Nanostructure Chem 12:129–140. https://doi.org/10.1007/s40097-021-00407-8

Shankar SS, Rai A, Ahmad A, Sastry M (2004) Rapid synthesis of Au, Ag, and bimetallic Au core–Ag shell NMs using neem (Azadirachta indica) leaf broth. J Colloid Interface Sci 275:496–502. https://doi.org/10.1016/j.jcis.2004.03.003

Sharma VK (2009) Aggregation and toxicity of titanium dioxide NMs in aquatic environment—a review. J Environ Sci Health A 44:1485–1495. https://doi.org/10.1080/10934520903263231

Sharma S, Kumari P, Shandilya M, Thakur S, Perveen K, Sheikh I, Ahmed Z, Sayyed R, Mastinu A (2025) The combination of α-Fe_2O_3NP and *Trichoderma* sp. improves antifungal activity against *Fusarium* wilt. J Basic Microbiol65(4):e2400613 https://doi.org/10.1002/jobm.202400613

Shen Z, Chen Z, Hou Z et al (2015) Ecotoxicological effect of zinc oxide NMs on soil microorganisms. Front Environ Sci Eng 9:912–918. https://doi.org/10.1007/s11783-015-0789-7

Siddique S, Chow JCL (2020) Application of nanomaterials in biomedical imaging and cancer therapy. Nano 10:1700. https://doi.org/10.3390/nano10091700

Singaravelu G, Arockiamary JS, Kumar VG, Govindaraju K (2007) A novel extracellular synthesis of monodisperse gold nanoparticles using marine alga, Sargassum wightii Greville. Colloids Surf B Biointerfaces 57:97–101. https://doi.org/10.1016/j.colsurfb.2007.01.010

Starowicz M, Stypuła B (2008) Electrochemical synthesis of ZnO NMs. Eur J Inorg Chem 2008:869–872. https://doi.org/10.1002/ejic.200700989

Subramanian KS, Rajkishore SK (2018) Regulatory framework for nanomaterials in Agri-food systems. In: Nanomaterials: ecotoxicity, safety, and public perception. Springer International Publishing, Cham, pp 319–342

Sundararajan N, Habeebsheriff HS, Dhanabalan K et al (2024) Mitigating global challenges: harnessing green synthesized nanomaterials for sustainable crop production systems. Global Chall 8:2300187. https://doi.org/10.1002/gch2.202300187

Tam SK, Ng KM (2015) High-concentration copper NMs synthesis process for screen-printing conductive paste on flexible substrate. J Nanopart Res 17:466. https://doi.org/10.1007/s11051-015-3277-x

Tang W, Zhang X, Hong H et al (2024) Computational nanotoxicology models for environmental risk assessment of engineered nanomaterials. Nano 14:155. https://doi.org/10.3390/nano14020155

Tao Z, Zhou Q, Zheng T et al (2023) Iron oxide NMs in the soil environment: adsorption, transformation, and environmental risk. J Hazard Mater 459:132107. https://doi.org/10.1016/j.jhazmat.2023.132107

Tasnim A, Roy A, Akash SR, Ali H, Habib MR, Barasarathi J, Muthukumaran M, Sayyed RZ, Yeasmin T (2024) Hibiscus sabdariffa L. petal biomass: a green source of nanoparticles of multifarious potential. Open Agric 9:20220332. https://doi.org/10.1515/opag-2022-0332

Thakur P, Thakur S, Kumari P, Shandilya M, Sharma S, Poczai P, Alarfaj AA, Sayyed RZ (2022) Nano-insecticide: synthesis, characterization, and evaluation of insecticidal activity of ZnO NPs against *Spodoptera litura* and *Macrosiphum euphorbiae*. Appl Nanosci 12:3835–3850. https://doi.org/10.1007/s13204-022-02530-6

Truong TT, Mondal S, Doan VHM et al (2024) Precision-engineered metal and metal-oxide NMs for biomedical imaging and healthcare applications. Adv Colloid Interf Sci 332:103263. https://doi.org/10.1016/j.cis.2024.103263

Usman M, Farooq M, Wakeel A et al (2020) Nanotechnology in agriculture: current status, challenges and future opportunities. Sci Total Environ 721:137778. https://doi.org/10.1016/j.scitotenv.2020.137778

Vimercati L, Cavone D, Caputi A et al (2020) NMs: an experimental study of zinc NMs toxicity on marine crustaceans. General overview on the health implications in humans. Front Public Health 8:499190. https://doi.org/10.3389/fpubh.2020.00192

Wang D, Buqa H, Crouzet M et al (2009) High-performance, nano-structured LiMnPO4 synthesized via a polyol method. J Power Sources 189:624–628. https://doi.org/10.1016/j.jpowsour.2008.09.077

Wang Y, Zhang X, Wang A et al (2014) Synthesis of ZnO NMs from microemulsions in a flow type microreactor. Chem Eng J 235:191–197. https://doi.org/10.1016/j.cej.2013.09.020

Wang Q, Yu Y, Liu J (2018) Preparations, characteristics and applications of the functional liquid metal materials. Adv Eng Mater 20:1700781. https://doi.org/10.1002/adem.201700781

Warheit DB, Donner EM (2015) Risk assessment strategies for nanoscale and fine-sized titanium dioxide particles: recognizing hazard and exposure issues. Food Chem Toxicol 85:138–147. https://doi.org/10.1016/j.fct.2015.07.001

Zhang Y, Erkey C (2006) Preparation of supported metallic NMs using supercritical fluids: a review. J Supercrit Fluids 38:252–267. https://doi.org/10.1016/j.supflu.2006.03.021

Zhang J, Ahmadi M, Fargas G et al (2022) Silver NMs for conductive inks: from synthesis and ink formulation to their use in printing technologies. Metals (Basel) 12:234. https://doi.org/10.3390/met12020234

Consumer Acceptance of the Use of Nanomaterial-Based Products

14

Aditya Pratap Singh, Laishram Olivia Devi, Athikho Kayia Alice, Tasso Yatung, Veluru Bhargav, and Siddhartha Singh

Abstract

Nanomaterials are materials that display peculiarities at nanoscale dimensions and have extensive applications in medicine, electronics, energy, and environmental protection. Nanomaterials open the door to many advances in terms of targeted drug delivery, enhanced material strength, and better energy storage, among others. However, there are concerns about their safety and risk to human health and the environment. The environmental fate of Nanomaterials, principally their persistence and the likelihood of environmental accumulation, is still a significant concern. Despite their increasing utilization, the safety regulation development for nanomaterials is still in its infancy. Currently, regulations are mostly driven by the properties associated with conventional materials and do not consider the distinctive properties and behavior of nanomaterials, thus creating a void in their risk management. Furthermore, consumer acceptance of nanomaterials, particularly in agricultural products, varies, influenced by perceived benefits like improved safety and yield, and concerns about health and environmental risks. Trust in regulatory bodies, clear labeling, and education can boost acceptance. In general, consumers are less concerned when nanomaterials are used in production, like fertilizers, rather than directly in food, but long-term effects remain a concern. Even though nanomaterials are very promising, their

A. P. Singh
Faculty of Agriculture, Bidhan Chandra Krishi Viswavidyalaya, Nadia, India

L. O. Devi · S. Singh (✉)
College of Horticulture & Forestry, Central Agricultural University (I), Pasighat, India

A. K. Alice
College of Agriculture, Central Agricultural University (I), Pasighat, India

T. Yatung · V. Bhargav
MTTC &VTC, Central Agricultural University (I), West Kameng, India

R. Sayyed et al. (eds.), *Nanobiostimulants in Innovative Agriculture*,
https://doi.org/10.1007/978-981-95-2477-8_14

usage needs up-to-date guidelines in regulatory and safety frameworks to counterbalance potential threats and ensure responsible development and use of these newest types of materials across different sectors.

Keywords

Consumer acceptance · Nanomaterials · Regulatory concerns · Safety guidelines · Toxicity evaluation · Health hazard · Environmental impact · Risk assessment

1 Introduction

Nanotechnology, a multidisciplinary field that operates on the nanometer scale, has emerged as the latest field generating revolutionary discoveries that evoke significant changes across various sectors. The versatility of nanotechnology arises from the ability to manipulate materials on the atomic and molecular scale. Nanomaterials (NM) are divided into two large categories: natural and synthetic, with the latter choosing specific applications for their design (Hochella Jr et al. 2019). Natural nanoparticles include those that have been formed during volcanic eruptions, during weathering of rocks, etc., or because of photochemical reactions. Examples of engineered nanoparticles include fullerenes and carbon-based nanotubes that are being used for various purposes (Jeevanandam et al. 2018). They can be further figured into a handful of compositions, namely, carbon, inorganic, organic, and composite-based NMs. Each one exhibits unique properties (Fig. 15.1) that have application within diverse areas; these include medicine and electronics (Kumar and Lal 2014).

NMs render unique chemical, physical, and biological properties at the nanoscale, thus enabling their use in electronics, cosmetics, energy storage, and healthcare (Mihranyan et al. 2012; Yeh 2013; Min et al. 2015; Hanigan et al. 2018; Pomerantseva et al. 2019). In medicine, nanotechnology facilitates innovative solutions for drug delivery, diagnostics, and treatment, leveraging the nanoscale size of biologically significant molecules like antibodies, enzymes, and receptors (Seigneuric et al. 2010). Despite its promising potential, nanomedicine faces challenges such as limited pharmacokinetic and toxicity data, regulatory uncertainties, and infancy in clinical translation, underscoring the need for robust guidelines to ensure legal and scientific clarity for stakeholders (Ventola 2012; He et al. 2019; Rasouli et al. 2019; Martins et al. 2020). Beyond healthcare, nanotechnology addresses energy needs, improves industrial products, and strengthens security systems, showcasing its wide-reaching benefits (Santamaria and Sayes 2010). However, integrating NMs into products like cosmetics, pharmaceuticals, and food raises concerns about their potential health and environmental impacts. Parallel to medical applications, nanotechnology's potential in agriculture addresses critical global challenges such as food security, climate change, and diminishing arable land. By 2050, the world population is projected to rise by 34%, necessitating innovative strategies to enhance crop yields and sustainability (Rosenzweig et al. 2020). Conventional plant breeding techniques, while foundational, fail to introduce novel traits absent in plant species (Arya et al. 2020). Advancements in genetic engineering, including RNA

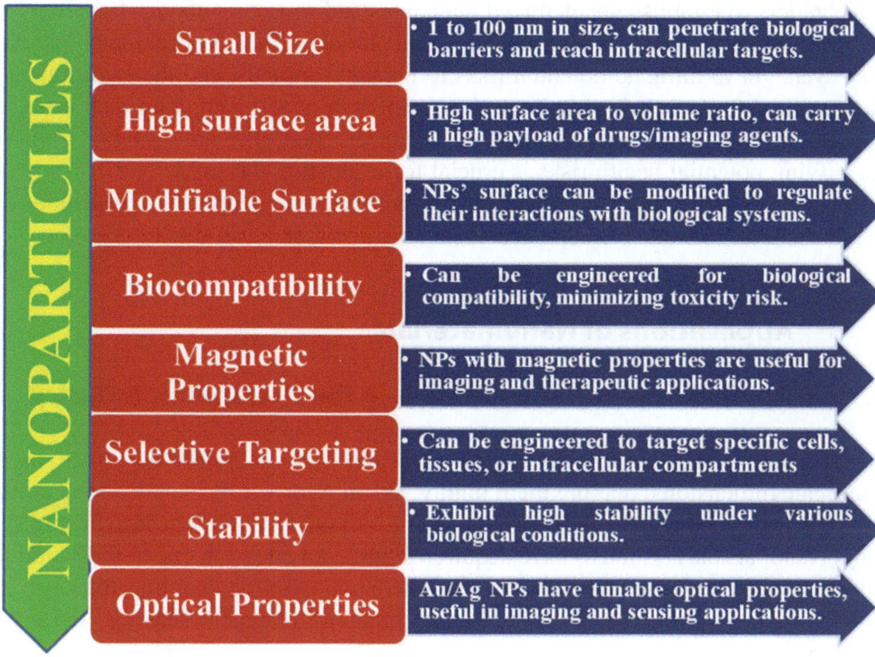

Fig. 15.1 Multifaceted properties of nanoparticles

interference and CRISPR/Cas genome editing, have enhanced plant resilience against biotic and abiotic stresses, significantly advancing agriculture. Nanobiotechnology complements these efforts by employing nanocarriers to deliver biomolecules into plant cells, overcoming barriers like the cell wall's exclusion limit of 5–20 nm (Zhang et al. 2019; Arya et al. 2021; Shahbaz et al. 2023). This approach has led to the successful modification of crops such as rice, maize, and spinach, demonstrating nanotechnology's transformative impact on agriculture (Demirer et al. 2019). Nanotechnology's contributions to agriculture extend to nanopesticides, nanobiofertilizers, and nanobiosensors, enhancing crop productivity while reducing environmental impacts (Sonawane et al. 2021; An et al. 2022; Akhtar et al. 2022; Sharma et al. 2025). These NMs optimize nutrient use, enhance pest control, and improve soil decontamination, aligning with sustainable agriculture goals (Usman et al. 2020; Kumari et al. 2023).

One of the primary challenges in nanotechnology is the design of efficient nanocarriers (NCs) with broad host ranges and high cargo-loading capacities. These carriers must navigate plant or human systems without mechanical assistance, facilitating passive and targeted delivery of biomolecules (Cunningham et al. 2018). Recent studies highlight the potential of NC-based genetically modified crops and nanomedicines to address pressing global issues, provided safety protocols are rigorously adhered to (Burlaka et al. 2015; Demirer et al. 2019). However, its advancement demands a multidisciplinary approach to address health and environmental risks. Ensuring safety requires collaboration among scientists, policymakers, and industry stakeholders to establish standardized guidelines and robust regulatory frameworks. Further, the deployment of engineered nanomaterials (ENMs) necessitates regulatory oversight to mitigate potential risks to humans, animals, and the environment. Hence, the safety of NMs remains a significant area of research, with a knowledge gap persisting between technological advancements and nano-safety studies. Identifying potential health risks is crucial for evaluating the safety and efficacy of new materials and products (Ventola 2012). This chapter explores the applications of nanomaterials, consumer acceptance, associated risks, and evolving regulations.

2 Applications of Nanomaterials

2.1 Applications of Nanomaterials in the Therapeutic Industry

Nanoparticles (NPs) possess distinct physical and chemical properties due to their nanoscale dimensions, making them highly versatile for various applications in the medical field, such as drug delivery, diagnostics, tissue engineering, etc. (Fig. 15.2). Gold Nanoparticles (AuNPs) have acquired significant attention due to their unique optical properties, ease of synthesis, and chemical stability, enabling their use in biomedical applications such as biological imaging, chemical sensing, and drug delivery.

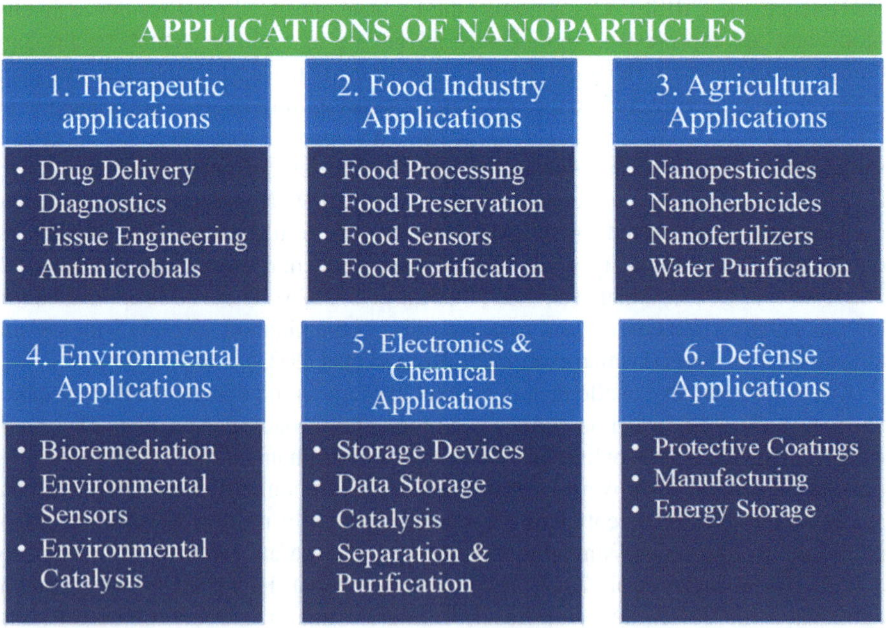

Fig. 15.2 Applications of nanoparticles

Research suggests that nanoparticles enable targeted drug delivery to specific sites of the cell, enhancing their therapeutic precision and minimizing side effects (Siddique and Chow 2020). Silver nanoparticles (AgNPs) are promising candidates for drug delivery due to their stability and ability to accumulate in cancerous tumors (Siddique and Chow 2020). Zinc oxide nanoparticles (ZnONPs) have selective targeting capabilities for cancerous cells (Anjum et al. 2021), and copper nanoparticles (CuNPs) are being investigated for bacterial infection treatments due to their antimicrobial properties (Yuan et al. 2018).

AgNPs are integrated into wound dressings, bone cement, and implants to enhance healing processes (Schröfel et al. 2014), and along with CuNPs, they exhibit potent antimicrobial properties in medical products (Hoseinzadeh et al. 2017). Iron oxide (Fe_3O_4) nanoparticles are widely used as MRI contrast agents (Menon et al. 2013), and AuNPs are explored for diagnostic purposes (Siddique and Chow 2020). Titanium dioxide (TiO_2) nanoparticles are effective in stimulating bone cell growth for tissue engineering (Kim et al. 2014). Nanoparticles hold immense potential in the medical industry, but their safe use must be ensured.

2.2 Food Industry Applications

The versatile nature of nanoparticles has made them a tool of immense potential in the food industry (Fig. 15.2). They enhance food processing efficiency and preservation techniques. Silver nanoparticles (AgNPs) are used as natural antimicrobial agents in food processing to ensure food safety and extend shelf life (Dangi and Verma 2021). Nanoparticles are integrated into food packaging materials to enhance barrier properties against moisture and gases.

Nanoparticles play a critical role in food fortification by delivering essential nutrients like iron via Fe_2O_3 NPs and copper via CuNPs for iron metabolism (Kopittke et al. 2019). They are also used in advanced food sensors for improved safety monitoring and quality assurance (Yadav et al. 2010). Their incorporation offers improved food safety, extended shelf life, and enhanced nutritional value.

2.3 Agricultural Industry Applications

Nanoparticles (NPs) present significant potential in agriculture for enhancing efficiency, minimizing environmental impacts, and improving crop and food safety. They are used in nanopesticides, nanoherbicides, and nanofertilizers. NPs enable precise delivery of pesticides and herbicides, reducing chemical usage and environmental risks (Khan et al. 2019). Silver (AgNPs) and copper (CuNPs) exhibit antimicrobial properties for pest control (Hoseinzadeh et al. 2017; Dangi and Verma 2021). Nano-fertilizers improve plant nutrition with efficient nutrient release. NPs like Ag, Zn, Cu, Au, Al, and Fe-based enhance crop yield (Jaskulski et al. 2022; Song and Kim 2020; Kopittke et al. 2019). NPs enhance food safety by detecting and eliminating pathogens and purifying irrigation water (Zhuang and Gentry 2011). Further research is needed on their long-term impacts.

2.4 Environmental Industry Applications

Nanoparticles (NPs) possess unique properties for environmental applications like bioremediation, sensors, and catalysis. They remove pollutants like heavy metals and organic contaminants. Silver NPs degrade organic dyes (Zhuang and Gentry 2011; Altammar 2023), and nanoscale zerovalent iron (nZVI) remediates chlorinated solvents (Elliott et al. 2013; Altammar 2023). Iron oxide NPs are efficient for heavy metals (Elliott et al. 2013).

Nanotechnology improves water quality and clean-up efforts (Pradeep 2009). Nanosensors detect pollutants and pathogens (Yadav et al. 2010; Theron et al. 2010). NPs act as catalysts in biofuel production and sensing (Lam and Luong 2014; Kora and Rastogi 2018). Thus, NPs offer versatile solutions for environmental challenges.

2.5 Electronics, Chemical, and Defense Industry Applications

NPs are revolutionizing various industries with their unique properties. In electronics, they enhance display systems, energy storage devices like batteries and supercapacitors, and data storage technology. Silver (Ag) and gold (Au) NPs improve LCDs and OLEDs (Park and Choi 2019; Bahadur et al. 2021; Gwynne 2020). Zinc oxide (ZnO) NPs boost energy storage efficiency (Singh et al. 2011).

Magnetic NPs advance data storage capacity (Ahmad et al. 2021). In the chemical industry, NPs are used in catalysis and separation processes. Platinum (Pt), palladium (Pd), and nickel (Ni) NPs are catalysts (Lara and Philippot 2014; Bhavani et al. 2021; Pérez-Lorenzo 2012; Salem and Fouda 2021). Iron oxide (Fe_2O_3), silver (Ag), gold (Au), and aluminum (Al) NPs aid in separation and purification (Pradeep 2009; Siddique and Chow 2020; Zhuang and Gentry 2011). In defense, NPs improve sensors for threat detection (Zheng et al. 2010), protective coatings (Rangel-Olivares et al. 2021), armor materials (Ye et al. 2020), energy storage (Morsi et al. 2022), and manufacturing materials (Khan et al. 2019).

3 Consumer Acceptance of the Use of NM-Based Agri-Products

Consumer acceptance (of NMs) is defined as the willingness to purchase and consume agricultural products that have been produced using NMs or contain NMs. This includes final food products grown with nanofertilizers or treated with nanopesticides, as well as products where NMs might be present as residues (Thakur et al. 2022). The context is critical, as acceptance varies based on whether NMs are used in production processes or are directly part of the consumable product, reflecting differing levels of perceived proximity and risk.

Consumer acceptance of agricultural products with NMs is key to their success, affecting market adoption and societal impact. Consumer acceptance is shaped by socio-psychological, cultural, and informational factors. The acceptance of NM-based agricultural products is not a static attribute but varies dynamically based on perceived risks and benefits, institutional trust, cultural predispositions, knowledge levels, specific applications, economic considerations, transparency, and concerns regarding health and environmental safety. Furthermore, distinctions in application type—whether NMs are directly present in the final product or utilized solely in production processes—along with demographic and regional variations, add additional layers of complexity to this issue.

3.1 Factors Affecting Consumer Acceptance

A primary factor influencing consumer acceptance is the perceived balance between risks and benefits associated with NM-based agricultural products. Research has consistently demonstrated that acceptance increases when consumers perceive tangible advantages, such as enhanced food safety, increased crop yields, and reduced environmental impact through lower chemical usage (Siegrist et al. 2007). Conversely, apprehensions about potential health risks, particularly those with unknown long-term implications, can significantly diminish acceptance (Frewer et al. 2014). For instance, a study conducted by Ho et al. (2010) revealed that when benefits such as improved nutrient delivery are effectively communicated, they often outweigh perceived risks, thereby bolstering consumer support. This duality underscores the importance of framing NM applications in a manner that highlights their practical advantages while addressing safety concerns transparently.

Trust in regulatory bodies, governmental agencies, and scientific institutions constitutes another pivotal determinant of consumer acceptance. Consumers exhibit greater willingness to embrace nanotechnology in agriculture when they are confident that these entities are vigilantly overseeing its safe implementation. A survey conducted in Germany by Siegrist et al. (2007) illustrated that trust in food safety authorities markedly enhances acceptance of nanotechnology in food production, suggesting that robust institutional credibility serves as a foundational pillar for public reassurance. This trust mitigates uncertainties and fosters a sense of security, particularly in contexts where the technology's novelty might otherwise engender skepticism.

Cultural worldviews further shape consumer attitudes toward NMs, influencing risk perception and acceptance across diverse populations. Kahan et al. (2007) observed that individuals with egalitarian worldviews tend to perceive heightened risks associated with nanotechnology, whereas those with individualistic perspectives are more inclined to embrace it. Such variations are particularly pronounced in agricultural contexts, where traditional farming methods hold significant cultural value, potentially amplifying resistance among certain demographic groups. This cultural lens highlights the necessity of considering societal values when assessing and promoting NM adoption.

The level of consumer knowledge and access to information about nanotechnology also plays a critical role, albeit with nuanced effects. Generally, public understanding of nanotechnology remains limited, which can lead to divergent outcomes depending on the nature of the information provided. Cacciatore et al. (2011) noted that media exposure and scientific literacy shape attitudes, with more informed consumers occasionally expressing heightened concern due to increased awareness of potential risks. In contrast, Cobb and Macoubrie (2004) emphasized the value of educational initiatives, finding that well-informed consumers are more likely to weigh benefits against risks judiciously, suggesting that targeted knowledge dissemination can enhance acceptance by fostering informed decision-making.

Acceptance also varies significantly depending on the specific application of NMs in agriculture. Consumers tend to exhibit greater receptivity when nanotechnology is employed in production processes, such as nanofertilizers, rather than when it is directly incorporated into food products. Gupta et al. (2012) reported higher acceptance for nanotechnology in food packaging—owing to indirect benefits like extended shelf life compared to its use as a direct additive, a distinction particularly relevant to agricultural products where NMs are predominantly utilized in farming inputs rather than the final consumable. This differentiation reflects a preference for applications perceived as less invasive to human health.

Economic considerations, particularly willingness to pay, further influence consumer acceptance. Some consumers are prepared to pay a premium for NM-enhanced agricultural products, especially when clear benefits such as enhanced safety or quality are evident. Yue et al. (2015) found that US consumers expressed a willingness to pay more for food packaging incorporating nanotechnology to improve safety, highlighting a value-driven economic incentive. However, price sensitivity can curtail acceptance if costs are deemed unjustified, as noted in a report by the Food Standards Agency (2020), underscoring the need to align perceived value with affordability.

Transparency, particularly through clear labeling, emerges as a vital component in fostering consumer trust and acceptance. Consumers value the ability to make informed choices, and proper labeling reduces uncertainty while building confidence in NM-based products. Kuzma et al. (2009) emphasized that labeling enhances trust, particularly for agricultural goods where nanotechnology use might otherwise go unnoticed, whereas a lack of transparency can engender skepticism, as observed in focus groups conducted by Frewer et al. (2011). This transparency is essential to bridge the gap between technological innovation and consumer perception.

Concerns regarding potential health and environmental risks represent significant barriers to acceptance, particularly with respect to long-term effects. Rollin et al. (2011) documented that French consumers expressed apprehension about health risks linked to nanotechnology in food production, reflecting broader public unease. Similarly, environmental risks, such as soil and water contamination from nanofertilizers—impact acceptance, as highlighted in reviews by Parisi et al. (2015), emphasizing the need for comprehensive risk assessments to address these legitimate concerns.

The distinction between direct presence and production process applications of NMs further delineates consumer acceptance patterns. When NMs are directly incorporated into food products, such as nano-encapsulated nutrients, consumers exhibit heightened concern due to perceived direct exposure risks, a finding corroborated by increased risk perception in studies like Siegrist et al. (2007). In contrast, when NMs are confined to production processes, such as fertilizers or pesticides and are not present in the final product, acceptance tends to be higher, provided benefits and safety measures are clearly communicated (Gupta et al. 2012). Nonetheless, residual concerns about environmental impact or unintended residues persist, as noted in environmental risk assessments by Iavicoli et al. (2017), necessitating ongoing monitoring and reassurance.

Demographic variations also contribute to differing levels of acceptance. Research from the National Institute of Food and Agriculture (NIFA 2017) indicated that men are more likely than women to accept nanotechnology in food, based on taste tests and surveys at Rutgers University, while Cobb and Macoubrie (2004) observed that younger, more educated consumers often display greater openness, reflecting the influence of age and education on receptivity. These demographic insights are crucial for tailoring outreach efforts to specific population segments.

3.2 Strategies to Enhance Consumer Acceptance

To enhance consumer acceptance of NM-based agricultural products, several evidence-based strategies have been proposed. Transparent communication, delivering clear and accurate information about NM use, benefits, and safety assessments through public campaigns and media engagement, is advocated by Macoubrie (2006) as a means to demystify the technology. Clear labeling, indicating nanotechnology use, allows consumers to make informed choices and reduces uncertainty, thereby building trust (Kuzma et al. 2009). Robust regulatory frameworks, ensuring safe NM application in agriculture, are critical to address safety concerns, as underscored in a Food Standards Agency report (2018), and their communication to the public enhances credibility. Educational initiatives, aimed at increasing public understanding through programs and media, are deemed essential by Parisi et al. (2015) to shape positive attitudes. Finally, highlighting benefits, such as increased crop yields, reduced environmental impact, and enhanced food safety can offset concerns, as demonstrated by Ho et al. (2010), aligning NM adoption with consumer priorities.

Regional and cultural considerations further modulate acceptance, with studies indicating higher receptivity in the US compared to Europe, where stricter regulatory scrutiny prevails. Siegrist et al. (2007) found lower acceptance in Germany due to elevated risk perceptions, reflecting cultural differences, whereas Yue et al. (2015) documented a willingness to pay for benefits in the US, suggesting a more favorable outlook. These disparities underscore the importance of context-specific approaches to NM promotion.

Consumer acceptance of NM-based agricultural products is a complex, multifaceted issue influenced by perceived benefits, trust, knowledge, application types, and demographic and regional factors (Siimes et al. 2022). By addressing consumer concerns through transparent communication, clear labeling, robust regulation, educational initiatives, and benefit-focused messaging, stakeholders can enhance acceptance, facilitating the sustainable integration of nanotechnology into agriculture while aligning with consumer expectations and preferences. This approach ensures that the transformative potential of NMs is realized in a manner that garners public support and trust, paving the way for their widespread adoption in agricultural systems globally.

4 Safety Concerns of NMs

4.1 Environmental Concerns

The environmental impacts of NMs remain challenging to assess due to the complexity of their interactions with various environmental variables. These include difficulties in NM identification, their low detection limits, and uncertainties surrounding their environmental concentrations. Such intricacies hinder definitive conclusions regarding their ecological effects and environmental stability. Even slight modifications in the chemical structure of NMs can significantly alter their properties, potentially transforming them into toxic substances. The United States Environmental Protection Agency (USEPA) highlights that NMs exhibit unique chemical properties, high reactivity, and resistance to dissolution in liquids, complicating toxicity assessments (USEPA 2016). Furthermore, the reactive nature of NMs can result in changes to their properties between sample collection and analysis (USEPA 2016). Determining the origin of NMs in environmental samples is also problematic. Recent research has explored the ecological effects of specific nanoparticles, such as Ag_2S nanoparticles. Studies indicate that plants like cucumber (*Cucumis sativus*) and wheat (*Triticum aestivum L.*) can absorb nanosilver from soil, leading to gene upregulation associated with ethylene signaling pathways and reduced plant growth. Additionally, NMs activate plant defense mechanisms, further suppressing growth. The accumulation of Ag_2S nanoparticles in plant leaves raises concerns about trophic transfer through the food chain (Wang et al. 2017). Similarly, nanoparticles like TiO_2 have been observed to induce genotoxic effects, causing DNA damage at higher concentrations in plants such as *Allium cepa* and *Nicotiana tabacum*. For instance, malondialdehyde levels in *A. cepa* increased in the presence of TiO_2, suggesting potential oxidative stress (Ghosh et al. 2010).

The toxicity of NMs often depends on their dissolved components or leached metals. For example, nano-ZnO exhibited significantly higher toxicity at acidic pH (4.5) due to rapid dissolution, while minimal effects were observed at neutral or coated conditions (Chen et al. 2016). Phytotoxicity also varies among NM types; nano-Zn and nano-ZnO negatively impact seed germination and growth in ryegrass and corn, respectively, while multi-walled carbon nanotubes and aluminum-based NMs showed negligible effects (Lin and Xing 2007). In aquatic systems, NMs pose a spectrum of toxicity risks. Nano-Cd demonstrated dose-dependent toxicity in *Daphnia magna*, where low to medium concentrations stimulated antioxidative enzymes, but higher levels caused oxidative damage. Carbon nanotubes (CNTs) also exert toxic effects on marine organisms, such as *Thalassiosira pseudonana*, *Tigriopus japonicus*, and *Oryzias melastigma*, with half-maximal effective concentrations (EC50) ranging from 0.1 to 10 mg/L. CNTs induce oxidative and physical stress, adversely affecting marine larvae and increasing mortality in some freshwater species (Kwok et al. 2010; Hussain et al. 2016). Similarly, nano-ZnO was reported to inhibit algal growth, while nano-TiO_2 and nano-ZnO damaged bacterial membranes in aquatic bacteria like *Escherichia coli* and *Aeromonas hydrophila* through oxidative stress mechanisms (Lok et al. 2006; Tong et al. 2015; Das et al. 2017).

Soil ecosystems are also vulnerable to NM toxicity. Nano-CeO_2 caused oxidative stress in *Caenorhabditis elegans*, reducing their lifespan by 12% at 1 nM and 18% at 100 nM concentrations due to reactive oxygen species accumulation (Zhang et al. 2011). In addition, airborne NMs can contribute to environmental degradation by forming dust clouds, impacting air quality (Turkevich et al. 2015). The adverse effects of NMs span multiple environmental media, including soil, water, and air, influencing terrestrial and aquatic life and degrading overall environmental quality (Doshi et al. 2008; Ghosh et al. 2010; Clément et al. 2013; Tong et al. 2015; Chen et al. 2016; Kabir et al. 2018). These findings emphasize the necessity for comprehensive assessments of NMs, both individually and in mixtures, to understand and mitigate their ecological impacts effectively.

4.2 Health Concerns

The rising concerns regarding the toxicity and exposure pathways of nanomaterials stem from their unique ability to penetrate and be absorbed by mammalian cell membranes. This absorption is contingent on the size, aggregation, and sedimentation properties of the NMs (Binderup et al. 2013). Cellular uptake of NMs primarily occurs through endocytosis or phagocytosis mechanisms (Mallapragada et al. 2015). The exposure routes of NMs are varied and include oral, dermal, inhalation, and gastrointestinal pathways. These routes are often associated with the use of consumer products such as sunscreens, skincare items, paints, coatings, food additives, and health supplements (Ma et al. 2015; Mackevica and Foss Hansen 2016). Furthermore, engineered nanomaterials (ENMs) are sometimes intentionally administered into the human body for therapeutic and medical purposes. Within the body, NMs generate reactive oxygen species (ROS), leading to oxidative stress and subsequent adverse health effects (Liou et al. 2012). In particular, NMs have been identified as potential allergens during early developmental stages, potentially triggering immune responses such as allergic inflammation in later life stages (Sly and Schüepp 2012). Studies suggest that neonatal exposure to NMs correlates with increased asthma exacerbation, reduced lung function, and respiratory symptoms like wheezing and coughing in the absence of infections (Meldrum et al. 2017). Despite these risks, nanotechnology holds promise in pediatric disease treatment, especially in diagnostic imaging and therapeutic applications (Zare-Zardini et al.

2015). Inhaled NMs have systemic effects, as they can disseminate throughout the body via the bloodstream (Terzano et al. 2010). Epidemiological data indicate that NMs can impact cardiovascular health by altering blood coagulation, which may lead to changes in cardiac rhythm and function (Araujo and Nel 2009; Quan et al. 2010; Liou et al. 2012). Chronic exposure to NMs has been linked to reduced lung function parameters, including forced expiratory volume (FEV1) and forced vital capacity (FVC) (Bonner 2010). Animal studies have revealed that exposure to carbon nanotubes and nanoscale TiO_2 particles can induce tumor formation (Becker et al. 2011). Additionally, NMs have been shown to exert toxic effects on fibroblasts and keratinocytes, potentially altering gene and protein expression (Haliullin et al. 2015).

Specific studies on human cells demonstrate the harmful impacts of NMs. For example, cadmium telluride quantum dots (CdTe QDs) have been found to disrupt mitochondrial morphology and structure in human hepatocellular carcinoma (HepG2) cells, leading to reduced mitochondrial function, altered biogenesis, and increased intracellular calcium levels. These disruptions also impair cellular respiration and adenosine triphosphate (ATP) synthesis, underscoring the severe health risks posed by such NMs (Nguyen et al. 2015). NMs, through oxidative stress, are also implicated in interstitial fibrosis and airway inflammation in animal models (Smita et al. 2012). They can induce ROS production by disrupting mitochondrial function or initiating cell signaling pathways (Liu et al. 2013). Although studies on silver NMs suggest potential reproductive and developmental toxicity, these studies often use unrealistically high doses (15–1000 mg/kg) and large particle sizes (6.45–323 nm) that do not reflect typical environmental exposures (Ema et al. 2017).

Several investigations highlight the toxic effects of NMs, such as oxidative stress, DNA damage, cell membrane disruption, glutathione depletion, and inflammatory responses (Sayes et al. 2006; Li et al. 2008; Lin et al. 2009; Akhtar et al. 2010; Mittal and Pandey 2014; Chen et al. 2015; Fede et al. 2015; Ng et al. 2017). Nano-ZnO exposure in human bronchoalveolar carcinoma-derived cells (A549) elevates ROS levels, causing oxidative stress, lipid peroxidation, and DNA damage. Higher concentrations of nano-ZnO have been reported to decrease the cell viability of A549 cells (Lin et al. 2009). Beyond engineered NMs, incidental nanoparticles, such as nano-plastics, are emerging as environmental pollutants due to plastic overuse and poor waste management. These nano-plastics contaminate aquatic and atmospheric systems, entering the food chain and potentially impacting human health by releasing harmful monomers and accumulating in the body (Bouwmeester et al. 2015; Galloway 2015; Prata 2018). While indirect studies on human cell lines and animal models offer insights into NMs' toxic effects, direct evidence on humans and higher animals is limited, hindering precise risk assessment. However, current toxicological research emphasizes the role of oxidative stress and DNA damage in mediating these adverse effects. Despite substantial research, limitations persist, such as small study populations, dose-response inconsistencies, and exposure misclassification. Consequently, the toxicological evaluation of NMs requires more robust methodologies, including unbiased in vivo studies employing advanced technologies.

5 Conclusion

The nanoscale manipulation capabilities of NMs have a revolutionary effect in the fields of medicine, agriculture, environment, chemical industry, defense sector, etc. They have been pivotal in providing transformative solutions to problems such as targeted drug delivery, sustainable farming, pollution control, and advanced materials. However, despite its vast potential, many challenges remain unaddressed regarding our understanding of long-term health and environmental impacts, thus requiring well-structured risk assessment and regulatory systems. The future of NM use is geared toward the realization of precision medicine with sustainable agricultural production and shaping a cleaner environment, with the safe and sustainable addition of new material and processes into the industrialization process, and interdisciplinary collaboration among researchers, policymakers, and industry stakeholders will be the key to unlocking its full potential responsibly.

References

Ahmad AA, Alsaad AM, Al-Bataineh QM, Al-Akhras MAH, Albataineh Z, Alizzy KA, Daoud NS (2021) Synthesis and characterization of ZnO NPs-doped PMMA-BDK-MR polymer-coated thin films with UV curing for optical data storage applications. Polym Bull 78:1189–1211

Akhtar MJ, Ahamed M, Kumar S, Siddiqui H, Patil G, Ashquin M, Ahmad I (2010) Nanotoxicity of pure silica mediated through oxidant generation rather than glutathione depletion in human lung epithelial cells. Toxicology 276(2):95–102

Akhtar N, Ilyas N, Meraj TA, Aboughadareh AP, Sayyed RZ, Mashwani Z, Poczai P (2022) Improvement of plant responses by nanobiofertilizer: a step towards sustainable agriculture. Nano 12:965. https://www.mdpi.com/2079-4991/12/6/965

Altammar KA (2023) A review on nanoparticles: characteristics, synthesis, applications, and challenges. Front Microbiol 14:1155622

An C, Sun C, Li N, Huang B, Jiang J, Shen Y et al (2022) Nanomaterials and nanotechnology for the delivery of agrochemicals: strategies towards sustainable agriculture. J Nanobiotechnol 20(1):11

Anjum S, Hashim M, Malik SA, Khan M, Lorenzo JM, Abbasi BH, Hano C (2021) Recent advances in zinc oxide nanoparticles (ZnO NPs) for cancer diagnosis, target drug delivery, and treatment. Cancer 13(18):4570

Araujo JA, Nel AE (2009) Particulate matter and atherosclerosis: role of particle size, composition and oxidative stress. Part Fibre Toxicol 6:1–19

Arya SS, Mahto BK, Ramkumar TR, Lenka SK (2020) Sharpening gene editing toolbox in Arabidopsis for plants. J Plant Biochem Biotechnol 29:769–784

Arya SS, Lenka SK, Cahill DM, Rookes JE (2021) Designer nanoparticles for plant cell culture systems: mechanisms of elicitation and harnessing of specialized metabolites. BioEssays 43(11):2100081

Bahadur PS, Jaiswal S, Srivastava R, Kumar A (2021) Advanced application of nanotechnology in engineering. In: 2021 international conference on technological advancements and innovations (ICTAI). IEEE, pp 92–95 Tashkent, Uzbekistan

Becker H, Herzberg F, Schulte A, Kolossa-Gehring M (2011) The carcinogenic potential of nanomaterials, their release from products and options for regulating them. Int J Hyg Environ Health 214(3):231–238

Bhavani KS, Anusha T, Brahman PK (2021) Platinum nanoparticles decorated on graphitic carbon nitride-ZIF-67 composite support: an electrocatalyst for the oxidation of butanol in fuel cell applications. Int J Hydrog Energy 46(13):9199–9214

Binderup ML, Bredsdorff L, Beltoft VM, Mortensen A, Löschner K, Larsen EH, Eriksen FD (2013) Systemic absorption of nanomaterials by oral exposure: part of the "better control of nano" initiative 2012–2015

Bonner JC (2010) Mesenchymal cell survival in airway and interstitial pulmonary fibrosis. Fibrogenesis & Tissue Repair. 3(1):15

Bouwmeester H, Hollman PC, Peters RJ (2015) Potential health impact of environmentally released micro-and nanoplastics in the human food production chain: experiences from nanotoxicology. Environ Sci Technol 49(15):8932–8947

Burlaka OM, Pirko YV, Yemets AI, Blume YB (2015) Plant genetic transformation using carbon nanotubes for DNA delivery. Cytol Genet 49(6):349–357

Cacciatore MA, Scheufele DA, Corley EA (2011) From enabling technology to applications: the evolution of risk perceptions about nanotechnology. Public Underst Sci 20(3):385–404

Chen LQ, Fang L, Ling J, Ding CZ, Kang B, Huang CZ (2015) Nanotoxicity of silver nanoparticles to red blood cells: size dependent adsorption, uptake, and hemolytic activity. Chem Res Toxicol 28(3):501–509

Chen X, O'Halloran J, Jansen MA (2016) The toxicity of zinc oxide nanoparticles to Lemna minor (L.) is predominantly caused by dissolved Zn. Aquat Toxicol 174:46–53

Clément L, Hurel C, Marmier N (2013) Toxicity of TiO_2 nanoparticles to cladocerans, algae, rotifers and plants–effects of size and crystalline structure. Chemosphere 90(3):1083–1090

Cobb MD, Macoubrie J (2004) Public perceptions about nanotechnology: risks, benefits and trust. J Nanopart Res 6:395–405

Cunningham FJ, Goh NS, Demirer GS, Matos JL, Landry MP (2018) Nanoparticle-mediated delivery towards advancing plant genetic engineering. Trends Biotechnol 36(9):882–897

Dangi K, Verma AK (2021) Efficient & eco-friendly smart nano-pesticides: emerging prospects for agriculture. Mater Today Proc 45:3819–3824

Das B, Dash SK, Mandal D, Ghosh T, Chattopadhyay S, Tripathy S et al (2017) Green synthesized silver nanoparticles destroy multidrug resistant bacteria via reactive oxygen species mediated membrane damage. Arab J Chem 10(6):862–876

Demirer GS, Zhang H, Goh NS, González-Grandío E, Landry MP (2019) Carbon nanotube–mediated DNA delivery without transgene integration in intact plants. Nat Protoc 14(10):2954–2971

Doshi R, Braida W, Christodoulatos C, Wazne M, O'Connor G (2008) Nano-aluminum: transport through sand columns and environmental effects on plants and soil communities. Environ Res 106(3):296–303

Elliott JA, Shibuta Y, Amara H, Bichara C, Neyts EC (2013) Atomistic modelling of CVD synthesis of carbon nanotubes and graphene. Nanoscale 5(15):6662–6676

Ema M, Okuda H, Gamo M, Honda K (2017) A review of reproductive and developmental toxicity of silver nanoparticles in laboratory animals. Reprod Toxicol 67:149–164

Fede C, Fortunati I, Weber V, Rossetto N, Bertasi F, Petrelli L et al (2015) Evaluation of gold nanoparticles toxicity towards human endothelial cells under static and flow conditions. Microvasc Res 97:147–155

Food Standards Agency (2018) Applications of nanotechnology for food contact materials and food safety. Retrieved from https://www.food.gov.uk/research/chemical-hazards-in-food-and-feed/applications-of-nanotechnology-for-food-contact-materials-and-food-safety

Food Standards Agency (2020) Consumer attitudes towards emerging technologies. Retrieved from https://www.food.gov.uk/research/behaviour-and-perception/consumer-attitudes-towards-emerging-technologies-0

Frewer LJ, Howard C, Hedderley D, Shepherd R (2011) Consumer response to novel agri-food technologies: implications for predicting consumer acceptance of emerging food technologies. Trends Food Sci Technol 22(8):442–456. https://doi.org/10.1016/j.tifs.2011.05.005

Frewer LJ, Gupta N, George S, Fischer ARH, Giles EL, Coles D (2014) Consumer attitudes towards nanotechnologies applied to food production. Trends Food Sci Technol 40(2):211–225. https://doi.org/10.1016/j.tifs.2014.06.005

Galloway TS (2015) Micro-and nano-plastics and human health. In: Marine anthropogenic litter, pp 343–366 Springer Cham Heidelberg New York Dordrecht London

Ghosh M, Bandyopadhyay M, Mukherjee A (2010) Genotoxicity of titanium dioxide (TiO_2) nanoparticles at two trophic levels: plant and human lymphocytes. Chemosphere 81(10):1253–1262

Gupta N, Fischer ARH, Frewer LJ (2012) Socio-psychological determinants of public acceptance of technologies: a review. Public Underst Sci 21(7):782–795. https://doi.org/10.1177/0963662511423334

Gwynne KM (2020) Enhancement of the photostability of blue phosphorescence using plasmonic surfaces. Master's thesis, Rutgers The State University of New Jersey, School of Graduate Studies

Haliullin TO, Zalyalov RR, Shvedova AA, Tkachov AG (2015) Hygienic evaluation of multilayer carbon nanotubes. Med Tr Prom Ekol (7):37–42

Hanigan D, Truong L, Schoepf J, Nosaka T, Mulchandani A, Tanguay RL, Westerhoff P (2018) Trade-offs in ecosystem impacts from nanomaterial versus organic chemical ultraviolet filters in sunscreens. Water Res 139:281–290

He H, Liu L, Morin EE, Liu M, Schwendeman A (2019) Survey of clinical translation of cancer nanomedicines—lessons learned from successes and failures. Acc Chem Res 52(9):2445–2461

Ho SS, Scheufele DA, Corley EA (2010) Making sense of policy choices: understanding the roles of value predispositions, mass media, and cognitive processing in public attitudes toward nanotechnology. J Nanopart Res 12:2703–2715

Hochella MF Jr, Mogk DW, Ranville J, Allen IC, Luther GW, Marr LC et al (2019) Natural, incidental, and engineered nanomaterials and their impacts on the earth system. Science 363(6434):eaau8299

Hoseinzadeh E, Makhdoumi P, Taha P, Hossini H, Stelling J, Amjad Kamal M, Md. Ashraf G (2017) A review on nano-antimicrobials: metal nanoparticles, methods and mechanisms. Curr Drug Metab 18(2):120–128

Hussain M, Khalid P, Suman VB, Arun AB (2016) Toxicology of carbon nanotubes-a review. Int J Appl Eng Res 11(1):159–168

Iavicoli I, Leso V, Beezhold DH, Shvedova AA (2017) Nanotechnology in agriculture: opportunities, toxicological implications, and occupational risks. Toxicol Appl Pharmacol 329:96–111

Jaskulski D, Jaskulska I, Majewska J, Radziemska M, Bilgin A, Brtnicky M (2022) Silver nanoparticles (AgNPs) in urea solution in laboratory tests and field experiments with crops and vegetables. Materials 15(3):870

Jeevanandam J, Barhoum A, Chan YS, Dufresne A, Danquah MK (2018) Review on nanoparticles and nanostructured materials: history, sources, toxicity and regulations. Beilstein J Nanotechnol 9(1):1050–1074

Kabir E, Kumar V, Kim KH, Yip AC, Sohn JR (2018) Environmental impacts of nanomaterials. J Environ Manag 225:261–271

Kahan DM, Slovic P, Braman D, Gastil J, Cohen G (2007) Nanotechnology risk perceptions: the influence of affect and values. Cultural Cognition Project at Yale Law School, New Haven, Connecticut, USA

Khan I, Saeed K, Khan I (2019) Nanoparticles: properties, applications and toxicities. Arab J Chem 12(7):908–931

Kim JH, Sheikh FA, Ju HW, Park HJ, Moon BM, Lee OJ, Park CH (2014) 3D silk fibroin scaffold incorporating titanium dioxide (TiO_2) nanoparticle (NPs) for tissue engineering. Int J Biol Macromol 68:158–168

Kopittke PM, Lombi E, Wang P, Schjoerring JK, Husted S (2019) Nanomaterials as fertilizers for improving plant mineral nutrition and environmental outcomes. Environ Sci Nano 6(12):3513–3524

Kora AJ, Rastogi L (2018) Peroxidase activity of biogenic platinum nanoparticles: a colorimetric probe towards selective detection of mercuric ions in water samples. Sensors Actuators B Chem 254:690–700

Kuzma J, Najmaie P, Larson J (2009) Evaluating oversight systems for emerging technologies: a case study of genetically engineered organisms. Journal of Law, Medicine & Ethics. 37(4):546–586

Kumar R, Lal S (2014) Synthesis of organic nanoparticles and their applications in drug delivery and food nanotechnology: a review. J Nanomater Mol Nanotechnol 3:4. of, 11, 2

Kumari R, Suman K, Karmakar S, Mishra V, Lakra SG, Saurav GK, Mahto BK (2023) Regulation and safety measures for nanotechnology-based agri-products. Front Genome Ed 5:1200987

Kwok KW, Leung KM, Flahaut E, Cheng J, Cheng SH (2010) Chronic toxicity of double-walled carbon nanotubes to three marine organisms: influence of different dispersion methods. Nanomedicine 5(6):951–961

Lam E, Luong JH (2014) Carbon materials as catalyst supports and catalysts in the transformation of biomass to fuels and chemicals. ACS Catal 4(10):3393–3410

Lara P, Philippot K (2014) The hydrogenation of nitroarenes mediated by platinum nanoparticles: an overview. Cat Sci Technol 4(8):2445–2465

Li SQ, Zhu RR, Zhu H, Xue M, Sun XY, Yao SD, Wang SL (2008) Nanotoxicity of TiO2 nanoparticles to erythrocyte in vitro. Food Chem Toxicol 46(12):3626–3631

Lin D, Xing B (2007) Phytotoxicity of nanoparticles: inhibition of seed germination and root growth. Environ Pollut 150(2):243–250

Lin W, Xu Y, Huang CC, Ma Y, Shannon KB, Chen DR, Huang YW (2009) Toxicity of nano-and micro-sized ZnO particles in human lung epithelial cells. J Nanopart Res 11:25–39

Liou SH, Tsou TC, Wang SL, Li LA, Chiang HC, Li WF et al (2012) Epidemiological study of health hazards among workers handling engineered nanomaterials. J Nanopart Res 14:1–15

Liu G, Gao J, Ai H, Chen X (2013) Applications and potential toxicity of magnetic iron oxide nanoparticles. Small 9(9–10):1533–1545

Lok CN, Ho CM, Chen R, He QY, Yu WY, Sun H et al (2006) Proteomic analysis of the mode of antibacterial action of silver nanoparticles. J Proteome Res 5(4):916–924

Ma J, Mercer RR, Barger M, Schwegler-Berry D, Cohen JM, Demokritou P, Castranova V (2015) Effects of amorphous silica coating on cerium oxide nanoparticles induced pulmonary responses. Toxicol Appl Pharmacol 288(1):63–73

Mackevica A, Foss Hansen S (2016) Release of nanomaterials from solid nanocomposites and consumer exposure assessment–a forward-looking review. Nanotoxicology 10(6):641–653

Macoubrie J (2006) Nanotechnology: public concerns, reasoning and trust in government. Public Underst Sci 15(2):221–241

Mallapragada SK, Brenza TM, McMillan JM, Narasimhan B, Sakaguchi DS, Sharma AD et al (2015) Enabling nanomaterial, nanofabrication and cellular technologies for nanoneuromedicines. Nanomedicine 11(3):715–729

Martins JP, Das Neves J, de la Fuente M, Celia C, Florindo H, Günday-Türeli N et al (2020) The solid progress of nanomedicine. Drug Deliv Transl Res 10:726–729

Meldrum K, Guo C, Marczylo EL, Gant TW, Smith R, Leonard MO (2017) Mechanistic insight into the impact of nanomaterials on asthma and allergic airway disease. Part Fibre Toxicol 14:1–35

Menon JU, Jadeja P, Tambe P, Vu K, Yuan B, Nguyen KT (2013) Nanomaterials for photo-based diagnostic and therapeutic applications. Theranostics 3(3):152

Mihranyan A, Ferraz N, Strømme M (2012) Current status and future prospects of nanotechnology in cosmetics. Prog Mater Sci 57(5):875–910

Min Y, Caster JM, Eblan MJ, Wang AZ (2015) Clinical translation of nanomedicine. Chem Rev 115(19):11147–11190

Mittal S, Pandey AK (2014) Cerium oxide nanoparticles induced toxicity in human lung cells: role of ROS mediated DNA damage and apoptosis. Biomed Res Int 2014(1):891934

Morsi MA, Abdelrazek EM, Ramadan RM, Elashmawi IS, Rajeh A (2022) Structural, optical, mechanical, and dielectric properties studies of carboxymethyl cellulose/polyacrylamide/

lithium titanate nanocomposites films as an application in energy storage devices. Polym Test 114:107705
National Institute of Food and Agriculture (NIFA) (2017) Food nanotechnology: Expanding the parameters of consumer acceptance. Retrieved from https://www.nifa.usda.gov/food-nanotechnology-expanding-parameters-consumer-acceptance
Ng CT, Yong LQ, Hande MP, Ong CN, Yu LE, Bay BH, Baeg GH (2017) Zinc oxide nanoparticles exhibit cytotoxicity and genotoxicity through oxidative stress responses in human lung fibroblasts and Drosophila melanogaster. Int J Nanomedicine:1621–1637
Nguyen KC, Rippstein P, Tayabali AF, Willmore WG (2015) Mitochondrial toxicity of cadmium telluride quantum dot nanoparticles in mammalian hepatocytes. Toxicol Sci 146(1):31–42
Parisi C, Vigani M, Rodríguez-Cerezo E (2015) Agricultural nanotechnologies: what are the current possibilities? Nano Today 10(2):124–127
Park CY, Choi B (2019) Enhanced light extraction from bottom emission oleds by high refractive index nanoparticle scattering layer. Nano 9(9):1241
Pérez-Lorenzo M (2012) Palladium nanoparticles as efficient catalysts for Suzuki cross-coupling reactions. J Phys Chem Lett 3(2):167–174
Pomerantseva E, Bonaccorso F, Feng X, Cui Y, Gogotsi Y (2019) Energy storage: the future enabled by nanomaterials. Science 366(6468):eaan8285
Pradeep T (2009) Noble metal nanoparticles for water purification: a critical review. Thin Solid Films 517(24):6441–6478
Prata JC (2018) Airborne microplastics: consequences to human health? Environ Pollut 234:115–126
Quan C, Sun Q, Lippmann M, Chen LC (2010) Comparative effects of inhaled diesel exhaust and ambient fine particles on inflammation, atherosclerosis, and vascular dysfunction. Inhal Toxicol 22(9):738–753
Rangel-Olivares FR, Arce-Estrada EM, Cabrera-Sierra R (2021) Synthesis and characterization of polyaniline-based polymer nanocomposites as anti-corrosion coatings. Coatings 11(6):653
Rasóuli H, Djordjević JP, Sayyed RZ, Zarayneh S, Jafari M, Fazeli-Nasab B (2019) Nanoparticles: a new threat to crop plants and soil rhizobia? In: Hayat S, Pichtel J, Faizan M, Fariduddin Q (eds) Sustainable agriculture reviews, vol 41. Springer, Switzerland AG pp 201–214. https://doi.org/10.1007/978-3-030-33996-8_11
Rollin F, Kennedy J, Wills J (2011) Consumers and new food technologies. Trends Food Sci Technol 22(2–3):99–111
Rosenzweig C, Mbow C, Barioni LG, Benton TG, Herrero M, Krishnapillai M et al (2020) Climate change responses benefit from a global food system approach. Nature Food 1(2):94–97
Salem SS, Fouda A (2021) Green synthesis of metallic nanoparticles and their prospective biotechnological applications: an overview. Biol Trace Elem Res 199(1):344–370
Santamaria AB, Sayes CM (2010) Toxicological studies with nanoscale materials. In: Nanotechnology environmental health and safety. William Andrew Publishing,Amsterdam, The Netherlands, pp 3–47
Sayes CM, Wahi R, Kurian PA, Liu Y, West JL, Ausman KD et al (2006) Correlating nanoscale titania structure with toxicity: a cytotoxicity and inflammatory response study with human dermal fibroblasts and human lung epithelial cells. Toxicol Sci 92(1):174–185
Schröfel A, Kratošová G, Šafařík I, Šafaříková M, Raška I, Shor LM (2014) Applications of biosynthesized metallic nanoparticles–a review. Acta Biomater 10:4023–4042
Seigneuric R, Markey L, Nuyten DS, Dubernet C, Evelo CT, Finot E, Garrido C (2010) From nanotechnology to nanomedicine: applications to cancer research. Curr Mol Med 10(7):640–652
Shahbaz M, Akram A, Mehak A, Haq E, Fatima N, Wareen G, Aamir S, Hassan H, Poczai P, Sayyed RZ, Ilyas N (2023) Evaluation of selenium nanoparticles in inducing disease resistance against spot blotch disease and promoting growth in wheat under biotic stress. Plants 12(4):761. https://doi.org/10.3390/plants12040761
Sharma S, Kumari P, Shandilya M, Thakur S, Perveen K, Sheikh I, Ahmed Z, Sayyed R, Mastinu A (2025) The combination of α-Fe_2O_3NP and *Trichoderma* sp. improves antifungal activity against *Fusarium* wilt. J Basic Microbiol 65(4):e2400613. https://doi.org/10.1002/jobm.202400613

Siddique S, Chow JC (2020) Gold nanoparticles for drug delivery and cancer therapy. Appl Sci 10(11):3824
Siegrist M, Cousin ME, Kastenholz H, Wiek A (2007) Public acceptance of nanotechnology foods and food packaging: the influence of affect and trust. Appetite 49(2):459–466
Siimes N, Sharp EL, Lewis N, Kah M (2022) Determining acceptance and rejection of nano-enabled agriculture: a case study of the New Zealand wine industry. NanoImpact 28:100432
Singh RP, Shukla VK, Yadav RS, Sharma PK, Singh PK, Pandey AC (2011) Biological approach of zinc oxide nanoparticles formation and its characterization. Adv Mater Lett 2(4):313–317
Sly PD, Schüepp K (2012) Nanoparticles and children's lungs: is there a need for caution? Paediatr Respir Rev 2(13):71–72
Smita S, Gupta SK, Bartonova A, Dusinska M, Gutleb AC, Rahman Q (2012) Nanoparticles in the environment: assessment using the causal diagram approach. Environ Health 11:1–11
Sonawane H, Arya S, Math S, Shelke D (2021) Myco-synthesized silver and titanium oxide nanoparticles as seed priming agents to promote seed germination and seedling growth of Solanum lycopersicum: a comparative study. Int Nano Lett 11(4):371–379
Song U, Kim J (2020) Zinc oxide nanoparticles: a potential micronutrient fertilizer for horticultural crops with little toxicity. Hortic Environ Biotechnol 61(3):625–631
Terzano C, Di Stefano F, Conti V, Graziani E, Petroianni A (2010) Air pollution ultrafine particles: toxicity beyond the lung. Eur Rev Med Pharmacol Sci 14(10):809–821
Thakur P, Thakur S, Kumari P, Shandilya M, Sharma S, Poczai P, Alarfaj AA, Sayyed RZ (2022) Nano-insecticide: synthesis, characterization, and evaluation of insecticidal activity of ZnO NPs against *Spodoptera litura* and *Macrosiphum euphorbiae*. Appl Nanosci 12:3835–3850. https://doi.org/10.1007/s13204-022-02530-6
Theron J, Eugene Cloete T, de Kwaadsteniet M (2010) Current molecular and emerging nano-biotechnology approaches for the detection of microbial pathogens. Crit Rev Microbiol 36(4):318–339
Tong T, Wilke CM, Wu J, Binh CTT, Kelly JJ, Gaillard JF, Gray KA (2015) Combined toxicity of nano-ZnO and nano-TiO_2: from single-to multinanomaterial systems. Environ Sci Technol 49(13):8113–8123
Turkevich LA, Dastidar AG, Hachmeister Z, Lim M (2015) Potential explosion hazard of carbonaceous nanoparticles: explosion parameters of selected materials. J Hazard Mater 295:97–103
U.S. Environmental Protection Agency (USEPA) (2016) Nanotechnology White Paper. Available at: https://www.epa.gov/osa/nanotechnology-white-paper
Usman M, Farooq M, Wakeel A, Nawaz A, Cheema SA, ur Rehman H et al (2020) Nanotechnology in agriculture: current status, challenges and future opportunities. Sci Total Environ 721:137778
Ventola CL (2012) The nanomedicine revolution: part 1: emerging concepts. Pharm Ther 37(9):512
Wang P, Lombi E, Sun S, Scheckel KG, Malysheva A, McKenna BA et al (2017) Characterizing the uptake, accumulation and toxicity of silver sulfide nanoparticles in plants. Environ Sci Nano 4(2):448–460
Yadav R, Dwivedi S, Kumar S, Chaudhury A (2010) Trends and perspectives of biosensors for food and environmental virology. Food Environ Virol 2:53–63
Ye Q, Chen W, Huang H, Tang Y, Wang W, Meng F et al (2020) Iron and zinc ions, potent weapons against multidrug-resistant bacteria. Appl Microbiol Biotechnol 104:5213–5227
Yeh NC (2013) Nanotechnology for electronics & photonics. Technovation 33(4–5):108–108
Yuan P, Ding X, Yang YY, Xu QH (2018) Metal nanoparticles for diagnosis and therapy of bacterial infection. Adv Healthc Mater 7(13):1701392
Yue C, Zhao S, Cummings C, Kuzma J (2015) Investigating factors influencing consumer willingness to buy GM food and nano-food. Journal of Nanoparticle Research. 17(7):283
Zare-Zardini H, Amiri A, Shanbedi M, Taheri-Kafrani A, Sadri Z, Ghanizadeh F et al (2015) Nanotechnology and pediatric cancer: prevention, diagnosis and treatment. Iran J Pediatr Hematol Oncol 5(4):233
Zhang H, Demirer GS, Zhang H, Ye T, Goh NS, Aditham AJ, ..Landry MP (2019) DNA nanostructures coordinate gene silencing in mature plants. Proceedings of the National Academy of Sciences. 116(15):7543–7548

Zhang H, He X, Zhang Z, Zhang P, Li Y, Ma Y, Kuang Y, Zhao Y, Chai Z (2011) Nano-CeO2 exhibits adverse effects at environmental relevant concentrations. Environmental Science & Technology. 45(8):3725–3730

Zheng Z, Zhang X, Carbo D, Clark C, Nathan CA, Lvov Y (2010) Sonication-assisted synthesis of polyelectrolyte-coated curcumin nanoparticles. Langmuir 26(11):7679–7681

Zhuang J, Gentry RW (2011) Environmental application and risks of nanotechnology: a balanced view. In: Biotechnology and nanotechnology risk assessment: minding and managing the potential threats around us. American Chemical Society, USA, pp 41–67

Nano-biofertilizers and Formulations for Stress Mitigation

15

Mohamed Ferioun, Ilham Zouitane, Nouhaila Zouine, Khalid Derraz, Said Louahlia, Riyaz Sayyed, and Naïma El Ghachtouli

Abstract

Environmental stresses are among the principal factors that limit plant growth and yield production. With demographic increases and the intensification of climate change, finding suitable technologies to enhance plant production under various environmental conditions has become urgent. Here, we aim to highlight the use of nanoparticles (NPs) as a powerful fertilization technology that can promote plant growth and stress tolerance. NPs may be inoculated in the plant

Mohamed Ferioun and Ilham Zouitane contributed equally with all other contributors.

M. Ferioun (✉)
Microbial Biotechnology and Bioactive Molecules Laboratory, Sciences, and Technology Faculty, Sidi Mohamed Ben Abdellah University, Fez, Morocco

Natural Resources and Environmental Laboratory, Taza Polydisciplinary Faculty, Sidi Mohamed Ben Abdellah University, Fez, Morocco
e-mail: mohamed.ferioun@usmba.ac.ma

I. Zouitane · N. Zouine · N. El Ghachtouli
Microbial Biotechnology and Bioactive Molecules Laboratory, Sciences, and Technology Faculty, Sidi Mohamed Ben Abdellah University, Fez, Morocco
e-mail: naima.elghachtouli@usmba.ac.ma

K. Derraz
Laboratory of Functional Ecology and Environment, Sciences and Technology Faculty, Sidi Mohamed Ben Abdellah University, Fez, Morocco

S. Louahlia
Natural Resources and Environmental Laboratory, Taza Polydisciplinary Faculty, Sidi Mohamed Ben Abdellah University, Fez, Morocco

R. Sayyed
Department of Biological Sciences and Chemistry, College of Arts and Science, University of Nizwa, Nizwa, Sultanate of Oman

R. Sayyed et al. (eds.), *Nanobiostimulants in Innovative Agriculture*,
https://doi.org/10.1007/978-981-95-2477-8_15

rhizosphere or sprayed on plant leaves. NPs show an interesting plant growth-promoting ability by increasing plant biomass, enhancing water and nutrient uptake, supporting membrane stability, promoting photosynthesis efficiency, and limiting induced oxidative stress. This provides plants with robust tolerance enhancements against various stresses such as drought, salinity, heavy metals, and other stressors. Constructing fertilizers based on NPs could be an effective technology for enhancing plant stress tolerance and maintaining crop yields under various environmental conditions.

Keywords

Climate change · Drought · Fertilization technology · Heavy metals · Nanoparticles (NPs) · Nutrient uptake · Oxidative stress

1 Introduction

The increase in global food needs is a pressing issue driven by rapid demographic development (Faizan et al. 2024; Fizzah Zulfiqar et al. 2024). With the world's population projected to reach up to 9.6 billion by 2050, there is an anticipated 70% increase in food demand (Rajput et al. 2021; Khalid et al. 2022; Rehman et al. 2024). This situation necessitates a revolution in conventional agricultural practices to achieve sustainable food production and the UN's "Zero Hunger" goal by 2030 (Rajput et al. 2021). Concurrently, the global agricultural system has been unfavorably impacted by adverse environmental changes. These changes in the climate, in turn, have altered the abiotic conditions of plants (Al-Khayri et al. 2023; Elouattassi et al. 2024; Elouattassi et al. 2023). Abiotic stresses in plants, such as drought, salinity, soil erosion, temperature extremes, heavy metals, and nutrient depletion, are the main obstacles to global agricultural production and food security (Ferioun et al. 2025; Bright et al. 2025; Khan and Upadhyaya 2019; Bhushan et al. 2020; Iqbal et al. 2020; Jithendar et al. 2024; Elouattassi et al. 2023). They play a significant role in yield losses, accounting for up to 50% of losses in the yield of major crops (Al-Khayri et al. 2023). Different physiological, biochemical and molecular changes occur in plants during stress, resulting in disruptions in plant growth, development, and productivity (Vafa et al. 2024; Fizzah Zulfiqar et al. 2024; Khan and Upadhyaya 2019; Khalid et al. 2022; Ferioun et al. 2024a; Aktar et al. 2025; Chaithra et al. 2025). Moreover, chemical fertilizers, while essential for providing nutrients for plant growth, often lead to crop damage, groundwater contamination, and soil degradation when used in high concentrations (Hadri et al. 2024; Bouhraoua et al. 2023). Fertilizer efficiency is a significant concern because traditional fertilizers often fail to deliver nutrients efficiently, leading to nutrient losses through leaching and volatilization, which exacerbates environmental impacts and increases production costs for farmers (Kumar et al. 2023; Faizan et al. 2024). To address these challenges, there is a pressing need for robust fertilizers that not only provide essential nutrients but also improve plant stress tolerance (Yadav et al. 2024). Nanoparticles

(NPs) represent a significant advancement in this regard. Nanotechnology has emerged as one such novel approach to improve crop production, through the utilization of nanoscale products, such as nanofertilizers, nanofungicides (Faridvand et al. 2021a), nanoherbicides (Thakur et al. 2022) and nanopesticides (Khalid et al. 2022). Their ability to cross cellular barriers makes nanoparticles suitable for their application in agriculture (Rajput et al. 2021; Al-Khayri et al. 2023). Nanoparticles are small materials ranging in size from 1 to 100 nm and interact with plants, causing many morphological and physiological changes depending on their properties (Chen et al. 2024). The efficacy of NPs is determined by their chemical composition, size, surface covering, reactivity, and, most importantly, the dose at which they are effective, as well as plant species (Siddiqui et al. 2015b; Fizzah Zulfiqar et al. 2024). The use of nanoparticle-based fertilizers reduces soil pollution and leaching, supports sustainable agriculture by improving crop yield and quality through better plant nutrition and enhanced photosynthesis, and boosts plant metabolism due to their physicochemical properties. These fertilizers also enhance nutrient absorption, reduce toxicity, and minimize the use of agrochemicals (Avila-Quezada et al. 2022; Hadri et al. 2024). Nanoparticles used to treat plants and overcome environmental deficits and abiotic stresses include titanium dioxide (TiO_2), zinc (Zn), zinc oxide (ZnO), cesium (Ce), cobalt (Co), selenium (Se), silver (Ag), silicon (Si), silicon oxide (SiO_2), calcium carbonate ($CaCO_3$), magnesium (Mg), magnesium oxide (MgO), manganese (Mn), molybdenum (Mo), aluminum oxide (AlO_4), and carbon nanotubes (CNTs) (Khalid et al. 2022; Chaffai et al. 2024; Jafir et al. 2024). Additionally, phosphorous nanoparticles ($Ca_5(PO_4)^3OH$), (Ahmed et al. 2021), iron oxide nanoparticles (Fe_2O_3), copper nanoparticles (CuO), molybdenum nanoparticles (Mo), and aluminum oxide (AlO_4) CNTs are also used (Rajput et al. 2021). Nanoparticle-based fertilizers offer numerous advantages for agriculture. Their high surface area and small size enhance nutrient transfer to plants, improving crop yield and quality (Rajput et al. 2021). NPs can withstand severe atmospheric conditions and biotic-abiotic stress, contributing positively to plant development by acting at enzymatic and hormonal levels (Rasouli et al. 2020; Fazeli-Nasab et al. 2021; Betina et al. 2023; Faizan et al. 2024). They also strengthen plant defense against pathogens by providing essential microelements, although some elements may reduce nutrient absorption in alkaline soils. Additionally, NPs reduce oxidative stress by increasing antioxidant activity and regulating osmolytes and hormones (Hossain et al. 2015). They improve plant performance under drought conditions by up-regulating proline and sugar production, preserving cellular integrity (Rehman et al. 2024). NPs also play a crucial role in alleviating abiotic stress-induced toxicity, enhancing soil characteristics, and managing heavy metal contamination (Mitra et al. 2023). Their controlled release mechanisms, targeted activity, and reduced eco-toxicity make them more efficient than traditional fertilizers (Channab et al. 2024). Furthermore, NPs can act as carriers for bioactive compounds and exhibit antimicrobial properties, aiding in pathogen inhibition and disease management (Yadav et al. 2024). This chapter aims to address the knowledge gap regarding the long-term effects and mechanisms of nanoparticles (NPs) in enhancing plant tolerance against abiotic stresses. Despite their potential, much remains unknown about how NPs interact with plants and the environment. Studies show that NPs can change shape and structure when exposed to plants, leading to varied effects depending on the NP type and plant species (Rehman et al. 2024). For example, ZnO NPs release Zn ions in wheat root cells, while CuO NPs transform into $Cu(OH)_2$ and Cu_2O in soybean roots, affecting their efficacy and safety (Kim et al. 2017; Ali et al. 2021). This review evaluates the physiological, morphological, and molecular impacts of NPs on plants, assesses their role in enhancing tolerance to stresses like drought and salinity, investigates potential phytotoxicity, and identifies research gaps to guide future studies towards safer and more effective agricultural practices.

2 Effects of Nanoparticles on Plants

In literature, many studies revealed a significant positive impact of NPs on plant physiology (Ahmad et al. 2022; Sami et al. 2020; Venzhik et al. 2021). These impacts were revealed in plant water status, osmotic balance, oxidative status, membrane integrity, photosynthesis efficiency, hormonal balance, and nutrient uptake (Du et al. 2017). Consequently, these impacts extend to affect plant biomass, morphology, and yield production (Hafeez et al. 2015).

2.1 Water Status

Plants' water status is a critical trait reflecting tissue hydration under various environmental conditions. Several studies have shown a considerable ability of nanoparticles (NPs) to promote plant relative water content (Ashkavand et al. 2015; Martínez-Fernández et al. 2016; Van Nguyen et al. 2022), highlighting the role these particles play in ameliorating plant water status. The small size of NPs ensures a high interaction surface with soil particles, leading to improved soil moisture (Van Nguyen et al. 2022). Furthermore, NPs can increase the uptake of essential nutrients and bioactive compounds, enhancing water uptake (Dimkpa et al. 2018). NPs might also intervene in cell osmotic pressure regulation, maintaining plant cell turgor under various environmental conditions (Zahedi et al. 2024). Recently, some studies have revealed the role of NPs in the activation of aquaporin genes, which regulate water transport within plant tissues (Akdemir 2021) (Fig. 15.1).

2.2 Osmotic Status

When it comes to improving the osmotic state of plant cells, which is essential for preserving cellular turgor pressure and general plant health, NPs have shown great promise. These particles can enhance osmotic adjustment by promoting the absorption and assimilation of vital nutrients and the accumulation of osmo-regulators, including proline, glycine, betaine, and carbohydrates (Gürkan 2018). When used as fertilizers, nanoparticles can improve the effectiveness of nutrient transport to

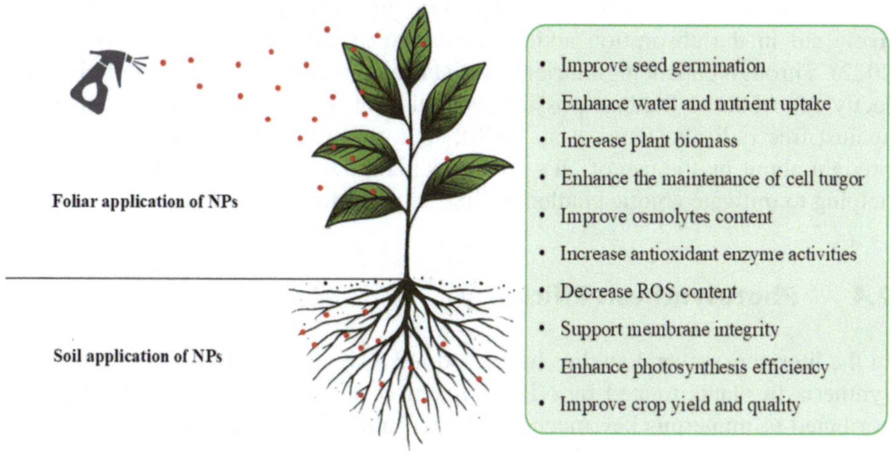

Fig. 15.1 Effects of nanoparticles (NPs) on plant growth and physiology

plant cells (Dimkpa et al. 2018). This guarantees the availability of vital ions like potassium and calcium, which are important for osmotic balance (Zahedi et al. 2024). Furthermore, osmoprotectants, which stabilize proteins and membranes against osmotic stress, may be synthesized with the aid of NPs (Kaya et al. 2016; Masouleh et al. 2017). Because of their enhanced osmotic condition, plants can better control the absorption and retention of water, preserving turgor pressure and cellular hydration even in the face of saline stress and drought (Ashkavand et al. 2015) (Fig. 15.1).

2.3 Oxidative Status

NPs have emerged as a potential technology that can enhance plant cells' oxidative status, which has shown great promise in this regard (Sami et al. 2020). Because of their special physicochemical characteristics, these particles can interact at the molecular level and affect many metabolic processes. Antioxidant enzymes, including superoxide dismutase (SOD), catalase (CAT), and peroxidase (POD), are among the main ways that NPs enhance oxidative status (Liu et al. 2020). These antioxidant enzymes are essential for preserving cellular redox equilibrium and scavenging reactive oxygen species (ROS) (Bouhraoua et al. 2024; Ferioun et al. 2024b). For example, it has been demonstrated that nanoparticles such as nano-silver and nano-titanium dioxide may, at low concentrations, promote the creation of ROS, which can function as signaling molecules to activate the antioxidant defense system of the plant (Manaf et al. 2021; Zia-ur-Rehman et al. 2023). Moreover, NPs might promote plant oxidative status by upregulating the expression of genes involved in antioxidant defense. This regulated induction aids in improving the plant's resistance to oxidative stress (Metryka et al. 2022). Indeed, as iron and zinc are necessary minerals and cofactors for several antioxidant enzymes, NPs can effectively aid in the absorption and translocation of these nutrients (Metryka et al. 2022). Through enhancing nutrient availability and absorption, nanoparticles indirectly contribute to the best possible operation of the plant's defense mechanisms against free radicals (Sami et al. 2020). As a result, the improved oxidative state brought about by the use of NPs enhances plant health and production while also helping to mitigate abiotic challenges like salt and drought (Fig. 15.1).

2.4 Photosynthesis Efficiency

In the literature, several studies have reported a significant improvement in photosynthesis in plants treated by NPs (Tripathi et al. 2016). These effects might be attributed to numerous key mechanisms: (i) Light absorption enhancement: It has been demonstrated that metal oxide nanoparticles, such as ZnO and TiO_2, enhance plant leaves' ability to absorb light. TiO_2 NPs, for instance, can function as a photocatalyst and enhance light energy absorption, enabling more effective solar energy use. This is especially advantageous in situations with less light, as NPs can enhance photosynthetic activity (Kataria et al. 2019). (ii) Stomatal regulation: By modifying stomatal activity, NPs can improve the balance of the CO_2 intake and water loss. NPs alter the stomatal opening, improving CO_2 incorporation and raising the substrate available for photosynthesis while decreasing transpiration rates (Liang et al. 2023; Tighe-Neira et al. 2018). (iii) Chloroplast functioning enhancement: Chloroplasts and NPs can interact directly to enhance the structural integrity and functional efficiency of chloroplasts (Liang et al. 2023; Tighe-Neira et al. 2018). NPs incorporate into the chloroplast membrane, stabilizing it and improving its tolerance to external stresses (Torres et al. 2018) (Fig. 15.1). Furthermore, NPs can improve the electron transport chain, a vital step in the photosynthetic process, inside chloroplasts (Poddar et al. 2020). NPs contribute to the maintenance of a high rate of ATP and NADPH production, which are necessary for the Calvin cycle and the subsequent synthesis of sugars, by enabling quicker and more efficient electron transport (Torres et al. 2018).

3 Impacts of NPs on Stress Tolerance of Plants

The application of nanoparticles in agriculture represents a significant advancement in enhancing crop resilience to climatic and biological stresses while improving agricultural efficiency and sustainability, thus contributing to food security (Akhtar et al. 2022; Fincheira et al. 2020). Nanotechnology, as an innovative approach, mitigates the adverse effects of stressful conditions and reduces the reliance on chemical fertilizers (Ingle 2021). Stress-modulating nanoparticles hold substantial potential to improve the quantity and quality of crop yields under stress conditions. Upon contact with plant cells, these nanoparticles are absorbed and transported through plasmodesmatal, apoplastic, and symplastic pathways, thereby enhancing photosynthesis, biomass, chlorophyll content, sugars, proteins, and the accumulation of osmolytes and antioxidants. They also optimize nitrogen metabolism and regulate the expression of genes associated with abiotic and biotic stress responses (Siddiqui et al. 2015a; Singh et al. 2021). Moreover, nanotechnology has been applied to various monocotyledonous and dicotyledonous plants and has proven beneficial in enhancing the growth and nutritional values of vegetables and fruits (Fincheira et al. 2020; Zahedi et al. 2020). Due to their large surface area, nanoparticles significantly increase their functionality and biological activity, positively influencing various stages of plant development and growth (Dubchak et al. 2010).

3.1 Drought Stress

One of the primary constraints on agricultural production in arid and semi-arid regions is the scarcity of water at various stages of plant development (Al-Ashkar et al. 2021; Harrison et al. 2014; Roy et al. 2021). The application of nanoparticles in agriculture represents an effective strategy to mitigate the adverse effects of water stress (Table 15.1). The results of several studies indicate that the biofertilization of cereals using nanoparticles can significantly enhance plant growth and tolerance to water stress. The addition of chitosan nanoparticles to the soil and through spraying showed a positive effect on yield, biomass, leaf area, relative water content, chlorophyll content, catalase, and superoxide dismutase of *Triticum aestivum* (Behboudi et al. 2019). However, TiO_2 nanoparticles improved productivity, seed gluten and starch content (Jaberzadeh et al. 2013), seedling length and dry weight, relative water content, catalase activity, ascorbate peroxidase activity, proline content, total chlorophyll, carotenoids, stomatal conductance, transpiration, superoxide dismutase activity, total soluble protein, net photosynthetic rate, intercellular CO_2 concentration, and decreased hydrogen peroxide and malondialdehyde content in *T. aestivum* (Faraji and Sepehri 2020) (Fig. 15.2).

The application of chitosan nanoparticles in *Hordeum vulgare* significantly improved leaf area, leaf color, the number of grains per spike, grain yield, relative water content, seed protein content, proline content, catalase and superoxide dismutase activities in barley plants under water stress (Behboudi et al. 2018). Nanoparticles are not limited to improving growth and stress tolerance in cereals but also demonstrate significant benefits for legumes and other crops. For example, in a study on *Cucumis sativus*, silicon nanoparticles (SiNP) enhanced various aspects of plant growth, including height, chlorophyll, and fruit yield, while increasing nitrogen, potassium, and silicon contents in different parts of the plant (Alsaeedi et al. 2019). In the case of *Lens culinaris*, silver nanoparticles (AgNP) significantly increased germination percentage and rate, as well as improved the length and fresh weight of radicles under water stress, demonstrating their potential to mitigate the negative effects of drought on lentil seed germination and growth (Hojjat and Kamyab 2017). Similarly, for *Zea mays*, ZnO nanoparticles promoted melatonin synthesis and activated enzymatic antioxidant systems, thereby alleviating drought-induced damage to leaf mitochondria and chloroplasts (Sun et al. 2020). Another

Table 15.1 Effects of various nanoparticles (NP) used in different plant species to study their response under drought stress

Plant	NP	Application method	Effect on plant	References
Cucumis sativus	SiO_2	Suspension of SiO_2 was added to soil	Improved growth and productivity via alerting nutrient uptake (nitrogen and potassium), plant height, chlorophyll and fruit yield	Alsaeedi et al. (2019)
Linum usitatissimum	TiO_2	Used in irrigation water	Improved morphological and physiological traits of the plant by improving chlorophyll and carotenoid contents, seed oil, protein contents and decreasing lipid peroxidation (decreasing MDA and H_2O_2)	Aghdam et al. (2016)
Gossypium barbadense	TiO_2 and SiO_2	Foliar application	Improved plant yield and increased pigment content, carbohydrates, phenolic compounds, soluble proteins, total free amino acids, proline content, total reducing power, antioxidant capacity and antioxidant enzyme activities	Shallan et al. (2016)
Lens culinaris	Ag	Suspension of silver was used for irrigation	Improved germination percentage and rate, length and fresh weight of radicle	Hojjat and Kamyab (2017)
Zea mays	Cu	Seeds immersed in a solution of copper nanoparticles	Improved leaf water content, plant biomass, grain yield, anthocyanin, chlorophyll and carotenoid content. Decreased the accumulation of reactive oxygen species	Nguyen et al. (2020)
	ZnO	Suspension was added 7 days after emergence	Promoted melatonin synthesis and activated the antioxidant enzyme system	Sun et al. (2020)

(continued)

Table 15.1 (continued)

Plant	NP	Application method	Effect on plant	References
Triticum aestivum	Chitosan	Used in soil and foliar application	Improved yield, biomass, leaf area, relative water content, chlorophyll content, catalase and superoxide dismutase	Behboudi et al. (2019)
	TiO_2	Foliar application	Improved productivity, seed gluten, and starch content	Jaberzadeh et al. (2013)
	TiO_2	Powders added and mixed directly in soil	Improved seedling dry weight, relative water content, catalase activity, ascorbate peroxidase activity, proline content, total chlorophyll, carotenoids, stomatal conductance, transpiration, seedling length, superoxide dismutase activity, total soluble protein, net photosynthetic rate and intercellular CO_2 concentration Decreased hydrogen peroxide and malondialdehyde content	Faraji and Sepehri (2020)
Hordeum vulgare	Chitosan	Used in soil and foliar application	Increased leaf area, SPAD index, grain yield and weight, relative water content, seed protein content, proline content, catalase and superoxide dismutase activities	Behboudi et al. (2018)
Sorghum bicolor	Zn	Zn added to pot soil and mixed	Increased crop development, grain yield while fortifying edible plant parts with Zn and increased N acquisition	Dimkpa et al. (2019)
Mentha piperita	TiO_2	Mixed with AMF and soil	Improved nutrient absorption, chlorophyll content, relative water content, antioxidant activity, quantity and quality of peppermint essential oil	Ostadi et al. (2023)
Glycine max	Zn	Suspension was added to seeds	Increased the germination rate and percentage, root length, root fresh and dry weight, and seed residual fresh and dry weight	Sedghi et al. (2013)
Crataegus sp.	Si	Added to nutrient solution of seedling	Improved photosynthesis rate, stomatal conductance, plant biomass, xylem water potential and MDA content	Ashkavand et al. (2016)

(continued)

Table 15.1 (continued)

Plant	NP	Application method	Effect on plant	References
Ocimum basilicum	TiO_2	Used in irrigation water	Decreased the negative effects of drought stress	Kiapour et al. (2015)
Dracocephalum moldavica	TiO_2	Spraying on shoot	Increased proline and reduced H_2O_2 and MDA levels	Mohammadi et al. (2016)
Arabidopsis thaliana	Fe	Used in nutrient media	Promoted HC-ATPase activity, maintained stomatal opening and closure, increased leaf area, biomass, photosynthetic pigments, and internal CO_2	Kim et al. (2015)
Musa acuminata	SiO_2	Used in nutrient media	Improved shoot growth and chlorophyll content and reduced malondialdehyde content and electrolyte leakage	Mahmoud et al. (2020)
Coriandrum Sativum	Si	Foliar application	Improved plant growth and yield, relative water content, total soluble sugar, total phenolic and flavonoid contents, essential oil percentage and yield of coriander	Afshari et al. (2021)
Cicer arietinum	Si	Mixed with soil	Improved the growth, Si concentrations, relative water content and the non-enzymatic antioxidant activity Decreased H_2O_2, proline, malonaldehyde and lipoxygenase activity	Gunes et al. (2007)
Solanum melongena	ZnO	Foliar application	Improved stem and leaf anatomical structures, relative water content, membrane stability index, photosynthetic efficiency and fruit yield	Semida et al. (2021)
Thymus vulgaris	TiO_2	Foliar application	Increased plant height, shoot fresh and dry weights and root dry weight	Fazeli-Nasab et al. (2018)
Lallemantia iberica	TiO_2	Foliar application	Increased the phenolic content, total flavonoid and antioxidant activity, catalase and superoxide dismutase activities	Sattari et al. (2020)

study by Nguyen et al. (2020) demonstrated that copper nanoparticles could enhance leaf water content, plant biomass, grain yield, anthocyanin, chlorophyll, and carotenoid content and decrease the accumulation of reactive oxygen species in maize. In

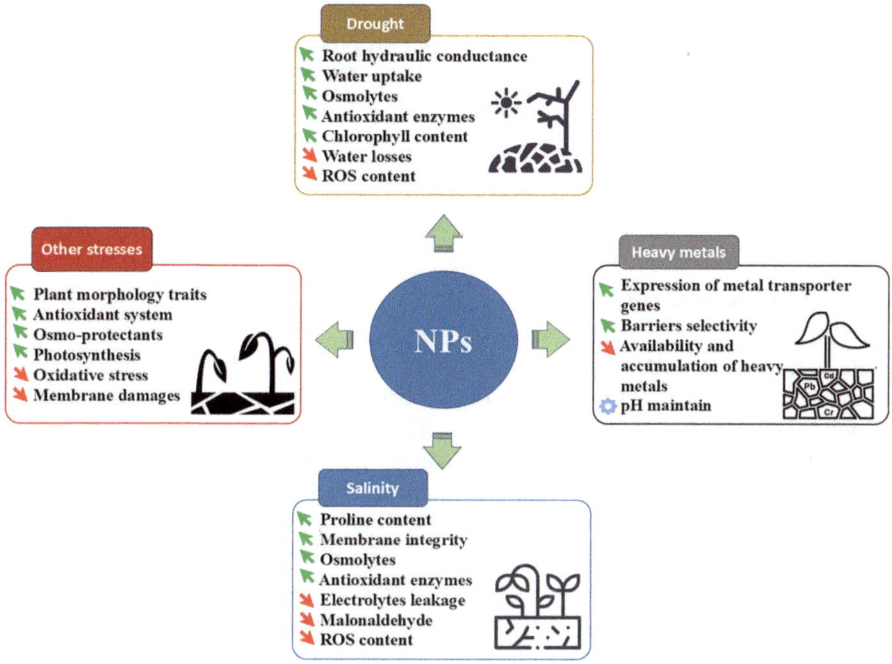

Fig. 15.2 Nanoparticles involved in enhancing plant stress tolerance

chickpea (*Cicer arietinum*) crops, the application of silicon improved the relative water content of leaves and reduced oxidative damage induced by drought by increasing antioxidant enzyme activities (Gunes et al. 2007). *Solanum melongena* (Eggplants) grown under water stress also benefited from the application of ZnO nanoparticles, which improved relative water content, membrane stability, and photosynthetic efficiency, thereby increasing growth and fruit yield despite drought stress (Semida et al. 2021). The improvement in productivity and the physiological and biochemical characteristics of plants can be attributed to the increased accessibility of nutrients. This leads to an increase in leaf area and plant height, which ensures an improvement in the photosynthesis rate by enhancing sunlight absorption (Begum et al. 2019; Silva et al. 2022). Additionally, the increase in phenol and flavonoid content may be due to the availability of nutrients that enhance the activity of enzymes involved in the biosynthesis of phenolic and flavonoid compounds (Duc et al. 2021). Since these enzymes are proteinaceous in nature, their activities increase due to the accessibility of nutrients, particularly nitrogen (Javanmard et al. 2022). Other studies have shown that these nanoparticles have improved the quantity and quality of essential oils in certain plants classified as aromatic and medicinal plants, including *Mentha piperita*, *Coriandrum sativum*, and *Thymus vulgaris* (Afshari et al. 2021; Fazeli-Nasab et al. 2018; Ostadi et al. 2023). In the case of *M. piperita*, biofertilization with TiO_2 improved nutrient absorption, which can directly or indirectly contribute to the enhancement of essential oil quality directly by improving the biosynthesis of essential oils through accessible nutrients that play a crucial role in the composition of essential oil precursors and intermediate compounds such as acetyl coenzyme A, NADPH, and ATP (Faridvand et al. 2021) and indirectly by increasing the photosynthesis rate resulting from better nutrient accessibility and also improving carbohydrate content for cell growth and the production of essential oil-secreting glands, thus increasing the essential oil content of the plant (Rostaei et al. 2018) (Fig. 15.2).

3.2 Salinity Stress

Nanoparticles have shown promising results in improving the resilience of cereal crops to salinity stress (Table 15.2). In *T. aestivum* (wheat), seed priming with Ag nanoparticles increased shoot fresh and dry weight, soluble sugars, proline, and peroxidase activity while decreasing catalase activity under salinity conditions (Mohamed et al. 2017). Foliar application of ZnO and Fe_2O_3 nanoparticles in wheat enhanced plant height, leaf area, shoot dry weight, and the concentration of Fe and Zn. Similarly, seed priming with Ag nanoparticles mitigated the damaging impacts of salinity stress in wheat (Ismail and Abou-Zeid 2018). In *Z. mays* (maize), spray treatment with ZnO and Fe_3O_4 nanoparticles increased Fe and Zn concentrations in leaves, root growth, net carbon dioxide assimilation rate, and substomatal CO_2 concentration (Fathi et al. 2017). The addition of Si to the nutrient solution also increased Si concentration in shoots and roots of maize plants, helping them cope with salinity stress (Gao et al. 2006). In *Oryza sativa* (rice), soaking and foliar application of Si and Se improved root thickness and volume, relative water content, dry matter, leaf area index, grain yield, Na^+ leaf content, and Na^+/K^+ ratio (Badawy et al. 2021).

Leguminous plants also benefit significantly from nanoparticle applications under salinity stress. For instance, in *P. vulgaris* (common bean), seed soaking with SiO_2 nanoparticles improved germination percentage, vigor index, germination rate, shoot and root lengths, and shoot and root dry weights (Alsaeedi et al. 2017). Foliar application of Si nanoparticles in common bean enhanced plant growth, chlorophylls, carotenoids, transpiration rate, net photosynthetic rate, stomatal conductance, membrane stability index, relative water content, free proline, total soluble sugars, and the nutrient contents of N, P, K, Ca_2^+, and K^+/Na^+ ratio, while decreasing electrolyte leakage, malondialdehyde, H_2O_2, O_2^-, Na^+, Pb, Cd, and Ni in leaves and pods of plants (El-Saadony et al. 2021a). In *T. foenum-graecum* (fenugreek), foliar application of ZnO nanoparticles increased protein, proline, and trigonelline contents, promoted antioxidant activity, and reduced H_2O_2 and MDA levels (Noohpisheh et al. 2021). Furthermore, Ag nanoparticles used in irrigation water improved seed germination in fenugreek (Hojjat and Kamyab 2017). In *V. faba* (faba bean), foliar application of Si nanoparticles increased the activity of ascorbate peroxidase, catalase, and peroxidase in plant leaves (Qados 2015), and also enhanced germination percentage, growth, relative water content, height of plant, fresh and dry weights, total yield, and seed quality (Qados and Moftah 2015). Additionally, foliar

Table 15.2 Effects of various nanoparticles (NP) used in different plant species to study their response under salinity stress

Plant	NP	Application method	Effect on plant	References
Phaseolus vulgaris	SiO_2	Seed soaking	Improved germination percentage, vigor index, germination rate for seeds, shoot and root lengths, shoot and root dry masses of common bean	Alsaeedi et al. (2017)
	Si	Foliar application	Enhanced plant growth and production, chlorophylls, carotenoids, transpiration rate, net photosynthetic rate, stomatal conductance, membrane stability index, relative water content, free proline, total soluble sugars, N, P, K, Ca_2^+, K^+/Na^+, and the activities of peroxidase, catalase, ascorbic peroxidase and superoxide dismutase. Decreased electrolyte leakage, malondialdehyde, H_2O_2, O_2^-, Na^+, Pb, Cd, and Ni in leaves and pods of plants	El-Saadony et al. (2021a)
Arabidopsis thaliana	CeO_2	Suspension was added to soil	Enhanced leaf mesophyll K^+ retention, chlorophyll content, biomass, and photosynthesis	Wu et al. (2018)
Brassica napus	CeO_2	Used in irrigation water	Modified the formation of the apoplastic barriers in roots; hence allowed more Na^+ transport to shoots and less accumulation of Na^+	Rossi et al. (2017)
		Mixed with soil	Enhanced plant biomass, photosynthesis and reduced stress	Rossi et al. (2016)
Trigonella foenumgraecum	ZnO	Foliar application	Increased protein, proline and trigonelline contents, promoted antioxidant activity, and reduced H_2O_2 and MDA levels	Noohpisheh et al. (2021)
	Ag	Used in irrigation water	Improved seed germination	Hojjat and Kamyab (2017)

(continued)

Table 15.2 (continued)

Plant	NP	Application method	Effect on plant	References
Solanum lycopersicum	ZnO	Used with PGPB	Alteration in cytosine methylation that induced a positive anti-genotoxic effect	Hosseinpour et al. (2020)
Helianthus annuus	ZnO	Foliar application	Improved SOD activity, shoot dry weight and biomass production of sunflower plants	Torabian et al. (2016)
			Improved leaf area, shoot dry weight, net carbon dioxide assimilation rate, sub-stomatal CO_2 concentration, chlorophyll content, Fv/Fm, Zn concentration and decreased Na content in leaves.	Torabian et al. (2015)
Lupinus termis	ZnO	Seeds priming	Enhanced photosynthetic pigments, organic solutes, total phenols, ascorbic acid, Zn, activities of SOD, CAT, APX and POD. Decreased MDA and Na^+ contents	Abdel Latef et al. (2016)

Lycopersicon esculentum	ZnO	Added to culture media	Had a positive response on plant metabolism by enhancing the expression of mRNA in the salt-tolerant cultivars	Alharby et al. (2016)
	Ag	Seeds soaking	Improved germination percentage, germination rate, root length and seedling fresh and dry weight. Up-regulation of salt stress genes AREB, MAPK2, P5CS and CRK1 and down-regulation of three salt stress genes TAS14, DDF2 and ZFHD1	Almutairi (2015)
	ZnO	Foliar application	Increased shoot and root length, biomass, leaf area, chlorophyll content, photosynthetic pigment content, protein content and antioxidative enzyme activity such as peroxidase, superoxide dismutase and catalase	Faizan et al. (2021)
	Si	Foliar application	Increased germination percentage, germination rate, root length and dry weight	(Haghighi et al. 2012)
	Si	Foliar application	Increased the fresh and dry weights, root volume, and chlorophyll concentration, photosynthetic rate, mesophyll conductance and the leaf water content	Haghighi and Pessarakli (2013)
Zea mays	ZnO and Fe_3O_4	Spray treatment	Increased Fe and Zn concentration in leaf, root growth, net carbon dioxide assimilation rate and substomatal CO_2 concentration of maize	Fathi et al. (2017)
	Si	Added to the nutrient solution	Increased Si concentration in shoots and roots of maize plants	Gao et al. (2006)

(continued)

Table 15.2 (continued)

Plant	NP	Application method	Effect on plant	References
Trachyspermum ammi	Fe_2O_3	Used in irrigation water	Enhanced K^+ uptake, K^+ /Na^+ ratio, Fe concentration, endogenous level of salicylic acid, superoxide dismutase, catalase, peroxidase, and polyphenol oxidase activities, membrane stability index, leaf water content, leaf pigments, root and shoot growth, and seed yield	Abdoli et al. (2020)
Dracocephalum moldavica	Fe_2O_3	Foliar application	Enhanced the activity of PAL, SOD and PPO enzymes and improved gene expression of tyrosine aminotransferase, rosmarinic acid synthase and rosmarinic acid content	Moradbeygi et al. (2020a)
			Increased leaf area, leaf length, fresh and dry weight of both shoot and root, total phenolic, flavonoid and anthocyanin content, as well as the activity of guaiacol peroxidase, ascorbate peroxidase, catalase and glutathione reductase enzymes	Moradbeygi et al. (2020b)
Sorghum bicolor	Fe_2O_3	Seed soaking	Enhanced the seedling vigor index, growth, photosynthesis, photosystem II efficiency, relative water content, and decreasing membrane damage	Maswada et al. (2018)
Grape	Fe	In nutrient media	Increased total protein content and reduced proline, enzymatic antioxidant activity and hydrogen peroxide, K^+ and decreased Na^+ contents	Mozafari et al. (2018)
Lathyrus sativus	Ag	Used in irrigation water	Improved the germination percentage, shoot and root length, seedling fresh weight, seedling dry weight and seedling dry contents of grass pea seedlings	Saeid Hojjat (2019)

Triticum aestivum	Ag	Seeds priming	Salinity increased shoot fresh and dry weight, soluble sugars, proline and peroxidase activity and decreased catalase activity	Mohamed et al. (2017)
	ZnO and Fe_2O_3	Foliar application	Improved plant height, leaf area, shoot dry weight, and the concentration of Fe and Zn	Fathi et al. (2017)
	Ag	Seeds priming	Mitigated the damaging impacts of salinity stress	Ismail and Abou-Zeid (2018)
Capsicum annuum	Mn	Seeds priming	Improved seed germination, root elongation and alteration in gene expression (MnSOD)	Ye et al. (2020)
	Si	Added in irrigation water	Enhanced plant height, number of branches, leaves fresh and dry weights and yield.	Tantawy et al. (2015)
Cucumis sativus	Mn_3O_4	Foliar application	Increased leaf photosynthetic pigment content, net photosynthesis, and biomass. Alleviate oxidative stress and alter the metabolism of plants, including the upregulation of phenolics and other antioxidant compounds	Lu et al. (2020)
	Si	Applied in irrigation water	Improved seed germination, seedlings growth, vigor index, germination rate index, fresh mass, shoot and root dry mass, and plant height. Decreased Na^+ uptake and improved high K^+/Na^+ ratio	Alsaeedi et al. (2018)
	SiO_2	Suspension of SiO_2 was added to soil	Improved growth and productivity via alerting nutrient uptake (nitrogen and potassium), plant height, chlorophyll and fruit yield	Alsaeedi et al. (2019)

(continued)

Table 15.2 (continued)

Plant	NP	Application method	Effect on plant	References
Mangifera indica	ZnO and Si	Foliar application	Improved plant growth, nutrient uptake, carbon assimilation, annual fruit yield and physiochemical characteristics	Elsheery et al. (2020)
Vicia faba	Si	Foliar application	Increased the activity of ascorbate peroxidase, catalase and peroxidase in plant leaves	Qados (2015)
			Enhanced the germination percentage, germination rate, mean germination time, growth, relative water content, height of plant, fresh and dry weights, total yield and seeds quality represented by nutrient elements	Qados and Moftah (2015)
	TiO_2	Foliar application	Increased the antioxidant enzymatic activities, soluble sugars, amino acids and proline contents.	Abdel et al. (2018)
Ocimum basilicum	Si	Foliar application	Increased leaf dry and fresh weight and the chlorophyll content	Kalteh et al. (2014)
Lens culinaris	SiO_2	Added to nutrient solution	Increased seed germination and seedling growth	Sabaghnia and Janmohammadi (2014)
Cucurbita pepo	SiO_2	Added to petri dishes containing seeds	Improved seed germination, net photosynthetic rate, stomatal conductance, transpiration rate, and water use efficiency. Decreased chlorophyll degradation, malondialdehyde and hydrogen peroxide levels as well as electrolyte leakage	Siddiqui et al. (2014)
Prunus mahaleb	SiO_2	Added in irrigation water	Improved gas exchange, nutritional status of the plants as the concentration of N, P, and K	Ashkavand et al. (2018)

Jatropha integerrima	Si	Foliar application	Enhanced growth parameters and chemical constituents, meanwhile decreasing the accumulation of Na%, total phenolic and flavonoids in leaves	Ashour and Mahmoud (2017)
Glycine max	Si	Foliar application	Improved shoot and root growth of seedlings, antioxidant activities, non-enzymatic compounds, K^+ concentration, and decreased Na^+ concentration, lipid peroxidation, and reactive oxygen species production	Farhangi-Abriz and Torabian (2018)
Oryza sativa	Si and Se	Soaking and foliar application	Improved root thickness and volume, relative water content, dry matter, leaf area index, grain yield, Na^+ leaf content and Na^+/K^+ ratio	Badawy et al. (2021)
Mentha piperita	Fe_2O_3	Used in water irrigation	Increased the leaf fresh and dry weights, P, K, Fe, Zn, and Ca contents. Decreased lipid peroxidation and proline contents	Askary et al. (2017)

application of TiO_2 nanoparticles increased antioxidant enzymatic activities, soluble sugars, amino acids, and proline contents in faba bean (Abdel et al. 2018). In *G. max* (soybean), foliar application of Si nanoparticles improved shoot and root growth, antioxidant activities, non-enzymatic compounds, K^+ concentration, and decreased Na^+ concentration, lipid peroxidation, and reactive oxygen species production (Farhangi-Abriz and Torabian 2018).

Vegetable crops also show improved tolerance to salinity stress with the application of nanoparticles. In *L. esculentum* (tomato), ZnO nanoparticles added to culture media had a positive effect on plant metabolism by enhancing the expression of mRNA in salt-tolerant cultivars (Alharby et al. 2016). Seeds soaking with Ag nanoparticles improved germination percentage, germination rate, root length, and seedling fresh and dry weight, along with the up-regulation of salt stress genes *AREB*, *MAPK2*, *P5CS*, and *CRK1*, and down-regulation of salt stress genes *TAS14*, *DDF2*, and *ZFHD1* in plants (Almutairi 2015). Foliar application of ZnO nanoparticles in tomatoes increased shoot and root length, biomass, leaf area, chlorophyll content, photosynthetic pigments, protein content, and antioxidative enzyme activity such as peroxidase, superoxide dismutase, and catalase (Faizan et al. 2021). Si nanoparticles applied as foliar spray in tomatoes increased the fresh and dry weights, root volume, chlorophyll concentration, photosynthetic rate, mesophyll conductance, and leaf water content (Haghighi and Pessarakli 2013). In *S. lycopersicum*, foliar application of Si nanoparticles increased germination percentage, germination rate, root length, and dry weight (Haghighi et al. 2012). Combined with PGPB (plant growth-promoting bacteria), ZnO nanoparticles induced positive antigenotoxic effects in *S. lycopersicum* (Hosseinpour et al. 2020). ZnO nanoparticles alone improved leaf area, shoot dry weight, net carbon dioxide assimilation rate, substomatal CO_2 concentration, chlorophyll content, Fv/Fm, Zn concentration, and decreased Na^+ content in leaves (Torabian et al. 2015). In *M. piperita* (peppermint), Fe_2O_3 nanoparticles used in irrigation water increased the leaf fresh and dry weights, and the contents of P, K, Fe, Zn, and Ca, while decreasing lipid peroxidation and proline contents (Askary et al. 2017).

Several other crops have also demonstrated enhanced tolerance to salinity stress by nanoparticles. In *H. annuus* (sunflower), foliar application of ZnO nanoparticles improved SOD activity, shoot dry weight, and biomass production (Torabian et al. 2016). In *S. bicolor* (sorghum), seed soaking with Fe_2O_3 nanoparticles enhanced seedling vigor index, growth, photosynthesis, photosystem II efficiency, relative water content, and decreased membrane damage (Maswada et al. 2018). In *C. sativus* (cucumber), SiO_2 nanoparticles added to soil improved growth and productivity via altering nutrient uptake, plant height, chlorophyll, and fruit yield (A. Alsaeedi et al. 2019). Foliar application of Mn_3O_4 nanoparticles in cucumber increased leaf photosynthetic pigment content, net photosynthesis, and biomass, alleviated oxidative stress, and altered plant metabolism by upregulating phenolics and other antioxidant compounds (Lu et al. 2020) (Fig. 15.2).

3.3 Heavy Metal Stress

Nanoparticles have shown significant potential in mitigating the effects of heavy metals on leguminous plants, enhancing growth and reducing metal accumulation (Table 15.3). The addition of SiO_2 nanoparticles to the nutrient medium alleviated growth issues, improved chlorophyll content and antioxidant enzyme activity, and reduced the accumulation and translocation of mercury in soybean seedlings (Li et al. 2020). Similarly, foliar application of ZnO nanoparticles in soybean plants decreased arsenic concentration in roots and shoots while increasing shoot and root lengths, net photosynthetic rate, transpiration, stomatal conductance, photochemical yield, and activities of several antioxidant enzymes (Ahmad et al. 2020). In *P. sativum* (pea), the addition of Si nanoparticles to the nutrient solution reduced chromium accumulation in roots and shoots, decreased reactive oxygen species levels, and upregulated the antioxidant defense system and mineral nutrient levels, providing protection against oxidative stress (Tripathi et al. 2015). *V. radiata* (mung bean) treated with TiO_2 nanoparticles showed improved growth traits, membrane stability index, and protein content, along with reduced levels of reactive oxygen species and malonaldehyde by increasing the activities/gene expressions of antioxidant enzymes under arsenic stress (Katiyar et al. 2020). In *L. luteus* (yellow lupin), soil mixed with Ag nanoparticles improved glutathione peroxidase activity and metallothionein expression when exposed to cadmium, lead, zinc and nickel (Jaskulak et al. 2019).

Table 15.3 Effects of various nanoparticles (NP) used in different plant species to study their response under heavy metal stress

Plant	Heavy metal	NP	Application method	Effect on plant	References
Glycine max	Mercury	SiO_2	SiO_2 added to nutrient medium	Alleviated the growth, chlorophyll, antioxidant enzyme and reduced the accumulation and translocation of Hg in soybean seedlings	Li et al. (2020)
	Arsenic	ZnO	Foliar application	Decreased arsenic concentration in the root and shoot of soybean plants. Increased the lengths of shoot and root, net photosynthetic rate, transpiration, stomatal conductance, photochemical yield, superoxide dismutase, catalase, ascorbate peroxidase and glutathione reductase activities	Ahmad et al. (2020)

(continued)

Table 15.3 (continued)

Plant	Heavy metal	NP	Application method	Effect on plant	References
Triticum aestivum	Cadmium	Si	Seed priming	Improved growth, chlorophyll content, antioxidant enzyme, Si concentration in plants and decreased the oxidative stress and the Cd concentrations in wheat, especially in grains	Hussain et al. (2019)
			Applied in soil and foliar spray	Increased the dry biomass of shoots, roots, spikes and grains, the leaf gas exchange attributes, chlorophyll a and b concentrations, and Si concentration in shoots and roots. Diminished the oxidative stress by reducing the electrolyte leakage and enhancing the superoxide dismutase and peroxidase activities in leaf. Also, reduced Cd contents in shoots, roots, and grains	Rizwan et al. (2019)
		ZnO	Mixed with soil	Increased the tissue dry weight and minimized the oxidative. Reduced the Cd accumulation in tissues and grains by reducing the soil bioavailable Cd and its accumulation by roots.	Khan et al. (2019)
		ZnO and Fe_3O_4	Seed priming	Increased plant height, spike length, dry weights of shoots, roots, spikes, and grains Reduced the electrolyte leakage, superoxide dismutase and peroxidase activities, the concentrations of Cd in roots, shoots, and grains	Rizwan et al. (2019)
	Lead, zinc, cadmium and copper	Fe_3O_4	Added to seedling germination medium	Decreased the growth inhibition and activated protective mechanisms to reduce oxidative stress by increasing peroxidase and superoxide dismutase activities	Konate et al. (2017)

(continued)

Table 15.3 (continued)

Plant	Heavy metal	NP	Application method	Effect on plant	References
Oryza sativa	Cadmium	ZnO	Foliar application	Improved the biomass, photosynthesis, and Zn concentration of rice plants and diminished the Cd concentration in shoots and roots	Rizwan et al. (2019)
	Arsenic	ZnO	Added to seedling nutrient solution	Enhanced the growth and photosynthesis of rice seedlings. Decreased the concentration of arsenic in the shoots and roots and upregulated the phytochelatin content	Yan et al. (2021)
	Cadmium, lead, copper and zinc	Si	Foliar application	Improved biomass and yield. Reduced the concentrations of heavy metals in rice (reduced bioconcentration factor of the heavy metals in grains and decreased the translocation factor of heavy metals from roots to shoots and/or from shoots to grains)	Wang et al. (2016)
Lupinus luteus	Cadmium, lead, zinc and nickel	Ag	Mixed with soil	Improved GPX activity and metallothionein expression	Jaskulak et al. (2019)
Pisum sativum	Chromium	Si	Added to nutrient solution	Reduced the accumulation of Cr in roots and shoots, levels of reactive oxygen species, upregulation of antioxidant defense system and mineral nutrients, which provide protection against oxidative stress	Tripathi et al. (2015)
Zea mays	Arsenic	Si	Seed soaking	Decreased the accumulation of arsenic and reactive oxygen species, and enhanced levels of superoxide dismutase, ascorbate peroxidase, glutathione reductase, dehydrateascorbate reductase, reduced ascorbate, reduced glutathione, which re-establish the redox status of the cell and thus mitigate arsenic toxicity	Tripathi et al. 2016)

(continued)

Table 15.3 (continued)

Plant	Heavy metal	NP	Application method	Effect on plant	References
Vigna radiata	Arsenic	TiO_2	Exogenous application	Improved growth traits, membrane stability index and protein content. Reduced the levels of reactive oxygen species and malonaldehyde via increasing the activities/gene expressions of antioxidant enzymes (superoxide dismutase and catalase)	Katiyar et al. (2020)
Brassica juncea	Chromium	Fe	Seed soaking	Enhanced the reduction of Cr (VI) to Cr (III) and decreased the bioavailability of Cr (VI) for plant uptake	Vemula et al. (2013)

Cereal grains have also benefited from nanoparticle treatments under heavy metal stress, showing improvements in growth, yield, and reduction in metal accumulation. *T. aestivum* (wheat) seed priming with Si nanoparticles improved growth, chlorophyll content, antioxidant enzyme activity, Si concentration in plants, and decreased oxidative stress and Cd concentrations, especially in grains (Hussain et al. 2019). Applying Si nanoparticles to soil and as a foliar spray increased dry biomass of shoots, roots, spikes, and grains, leaf gas exchange attributes, chlorophyll concentrations, and Si concentration in shoots and roots, while diminishing oxidative stress and reducing Cd contents in wheat plants (Ali et al. 2019a). Mixing ZnO nanoparticles with soil increased tissue dry weight and minimized oxidative stress by reducing Cd accumulation in wheat (Khan et al. 2019). Seed priming with ZnO and Fe_3O_4 nanoparticles increased plant height, spike length, dry weights, and reduced electrolyte leakage, superoxide dismutase, and peroxidase activities, as well as Cd concentrations in *T. aestivum* (Rizwan et al. 2019). Additionally, Fe_3O_4 nanoparticles added to the seedling germination medium of wheat decreased growth inhibition and activated protective mechanisms to reduce oxidative stress (Konate et al. 2017). *O. sativa* (rice) also exhibits enhanced tolerance to heavy metal stress with nanoparticle applications. Foliar application of ZnO nanoparticles improved biomass, photosynthesis, Zn concentration, and reduced Cd concentration in shoots and roots (Ali et al. 2019b). Adding ZnO nanoparticles to the seedling nutrient solution enhanced growth and photosynthesis, decreased arsenic concentration in shoots and roots, and upregulated phytochelatin content of rice plants (Yan et al. 2021). Foliar application of Si nanoparticles improved biomass and yield, reduced the concentrations of multiple heavy metals (Cd, Pb, Cu, and Zn) in rice, lowered the bioconcentration factor of these metals in grains, and decreased their translocation from roots to shoots and grains (Wang et al. 2016). In *Z. mays* (maize), seed soaking in Si nanoparticles decreased arsenic accumulation and reactive oxygen species levels, enhanced activities of superoxide dismutase, ascorbate peroxidase, glutathione reductase, dehydrascorbate reductase, reduced ascorbate, and glutathione, thus mitigating arsenic toxicity and re-establishing the redox status of cells.

Various other plants also benefit from nanoparticle applications, showing improvements in growth, antioxidant activity, and reduction in metal accumulation. *M. piperita* (peppermint) treated with Fe_2O_3 nanoparticles in water irrigation showed increased leaf fresh and dry weights, phosphorus (P), potassium (K), iron (Fe), zinc (Zn), and calcium (Ca) contents, and decreased lipid peroxidation and proline contents (Askary et al. 2017). In *B. juncea* (Indian mustard), seed soaking with Fe nanoparticles enhanced the reduction of Cr(VI) to Cr(III) and decreased the bioavailability of Cr(VI) for plant uptake, providing a protective effect against Cr stress (Vemula et al. 2013) (Fig. 15.2).

3.4 Other Stresses

Titanium dioxide nanoparticles (TiO_2 NPs) have shown significant beneficial effects on plants under cold stress (Table 15.4). In *G. glabra*, the addition of TiO_2 to the medium led to an increase in phenolics, total proteins, and osmolytes, while reducing oxidative damage by lowering the content of malondialdehyde and hydrogen peroxide (Ghabel and Karamian 2020). Additionally, in *C. arietinum*, the addition of TiO_2 to soil and irrigation water increased the transcription level of transcript-derived fragments, crucial for developing cold tolerance by decreasing electrolyte leakage index content (Amini et al. 2017). Another study on *C. arietinum* showed that adding TiO_2 to the nutrient solution improved the redox status of chickpea genotypes and reduced electrolyte leakage index and malondialdehyde content (Mohammadi et al. 2013). Furthermore, TiO_2 suspension added to *C. arietinum* seeds increased the transcription levels of *CaLRubisco*, *CaSRubisco*, and *Cachlorophyll* a/b-binding protein genes, enhancing photosynthetic activity, increasing phosphoenolpyruvate carboxylase activity, and decreasing H_2O_2 content (Hasanpour et al. 2015). In *A. elongatum*, SiO_2 added to petri dishes containing seeds prevented seed dormancy and improved seed germination, dry weight of shoots, roots, and seedlings (Azimi et al. 2014).

Nanoparticles have also demonstrated significant benefits under heat stress (Table 15.4). In *T. aestivum*, a colloidal emulsion of ZnO and Fe_3O_4 applied to shoots increased yield quantity and the activity of antioxidant enzymes such as glutathione S-transferase, superoxide dismutase, peroxidase, and catalase, while decreasing the lipid peroxidation product (malondialdehyde) (Hassan et al. 2018). Additionally, a solution of Ag added to *T. aestivum* soil improved root and shoot lengths, root number, plant fresh and dry weights, leaf area, and leaf number (Iqbal et al. 2019). In *L. esculentum*, spraying TiO_2 increased the photosynthetic rate, conductance to H_2O, and transpiration rate of tomato leaves. It also regulated photosystem II (PS II) energy dissipation and decreased non-regulated PS II energy dissipation, minimum chlorophyll fluorescence, and relative electron transport in leaves (Qi et al. 2013).

Nanoparticles also play a role in mitigating the effects of flooding stress (Table 15.4). In *G. max*, the addition of Al_2O_3 to the medium affected mitochondrial proteins by regulating membrane permeability and tricarboxylic acid cycle activity. It also increased root length, regulated the ascorbate/glutathione pathway, and suppressed proteins related to glycolysis (Mustafa and Komatsu 2016). The use of Ag nanoparticles in the irrigation of *G. max* improved plant growth, root length/weight, and hypocotyl length/weight. The accumulation of calnexin/calreticulin and glycoproteins increased, and protein quality control for misfolded proteins in the endoplasmic reticulum was upregulated (Hashimoto et al. 2020). Additionally, soaking *C. sativus* seeds in Ag blocked ethylene signaling, promoted root number and length, and increased leaf dry weight (Rezvani et al. 2012).

To mitigate UV radiation stress, Si applied as a foliar spray in *T. aestivum* improved the protection of wheat seedlings via NO-mediated antioxidant defense, consequently counterbalancing toxic reactive oxygen species (ROS) generation (Tripathi et al. 2017). In *S. oleracea*, foliar application of TiO_2 increased the activities of superoxide dismutase, catalase, ascorbate peroxidase, and guaiacol peroxidase, while elevating oxygen evolution rate. It also decreased the accumulation of superoxide radicals, hydrogen peroxide, and malondialdehyde content (Lei et al. 2008) (Fig. 15.2).

Silver nanoparticles (Ag) have also shown beneficial effects on plants under dark stress. In *P. zonale*, Ag spray increased petal longevity, chlorophyll, carotenoids, ascorbate peroxidase, superoxide dismutase, and guaiacol peroxidase activities, while decreasing malondialdehyde content and petal abscission (Hatami and Ghorbanpour 2013, 2014).

To mitigate light stress, soaking *S. oleracea* seeds in TiO_2 increased the activities of superoxide dismutase, catalase, and peroxidase. It also decreased the accumulation of reactive oxygen free radicals and malondialdehyde levels, while maintaining the stability of the chloroplast membrane structure under light (Hong et al. 2005).

Finally, in *C. morifolium*, spraying Ag enhanced the vitality and succulence of cut flowers and reduced fresh weight loss and stem bacterial count (Kazemipour et al. 2013).

In summary, nanoparticles offer promising potential to mitigate various abiotic stresses in plants, thereby improving their growth, development, and productivity. Their application has been shown to enhance plant resilience against drought, salinity, heavy metals, cold, heat, flooding, UV radiation, dark, and light stresses. As a perspective, nanoparticles can also be explored for mitigating biotic stresses such as

Table 15.4 Effects of various nanoparticles (NP) used in different plant species to study their response under abiotic stresses

Stress type	Plant	NP	Application method	Effect on plant	References
Cold stress	*Glycyrrhiza glabra*	TiO_2	Added to the medium	Increased the phenolics, total protein and osmolytes contents. Decreased the level of oxidative damage by reducing malondialdehyde and hydrogen peroxide contents	Ghabel and Karamian (2020)
	Cicer arietinum	TiO_2	Added in soil and water irrigation	Increased the transcription level of transcript-derived fragments and was crucial for developing cold tolerance by decreasing electrolyte leakage index content	Amini et al. (2017)
			Added in nutrient solution	Improved the redox status of chickpea genotypes and reduced the electrolyte leakage index and malondialdehyde content	Mohammadi et al. (2013)
			Suspension was added to seeds	Increased transcription levels of *CaLRubisco, CaSRubisco*, and *Cachlorophyll* a/b-binding protein genes, enhancing photosynthetic activity. TiO_2-NPs increased the activity of phosphoenolpyruvate carboxylase and decreased the H_2O_2 content	Hasanpour et al. (2015)
	Agropyron elongatum	SiO_2	Added to petri dishes containing seeds	Averted seed dormancy and improved seed germination, dry weight of shoot, root and seedling of tall wheatgrass	Azimi et al. (2014)

(continued)

Table 15.4 (continued)

Stress type	Plant	NP	Application method	Effect on plant	References
Heat stress	*Triticum aestivum*	ZnO and Fe_3O_4	Colloidal Emulsion solution applied to shoot	Increased yield quantity, antioxidant enzymes activities such as glutathione S transferase, superoxide dismutase, peroxidase, catalase and decreased the lipid peroxidation product malondialdehyde	Hassan et al. (2018)
		Ag	Solution added to soil	Improved plant root and shoot lengths, root number, plant fresh and dry weights, leaf area, and leaf number	Iqbal et al. (2019)
	Lycopersicon esculentum	TiO_2	Spraying	Increased photosynthetic rate, conductance to H_2O and transpiration rate of tomato leaves. Also, regulated photosystem II (PS II) energy dissipation and decreased non-regulated PS II energy dissipation, the minimum chlorophyll fluorescence and relative electron transport in leaves	Qi et al. (2013)

Flooding stress	*Glycine max*	Al_2O_3	Added to medium	Affected the mitochondrial proteins by regulating membrane permeability and tricarboxylic acid cycle activity. Increased the length of the root, regulated ascorbate/glutathione pathway and suppressed the proteins related to glycolysis	Mustafa and Komatsu (2016)
		Ag	Used in water irrigation	Improved plant growth, root length/weight and hypocotyl length/weight of soybean. Accumulation of calnexin/calreticulin and glycoproteins was increased. The protein quality control for the mis-folded proteins in the endoplasmic reticulum was up-regulated	Hashimoto et al. (2020)
	Crocus sativus	Ag	Seed soaking	Blocked ethylene signaling, promoted root number and length, and increased leaf dry weight.	Rezvani et al. (2012)
UV radiation stress	*Triticum aestivum*	Si	Foliar application	Improved protection of wheat seedlings via NO-mediated antioxidant defense, consequently counterbalancing toxic reactive oxygen species generation	Tripathi et al. (2017)
	Spinacia oleracea	TiO_2	Foliar application	Increased the superoxide dismutase, catalase, ascorbate peroxidase, and guaiacol peroxidase activities, and elevated evolution oxygen rate. Decreased the accumulation of superoxide radicals, hydrogen peroxide and malonyldialdehyde content	Lei et al. (2008)

(continued)

Table 15.4 (continued)

Stress type	Plant	NP	Application method	Effect on plant	References
Dark stress	*Pelargonium zonale*	Ag	Spray	Increased petal longevity, chlorophyll, carotenoids, ascorbate peroxidase, superoxide dismutase and guaiacol peroxidase activities. Decreased malondialdehyde content and petal abscission.	Hatami and Ghorbanpour (2013, 2014)
Light stress	*Spinacia oleracea*	TiO_2	Seed soaking	Increased the activities of superoxide dismutase, catalase and peroxidase. Decreased the accumulation of reactive oxygen free radicals and the level of malondialdehyde and maintained the stability of the membrane structure of chloroplast under light	Hong et al. (2005)
Post-harvest stress	*Chrysanthemum morifolium*	Ag	Spraying	Enhanced the vitality and succulence of cut flowers and reduced fresh weight loss and stem bacterial count	Kazemipour et al. (2013)

fungal infections. For example, the application of selenium (Se) nanoparticles in soil and as a foliar spray in *T. aestivum* has maintained leaf water status, chlorophyll, and carotenoid contents, thereby enhancing plant growth and increasing biomass under fungal infection stress by *Fusarium* species (*F. culmorum, F. graminearum, F. avenaceum, F. poae, F. sporotrichioides*, and *F. cerealis*) (El-Saadony et al. 2021b). This indicates the potential for nanoparticles to offer comprehensive stress protection, both abiotic and biotic, leading to more resilient and productive crop systems.

4 Nanoparticle Phytotoxicity

Although nanofertilizers offer numerous benefits, they also pose significant risks. Nano-sized particles can be hazardous to health, potentially damaging vital organs and tissues if they enter the human body. They may contribute to environmental issues such as eutrophication and delayed nutrient release, complicating their management. Additionally, the accumulation of metal-based and carbon-based engineered nanomaterials in soil and plants can negatively impact ecosystem dynamics and soil health (Faizan et al. 2024). Concerns about the toxicological and environmental impacts of nanofertilizers are prevalent in the research community, with reports indicating that excessive use can lead to physiological abnormalities, oxidative stress, reduced antioxidant activities, and disrupted cell division and root growth in plants. This highlights the need for further research to fully understand the potential adverse effects of nanofertilizers (Fizzah Zulfiqar et al. 2024).

5 Conclusion

Nanoparticles have demonstrated considerable potential in enhancing plant tolerance to abiotic stresses, which is a critical requirement today due to the increasing global food demand and severe climate-related stresses. NPs enhance water retention, osmotic balance, photosynthesis, and overall plant growth under stress conditions, reducing the impacts of drought, salinity, heavy metals, and other environmental stresses. Despite their promising results, significant concerns persist regarding the phytotoxicity and potential negative impacts of improper nanoparticle use. More studies are crucial to elucidate their mechanisms of action and to develop safer and more effective formulations. Future research should aim to create nanoparticles that are non-toxic, ecologically safe, and self-degradable, facilitating their integration into sustainable agricultural practices without creating any harmful effects on the environment. Moreover, it is essential to investigate the molecular mechanisms behind NP-plant interactions. This will be useful in applying the NPs more effectively by knowing how they act on plant signaling pathways, leading to gene expression for stress tolerance purposes. Research should be extended to long-term field trials in multiple crops and soil types. In the future, it may be necessary to study the implications of NPs treatments on plant growth and nutrition in relation to soil ecology for a better understanding of a sustainable ecosystem. Likewise, before adopting widely NPs for agricultural uses, it should be a priority to have complete toxicological assessments on model organisms, soil microbiota and non-target species, which will help in defining safe usage instructions as well as prevent possible risks. In addition, investigation of the synergistic effect of NPs with other sustainable agricultural technologies like precision farming and organic amendments may improve their effectiveness, avoiding or minimizing the need for synthetic fertilizers. The potential of NPs shows a positive future in the field that can promise to revolutionize agriculture through improving plant abiotic stress tolerance but needs strong efforts to address queries associated with their safety and sustainability. Future research and innovation will be critical in expressing the full capabilities of NPs to shape resilient, sustainable agriculture systems.

References

Abdel Latef AAH, Abu Alhmad MF, Abdelfattah KE (2016) The possible roles of priming with ZnO nanoparticles in mitigation of salinity stress in lupine (*Lupinus termis*) plants. J Plant Growth Regul 36(1):60–70. https://doi.org/10.1007/s00344-016-9618-x

Abdel A, Abdel H, Srivastava AK, El-sadek MSA, Kordrostami M, Tran LP (2018) Titanium dioxide nanoparticles improve growth and enhance tolerance of broad bean plants under saline soil conditions. Land Degrad Develop 1073(September 2017):1065–1073. https://doi.org/10.1002/ldr.2780

Abdoli S, Ghassemi-Golezani K, Alizadeh-Salteh S (2020) Responses of ajowan (*Trachyspermum ammi* L.) to exogenous salicylic acid and iron oxide nanoparticles under salt stress. Environ Sci Pollut Res 27(29):36939–36953. https://doi.org/10.1007/s11356-020-09453-1

Afshari M, Pazoki A, Sadeghipour O (2021) Foliar-applied silicon and its nanoparticles stimulate physio-chemical changes to improve growth, yield and active constituents of coriander (*Coriandrum Sativum* L.) essential oil under different irrigation regimes. SILICON 13(11):4177–4188. https://doi.org/10.1007/s12633-021-01101-8

Aghdam MTB, Mohammadi H, Ghorbanpour M (2016) Effects of nanoparticulate anatase titanium dioxide on physiological and biochemical performance of *Linum usitatissimum* (Linaceae) under well-watered and drought stress conditions. Rev Bras Bot 39(1):139–146. https://doi.org/10.1007/s40415-015-0227-x

Ahmad P, Alyemeni MN, Al-Huqail AA, Alqahtani MA, Wijaya L, Ashraf M, Kaya C, Bajguz A (2020) Zinc oxide nanoparticles application alleviates arsenic (As) toxicity in soybean plants by restricting the uptake of as and modulating key biochemical attributes, antioxidant enzymes, ascorbate-glutathione cycle and glyoxalase system. Plants (Basel) 9(7). https://doi.org/10.3390/plants9070825

Ahmad A, Hashmi SS, Palma JM, Corpas FJ (2022) Influence of metallic, metallic oxide, and organic nanoparticles on plant physiology. Chemosphere 290:133329

Ahmed S, Choudhury A, Roy S, Choi J, Sayyed R, Sa T (2021) Biomolecular painstaking utilization and assimilation of phosphorus under indigent stage in agricultural crops. In: Antioxidants in plant-microbe interaction, pp 565–588. https://doi.org/10.1007/978-981-16-1350-0_26

Akdemir H (2021) Evaluation of transcription factor and aquaporin gene expressions in response to Al_2O_3 and ZnO nanoparticles during barley germination. Plant Physiol Biochem 166:466–476
Akhtar N, Ilyas N, Meraj TA, Pour-Aboughadareh A, Sayyed RZ, Mashwani Z-U-R, Poczai P (2022) Improvement of plant responses by nanobiofertilizer: a step towards sustainable agriculture. Nanomaterials (Basel) 12(6). https://doi.org/10.3390/nano12060965
Aktar SN, Shah MH, Mohanty B, Pramanik K, Hosen MS, Ddoza RMB, Kumawat SM, Sharma NK, Sayyed RZ (2025) Beneficial microbes in plant health, immunity and resistance. In: Mathur P, Roy S (eds) Plant microbiome and biological control, Sustainability in plant and crop protection, vol 20. Springer, Cham, pp 1–17. https://doi.org/10.1007/978-3-031-75845-4_1
Al-Ashkar I, Al-Suhaibani N, Abdella K, Sallam M, Alotaibi M, Seleiman MF (2021) Combining genetic and multidimensional analyses to identify interpretive traits related to water shortage tolerance as an indirect selection tool for detecting genotypes of drought tolerance in wheat breeding. Plants 10(5):1–23. https://doi.org/10.3390/plants10050931
Alharby HF, Metwali EMR, Fuller MP, Aldhebiani AY (2016) The alteration of mRNA expression of SOD and GPX genes, and proteins in tomato (*Lycopersicon esculentum* Mill) under stress of NaCl and/or ZnO nanoparticles. Saudi J Biol Sci 23(6):773–781. https://doi.org/10.1016/j.sjbs.2016.04.012
Ali S, Rizwan M, Hussain A, Zia Ur Rehman M, Ali B, Yousaf B, Wijaya L, Alyemeni MN, Ahmad P (2019a) Silicon nanoparticles enhanced the growth and reduced the cadmium accumulation in grains of wheat (*Triticum aestivum* L.). Plant Physiol Biochem 140:1–8. https://doi.org/10.1016/j.plaphy.2019.04.041
Ali S, Rizwan M, Noureen S, Anwar S, Ali B, Naveed M, Abd Allah EF, Alqarawi AA, Ahmad P (2019b) Combined use of biochar and zinc oxide nanoparticle foliar spray improved the plant growth and decreased the cadmium accumulation in rice (*Oryza sativa* L.) plant. Environ Sci Pollut Res Int 26(11):11288–11299. https://doi.org/10.1007/s11356-019-04554-y
Ali S, Mehmood A, Khan N (2021) Uptake, translocation, and consequences of nanomaterials on plant growth and stress adaptation. J Nanomater 2021(1):6677616. https://doi.org/10.1155/2021/6677616
Al-Khayri JM, Rashmi R, Surya Ulhas R, Sudheer WN, Banadka A, Nagella P, Aldaej MI, Rezk AA-S, Shehata WF, Almaghasla MI (2023) The role of nanoparticles in response of plants to abiotic stress at physiological, biochemical, and molecular levels. Plants 12(2):292. https://doi.org/10.3390/plants12020292
Almutairi Z (2015) Influence of silver nano-particles on the salt resistance of tomato (Solanum lycopersicum L.) during germination. Int J Agric Biol 18. https://doi.org/10.17957/IJAB/15.0114
Alsaeedi AH, El-Ramady H, Alshaal T, El-Garawani M, Elhawat N, Almohsen M (2017) Engineered silica nanoparticles alleviate the detrimental effects of Na+ stress on germination and growth of common bean (Phaseolus vulgaris). Environ Sci Pollut Res 24(27):21917–21928. https://doi.org/10.1007/s11356-017-9847-y
Alsaeedi A, El-Ramady H, Alshaal T, El-Garawani M, Elhawat N, Al-Otaibi A (2018) Exogenous nanosilica improves germination and growth of cucumber by maintaining K+/Na+ ratio under elevated Na+ stress. Plant Physiol Biochem 125(December 2017):164–171. https://doi.org/10.1016/j.plaphy.2018.02.006
Alsaeedi A, El-Ramady H, Alshaal T, El-Garawany M, Elhawat N, Al-Otaibi A (2019) Silica nanoparticles boost growth and productivity of cucumber under water deficit and salinity stresses by balancing nutrients uptake. Plant Physiol Biochem 139(March):1–10. https://doi.org/10.1016/j.plaphy.2019.03.008
Amini S, Maali-Amiri R, Mohammadi R, Kazemi- Shahandashti SS (2017) cDNA-AFLP analysis of transcripts induced in chickpea plants by TiO_2 nanoparticles during cold stress. Plant Physiol Biochem 111:39–49. https://doi.org/10.1016/j.plaphy.2016.11.011
Ashkavand P, Tabari M, Zarafshar M, Tomášková I, Struve D (2015) Effect of SiO_2 nanoparticles on drought resistance in hawthorn seedlings. https://open.icm.edu.pl/handle/123456789/10837

Ashkavand P, Tabari M, Zarafshar M, Tomášková I, Struve D (2016) Effect of SiO_2 nanoparticles on drought resistance in hawthorn seedlings. Forest Research Papers / Leśne Prace Badawcze 76:350–359. https://doi.org/10.1515/frp-2015-0034

Ashkavand P, Zarafshar M, Tabari M, Mirzaie J, Nikpour A, Bordbar S, Struve D, Striker G (2018) Application of SiO_2 nanoparticles AS pretreatment alleviates the impact of drought on the physiological performance of Prunus Mahaleb (Rosaceae). Boletin de La Sociedad Argentina de Botanica 53:207–219. https://doi.org/10.31055/1851.2372.v53.n2.20578

Ashour HA, Mahmoud A (2017) Response of Jatropha integerrima plants irrigated with different levels of saline water to nano silicon and gypsum. J Agric Stud 5(4):136. https://doi.org/10.5296/jas.v5i4.12170

Askary M, Talebi S, Amini F, Bangan A (2017) Effects of iron nanoparticles on Mentha piperita L. under salinity stress. Biologija 63. https://doi.org/10.6001/biologija.v63i1.3476

Avila-Quezada GD, Ingle AP, Golińska P, Rai M (2022) Strategic applications of nano-fertilizers for sustainable agriculture: benefits and bottlenecks. Nanotechnol Rev 11(1):2123–2140. https://doi.org/10.1515/ntrev-2022-0126

Azimi R, Borzelabad MJ, Feizi H, Azimi A (2014) Interaction of SiO_2 nanoparticles with seed prechilling on germination and early seedling growth of tall wheatgrass (Agropyron Elongatum L.). Pol J Chem Technol 16(3):25–29. https://doi.org/10.2478/pjct-2014-0045

Badawy SA, Zayed BA, Bassiouni SMA, Mahdi AHA, Majrashi A, Ali EF, Seleiman MF (2021) Influence of nano silicon and nano selenium on root characters, growth, ion selectivity, yield, and yield components of rice (Oryza sativa L.) under salinity conditions. Plants (Basel, Switzerland) 10(8). https://doi.org/10.3390/plants10081657

Begum N, Qin C, Ahanger MA, Raza S, Khan MI, Ashraf M, Ahmed N, Zhang L (2019) Role of arbuscular mycorrhizal fungi in plant growth regulation: implications in abiotic stress tolerance. Front Plant Sci 10(September):1–15. https://doi.org/10.3389/fpls.2019.01068

Behboudi F, Sarvestani TT, Kassaee ZZ, Sanavi MM, Sorooshzadeh S, Ahmadi AB (2018) Evaluation of chitosan nanoparticles effects on yield and yield components of barley (Hordeum vulgare L.) under late season drought stress. J Water Environ Nanotechnol 3(1):22–39. https://doi.org/10.22090/jwent.2018.01.003

Behboudi F, Tahmasebi-Sarvestani Z, Kassaee MZ, Modarres-Sanavy SAM, Sorooshzadeh A, Mokhtassi-Bidgoli A (2019) Evaluation of chitosan nanoparticles effects with two application methods on wheat under drought stress. J Plant Nutr 42(13):1439–1451. https://doi.org/10.1080/01904167.2019.1617308

Betina C, Agaras, Shahbaz M, Akram A, Mehak A, Haq E, Fatima N, Wareen G, Fitriatin B, Sayyed R, Ilyas N, Sabullah M (2023) Evaluation of selenium nanoparticles in inducing disease resistance against spot blotch disease and promoting growth in wheat under biotic stress. Plants 12. https://doi.org/10.3390/plants12040761

Bhushan I, Singh VK, Tripathi DK (eds) (2020) Nanomaterials and environmental biotechnology. Springer, Cham, p 342. https://doi.org/10.1007/978-3-030-34544-0

Bouhraoua S, Ferioun M, Nassira S, Boussakouran A, Akhazzane M, Belahcen D et al (2023) Biomass partitioning and physiological responses of four Moroccan barley varieties subjected to salt stress in a hydroponic system. J Plant Biotechnol 50(1):115–126

Bouhraoua S, Ferioun M, Boussakouran A, Belahcen D, Srhiouar N, Hammani K, Louahlia S (2024) Physio-biochemical responses and yield performance of North African barley genotypes submitted to moderate and severe salinity. Cereal Res Commun. https://doi.org/10.1007/s42976-024-00549-9

Bright JP, Maheshwari HS, Thangappan S, Perveeen K, Bukhari NA, Mitra D, Sayyed RZ, Mantinu A (2025) Biofilmed-PGPR: a next-generation bioinoculant for plant growth promotion in rice (*Oryza sativa* L.) under changing climate. Rice Sci 2025(32):10.1016/j.rsci.2024.08.008

Chaffai R, Ganesan M, Cherif A (2024) Nanoparticles for sustainable agriculture and abiotic stress resistance. In: Plant adaptation to abiotic stress: from signaling pathways and microbiomes to molecular mechanisms. Springer Nature Singapore, Singapore, pp 195–208. https://doi.org/10.1007/978-981-97-0672-3_9

Chaithra M, Nishmitha K, Patel PS, Mitra D, Gangaraj R, Sayyed RZ (2025) Fungal bio-stimulants: cutting-edge bioinoculants for sustainable agriculture. In: Mathur P, Roy S (eds) Plant microbiome and biological control, Sustainability in plant and crop protection, vol 20. Springer, Cham, pp 289–307. https://doi.org/10.1007/978-3-031-75845-4_13

Channab B-E, Idrissi AE, Ammar A, Dardari O, Marrane SE, el Gharrak A, Akil A, Essemlali Y, Zahouily M (2024) Recent advances in nano-fertilizers: synthesis, crop yield impact, and economic analysis. Nanoscale 16(9):4484–4513. https://doi.org/10.1039/D3NR05012B

Chen L, Zhu L, Liu X, Chen L, Zhou H, Ma H, Sun G, Nyande A, Li Z, Wu H (2024) Recent omics progress in nanobiotechnology for plant abiotic stress tolerance improvement. Crop J. https://doi.org/10.1016/j.cj.2024.05.017

Dimkpa CO, Singh U, Adisa IO, Bindraban PS, Elmer WH, Gardea-Torresdey JL, White JC (2018) Effects of manganese nanoparticle exposure on nutrient acquisition in wheat (Triticum aestivum L.). Agronomy 8(9):158

Dimkpa CO, Singh U, Bindraban PS, Elmer WH, Gardea-Torresdey JL, White JC (2019) Zinc oxide nanoparticles alleviate drought-induced alterations in sorghum performance, nutrient acquisition, and grain fortification. Sci Total Environ 688:926–934. https://doi.org/10.1016/j.scitotenv.2019.06.392

Du W, Tan W, Peralta-Videa JR, Gardea-Torresdey JL, Ji R, Yin Y, Guo H (2017) Interaction of metal oxide nanoparticles with higher terrestrial plants: physiological and biochemical aspects. Plant Physiol Biochem 110:210–225

Dubchak S, Ogar A, Mietelski J, Turnau K (2010) Influence of silver and titanium nanoparticles on arbuscular mycorrhiza colonization and accumulation of radiocaesium in Helianthus annuus. Span J Agric Res. https://doi.org/10.5424/sjar/201008S1-1228

Duc NH, Vo AT, Haddidi I, Daood H, Posta K (2021) Arbuscular mycorrhizal fungi improve tolerance of the medicinal plant Eclipta prostrata (L.) and induce major changes in polyphenol profiles under salt stresses. Front Plant Sci 11(January):1–18. https://doi.org/10.3389/fpls.2020.612299

Elouattassi Y, Ferioun M, El Ghachtouli N, Derraz K, Rachidi F (2023) Agroecological concepts and alternatives to the problems of contemporary agriculture: monoculture and chemical fertilization in the context of climate change. J Agric Environ Int Dev 117(2):41–98

Elouattassi Y, Ferioun M, El Ghachtouli N, Derraz K, Rachidi F (2024) Enhancing onion growth and yield through agroecological practices: organic fertilization and intercropping. Ecol Front 44(3):547–557

El-Saadony MT, Desoky ESM, Saad AM, Eid RSM, Selem E, Elrys AS (2021a) Biological silicon nanoparticles improve Phaseolus vulgaris L. yield and minimize its contaminant contents on a heavy metals-contaminated saline soil. J Environ Sci (China) 106:1–14. https://doi.org/10.1016/j.jes.2021.01.012

El-Saadony MT, Saad AM, Najjar AA, Alzahrani SO, Alkhatib FM, Shafi ME, Selem E, Desoky ESM, Fouda SEE, El-Tahan AM, Hassan MAA (2021b) The use of biological selenium nanoparticles to suppress Triticum aestivum L. crown and root rot diseases induced by Fusarium species and improve yield under drought and heat stress. Saudi J Biol Sci 28(8):4461–4471. https://doi.org/10.1016/j.sjbs.2021.04.043

Elsheery NI, Helaly MN, El-Hoseiny HM, Alam-Eldein SM (2020) Zinc oxide and silicone nanoparticles to improve the resistance mechanism and annual productivity of salt-stressed mango trees. Agronomy 10(4). https://doi.org/10.3390/agronomy10040558

Faizan M, Bhat JA, Chen C, Alyemeni MN, Wijaya L, Ahmad P, Yu F (2021) Zinc oxide nanoparticles (ZnO-NPs) induce salt tolerance by improving the antioxidant system and photosynthetic machinery in tomato. Plant Physiol Biochem 161(February):122–130. https://doi.org/10.1016/j.plaphy.2021.02.002

Faizan M, Karabulut F, Khan I, Akhtar MS, Alam P (2024) Emergence of nanotechnology in efficient fertilizer management in soil. S Afr J Bot 164:242–249. https://doi.org/10.1016/j.sajb.2023.12.004

Faraji J, Sepehri A (2020) Exogenous nitric oxide improves the protective effects of TiO_2 nanoparticles on growth, antioxidant system, and photosynthetic performance of wheat seedlings under drought stress. J Soil Sci Plant Nutr 20(2):703–714. https://doi.org/10.1007/s42729-019-00158-0

Farhangi-Abriz S, Torabian S (2018) Nano-silicon alters antioxidant activities of soybean seedlings under salt toxicity. Protoplasma 255(3):953–962. https://doi.org/10.1007/s00709-017-1202-0

Faridvand S, Amirnia R, Tajbakhsh M, El Enshasy HA, Sayyed RZ (2021) The effect of foliar application of magnetic water and nano-fertilizers on phytochemical and yield characteristics of fennel. Horticulturae 7(11). https://doi.org/10.3390/horticulturae7110475

Fathi A, Zahedi M, Torabian S (2017) Effect of interaction between salinity and nanoparticles (Fe_2O_3 and ZnO) on physiological parameters of Zea mays L. J Plant Nutr 40(19):2745–2755. https://doi.org/10.1080/01904167.2017.1381731

Fazeli-Nasab B, Sirousmehr A-R, Azad H (2018) Effect of titanium dioxide nanoparticles on essential oil quantity and quality in *Thymus vulgaris* under water deficit. J Med Plants By-Prod 2:125–133

Fazeli-Nasab B, Sayyed R, Piri R, Rahmani A (2021) Biopriming and nanopriming: green revolution wings to increase plant yield, growth, and development under stress condition and forward dimensions. In: Antioxidants in plant-microbe interaction, pp 623–655. https://doi.org/10.1007/978-981-16-1350-0_29

Ferioun M, Bouhraoua S, Srhiouar N, Boussakouran A, Belahcen D, El Ghachtouli N et al (2024a) Advanced multivariate approaches for selecting Moroccan drought-tolerant barley (*Hordeum vulgare* L.) cultivars. Ecol Front 44(4):820–828

Ferioun M, Bouhraoua S, Belahcen D, Zouitane I, Srhiouar N, Louahlia S, El Ghachtouli N (2024b) PGPR consortia enhance growth and yield in barley cultivars subjected to severe drought stress and subsequent recovery. Rhizosphere 31:100926

Ferioun M, Zouitane I, Bouhraoua S, Elouattassi Y, Belahcen D, Errabbani A, Louahlia S, Sayyed R, El-Ghachtouli N (2025) Applying microbial biostimulants and drought-tolerant genotypes to enhance barley growth and yield under drought stress. Front Plant Sci 15:1494987. https://doi.org/10.3389/fpls.2024.1494987

Fincheira P, Tortella G, Duran N, Seabra AB, Rubilar O (2020) Current applications of nanotechnology to develop plant growth inducer agents as an innovation strategy. Crit Rev Biotechnol 40(1):15–30. https://doi.org/10.1080/07388551.2019.1681931

Fizzah Zulfiqar H, Afroze B, Shakoor S, Bhutta MS, Ahmed M, Hassan S, Batool F, Rashid B, Zulfiqar HF, Afroze B, Shakoor S, Bhutta MS, Ahmed M, Hassan S, Batool F, Rashid B (2024) Nanoparticles in agriculture: enhancing crop resilience and productivity against abiotic stresses. In: Abiotic stress in crop plants—ecophysiological responses and molecular approaches. IntechOpen. https://doi.org/10.5772/intechopen.114843

Gao X, Zou C, Wang L, Zhang F (2006) Silicon decreases transpiration rate and conductance from stomata of maize plants. J Plant Nutr 29(9):1637–1647. https://doi.org/10.1080/01904160600851494

Ghabel V, Karamian R (2020) Effects of TiO_2 nanoparticles and spermine on antioxidant responses of Glycyrrhiza glabra L. to cold stress. Acta Bot Croat 79. https://doi.org/10.37427/botcro-2020-025

Gunes A, Pilbeam DJ, Inal A, Bagci EG, Coban S (2007) Influence of silicon on antioxidant mechanisms and lipid peroxidation in chickpea (Cicer arietinum L.) cultivars under drought stress. J Plant Interact 2(2):105–113. https://doi.org/10.1080/17429140701529399

Gürkan M (2018) Effects of three different nanoparticles on bioaccumulation, oxidative stress, osmoregulatory, and immune responses of Carcinus aestuarii. Toxicol Environ Chem 100(8–10):693–716. https://doi.org/10.1080/02772248.2019.1579818

Hadri SH, Afzaal A, Saeed L, Arshad A, Nazeer S, Akram M (2024) Recent advances in the development of nanoparticle based fertilizers for different kinds of crops: a review. Biocatal Agric Biotechnol 58:103194. https://doi.org/10.1016/j.bcab.2024.103194

Hafeez A, Razzaq A, Mahmood T, Jhanzab HM (2015) Potential of copper nanoparticles to increase growth and yield of wheat. J Nanosci Adv Technol 1(1):6–11
Haghighi M, Pessarakli M (2013) Influence of silicon and nano-silicon on salinity tolerance of cherry tomatoes (Solanum lycopersicum L.) at early growth stage. Sci Hortic 161:111–117. https://doi.org/10.1016/j.scienta.2013.06.034
Haghighi M, Afifipour Z, Mozafarian M (2012) The effect of N-Si on tomato seed germination under salinity levels. J Biol Environ Sci 6(16) 87-90
Harrison MT, Tardieu F, Dong Z, Messina CD, Hammer GL (2014) Characterizing drought stress and trait influence on maize yield under current and future conditions. Glob Chang Biol 20(3):867–878. https://doi.org/10.1111/gcb.12381
Hasanpour H, Maali-Amir R, Zeinali H (2015) Effect of TiO_2 nanoparticles on metabolic limitations to photosynthesis under cold in chickpea. Russ J Plant Physiol 62(6):779–787. https://doi.org/10.1134/S1021443715060096
Hashimoto T, Mustafa G, Nishiuchi T, Komatsu S (2020) Comparative analysis of the effect of inorganic and organic chemicals with silver nanoparticles on soybean under flooding stress. Int J Mol Sci 21(4). https://doi.org/10.3390/ijms21041300
Hassan NS, Salah El Din TA, Hendawey MH, Borai IH, Mahdi AA (2018) Magnetite and zinc oxide nanoparticles alleviated heat stress in wheat plants. Curr Nanomater 3(1):32–43. https://doi.org/10.2174/2405461503666180619160923
Hatami M, Ghorbanpour M (2013) Effect of nanosilver on physiological performance of pelargonium plants exposed to dark storage. J Hortic Res 21(1):15–20. https://doi.org/10.2478/johr-2013-0003
Hatami M, Ghorbanpour M (2014) Defense enzyme activities and biochemical variations of Pelargonium zonale in response to nanosilver application and dark storage. Turk J Biol 38(1):130–139. https://doi.org/10.3906/biy-1304-64
Hojjat SS, Kamyab M (2017) The effect of silver nanoparticle on Fenugreek seed germination under salinity levels. Russ Agric Sci 43(1):61–65. https://doi.org/10.3103/s1068367417010189
Hong F, Yang F, Liu C, Gao Q, Wan Z, Gu F, Wu C, Ma Z, Zhou J, Yang P (2005) Influences of nano-TiO_2 on the chloroplast aging of spinach under light. Biol Trace Elem Res 104(3):249–260. https://doi.org/10.1385/BTER:104:3:249
Hossain Z, Mustafa G, Komatsu S (2015) Plant responses to nanoparticle stress. Int J Mol Sci 16(11):26644–26653. https://doi.org/10.3390/ijms161125980
Hosseinpour A, Haliloglu K, Cinisli KT, Ozkan G, Ozturk HI, Pour-Aboughadareh A, Poczai P (2020) Application of zinc oxide nanoparticles and plant growth promoting bacteria reduces genetic impairment under salt stress in tomato (Solanum lycopersicum l. 'Linda'). Agriculture (Switzerland) 10(11):1–16. https://doi.org/10.3390/agriculture10110521
Hussain A, Rizwan M, Ali Q, Ali S (2019) Seed priming with silicon nanoparticles improved the biomass and yield while reduced the oxidative stress and cadmium concentration in wheat grains. Environ Sci Pollut Res Int 26(8):7579–7588. https://doi.org/10.1007/s11356-019-04210-5
Ingle A (2021) Nanotechnology in plant growth promotion and protection. https://doi.org/10.1002/9781119745884
Iqbal M, Raja NI, Mashwani ZUR, Hussain M, Ejaz M, Yasmeen F (2019) Effect of silver nanoparticles on growth of wheat under heat stress. Iran J Sci Technol Trans A Sci 43(2):387–395. https://doi.org/10.1007/s40995-017-0417-4
Iqbal MS, Singh AK, Singh SP, Ansari MI (2020) Nanoparticles and plant interaction with respect to stress response. In: Bhushan I, Singh VK, Tripathi DK (eds) Nanomaterials and environmental biotechnology. Springer International Publishing, pp 1–15. https://doi.org/10.1007/978-3-030-34544-0_1
Ismail G, Abou-Zeid H (2018) The role of priming with biosynthesized silver nanoparticles in the response of Triticum aestivum L. to salt stress. Egypt J Bot. https://doi.org/10.21608/ejbo.2017.1873.1128

Jaberzadeh A, Moaveni P, Moghadam T, Zahedi H (2013) Influence of bulk and nanoparticles titanium foliar application on some agronomic traits, seed gluten and starch contents of wheat subjected to water deficit stress. Not Bot Horti Agrobo Cluj-Napoca 41:201–207. https://doi.org/10.15835/nbha4119093

Jafir M, Khan A, Ahmad A, Hussain K, ur Rehman MZ, Nazeer Ahmad SJ, Irfan M, Sabir MA, Khan TH, Zulfiqar U (2024) Zinc nanoparticles for enhancing plant tolerance to abiotic stress: a bibliometric analysis and review. J Soil Sci Plant Nutr 24(2):1704–1719. https://doi.org/10.1007/s42729-024-01733-w

Jaskulak M, Rorat A, Grobelak A, Chaabene Z, Kacprzak M, Vandenbulcke F (2019) Bioaccumulation, antioxidative response, and metallothionein expression in *Lupinus luteus* L. exposed to heavy metals and silver nanoparticles. Environ Sci Pollut Res Int 26(16):16040–16052. https://doi.org/10.1007/s11356-019-04972-y

Javanmard A, Ashrafi M, Morshedloo MR, Machiani MA, Rasouli F, Maggi F (2022) Optimizing phytochemical and physiological characteristics of Balangu (*Lallemantia iberica*) by foliar application of chitosan nanoparticles and myco-root inoculation under water supply restrictions. Horticulturae 8(8). https://doi.org/10.3390/horticulturae8080695

Jithendar B, Kumar R, Rana N (2024) Revolutionizing crop nutrition: exploring nano fertilizers in agriculture. Int J Plant Soil Sci 36(6):327–339. https://doi.org/10.9734/ijpss/2024/v36i64635

Kalteh M, Zarrin TA, Shahram A, Maryam MA, Alireza FN (2014) Effect of silica nanoparticles on basil (Ocimum basilicum) under salinity stress. J Chem Health Risks 4:49–55

Kataria S, Jain M, Rastogi A, Živčák M, Brestic M, Liu S, Tripathi DK (2019) Role of nanoparticles on photosynthesis: avenues and applications. In: Nanomaterials in plants, algae and microorganisms. Elsevier, pp 103–127. https://www.sciencedirect.com/science/article/pii/B9780128114889000068

Katiyar P, Yadu B, Korram J, Satnami ML, Kumar M, Keshavkant S (2020) Titanium nanoparticles attenuates arsenic toxicity by up-regulating expressions of defensive genes in Vigna radiata L. J Environ Sci (China) 92(February):18–27. https://doi.org/10.1016/j.jes.2020.02.013

Kaya H, Aydın F, Gürkan M, Yılmaz S, Ates M, Demir V, Arslan Z (2016) A comparative toxicity study between small and large size zinc oxide nanoparticles in tilapia (Oreochromis niloticus): organ pathologies, osmoregulatory responses and immunological parameters. Chemosphere 144:571–582

Kazemipour S, Hashemabadi D, Kaviani B (2013) Effect of silver nanoparticles on the vase life and quality of cut chrysanthemum (Chrysanthemum morifolium L.) flower. Pelagia Research Library European. J Exp Biol 3(3):298–302

Khalid MF, Iqbal Khan R, Jawaid MZ, Shafqat W, Hussain S, Ahmed T, Rizwan M, Ercisli S, Pop OL, Alina Marc R (2022) Nanoparticles: the plant saviour under abiotic stresses. Nano 12(21):3915. https://doi.org/10.3390/nano12213915

Khan Z, Upadhyaya H (2019) Impact of nanoparticles on abiotic stress responses in plants. In: Nanomaterials in plants, algae and microorganisms. Elsevier, pp 305–322. https://doi.org/10.1016/B978-0-12-811488-9.00015-9

Khan ZS, Rizwan M, Hafeez M, Ali S, Javed MR, Adrees M (2019) The accumulation of cadmium in wheat (Triticum aestivum) as influenced by zinc oxide nanoparticles and soil moisture conditions. Environ Sci Pollut Res Int 26(19):19859–19870. https://doi.org/10.1007/s11356-019-05333-5

Kiapour H, Moaveni P, Habibi D, Sani B (2015) Evaluation of the application of gibbrellic acid and titanium dioxide nanoparticles under drought stress on some traits of basil (Ocimum basilicum L.). Int J Agron Agric Res 6:138–150

Kim JH, Oh Y, Yoon H, Hwang I, Chang YS (2015) Iron nanoparticle-induced activation of plasma membrane H+-ATPase promotes stomatal opening in Arabidopsis thaliana. Environmental science & technology, 49(2):1113–1119

Kim DH, Gopal J, Sivanesan I (2017) Nanomaterials in plant tissue culture: the disclosed and undisclosed. RSC Adv 7(58):36492–36505. https://doi.org/10.1039/C7RA07025J

Konate A, He X, Zhang Z, Ma Y, Zhang P, Alugongo GM, Rui Y (2017) Magnetic (Fe_3O_4) nanoparticles reduce heavy metals uptake and mitigate their toxicity in wheat seedling. Sustainability (Switzerland) 9(5):1–16. https://doi.org/10.3390/su9050790

Kumar N, Samota SR, Venkatesh K, Tripathi SC (2023) Global trends in use of nano-fertilizers for crop production: advantages and constraints – a review. Soil Tillage Res 228:105645. https://doi.org/10.1016/j.still.2023.105645

Lei Z, Mingyu S, Xiao W, Chao L, Chunxiang Q, Liang C, Hao H, Xiaoqing L, Fashui H (2008) Antioxidant stress is promoted by nano-anatase in spinach chloroplasts under UV-B radiation. Biol Trace Elem Res 121(1):69–79. https://doi.org/10.1007/s12011-007-8028-0

Li Y, Li Y, Zhu N, Liang X, Bai X, Zheng L, Zhao J, Li YF, Zhang Z, Gao Y (2020) Silica nanoparticles alleviate mercury toxicity: via immobilization and inactivation of Hg(ii) in soybean (Glycine max). Environ Sci Nano 7(6):1807–1817. https://doi.org/10.1039/d0en00091d

Liang Y, Liu H, Fu Y, Li P, Li S, Gao Y (2023) Regulatory effects of silicon nanoparticles on the growth and photosynthesis of cotton seedlings under salt and low-temperature dual stress. BMC Plant Biol 23(1):504. https://doi.org/10.1186/s12870-023-04509-z

Liu W, Worms I, Slaveykova VI (2020) Interaction of silver nanoparticles with antioxidant enzymes. Environ Sci Nano 7(5):1507–1517

Lu L, Huang M, Huang Y, Corvini PFX, Ji R, Zhao L (2020) Mn_3O_4 nanozymes boost endogenous antioxidant metabolites in cucumber (: Cucumis sativus) plant and enhance resistance to salinity stress. Environ Sci Nano 7(6):1692–1703. https://doi.org/10.1039/d0en00214c

Mahmoud LM, Dutt M, Shalan AM, El-Kady ME, El-Boray MS, Shabana YM, Grosser JW (2020) Silicon nanoparticles mitigate oxidative stress of in vitro-derived banana (Musa acuminata 'Grand Nain') under simulated water deficit or salinity stress. S Afr J Bot 132:155–163. https://doi.org/10.1016/j.sajb.2020.04.027

Manaf A, Wang X, Tariq F, Jhanzab HM, Bibi Y, Sher A, Razzaq A, Fiaz S, Tanveer SK, Qayyum A (2021) Antioxidant enzyme activities correlated with growth parameters of wheat sprayed with silver and gold nanoparticle suspensions. Agronomy 11(8):1494

Martínez-Fernández D, Barroso D, Komárek M (2016) Root water transport of Helianthus annuus L. under iron oxide nanoparticle exposure. Environ Sci Pollut Res 23(2):1732–1741. https://doi.org/10.1007/s11356-015-5423-5

Masouleh FF, Amiri BM, Mirvaghefi A, Ghafoori H, Madsen SS (2017) Silver nanoparticles cause osmoregulatory impairment and oxidative stress in Caspian kutum (*Rutilus kutum*, Kamensky 1901). Environ Monit Assess 189(9):448. https://doi.org/10.1007/s10661-017-6156-3

Maswada HF, Djanaguiraman M, Prasad PVV (2018) Seed treatment with nano-iron (III) oxide enhances germination, seeding growth and salinity tolerance of sorghum. J Agron Crop Sci 204(6):577–587. https://doi.org/10.1111/jac.12280

Metryka O, Wasilkowski D, Mrozik A (2022) Evaluation of the effects of Ag, Cu, ZnO and TiO_2 nanoparticles on the expression level of oxidative stress-related genes and the activity of antioxidant enzymes in Escherichia coli, Bacillus cereus and Staphylococcus epidermidis. Int J Mol Sci 23(9):4966

Mitra D, Adhikari P, Djebaili R, Thathola P, Joshi K, Pellegrini M, Adeyemi NO, Khoshru B, Kaur K, Priyadarshini A, Senapati A, Del Gallo M, Das Mohapatra PK, Nayak AK, Shanmugam V, Panneerselvam P (2023) Biosynthesis and characterization of nanoparticles, its advantages, various aspects and risk assessment to maintain the sustainable agriculture: emerging technology in modern era science. Plant Physiol Biochem 196:103–120. https://doi.org/10.1016/j.plaphy.2023.01.017

Mohamed AKSH, Qayyum MF, Abdel-Hadi AM, Rehman RA, Ali S, Rizwan M (2017) Interactive effect of salinity and silver nanoparticles on photosynthetic and biochemical parameters of wheat. Arch Agron Soil Sci 63(12):1736–1747. https://doi.org/10.1080/03650340.2017.1300256

Mohammadi R, Maali Amiri R, Abbasi A (2013) Effect of TiO_2 nanoparticles on chickpea response to cold stress. Biol Trace Elem Res 152. https://doi.org/10.1007/s12011-013-9631-x

Mohammadi H, Esmailpour M, Gheranpaye A (2016) Effects of TiO_2 nanoparticles and water-deficit stress on morpho-physiological characteristics of dragonhead (Dracocephalum moldavica L.) plants. Acta Agric Slov 107(2):385–396. https://doi.org/10.14720/aas.2016.107.2.11

Moradbeygi H, Jamei R, Heidari R, Darvishzadeh R (2020a) Fe_2O_3 nanoparticles induced biochemical responses and expression of genes involved in rosmarinic acid biosynthesis pathway in Moldavian balm under salinity stress. Physiol Plant 169(4):555–570. https://doi.org/10.1111/ppl.13077

Moradbeygi H, Jamei R, Heidari R, Darvishzadeh R (2020b) Investigating the enzymatic and non-enzymatic antioxidant defense by applying iron oxide nanoparticles in Dracocephalum moldavica L. plant under salinity stress. Sci Hortic 272(June):109537. https://doi.org/10.1016/j.scienta.2020.109537

Mozafari A a, Ghadakchiasl A, Ghaderi N (2018) Grape response to salinity stress and role of iron nanoparticle and potassium silicate to mitigate salt induced damage under in vitro conditions. Physiol Mol Biol Plants 24(1):25–35. https://doi.org/10.1007/s12298-017-0488-x

Mustafa G, Komatsu S (2016) Insights into the response of soybean mitochondrial proteins to various sizes of aluminum oxide nanoparticles under flooding stress. J Proteome Res 15(12). https://doi.org/10.1021/acs.jproteome.6b00572

Nguyen D, Huong N, Le N, Nguyen K, Nguyen H, Le H, Nguyen A, Dinh N, Hoang S, Ha C (2020) Copper nanoparticle application enhances plant growth and grain yield in maize under drought stress conditions. J Plant Growth Regul 41:364–375. https://doi.org/10.1007/s00344-021-10301-w

Noohpisheh Z, Amiri H, Mohammadi A, Farhadi S (2021) Effect of the foliar application of zinc oxide nanoparticles on some biochemical and physiological parameters of Trigonella foenum-graecum under salinity stress. Plant Biosyst 155(2):267–280. https://doi.org/10.1080/11263504.2020.1739160

Ostadi A, Javanmard A, Machiani MA, Kakaei K (2023) Optimizing antioxidant activity and phytochemical properties of peppermint (*Mentha piperita* L.) by integrative application of biofertilizer and stress-modulating nanoparticles under drought stress conditions. Plants 12(1):151. https://doi.org/10.3390/plants12010151

Poddar K, Sarkar D, Sarkar A (2020) Nanoparticles on photosynthesis of plants: effects and role. In: Patra JK, Fraceto LF, Das G, Campos EVR (eds) Green nanoparticles. Springer International Publishing, pp 273–287. https://doi.org/10.1007/978-3-030-39246-8_13

Qados A (2015) Mechanism of nanosilicon-mediated alleviation of salinity stress in faba bean (Vicia faba L.) plants. Am J Exp Agric 7:78–95. https://doi.org/10.9734/AJEA/2015/15110

Qados A, Moftah A (2015) Influence of silicon and nano-silicon on germination, growth and yield of faba bean (Vicia faba L.) under salt stress conditions. Am J Exp Agric 5(6):509–524. https://doi.org/10.9734/ajea/2015/14109

Qi M, Liu Y, Li T (2013) Nano-TiO_2 improve the photosynthesis of tomato leaves under mild heat stress. Biol Trace Elem Res 156(1–3):323–328. https://doi.org/10.1007/s12011-013-9833-2

Rajput VD, Minkina T, Kumari A, Harish, Singh VK, Verma KK, Mandzhieva S, Sushkova S, Srivastava S, Keswani C (2021) Coping with the challenges of abiotic stress in plants: new dimensions in the field application of nanoparticles. Plants 10(6):1221. https://doi.org/10.3390/plants10061221

Rasouli H, Popović-Djordjević J, Sayyed R, Zarayneh S, Jafari M, Fazeli-Nasab B (2020) Nanoparticles: a new threat to crop plants and soil rhizobia? In: Sustainable agriculture reviews 41. https://doi.org/10.1007/978-3-030-33996-8_11

Rehman A, Khan S, Sun F, Peng Z, Feng K, Wang N, Jia Y, Pan Z, He S, Wang L, Qayyum A, Du X, Li H (2024) Exploring the nano-wonders: unveiling the role of nanoparticles in enhancing salinity and drought tolerance in plants. Front Plant Sci 14:1324176. https://doi.org/10.3389/fpls.2023.1324176

Rezvani N, Sorooshzadeh A, Farhadi N (2012) Effect of nano-silver on growth of saffron in flooding stress. World Acad Sci Eng Technol 6:519–524

Rizwan M, Ali S, Ali B, Adrees M, Arshad M, Hussain A, Zia Ur Rehman M, Waris AA (2019) Zinc and iron oxide nanoparticles improved the plant growth and reduced the oxidative stress and cadmium concentration in wheat. Chemosphere 214:269–277. https://doi.org/10.1016/j.chemosphere.2018.09.120

Rossi L, Zhang W, Lombardini L, Ma X (2016) The impact of cerium oxide nanoparticles on the salt stress responses of Brassica napus L. Environ Pollut 219:28–36. https://doi.org/10.1016/j.envpol.2016.09.060

Rossi L, Zhang W, Ma X (2017) Cerium oxide nanoparticles alter the salt stress tolerance of Brassica napus L. by modifying the formation of root apoplastic barriers. Environ Pollut 229:132–138. https://doi.org/10.1016/j.envpol.2017.05.083

Rostaei M, Fallah S, Lorigooini Z, Abbasi Surki A (2018) The effect of organic manure and chemical fertilizer on essential oil, chemical compositions and antioxidant activity of dill (Anethum graveolens) in sole and intercropped with soybean (Glycine max). J Clean Prod 199:18–26. https://doi.org/10.1016/j.jclepro.2018.07.141

Roy R, Núñez-Delgado A, Sultana S, Wang J, Munir A, Battaglia ML, Sarker T, Seleiman MF, Barmon M, Zhang R (2021) Additions of optimum water, spent mushroom compost and wood biochar to improve the growth performance of Althaea rosea in drought-prone coal-mined spoils. J Environ Manag 295(May). https://doi.org/10.1016/j.jenvman.2021.113076

Sabaghnia N, Janmohammadi M (2014) Effect of nano-silicon particles application on salinity tolerance in early growth of some lentil genotypes. Ann UMCS Biol 69(2):39–55. https://doi.org/10.1515/umcsbio-2015-0004

Saeid Hojjat S (2019) Effect of interaction between Ag nanoparticles and salinity on germination stages of lathyrus sativus l. Open Access J Environ Soil Sci 2(2):193–198. https://doi.org/10.32474/oajess.2019.02.000132

Sami F, Siddiqui H, Hayat S (2020) Impact of silver nanoparticles on plant physiology: a critical review. In: Hayat S, Pichtel J, Faizan M, Fariduddin Q (eds) Sustainable agriculture reviews 41, vol 41. Springer International Publishing, pp 111–127. https://doi.org/10.1007/978-3-030-33996-8_6

Sattari NS, Jamei R, Eslam BP, Lisar SYS (2020) Titanium dioxide nanoparticles increase resistance of Lallemantia iberica to drought stress due to increased accumulation of protective antioxidants. Iran J Plant Physiol 10(4):3343–3354. https://doi.org/10.22034/ijpp.2020.1901976.1233

Sedghi M, Hadi M, Toluie S (2013) Effect of nano zinc oxide on the germination of soybean seeds under drought stress. Ann West Univ Timişoara Ser Biol 16:73–78

Semida WM, Abdelkhalik A, Mohamed GF, Abd El-Mageed TA, Abd El-Mageed SA, Rady MM, Ali EF (2021) Foliar application of zinc oxide nanoparticles promotes drought stress tolerance in eggplant (*Solanum melongena* L.). Plants 10(2):1–18. https://doi.org/10.3390/plants10020421

Shallan M, Hassan H, Namich A, Ibrahim A (2016) Biochemical and physiological effects of TiO_2 and SiO_2 nanoparticles on cotton plant under drought stress. Res J Pharm, Biol Chem Sci 7:1540–1551

Siddiqui MH, Al-Whaibi MH, Faisal M, Al Sahli AA (2014) Nano-silicon dioxide mitigates the adverse effects of salt stress on Cucurbita pepo L. Environ Toxicol Chem 33(11):2429–2437. https://doi.org/10.1002/etc.2697

Siddiqui WA, Ahad A, Ahsan H (2015a) The mystery of BCL2 family: BCL-2 proteins and apoptosis: an update. Arch Toxicol 89(3):289–317. https://doi.org/10.1007/s00204-014-1448-7

Siddiqui MH, Al-Whaibi MH, Firoz M, Al-Khaishany MY (2015b) Role of nanoparticles in plants. In: Siddiqui M, Al-Whaibi M, Mohammad F (eds) Nanotechnology and plant sciences. Springer International Publishing, pp 19–35. https://doi.org/10.1007/978-3-319-14502-0_2

Silva S, Dias MC, Silva AMS (2022) Titanium and zinc based nanomaterials in agriculture: a promising approach to Deal with (a)biotic stresses? Toxics 10(4):1–30. https://doi.org/10.3390/toxics10040172

Singh A, Tiwari S, Pandey J, Lata C, Singh IK (2021) Role of nanoparticles in crop improvement and abiotic stress management. J Biotechnol 337(June):57–70. https://doi.org/10.1016/j.jbiotec.2021.06.022

Sun L, Song F, Guo J, Zhu X, Liu S, Liu F, Li X (2020) Nano-ZnO-induced drought tolerance is associated with melatonin synthesis and metabolism in maize. Int J Mol Sci 21(3):1–18. https://doi.org/10.3390/ijms21030782

Tantawy A, Salama Y, El-Nemr MA, Abdel-Mawgoud AMR (2015) Nano silicon application improves salinity tolerance of sweet pepper plants. Int J ChemTech Res. 8:11–17

Thakur P, Thakur S, Kumari P, Shandilya M, Sharma S, Poczai P, Alarfaj A, Sayyed R (2022) Nano-insecticide: synthesis, characterization, and evaluation of insecticidal activity of ZnO NPs against Spodoptera litura and Macrosiphum euphorbiae. Appl Nanosci 12. https://doi.org/10.1007/s13204-022-02530-6

Tighe-Neira R, Carmora E, Recio G, Nunes-Nesi A, Reyes-Diaz M, Alberdi M, Rengel Z, Inostroza-Blancheteau C (2018) Metallic nanoparticles influence the structure and function of the photosynthetic apparatus in plants. Plant Physiol Biochem 130:408–417

Torabian S, Zahedi M, Khoshgoftar AH (2015) Effects of foliar spray of two kinds of zinc oxide on the growth and ion concentration of sunflower cultivars under salt stress. J Plant Nutr 39(2):172–180. https://doi.org/10.1080/01904167.2015.1009107

Torabian S, Zahedi M, Khoshgoftarmanesh A (2016) Effect of foliar spray of zinc oxide on some antioxidant enzymes activity of sunflower under salt stress. J Agric Sci Technol 18(4)

Torres R, Diz VE, Lagorio MG (2018) Effects of gold nanoparticles on the photophysical and photosynthetic parameters of leaves and chloroplasts. Photochem Photobiol Sci 17(4):505–516. https://doi.org/10.1039/c8pp00067k

Tripathi DK, Singh VP, Prasad SM, Chauhan DK, Dubey NK (2015) Silicon nanoparticles (SiNp) alleviate chromium (VI) phytotoxicity in Pisum sativum (L.) seedlings. Plant Physiol Biochem 96(Vi):189–198. https://doi.org/10.1016/j.plaphy.2015.07.026

Tripathi DK, Singh S, Singh S, Dubey NK, Chauhan DK (2016) Impact of nanoparticles on photosynthesis: challenges and opportunities. Mater Focus 5(5):405–411

Tripathi DK, Singh S, Singh VP, Prasad SM, Dubey NK, Chauhan DK (2017) Silicon nanoparticles more effectively alleviated UV-B stress than silicon in wheat (Triticum aestivum) seedlings. Plant Physiol Biochem 110:70–81. https://doi.org/10.1016/j.plaphy.2016.06.026

Vafa ZN, Sohrabi Y, Heidari G, Rizwan M, Sayyed RZ (2024) Wheat growth and yield response are regulated by mycorrhizae application and supplemental irrigation. Chemosphere 364:143068. https://doi.org/10.1016/j.chemosphere.2024.143068

Van Nguyen D, Nguyen HM, Le NT, Nguyen KH, Nguyen HT, Le HM, Nguyen AT, Dinh NTT, Hoang SA, Van Ha C (2022) Copper nanoparticle application enhances plant growth and grain yield in maize under drought stress conditions. J Plant Growth Regul 41(1):364–375. https://doi.org/10.1007/s00344-021-10301-w

Vemula M, Prasad TNVKV, Vijay A, Vijaya Bhaskar Reddy A, Madhavi G (2013) Plant growth promoting potential of nano-bioremediation under Cr (VI) stress. 3:2277–4777

Venzhik YV, Moshkov IE, Dykman LA (2021) Gold nanoparticles in plant physiology: principal effects and prospects of application. Russ J Plant Physiol 68(3):401–412. https://doi.org/10.1134/S1021443721020205

Wang S, Wang F, Gao S, Wang X (2016) Heavy metal accumulation in different rice cultivars as influenced by foliar application of nano-silicon. Water Air Soil Pollut 227(7). https://doi.org/10.1007/s11270-016-2928-6

Wu H, Shabala L, Shabala S, Giraldo JP (2018) Hydroxyl radical scavenging by cerium oxide nanoparticles improves Arabidopsis salinity tolerance by enhancing leaf mesophyll potassium retention. Environ Sci Nano 5(7):1567–1583. https://doi.org/10.1039/c8en00323h

Yadav N, Bora S, Devi B, Upadhyay C, Singh P (2024) Nanoparticle-mediated defense priming: a review of strategies for enhancing plant resilience against biotic and abiotic stresses. Plant Physiol Biochem 213:108796. https://doi.org/10.1016/j.plaphy.2024.108796

Yan S, Wu F, Zhou S, Yang J, Tang X, Ye W (2021) Zinc oxide nanoparticles alleviate the arsenic toxicity and decrease the accumulation of arsenic in rice (Oryza sativa L.). BMC Plant Biol 21(1):1–11. https://doi.org/10.1186/s12870-021-02929-3

Ye Y, Cota-Ruiz K, Hernández-Viezcas JA, Valdés C, Medina-Velo IA, Turley RS, Peralta-Videa JR, Gardea-Torresdey JL (2020) Manganese nanoparticles control salinity-modulated molecular responses in Capsicum annuum L. through priming: a sustainable approach for agriculture. ACS Sustain Chem Eng 8(3):1427–1436. https://doi.org/10.1021/acssuschemeng.9b05615

Zahedi SM, Karimi M, Teixeira da Silva JA (2020) The use of nanotechnology to increase quality and yield of fruit crops. J Sci Food Agric 100(1):25–31. https://doi.org/10.1002/jsfa.10004

Zahedi SM, Harfi T, Marjani M, Vaculík M, Amini M, Sarikhani S (2024) Various forms of foliar applied zinc improve drought acclimation in pomegranate: response of photosynthesis, osmoregulation and antioxidant defense. J Soil Sci Plant Nutr 24(2):2694–2705. https://doi.org/10.1007/s42729-024-01694-0

Zia-ur-Rehman M, Anayatullah S, Irfan E, Hussain SM, Rizwan M, Sohail MI, Jafir M, Ahmad T, Usman M, Alharby HF (2023) Nanoparticles assisted regulation of oxidative stress and antioxidant enzyme system in plants under salt stress: a review. Chemosphere 314:137649

Nanotoxicology

16

Rosalba Salgado-Morales, Fernando Cuevas-Remigio, Armando Hernández-Mendoza, Gloria Sarahí Castañeda-Ramírez, Gilberto Villa-Rojas, and Edgar Dantán-González

Abstract

Nanotoxicology is a specialized field of toxicology focused on studying the harmful effects of nanomaterials. Nanomaterials have gained significant attention in sectors like agriculture due to their potential to enhance crop yields and sustainability. However, their fate and transport in the environment are key concerns, as their small size allows them to move easily through soil, water, and air, potentially impacting ecosystems. The mechanisms of nanomaterial toxicity are still under investigation, with evidence suggesting they can cause cellular damage, oxidative stress, and disruptions in biological processes. The environmental fate and transport of nanomaterials, their mechanisms of toxicity, and their impact on agriculture, medicine, and industry are crucial aspects of nanotoxicology research. The use of omics approaches, such as genomics and proteomics, is used to identify biomarkers and molecular responses to nanomaterial exposure, furthering our understanding of their toxicity. To mitigate these risks, strategies like safer design, toxicity testing, and regulatory frameworks are being developed. This work aims to examine the impact of nanomaterials on agriculture and the environment, emphasizing safe practices and regulatory measures for their use in various industries.

R. Salgado-Morales · G. S. Castañeda-Ramírez · G. Villa-Rojas · E. Dantán-González (✉)
Laboratorio de Estudios Ecogenómicos, Centro de Investigación en Biotecnología, Universidad Autónoma del Estado de Morelos, Morelos, Mexico
e-mail: edantan@uaem.mx

F. Cuevas-Remigio
Centro de Investigación en Biotecnología, Universidad Autónoma del Estado de Morelos, Morelos, Mexico

A. Hernández-Mendoza
Laboratorio de Bioinformática y Genómica Evolutiva, Centro de Investigación en Dinámica Celular, Universidad Autónoma del Estado de Morelos, Morclos, Mexico

R. Sayyed et al. (eds.), *Nanobiostimulants in Innovative Agriculture*,
https://doi.org/10.1007/978-981-95-2477-8_16

Keywords

Agricultural · Plants · Environmental · Omics approaches · Risk assessment · Regulatory Frameworks

1 Introduction

Nanotoxicology is an emerging and interdisciplinary field of research dedicated to understanding the potential toxic effects of nanomaterials, which are substances with structures at the nanoscale, typically ranging from 1 to 100 nanometers (Oberdörster et al. 2007; Oberdörster 2010; Zhao and Castranova 2011). Due to their unique properties—such as high surface area, reactivity, and small size—nanomaterials exhibit distinct behaviors compared to bulk materials. As the use of nanomaterials continues to grow in various sectors, particularly in agriculture, medicine, and industry, their potential effects on human health and the environment have become a crucial area of concern (Zhao and Castranova 2011). In agriculture, for instance, nanomaterials are being explored for their potential to enhance crop yields, improve pest control, and increase sustainability (Wei et al. 2024). However, their small size and high reactivity raise significant concerns about their behavior, transport, and persistence in the environment (Remédios et al. 2012). As these materials may be released into the environment during production, use, or disposal, it is important to understand how they interact with ecosystems and living organisms (Ray et al. 2009; Wei et al. 2024).

This chapter aims to provide a comprehensive exploration of nanotoxicology, focusing on the environmental fate and transport of nanomaterials, as well as their mechanisms of toxicity. One of the key areas of concern is how nanomaterials interact with plants and ecosystems. These materials can enter the soil, water, and air, potentially accumulating in plants and animals, leading to disruptions in biological processes and affecting biodiversity (Wei et al. 2024). Moreover, the mechanisms of nanomaterial toxicity remain an active area of investigation, with evidence suggesting that these materials can induce cellular damage, oxidative stress, and interfere with biological systems (Shukla et al. 2014; Ji et al. 2024). As such, understanding the underlying toxicological mechanisms is vital for assessing potential health risks and ensuring their safe use in various applications, including agriculture. An important aspect of nanotoxicology research is the use of advanced "omics" approaches, such as genomics, proteomics, and metabolomics. These technologies allow researchers to study the molecular and genetic responses of organisms exposed to nanomaterials, providing deeper insights into the pathways and biomarkers associated with nanotoxicity (Saarimäki et al. 2021). For example, genomics can help identify genetic changes in response to exposure, while proteomics can shed light on how nanomaterials influence protein expression and function. By applying these approaches, researchers can gain a better understanding of the effects of nanomaterials on living organisms, facilitating the identification of early warning signs of toxicity and potential health risks.

Furthermore, understanding the risks associated with nanomaterials is essential for developing effective regulatory frameworks to manage their safe use. Although nanomaterials hold great promise in various industries, their potential impact on health and the environment requires careful risk assessment. Regulatory measures must account for factors such as exposure levels, bioaccumulation potential, persistence in the environment, and possible toxicity to humans and wildlife. This chapter will also discuss current efforts to establish robust risk assessment protocols and regulatory guidelines that can effectively address the challenges posed by nanomaterials.

Lastly, strategies for mitigating the risks of nanotoxicity are crucial to ensuring the safe development and application of nanomaterials. Safer-by-design approaches are being developed, which focus on designing nanomaterials with reduced toxicity or less environmental impact. Additionally, toxicity testing protocols, including in vitro, in vivo, and in silico models, are essential to assess the potential risks associated with these materials. By integrating these testing methods and implementing effective regulatory frameworks, it is possible to ensure that nanomaterials are used safely across various sectors. This chapter will conclude by emphasizing the importance of a multidisciplinary approach to nanotoxicology, which combines scientific research with effective policy and regulatory frameworks to minimize potential risks to human health and the environment.

In summary, this chapter will cover the critical aspects of nanotoxicology, from understanding the environmental fate and transport of nanomaterials to exploring their toxicity mechanisms and evaluating the effectiveness of current regulatory frameworks. The goal is to provide a comprehensive overview of the research and strategies needed to ensure the safe and sustainable use of nanomaterials in diverse applications, particularly in agriculture.

2 Overview of Nanomaterials in Agriculture

Agriculture has been a vital activity since ancient times, not only for food production, but also because it plays a crucial role in the global economy, food security, and social welfare (Dethier and Effenberger 2012). Currently, agriculture faces significant challenges due to ongoing population growth and environmental deterioration caused by erosion and indiscriminate use of chemicals to improve agricultural production (Aktar et al. 2025; Chaithra et al. 2025). This underscores the need to develop new agrochemicals that can boost production while being more environmentally sustainable (Sethi et al. 2025; Bright et al. 2025; Ferioun et al. 2025. An alternative to this problem is the application of nanotechnology in agriculture using different nanoformulations and nanobiosensors. Nanoformulations include products such as fertilizers, herbicides, insecticides and fungicides, which offer benefits such as selective administration, slow degradation and controlled release of active ingredients, making them more effective than conventional formulations (Chakraborty et al. 2023). Nanobiosensors help in the early detection of abiotic and biotic stresses in plants, such as drought or disease (Shawon et al. 2020).

Nanofertilizers (NFs) or smart fertilizers are classified as inorganic NFs, mainly consisting of iron or zinc oxide nanoparticles (NPs) and organic NFs consisting of chitosan or humic acid NPs that are formulated in capsules, vesicles, and polymersomes. This novel technology improves soil structure and stimulates microbial activity, promoting environmentally friendly agriculture (Dimkpa and Bindraban 2017). Among the most widely used commercial NFs we find NanoGro™ and NanoFert Plus, both consisting of NPs that can penetrate the soil surface and reach plant roots, allowing more efficient absorption of the active ingredient (Pettersen and Aamlid 2019; Oancea 2022; Yadav et al. 2023). Some inorganic NFs have increased the supply of essential micronutrients, such as zinc, potassium, and phosphorus, improving the yield of rice and cucumber crops (Abraar et al. 2023; Méndez-López et al. 2024). Pharmks® organic nanofertilizers and humic acid treatments improved seed weight and plant size in *Nigella sativa* (Merghany et al. 2019). Another research showed that the combination of iron and potassium nanofertilizers increased pod and seed size in fava bean crops (Alhasany et al. 2021).

On the other hand, the constantly evolving agricultural environment suggests continued dependence on herbicides in the future, since weeds generate millions of dollars in losses in agricultural crops (Gharde et al. 2018). Many current herbicides only damage the aerial parts of weeds, ignoring the subway parts that may produce new weeds in the future. In contrast nanoherbicides can reach these subterranean parts of the weed and follow apoplastic and symplastic pathways to move within the plant (Adarsh et al. 2024; Wu and Li 2022). Nanoherbicide carrier systems typically consist of chitosan, tripolyphosphate, alginate, polyepsilon caprolactone, starch, rice husk, silica dioxide, and zinc oxide nanoparticles (Grillo et al. 2014). Several nanoherbicides have improved crop efficiency. For example, some containing H_2O_2, pendimethalin, and ZnO NPs enhanced the growth of bean (black gram) plants by optimizing auxin metabolism, facilitating Zn uptake, and establishing a weed-free environment (Vimalrajiv et al. 2018). A nanoemulsion of pretilachlor exhibited better weed control (Kumar et al. 2016), and nanoherbicides consisting of solid lipid nanoparticles presented higher efficacy against pest growth inhibition compared to commercial formulations (de Oliveira et al. 2015).

Nanotechnology in agriculture has also expanded to include the use of nanopesticides. Among the most widely commercialized are Banner MAXX, Subdue MAXX, Bifender FC, AZeroid, and Fenstro (Walker et al. 2017). Additionally, ZnO NPs have enhanced the insecticidal activity of thiamethoxam against tobacco cutworm *Spodoptera litura* larvae (Jameel et al. 2020). Likewise, SiO_2 is effective against cotton leafworm (*Spodoptera littoralis*), rice weevil (*Sitophilus oryzae*), wheat weevil (*Rhizopertha dominica*), red flour beetle (*Tribolium castaneum*), and grain beetle (*Oryzaephilus surinamensis*) (El-Naggar et al. 2020). Thus, most nanoinsecticides are formulated with NPs and active ingredients with insecticidal properties. However, nanopesticides cannot yet be considered fully environmentally friendly. An alternative is the use of organic bionanopesticides, which could reduce the adverse environmental impact of traditional nanopesticides (Manna et al. 2023). In these bionanoinsecticides, both the active ingredient and the carrier molecule are derived from biological sources, making them more effective for pest control and environmentally friendly. However, the options for biologically active ingredients and bionanotransporters are currently limited by the available resources.

Finally, bionanosensors are increasingly used in the agricultural industry for routine crop monitoring (Bodkhe et al. 2024). A novel application of nanobiosensors integrated with computer systems has been developed for crop improvement in rural areas (Alvarado et al. 2019). This system could help farmers in remote regions monitor herbicide and insecticide use, water levels, and both chemical and physical stresses (Shang et al. 2019). In addition, nanobiosensors enable the early detection of pathogens (nanodiagnostics). For instance, gold nanoparticle-based nanobiosensors have been used to identify plant diseases, track pathogen-plant interactions, sequence pathogen genomes, and monitor pathogen-host mutations (Khiyami et al. 2014). Another application of nanobiosensors is detecting pesticide residues in crops, helping prevent chemical overuse and ensuring responsible pesticide application, as these residues pose an environmental risk (Zhao et al. 2015). Some chemical compounds found in insecticides, such as methyl parathion, parathion, fenitrothion, and paraoxon, can be detected using surface plasmon resonance (SPR) biosensors (Alvarado et al. 2019). Another gold nanoparticle-based nanobiosensor detected carbofuran at 0.08 $\mu g/dm^3$ and chlorpyrifos at 0.06 $\mu g/dm^3$, respectively (Sun et al. 2014).

Although nanomaterials offer significant benefits in agriculture, such as crop improvement and environmental sustainability, they also present substantial challenges. These include long-term efficacy, increased production and storage costs, and regulatory challenges regarding their use in different countries. In addition, there is a potential threat that some nanoparticles can penetrate human skin and become airborne, which could lead to environmental problems as well as human and animal health issues (Pradhan et al. 2024). Therefore, it is crucial to address these challenges to maximize the benefits and minimize the negative impacts of such nanomaterials.

3 Environmental Fate and Transport of Nanomaterials

With the rapid advancement of nanotechnology across multiple sectors, nanomaterials (NMs) are becoming increasingly present in the environment. A consensus is emerging in the scientific community that NMs are considered to pose a relatively low environmental risk. However, it is important to note that this conclusion is still developing, as uncertainty remains about the complete understanding of how nanoparticles (NPs) within NMs interact with ecosystems and living organisms (Namasivayam et al. 2024). To assess this environmental damage, various studies have been conducted using soil samples, soybean seeds, zebrafish larvae, fish, and Daphnia magna hatchlings as biological indicators (Kumah et al. 2023). It should first be clarified that there are natural nanomaterials (NNMs) and engineered nanomaterials (ENMs), both of which can be released into the environment at any stage during their lifecycle. There are four typical ways this occurs: (a) release during the production of NPs and the manufacture of products containing NPs; (b) release during the use of NPs; (c) release after the disposal of products containing NPs; (d) natural emissions, such as volcanic eruptions, forest fires, and soil erosion. Once in the environment, NMs can reach water bodies, sediments, soil, sewage sludge, and air, where they can undergo various transformations (Mishra and Sundaram 2023b; Yang et al. 2023). In some cases, this can lead to the bioaccumulation of NPs in the food chain and their transfer from generation to generation (Thomas et al. 2013). The transformation of NPs in different ecosystems depends on factors such as shape, size, surface chemistry, and aggregation behavior of the particles. These factors also influence the stability, mobility, and persistence of nanoparticles in the environment. Additionally, the transformation and degradation of these materials are influenced by processes such as dissolution in the aqueous phase, oxidation, photodegradation, and surface coating or passivation by coexisting matter. Figure 16.1 illustrates the life cycle of ENMs, as well as the transport and fate of NMs in the environment (Rastogi et al. 2020). Life cycle analyses of ENMs indicate that the greatest potential environmental and health impacts stem from their production, rather than from accidental release during use (Huang et al. 2024).

Thus, NPs can be transported through various environmental systems. In aquatic environments, NPs may be suspended, sediment or bind to organic matter, while in soil they may move vertically by leaching or laterally by runoff. In soil, NPs can undergo biotransformations, affecting their bioavailability and toxicity (Rajput et al. 2020; Tran et al. 2023). It should be noted that the behavior and transport of NPs are influenced by environmental factors such as temperature, pH, and salinity, which affect their stability, solubility, and dispersion (Mishra and Sundaram 2023b). Likewise, environmental conditions give rise to redox reactions on the surface of NPs, in addition to other chemical transformations, as well as aggregation and

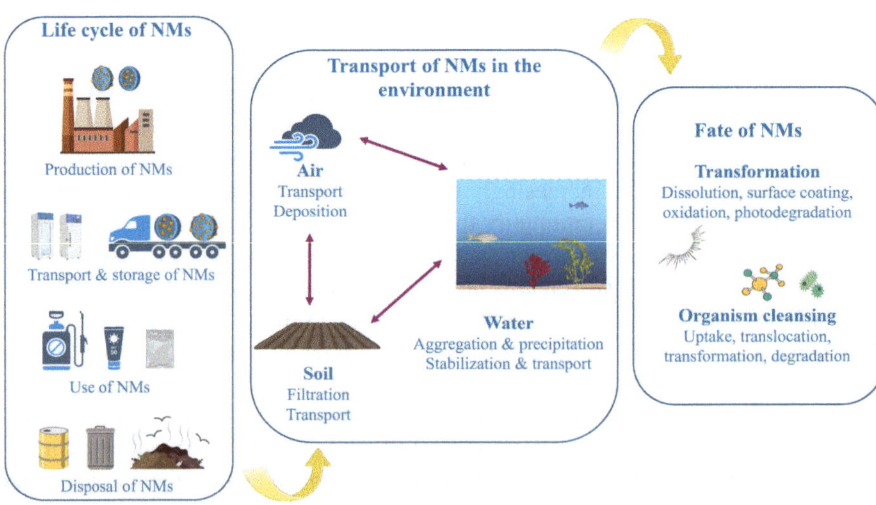

Fig. 16.1 Life cycle, transport, and fate of nanomaterials

disaggregation processes. These factors are crucial in the possible negative effects of NPs (Peijnenburg et al. 2015).

Regarding the atmospheric environment, NPs can be transported by air over long distances and end up deposited on land or in surface waters through wet or dry deposition. NPs can enter soil or water bodies through wind or rain (Hochella Jr et al. 2019). On the other hand, bioaerosols are airborne particles originating from biological sources and encompassing microorganisms (Hung et al. 2021). NPs can be associated with these types of particles and bring with them implications for human health. The silicon dioxide NPs have been identified as physicochemical markers of SARS-CoV-2, allowing the study of the transport of these bioaerosols in public spaces (Hildebrandt et al. 2022). These new interactions must be understood to subsequently evaluate their possible effects on human health and take measures to reduce the risks. Regarding the detection and quantification of NPs in the environment, studies have estimated that the presence of silver nanoparticles (Ag NPs) in sediments ranges between 0.19 and 470.6 μg kg^{-1} and in surface freshwater, they were estimated between 0.03 and 2.79 ng L^{-1} (Giese et al. 2018). Along WWTPs, it ranges between μg L^{-1} and ng L^{-1} (Cervantes-Avilés et al. 2019). Likewise, a study conducted in Brazil showed metallic NPs in the atmosphere (Souza et al. 2021). On the other hand, average levels of NPs in surface waters of around 1.5 ng L^{-1} for Ag NPs and 2.2 μg L^{-1} for TiO_2 NPs were quantified in the European Union (Sun et al. 2016). It is also estimated that copper nanoparticles released into the environment may eventually reach aquatic sediments and soils, reaching concentrations of up to 500 μg kg^{-1} (Keller et al. 2017). Finally, we show a comprehensive view on the toxicity of nanoparticles in plants, microorganisms, aquatic organisms, animals and humans. Some NPs affect plant growth and development. Specific examples are Ag NPs, which retard growth and activate plant immune systems, and titanium dioxide NPs, which cause genotoxic effects and damage to plant deoxyribonucleic acid (DNA) (Ghosh et al. 2010). Zinc and zinc oxide NPs inhibit seed germination and plant growth (Lin and Xing 2007).

Metal NPs can affect the microbiome by altering the homeostasis and functionality of the microorganisms that make it up. This occurs through mechanisms such as membrane disruption, generation of reactive oxygen species, and oxidative DNA damage (Abishagu et al. 2024). Aquatic organisms can be significantly harmed by NPs. Some specific cases are carbon nanotubes (CNTs), which have been shown to be toxic to several aquatic species, such as *Oryzias melastigma*, *Tigriopus japonicas* and *Thalassiosira pseudonana*. Also, exposure to CNTs can cause elevated mortality rates in freshwater crabs and adversely affect larvae of aquatic species (Khalid et al. 2015).

Regarding humans and animals, it has been observed that animals exposed to nanoparticles can suffer lung damage in the short or long term (Asmatulu et al. 2022). Also, accumulation of nanoparticles in the lungs is related to fibrosis (Qiao et al. 2015). Additionally, nanoparticles can cause oxidative stress, chromosomal mutations, and mitochondrial damage in human cells (Kabir et al. 2018) and lead to various health problems (Mishra and Sundaram 2023a), such as those mentioned below. In infancy, NPs could act as allergens (Sly and Schüepp 2012); NPs have been linked to increased severity of asthma and impaired lung function in the neonatal period (Meldrum et al. 2017); NPs are associated with cardiovascular problems, such as alterations in blood clotting and changes in cardiac function (Quan et al. 2010).

The widespread presence of NPs in the environment is of concern because of the limited understanding of their physicochemical properties and, consequently, their harmful effects. There is a pressing need for research to develop effective and affordable methods to assess the role of NMs in ecosystems and their potential risks to both the environment and human health. In addition, the development of biodegradable and environmentally sustainable NMs is necessary. This type of research will be crucial to develop strategies for managing and mitigating the negative effects of NMs.

4 Mechanisms of Nanomaterial Toxicity

Nanomaterials such as nanoparticles are particles with dimensions between 1 and 100 nanometers. Most authors define them as portions of material that are distinguishable from the medium in which they are found. These particles can form naturally, as in the case of volcanic eruptions. There are also incidental nanoparticles, which are generated accidentally through human processes, such as fuel emissions. In addition, there are manufactured nanoparticles, created for a specific purpose (Khan et al. 2019). On the other hand, the negative impacts that nanoparticles can have, including toxic and adverse effects on both living beings and the environment, have been studied. Research has been carried out to understand the mechanisms responsible for the toxicity of nanoparticles (Jane et al. 2021). It has been observed that, upon reaching the outer cell membrane, nanoparticles can interact with the outer cell membrane and enter the cell through endocytosis (Behzadi et al. 2017). However, endocytosis is not the only mechanism of action. Other processes, such as dissolution of nanoparticles, can release toxic ions and cause oxidative stress. In addition, the specific properties of nanoparticles influence their toxicity by determining how they interact with cells.

4.1 Properties of Nanoparticles

Nanoparticles have their own physical and chemical properties. These properties define their different applications, as well as determine their toxicity. Another important characteristic is their size, which is nanometric; they have a larger specific surface of action and thus can interact with cellular components or macromolecules (nucleic acids, proteins, lipids and carbohydrates). It is also likely that the size determines the ease with which the nanoparticle crosses the cell membrane, causing damage to it. Chemical composition is another important characteristic of nanoparticles as they can be made up of different materials, such as metals, carbon-based materials and some semiconductors. These chemical properties can determine their optical, electrical and magnetic properties (Mu et al. 2014). These varied characteristics can be important in the implementation of nanoparticles in different fields. The surface area and reactivity of nanoparticles are important in their interaction with cells. Modifying the reactive characteristics of the nanoparticle surface can improve cell interaction (Oberdörster et al. 2007). Solubility and stability are important characteristics. The effectiveness of nanoparticles is also determined by their solubility in the medium in which they are found; on the other hand, their stability will determine their efficacy and safety, since unstable nanoparticles tend to agglomerate or disintegrate, thus compromising their activity.

4.2 General Mechanisms of Nanoparticle Toxicity

As previously mentioned, nanoparticles, due to their size, react strongly with the surface and volume of the medium in which they are found. However, this same particularity facilitates interaction with biological systems and, in turn, induces toxicity. There are different mechanisms by which nanoparticles can cause damage and induce toxicity to different living organisms. These general mechanisms are described below.

4.3 Reactive Oxygen Species (ROS) Generation

One of the main mechanisms of toxicity of nanoparticles is the generation of reactive oxygen species (ROS). The high reactivity of ROS can damage cellular components such as proteins, nucleic acids and lipids, causing oxidative stress and inflammation (Oberdörster et al. 2007). Examples of nanoparticles that can catalyze the formation of ROS are gold and silver nanoparticles, due to their electronic properties and surface-to-volume ratio. The generation of ROS can cause damage to tissues and cells, altering function or generating chronic diseases such as cancer (Ray et al. 2012).

4.4 Interaction with Cell Membranes

The high surface reactivity of nanoparticles is another mechanism that involves interaction with cell membranes, which can induce damage to cell membranes, causing alterations in their function and permeability. Some nanoparticles have the ability to insert their structure into the membranes, causing rupture of the membrane and consequently the release of cellular content, triggering cell death by apoptosis or necrosis (Rodriguez-Hernandez et al. 2018). It has been observed that carbon nanotubes, due to their rigidity and elongated shape, can perforate the cell membrane, causing significant damage to it (Zhou et al. 2017).

4.5 Accumulation and Systemic Toxicity

The accumulation of nanoparticles in various organs and tissues of the body can cause systemic toxicity, altering the function of vital organs such as the liver, kidneys and lungs. This can lead to organ dysfunction and health problems. For example, nanoparticles that accumulate in the liver can hinder the metabolism and elimination of toxins, while those deposited in the lungs can cause chronic inflammation and respiratory diseases, according to Oberdörster et al. (2005). How nanoparticles are distributed and eliminated in the body depends on their characteristics, such as size, shape and chemical composition, which determines their potential to cause harm.

4.6 Immune System Modulation

Nanoparticles have the ability to influence the immune response by being recognized by the immune system as foreign entities, triggering an inflammatory reaction. This inflammatory response can be either acute or chronic and has the potential to trigger autoimmune diseases or worsen pre-existing conditions (Liu et al. 2017). Specifically, zinc oxide and silica nanoparticles have been associated with increased production of proinflammatory cytokines, suggesting a possible immunotoxic effect.

4.7 Mechanisms of Toxicity of Nanoparticles in Plants

The use of nanoparticles has experienced significant growth in several areas, including medical, agricultural and environmental applications. In the agricultural sector, they have been used to enhance beneficial characteristics in plants, such as growth and development. In addition, they have proven useful in addressing soil-related problems and have shown remediation properties. However, it is important to consider that the interaction of nanoparticles with biological systems can generate toxic effects, such as, for example, negatively affecting plant growth and health.

4.7.1 Oxidative Stress Generation

The toxicity of nanoparticles in plants is mainly attributed to the induction of oxidative stress, mediated by the generation of reactive oxygen species (ROS) such as hydrogen peroxide and free radicals, due to the high surface-to-volume ratio of nanoparticles (Risom et al. 2005). This process can trigger oxidative damage to essential biomolecules such as lipids, proteins and DNA, disrupting critical cellular functions. Induction of oxidative stress by nanoparticles can lead to lipid peroxidation, protein modification and genetic damage, negatively affecting plant growth and development (Mehrian and De Lima 2016).

4.7.2 Interference with the Nutrient Transport System

Exposure to nanoparticles can modify the uptake and distribution of essential nutrients in plants, altering their nutritional homeostasis. For example, silver and zinc nanoparticles can hinder the transport of ions and nutrients in roots, limiting the availability of crucial minerals such as iron and potassium (Lee et al. 2022). This disruption in nutrient uptake can lead to nutrient deficiencies, negatively impacting plant growth and development. In addition, the accumulation of nanoparticles in plant tissues can interfere with the binding of nutrients to their sites of action, reducing their bioavailability and affecting cell function (Al-Khayri et al. 2023).

4.7.3 Toxicity by Accumulation in Cellular Organelles

The accumulation of nanoparticles in vital cellular organelles, such as chloroplasts and mitochondria, can disrupt essential metabolic processes. The presence of nanoparticles in chloroplasts can compromise photosynthesis, damaging photosynthetic pigments and disrupting photosystem function (Vochita et al. 2019). Similarly, accumulation in mitochondria can hinder ATP production, affecting cellular energy and the plant's ability to perform critical metabolic processes (Tripathi et al. 2017).

4.7.4 Alteration of Water and Cell Membrane Homeostasis

Exposure to nanoparticles can alter water regulation and cell membrane integrity, affecting cell homeostasis. Cell membrane permeability is altered, allowing excessive flow of ions and water into the cell, which can lead to water stress and cell damage (Foyer and Noctor 2005). Nanoparticles can modify the structure and function of cell membranes, affecting cell-to-cell communication and signaling mechanisms, thus altering cell function.

4.7.5 Impact on Plant Health and Productivity

Exposure of plants to nanoparticles in agricultural environments can have significant detrimental effects on plant life and productivity. Due to their small size and high reactivity, nanoparticles can alter normal physiological processes in plants, as observed in several studies (Fig. 16.2). For example, the Journal of Nanoparticle Research has reported that metal nanoparticles can generate oxidative stress in plants, which negatively affects their growth and development (Singh et al. 2021). Additionally, a study published in Environmental Science & Technology indicates that nanoparticles can hinder plants' uptake of essential nutrients, leading to a decrease in productivity (Manke et al. 2013). These adverse effects not only compromise plant health but may also have significant implications for agricultural productivity and food security.

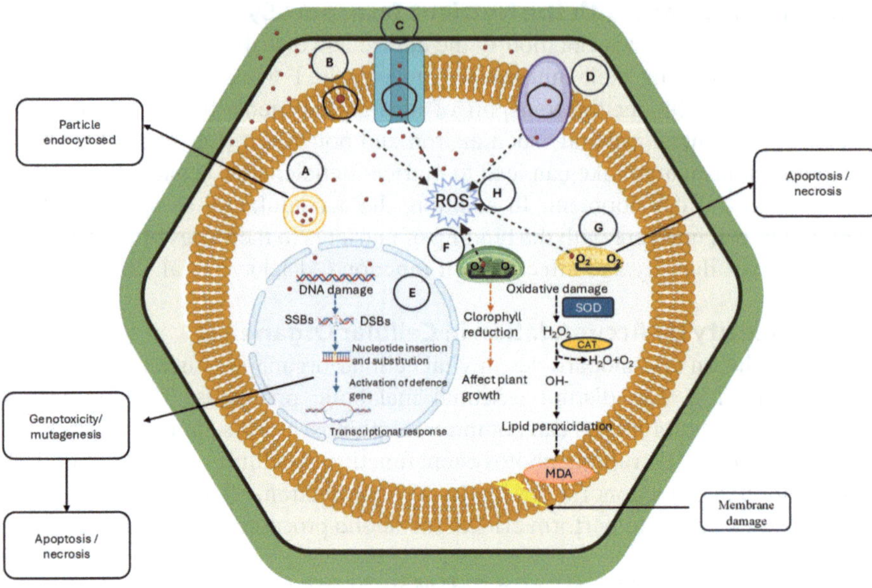

Fig. 16.2 Mechanisms of plant toxicity induced by nanoparticles (NPs). Mechanisms of nanoparticle entry into the cytoplasm (A, B, C, D). Free nanoparticles entering the nucleus through the nuclear pore, generating genotoxicity and mutagenesis (DSB: double-strand break / SSB: single-strand break) (E). Nanoparticles entering the chloroplast, generating oxidative damage and affecting growth (F). Introduction of NPs into mitochondria, generating oxidative stress and malondialdehyde production (G). ROS: reactive oxygen species (H)

5 Use of Omics Technologies in the Analysis of Nanotoxicity Effects in Plants

Nanotechnology has shown significant potential in agriculture, particularly in enhancing plant growth and development. However, while many studies have highlighted the positive effects of nanoparticles (NPs) and nanomaterials (NMs) on plants, evidence also exists of their toxic impacts at higher concentrations, leading to various harmful morphological, physiological, and genetic changes. These toxic effects are influenced by factors such as NP size, concentration, and chemical properties. The interaction of NPs with plants can lead to oxidative stress, disrupting cellular functions and causing phytotoxicity. NMs are classified based on their dimensions and chemical composition, including organic, inorganic, and carbon-based NPs. Organic NPs are typically biodegradable and less toxic, while inorganic NPs, such as metal and metal oxides, may pose greater risks. A comprehensive understanding of NP-based toxicity in vegetable crops is provided through the analysis of reports on the effects of NPs on plant toxicity (Sharma et al. 2022); this emphasizes the need to develop mitigation strategies to reduce phytotoxicity and its impact on human health. Nanotoxicity impedes plant growth, nutrient uptake, and biochemical processes at both physiological and molecular levels. Although research on the molecular mechanisms of NP toxicity is limited, studies on gene regulation may offer insights into both the beneficial and harmful effects of NPs on plants. Moreover, the bioaccumulation of NPs in plants can present serious health risks to humans through food chains (Sharma et al. 2022).

Mitigation strategies, including the use of antioxidants and surface modifications, are being explored to counteract the harmful effects of NPs. The review concludes by emphasizing the need for further research to optimize the use of NPs in agriculture and minimize their negative impacts. Omics sciences are essential in nanotoxicology, providing comprehensive molecular insights into the toxic effects of nanoparticles. By integrating multiple omics layers, researchers can achieve a more detailed and nuanced understanding of nanotoxicity, identify key biomarkers, and develop predictive models for risk assessment. These approaches are invaluable for advancing safety evaluations of nanoparticles in both environmental and biomedical contexts (Costa and Fadeel 2016). Nanotoxicology in plants has become a critical area of study, especially with the increasing use of engineered nanomaterials (ENMs) in agriculture. The integration of omics sciences—such as genomics, proteomics, metabolomics, and transcriptomics—provides a comprehensive approach to understanding the molecular and biochemical impacts of nanoparticles on plants. These advanced techniques enable the identification of specific genes, proteins, and metabolic pathways affected by nanoparticle exposure, offering insights into the mechanisms of toxicity, stress responses, and potential adaptations in crops. By leveraging omics data, researchers can more effectively assess the safety and environmental risks associated with ENM use, ultimately guiding the development of sustainable agricultural practices (Kumari et al. 2020). Plant diseases present a significant threat to the productivity and quality of food and fiber crops, putting increased pressure on agriculture to sustain the ever-growing global population. The worldwide loss due to plant pathogens varies significantly, with estimates ranging from 50% in wheat to over 80% in cotton. Although several management practices have been employed to mitigate these losses, the emergence of multidrug-resistant pathogens and the effects of climate change have exacerbated the challenge.

Silver nanoparticles (SNPs) have gained attention as potent antimicrobial agents, with established utility in medical applications. However, their application in plant disease management is still in its early stages.

In recent years, biogenic SNPs synthesized from biocontrol agents like *Trichoderma viride* have demonstrated enhanced antimicrobial activity compared to their chemically synthesized counterparts. For instance, SNPs synthesized from Trichoderma have been shown to effectively restrain phytopathogens by modulating pathogenic gene expression and inducing higher reactive oxygen species (ROS) generation. Despite these promising results, the plant's responses to these nanoparticles, particularly in terms of disease resistance, remain largely unexplored.

To address this gap, a study investigates the model pathosystem of *Arabidopsis thaliana* ecotype Columbia (col-0) and *Alternaria brassicicola* to assess the potential of biogenic silver nanoparticles in plant disease management. *A. brassicicola* is known to cause black spot disease, affecting nearly every important cultivated Brassica species. In the study, a combination of proteomic and metabolomic approaches was used, along with an evaluation of the plant's defense and morphological responses during the tripartite interaction of the plant, pathogen, and nanoparticles, to gain insights into the mechanistic aspects of this interaction (Kumari et al. 2020).

The results demonstrated that SNPs exhibit excellent antifungal activity in the *A. thaliana-A. brassicicola* pathosystem. The application of SNPs not only helped the host plant combat the disease but also enhanced phenolic levels and alleviated biotic stress, as evidenced by a reduction in hydrogen peroxide (H_2O_2), lipid peroxidation (LPX), and proline levels. These effects were achieved without causing any phytotoxicity.

The proteomic and metabolomic analysis revealed that numerous proteins and metabolites were differentially expressed during the tripartite interaction. Specifically, proteins related to bioenergy, metabolism, and plant defense were abundantly differentially expressed, leading to the differential regulation of sugar, steroid, and secondary metabolite synthesis. This study provides critical insights into the role of SNPs in plant disease management. The nanomaterials not only demonstrated their ability to kill the pathogen but also enhanced the plant's intrinsic antimicrobial properties and induced both primary and secondary immune responses by modulating the plant's proteome and metabolome (Kumari et al. 2020).

Nanotechnology has rapidly advanced over the past decade, with carbon nanotubes (CNTs) gaining significant attention due to their unique electro-mechanical and thermal properties, leading to widespread applications in industrial and biomedical sectors. The global production of CNTs has seen a substantial increase, driven by both single-walled (SWCNT) and multi-walled carbon nanotubes (MWCNT). However, this increased production raises concerns about the potential environmental release and bioaccumulation of CNTs. Studies on the toxicity and carcinogenic potential of MWCNTs' effects on plants remain limited, primarily focusing on physiological responses and germination rates (Ghosh et al. 2015). In a recent study using Allium cepa, the cytotoxic, genotoxic, and epigenetic effects of MWCNTs were evaluated (Ghosh et al. 2015), revealing significant oxidative stress, DNA damage, and global DNA hypermethylation induced by exposure to these nanoparticles. These findings highlight the importance of understanding not only the cytotoxic and genotoxic impacts but also epigenetic changes, such as DNA methylation, which may play a crucial role in plant stress tolerance and in adaptive responses. This study is among the first to investigate MWCNT-induced epigenetic changes in plants, suggesting the need for further research into histone modifications and other epigenetic alterations to fully characterize the toxicological profile of nanoparticles in plant systems.

Unlike larger materials, NPs can readily penetrate plant tissues, disrupting cellular functions and impairing growth. These tiny particles can accumulate in plant structures, interfering with vital processes like water transport. Agricultural concerns have arisen due to NPs' potential to reduce crop yields and contaminate the food chain. Evidence suggests that certain NPs, particularly carbon-based ones, can be absorbed, transported within plants, and even passed to subsequent generations. Furthermore, studies on *Arabidopsis thaliana* have revealed that different NPs, such as TiO_2, ZnO, and Ag, induce varying molecular responses, with some causing substantial alterations in gene expression (García-Sánchez et al. 2015). These findings underscore the complex nature of plant-nanoparticle interactions and highlight the need for further research to understand and mitigate their potential negative impacts. A study produced a set of 16 comparable transcriptome profiles to monitor early changes in gene expression upon NP and stress exposure (García-Sánchez et al. 2015). They examined the response of *A. thaliana* to different types (metallic and carbon-based) and sizes (4–80 nm) of NPs, in comparison with biotic and abiotic stress inducers that represent common environmental challenges for plants. Biotic stress was induced by infection with the necrotrophic fungus *Alternaria brassicicola* and the hemibiotrophic bacterium *Pseudomonas syringae pv. tomato* (Pst), which trigger plant defense responses through salicylic acid (SA) and jasmonic acid/ethylene pathways, respectively. Abiotic stresses included hypersalinity, drought, and mechanical wounding. The effects of abscisic acid (ABA), a key phytohormone in drought and osmotic stress responses, on the gene expression and phenotypes of NP-exposed plants were also assessed. The transcriptomic analysis revealed that the early response of *Arabidopsis* to NPs is characterized by significant transcriptional repression, with a common pattern observed across different NP types and sizes. Principal Component Analysis (PCA) grouped transcriptomes of NP-exposed plants together, distinguishing their impact from that of other stressors. Notably, NP exposure repressed genes involved in pathogen response, leading to increased bacterial survival during Pst infection. This repression was counteracted by SA signaling, which enhanced systemic acquired resistance (SAR) and reduced bacterial colonization. Furthermore, NP-induced repression affected genes related to phosphate starvation and root hair development, though supplementation with SA remedied the NP-induced root hair-less phenotype. In conclusion, the study provides critical insights into the molecular mechanisms underlying the early response of *Arabidopsis* to NPs, emphasizing the role of transcriptional repression as a common response regardless of NP composition (García-Sánchez et al. 2015). These findings contribute to our understanding of plant-nanoparticle interactions and highlight the potential risks posed by NP exposure in agricultural settings.

In the context of ZnO-NPs—a common type of metal oxide nanoparticle used in various products—transcriptomic analysis has proven essential. ZnO-NPs can induce oxidative stress and damage in plants, leading to a range of physiological and biochemical changes. By examining global gene expression changes and identifying stress-responsive microRNAs (miRNAs), transcriptomics provides a detailed view of how plants respond to NP stress at the molecular level. The recent study on maize (*Zea mays* L.) exposed to ZnO-NPs highlights the power of transcriptomics in uncovering the underlying mechanisms of NP-induced stress (Adhikari et al. 2020). This research utilized deep sequencing of small RNA libraries to identify both known and novel miRNAs that respond to ZnO-NP stress. These miRNAs play crucial roles in post-transcriptional regulation, affecting gene expression related to oxidative stress, growth, and development. By integrating transcriptomic data with physiological assessments—such as tissue Zn accumulation, ROS detection, and antioxidant enzyme activity—researchers can gain a comprehensive understanding of how ZnO-NPs impact maize. This approach revealed insights into the plant's defense strategies, including alterations in gene expression related to redox homeostasis and stress adaptation. The study's findings emphasize the importance of miRNAs in regulating stress responses and provide a detailed look at how plants adapt to NP exposure. Identifying differentially expressed miRNAs and their target genes sheds light on the complex regulatory networks involved in mitigating NP-induced damage. This knowledge not only enhances our understanding of plant stress biology but also informs strategies for improving crop resilience and managing the environmental impacts of nanotechnology.

In summary, transcriptomics is a powerful tool for dissecting the molecular effects of nanoparticles on plants. It enables researchers to explore gene expression changes, identify regulatory mechanisms, and develop strategies for mitigating NP-induced stress, ultimately contributing to the sustainable use of nanotechnology in agriculture and other sectors.

The proliferation of plastic particles in the environment, including microplastics (MPLs) and nanoplastics (NPLs), poses significant ecological challenges. MPs, ranging from 100 nm to 5 mm, and NPLs, smaller than 100 nm, have become ubiquitous pollutants due to their widespread use and persistence in the environment. Their small size and large surface area enable easy uptake by organisms, leading to potential ecological and health risks. Similarly, herbicide residues, such as Quinclorac (QNC), commonly used in agriculture, can adversely affect plant health and growth. The combined effects of MPLs, NPLs, and herbicides on plants, especially crops like rice, warrant detailed investigation to understand their impact and underlying mechanisms (Lu et al. 2023). Plants are crucial to terrestrial ecosystems and their exposure to MPLs, NPLs, and herbicides can significantly disrupt physiological and biochemical processes. These disruptions are often manifest in changes to plant biomass, growth parameters, photosynthetic activity, and nutrient transport. NPs, due to their smaller size, can penetrate plant cells more readily than MPs, leading to more pronounced physical toxicity and potential disruptions in nutrient and water transport. On the other hand, herbicide residues can impair photosynthesis and growth by affecting key metabolic pathways.

In recent studies, transcriptomics has been applied to investigate the effects of MPLs and NPLs on various plant species, revealing alterations in key biological processes and pathways. For instance, research has shown how MPLs and NPLs influence metabolic pathways, stress response mechanisms, and overall plant growth. Similarly, the impact of herbicide residues, such as QNC, has been analyzed using transcriptomic approaches to understand their effects on photosynthesis, growth, and stress adaptation. The study aims to bridge the knowledge gap by examining the combined effects of plastic particles and herbicides on rice, focusing on both individual and interactive impacts (Lu et al. 2023). By applying transcriptomics, we investigate how PS (polystyrene), a prevalent microplastic, and QNC herbicide, individually and in combination, influence rice growth and physiology. The hypothesis is that these stressors, whether alone or combined, will negatively affect rice through distinct mechanisms. The research utilizes transcriptomic analysis to explore the effects of PS and QNC on rice growth parameters, including biomass, photosynthetic pigments, and antioxidant enzyme activity (Lu et al. 2023). The findings reveal that PS and QNC, especially in combination, induce oxidative stress and disrupt key metabolic pathways, including the TCA cycle and photosynthesis. Transcriptomic data highlight how PS affects rice growth through alterations in the TCA cycle and plant-pathogen interactions, while QNC impacts photosynthesis and metabolic pathways. The combination of PS and QNC modulates a broader range of processes, including amino acid metabolism and antioxidant responses.

ENMs, particularly silver nanoparticles (AgNPs), are increasingly integrated into various applications due to their unique physicochemical properties. However, the environmental impact of these materials, especially their interactions with living organisms, is a growing concern. AgNPs, used in electronics, medicine, and agriculture, can have both beneficial and detrimental effects on plants. Their impact is influenced by particle size, concentration, and plant species. Primarily, AgNPs enter plants through their roots and can lead to several adverse effects, including reduced root and shoot length, decreased biomass, and diminished chlorophyll and carotenoid contents. At the cellular level, AgNPs can damage cytoplasmic membranes, cell walls, and organelles, causing plasmolysis, vacuolization, and malformations in chloroplasts and mitochondria. The toxic effects of AgNPs are often attributed to both their particulate nature and the silver ions (Ag^+) that leach from them. Ag^+ ions are known to produce reactive oxygen species (ROS), similar to the effects of AgNPs, suggesting that the toxicity might be due to Ag^+ rather than solely the nanoparticles themselves. To explore this, a study exposed *Arabidopsis thaliana* seedlings to AgNPs and $AgNO_3$, a source of Ag^+ ions, at equivalent concentrations (Mays et al. 2024). The results indicated that both AgNPs and $AgNO_3$ caused similar changes in gene expression and physiological functions, supporting the idea that Ag^+ ions play a significant role in toxicity. However, the particulate nature of AgNPs also contributes to their harmful effects. Two potential mechanisms of AgNP toxicity were proposed: physical damage to cell structures and localized release of Ag^+ ions inside the plant. These findings suggest that AgNP toxicity is associated with both the particulate nature of silver and the release of Ag^+ ions. Further research is necessary to fully understand these mechanisms and assess the risks of using nanosilver in agriculture, ensuring its safe application and preventing harm to non-target organisms.

Omics sciences, encompassing genomics, transcriptomics, proteomics, and metabolomics, are essential tools for analyzing nanotoxicity in plants. They provide comprehensive insights into the molecular, biochemical, and physiological responses of plants to nanoparticle exposure. By leveraging these high-throughput approaches, researchers can identify specific genes, proteins, and metabolites that are differentially expressed or regulated in response to nanoparticles. This holistic understanding is crucial for elucidating the complex mechanisms underlying nanoparticle-induced stress, enabling the development of strategies to mitigate potential adverse effects on plant health and ensuring the safe integration of nanotechnology in agriculture.

6 Nanotoxicology and Plants

Plants can absorb these nanoparticles through their roots, stems, and leaves, with their effects varying based on size, shape, and chemical composition. Nanoparticles can accumulate in different plant parts and affect processes like seed germination, root elongation, and overall growth, with outcomes that can be both beneficial and harmful (He et al. 2019). As global food production faces challenges from stressors such as extreme weather and pests, nanoagrochemicals offer promising solutions for pest management and plant growth enhancement. However, the unique structure and immune systems of plants make designing effective nanoparticle-based tools complex. Moreover, while nanotechnology offers significant potential, it also raises concerns about nanotoxicology, necessitating thorough investigations into the long-term effects of nanoparticles on plant health and the environment. Advanced methodologies, including omics technologies, are essential for understanding and managing these risks, ensuring the safe and sustainable application of nanotechnology in agriculture (Muzammil et al. 2023). Their widespread application in various industries, including agriculture, cosmetics, textiles, pharmaceuticals, ceramics, and electronics, has heightened worries about their potential hazards. Despite their benefits, these materials have introduced unforeseen problems, prompting researchers to focus on nanotoxicology to investigate NP toxicity on living organisms. NPs have been found to permeate the environment through multiple channels, accumulating in edible and crop plants and potentially reaching hazardous levels. Silver nanoparticles (AgNPs) are increasingly used across various industries, including electronics, medicine, and agriculture, due to their unique properties such as antimicrobial effects and enhanced crop production capabilities. However, their widespread use raises environmental concerns as AgNPs can be released into ecosystems through agricultural runoff, wastewater, and consumer waste. Their impact on plants can vary based on particle size, concentration, and plant species. Smaller AgNPs (<40 nm) can be absorbed by roots, penetrate cell walls, and spread throughout plants, potentially causing harmful effects such as reduced growth, decreased biomass, lower chlorophyll levels, and impaired photosynthesis. Cellular damage includes membrane disruption, plasmolysis, vacuolization, and organelle malformations, often accompanied by increased reactive oxygen species (ROS) production (Mays et al. 2024). The debate persists over whether these toxic effects are due to the nanoparticles themselves or the release of silver ions from their surface. One study investigates whether nanosilver phytotoxicity is linked to the particulate nature by exposing *Arabidopsis thaliana* seedlings to sub-morbid concentrations (4 μg/mL) of either AgNPs or $AgNO_3$ (Mays et al. 2024). The study controlled the pH to minimize ROS-induced stress and compared the effects of AgNPs and $AgNO_3$ on gene expression and physiological functions. By ensuring equal in planta Ag concentrations for both treatments and generating data from independently repeated experiments, this research aims to provide clarity on the role of AgNPs versus Ag^+ ions in phytotoxicity (Mays et al. 2024).

Copper oxide (II) nanoparticles (CuO NPs) are widely used in industries such as medicine, water treatment, and electronics, leading to increased environmental concentrations, with soil levels predicted to range from 0 to 0.54 mg/L. The impact of CuO NPs on plant growth is a growing concern, as their effects have been less studied compared to their known accumulation in cells. CuO NPs can inhibit root and leaf development, as well as seed germination, primarily due to the release of Cu^{2+} ions and oxidative stress. Elevated Cu^{2+} levels can disrupt phytohormone balance, leading to abnormal plant growth. Although CuO NPs can enter plant cells, form endosomes, and potentially alter gene expression, direct evidence of their effects on cell growth is limited. Additionally, the actin cytoskeleton, which is vital for cell division, growth, and response to stimuli, may be affected by these nanoparticles. The research of Jia et al. (2020) is the first to demonstrate that CuO NPs disrupt the actin cytoskeleton in plants, particularly in *Arabidopsis* root epidermal cells and root hairs. CuO NPs were found to directly interact with actin, causing cytoskeletal damage, growth defects, and cytotoxicity. This interaction impairs cell morphology and function, primarily affecting root cells due to the limited transfer ability of CuO NPs. The disruption of auxin transport, resulting from altered actin dynamics, causes secondary damage and is detectable in a dose-dependent manner. These findings provide a theoretical basis for further research into nano-toxicity in plants.

There is increasing interest in how nanoparticles (NPs) affect contaminated plants, with evidence showing both potential negative and positive impacts on plant health. NPs can impair germination, cell morphology, growth rate, and biomass, and induce oxidative stress. However, at low concentrations, certain NPs have shown beneficial effects on plants like *Avena sativa* and *Zea mays*, improving growth by enhancing water or nutrient uptake or mitigating toxic metal effects (Ekner-Grzyb et al. 2021). For instance, NaYF4: Yb3+, Er3+, Tm3+ nanoparticles improved mung bean growth at 10 μg/mL but inhibited it at 100 μg/mL. The impact of NPs varies with exposure time, surface contamination, size, and plant species. Lanthanide-doped nanocrystals (LDNCs), known for their luminescence and stability, are explored for applications in agriculture and biomedical fields. While gadolinium-based LDNCs have shown potential in imaging and drug delivery, concerns arise over gadolinium accumulation in water and animal tissues, which could affect vertebrate cells. Studies have found variable effects on cell types, with some research indicating no impact on blood cell health at certain concentrations (Ekner-Grzyb et al. 2021). Limited research exists on the phytotoxicity of LDNCs, especially vanadate gadolinium doped with europium ions. In a research by Ekner-Grzyb et al. (2021), the impact of GdVO4+ NPs on wheat seedlings was examined. After a 3-day incubation, LDNCs were detected inside the plants, but they had no significant effect on germination rate, seedling elongation, or biomass. The effect of NPs on cell viability was assessed using TTC and EB assays, which provided conflicting results. While TTC assays indicated no impact on cell viability, EB assays showed significant alterations in cell staining, suggesting possible loss of plasma membrane integrity. The unexpected finding that control cells had higher absorbance than treated cells may be due to decreased uptake of EB dye by cells incubated with LDNCs, potentially caused by NP aggregation blocking dye entry or interference with dye pumping mechanisms (Ekner-Grzyb et al. 2021).

In particular, the use of zinc oxide nanoparticles (ZnO-NPs) in agriculture and other sectors has raised concerns about their accumulation, translocation, and toxicity in plants. Zinc is a crucial micronutrient for plant growth but can be harmful at excessive concentrations. Low concentrations of ZnO-NPs can act as a Zn fertilizer, improving plant growth and development. However, higher doses can disrupt cellular homeostasis due to excessive Zn^{2+} release, leading to inhibited plant growth. Studies have shown that ZnO-NPs can enhance reactive oxygen species (ROS) generation, alter antioxidative enzyme activities, and affect pigment contents in plants. Furthermore, high doses of ZnO-NPs can cause DNA damage and increased genotoxicity, particularly in wheat seedlings, and induce biochemical changes in plants, such as elevated lipid peroxidation and protein degradation (Kumari et al. 2023). Despite plants' ability to adapt to harsh environments, distinguishing between adaptation and toxicity in the context of ZnO-NPs remains unclear. The variability in detoxifying processes among plants further complicates this understanding. This study of Kumari et al. (2023) aims to explore the genotoxic and phytotoxic effects of macro- and nano-forms of ZnO. It focuses on analyzing the genotoxic and cytotoxic behavior of these forms in pea roots and their phytotoxic responses, including germination, growth, and physiological states such as quantum yield and chlorophyll fluorescence in barley. Results indicate that both nano- and macro-dispersed ZnO induce genotoxicity and cytotoxicity in pea roots and inhibit barley seed germination and growth (Kumari et al. 2023). While the overall impact of both forms is similar, nano-ZnO causes more pronounced toxicity due to the accelerated release of Zn^{2+}. This research contributes to understanding ZnO-NP toxicity in plants and suggests that adjusting NP dosage and exposure duration can mitigate potential threats.

Adhikary et al. investigated the impact of zinc oxide nanoparticles (ZnO-NPs) on maize (*Zea mays* L.) seedlings, focusing on their molecular response to oxidative stress (Adhikari et al. 2020). ZnO-NPs, widely used in various industries, pose environmental threats due to their potential bioaccumulation in plants and subsequent entry into the food chain. The research highlights how ZnO-NPs induce oxidative stress, causing membrane damage, reduced biomass, and altered gene expression in maize. By analyzing microRNAs (miRNAs), the study uncovers their role in regulating stress-responsive genes, revealing that ZnO-NPs affect key cellular processes such as growth, development, and ROS homeostasis. Despite the upregulation of some defense-related genes, maize seedlings showed limited tolerance to ZnO-NPs, emphasizing the need for further research into adaptive mechanisms. This study provides novel insights into the post-transcriptional regulation of stress responses in plants exposed to ZnO-NPs, contributing to the broader understanding of nanomaterial-induced stress in crops.

The increasing use of metallic nanoparticles, such as silver (AgNPs) and gold nanoparticles (AuNPs), in everyday products, pesticides, and medical applications has led to their inevitable release into the environment, where they interact with crop plants. Understanding how plants uptake these nanoparticles is essential for assessing their accumulation and potential toxicity in the food chain. Recent research has revealed that the uptake of metallic NPs by plants is particle-specific, yet significant gaps remain, particularly regarding the mechanisms of uptake and the impact of environmental factors like dissolved oxygen. The biodissolution of NPs, such as AgNPs and AuNPs, can lead to simultaneous particulate and ionic uptake, complicating the interpretation of metal accumulation. Many studies have been limited by focusing on specific NP types and properties, with a lack of comparative assessments across different nanomaterials, which is necessary for a comprehensive understanding of NP interactions with the environment. Zhang et al. (2019) investigated the particle-specific uptake kinetics of silver (AgNPs) and gold nanoparticles (AuNPs) in wheat (*Triticum aestivum* L.) using hydroponic systems, with particle interactions characterized by transmission electron microscopy-energy dispersive spectrometry (TEM-EDS) and single particle inductively coupled plasma mass spectrometry (spICP-MS) (Zhang et al. 2019). The study provided insights into how these commonly used nanoparticles are absorbed by plants, finding that while the uptake rate constants were low, the model system successfully distinguished between particulate and dissolved metal uptake. The results showed similar uptake rates for AgNPs and AuNPs, suggesting comparable mechanisms potentially due to damage to cell membranes. However, the influence of particle size, coating, and core composition on bioavailability remains unclear. The research highlights the need for further studies on a broader range of nanoparticles to fully understand their uptake by plants.

The rise of nano-enabled agrochemicals, such as silver nanoparticles (AgNPs), aims to improve crop nutrition and protection sustainably. However, concerns arise over their bioavailability and the effects of Ag ion dissolution. Research comparing AgNPs to their dissolved forms or using thiol ligands has yielded mixed results, with some studies attributing effects solely to dissolved Ag while others emphasize the particles' role. The challenge is further complicated by the potential for dissolved Ag to reduce to nanoparticles within plants. Factors such as dissolved oxygen, root exudates, and reactive oxygen species can influence AgNP dissolution, potentially skewing uptake results. Understanding these interactions is crucial for accurate assessment of nanoparticle and ion bioavailability in plants and ensuring the sustainable use of nano-agrochemicals. The study of Dang et al. (2020) reexamines the bioavailability of silver nanoparticles (AgNPs) by differentiating between their dissolution in bulk suspension versus at the plant interface, and analyzing the uptake kinetics of particles versus dissolved Ag in wheat (*Triticum aestivum* L.). The research shows that despite controlling for AgNP dissolution, particle-specific uptake is notable, though uptake rate constants are low, indicating that particles are indeed taken up by plants. The study highlights that AgNP uptake might be underestimated, especially when dissolution is minimal. This research provides a quantitative distinction between particulate and dissolved Ag uptake and suggests that similar kinetic approaches could be useful for studying other metal-based nanoparticles (Dang et al. 2020). Further research with extended growth periods and realistic conditions is needed to fully understand AgNP interactions and applications.

The integration of engineered nanomaterials (ENMs) into various technologies is on the rise, including applications in agriculture, biotechnology, and medicine. Among these, carbon nanotubes (CNTs) are gaining attention for their potential agricultural uses due to their ability to influence plant growth, cross cell walls, and deliver nanopesticides. Despite their benefits, there is uncertainty about the effects of CNTs on plants, as their impact depends on factors such as type, size, concentration, and functionalization. Research is needed to understand these interactions and quantify potential nanotoxicological effects to ensure safe use. CNTs can induce cellular oxidative stress, influenced by their characteristics and the production of reactive oxygen species (ROS). They can affect plant physiological processes, including growth, hormone production, and gene expression. For instance, CNTs can impact nitric oxide (NO) levels, which play a role in stress responses and plant metabolism. Jordan et al focused on the effects of CNTs on tomato plants (Jordan et al. 2020), examining growth, flowering, fruiting, and changes in amino acids and phytohormones. The findings suggest that while CNT exposure may initially delay growth, plants might eventually adapt and perform better. However, CNTs do not significantly affect reproductive outcomes or fruit yield. The study also found no significant changes in phytohormone levels, indicating that CNT effects might be influenced by factors such as size, heterogeneity, and manufacturing impurities. Further research is needed to explore the impact of CNT aggregation on plant health and nutrient uptake, and to inform regulations ensuring safe ENM concentrations to protect ecosystems and human health.

Nanotechnology plays a crucial role in modern agricultural practices, with engineered nanoparticles (ENPs) being employed in various applications, including waste treatment, pesticide development, and fertilizer enhancement. Despite their benefits, concerns have been raised about the potential biosafety and environmental impact of ENPs, as they can inhibit plant growth, damage cellular structures, and induce oxidative stress. The research of Yuan et al. (2024) highlights the complex interactions between ENPs, such as graphene nanoplatelets (GNPs) and titanium dioxide nanoparticles (TiO_2 NPs), and their combined effects on lettuce, a sensitive crop. The findings suggest that the mixture of these nanoparticles results in less severe oxidative damage and influences critical molecular pathways related to stress responses, providing valuable insights into the broader implications of ENPs in agriculture. Understanding these interactions is essential for evaluating the safety and efficacy of ENPs in sustainable agricultural practices.

Nanotoxicology in plants is a rapidly emerging field of study, essential for understanding the complex interactions between engineered nanomaterials and plant systems. As nanotechnology continues to advance and its applications in agriculture and other sectors expand, the potential risks associated with nanoparticle exposure to plants cannot be overlooked. Plants play a critical role in ecosystems and serve as a primary entry point for nanoparticles into the food chain, making it crucial to assess their toxicological impact. Understanding how nanoparticles affect plant growth, development, and stress responses is vital for ensuring the safe and sustainable use of nanomaterials in agriculture and environmental applications. This knowledge not only helps in mitigating potential risks but also paves the way for harnessing the benefits of nanotechnology to improve crop resilience and productivity in a responsible manner

7 Risk Assessment and Regulatory Frameworks

The use of nanotechnology increased in the year 2000, when the use of nanomaterials (1–100 nm) was surprisingly discovered. These nanomaterials show very innovative properties, which allowed extensive research to be carried out and is still ongoing. In addition, there are already many everyday products with nanotechnology, such as titanium oxide (nanostructured), which is used as a pigment in paints, and other nanomaterials are found in personal care products. Nanotechnology is also on the European market, found in batteries, coatings, antibacterial clothing and cosmetics (European Commission 2024). Research also suggests future applications of this nanotechnology can be found in different areas: nanomedicine (nanomaterials, nano sensors, diagnosis and control); agriculture (bioremediation of soils to remove biopesticides or toxic fertilizers, indicators of pollutants, effects on phytopathogens); the construction industry (metallic compounds with special properties, such as nano-wires, carbon tubes, etc.). Among other sectors where these technologies are applied. The main innovation of nanotechnology is nano-size, which has its advantages and other implications; one of these implications is that nano-sized materials could easily pass through cells, engaging in beneficial or harmful effects. There are already studies of the cell membrane and the unspecificity of nanoparticles (Rodriguez-Hernandez et al. 2018).

Therefore, different international, national and regional organisms have sought to establish standards or guidelines for the safe use of this nanotechnology in living organisms and the surrounding environment (Saldívar Tanaka 2020). In 2002, nanomaterials were discussed at the joint meeting of the Versailles Project on Advanced Materials and Standards (VAMAS) (Kica and Bowman 2012). Following this meeting, China established the United Working Group for Nanomaterials Standardization in 2003. Subsequently, in 2004, the first report on the implications of nanotechnology was published by the Royal Society and the Royal Academy of Engineering in England. It highlights many of the benefits of these technologies in the future and the need for research on the environmental and health effects of nanoparticles (Royal 2004).

As these nanotechnologies began to be used, international organizations and national or regional committees began to establish and recommend standards, technical or legal guidelines for the use of nanotechnology. Currently, there are voluntary international standards established by the "International Organization for Standardization" (ISO), where there is already a list of more than 110 documents on nanomaterials information (ISO 2024). These standards started to be published by ISO from 2010 to date, starting with the explanation of definitions, terms, addressing the different fields of nanotechnology, classifications, and toxicology in different fields.

In the European Union, the Regulation on the Safe Use of Chemical Substances, REACH (Registration, Evaluation, Authorization and Restriction of Chemical Substances) came into force in 2007. In the case of nanomaterials, the European Regulation came into force, which establishes a high level of health protection for cosmetic products with nanomaterials, as well as the establishment of a Scientific

Committee for Consumer Safety (SCCS 2021). Where conclusions have been reached in “COMMISSION REGULATION (EU) 2024/858 of the OFFICIAL JOURNAL OF THE EUROPEAN UNION,” establishing collective physicochemical and toxicological tests of nano products, in order to avoid health risks. An example of the importance of these guidelines was determined by the Scientific Committee on Consumer Safety, where it indicates that colloidal (nano) copper could cause accumulation in different organs, as well as possible mutagenic/genotoxic effects of copper nanomaterials used in some cosmetics (SCCS 2020). This committee has also determined other nanomaterials and safe use concentrations in the cosmetics industry (SCCS 2023).

There are different agencies in the European Union that oversee establishing guidelines according to the type of nanomaterial being used. Such is the case of the “European Chemicals Agency,” which published the “Appendix for nanoforms applicable to the Guidance on Registration and Substance Identification” in 2022. The aim of this information is to comply with the obligations imposed by the REACH Regulation. In addition, there are also specific registration guidelines for those who manufacture or import nanoforms: “How to prepare registration dossiers covering nanoforms.” Other countries are developing and modifying their national standards based on ISO standards. There are country regions where standards are mandatory and contribute to the development of information in those countries. Despite all this effort, the fact that some standards are voluntary means that the industrial sector sometimes does not regulate (Saldívar Tanaka 2020). The aim of these guidelines or technical or legal standards is to preserve human health and the environment, study bioaccumulation, solubility, persistence, and study the risks from a multidisciplinary approach and promote nanotechnologies, taking care of the aspects.

8 Strategies for Mitigating Nanotoxicity

Toxicology is a science that studies the toxic levels that can be generated by an organism, compound, molecule, substance or material (Hernández et al. 2021). It is very necessary because it provides information that helps to establish parameters that indicate whether an environment has toxicity levels and for what types of organisms.

One of the sectors where toxicology can be observed in a more detailed way is in the study of medicines, these are studied in different stages (phases), establishing in vitro tests and then in vivo tests, each of these phases indicating the level of effectiveness of the medicine, pharmacokinetics, adverse effects, effects on health and the environment that can occur, all of these specifications are reported in international organizations. The development of new medicines is regulated and the process from inception of a medicine until it is available for sale to the public is a process that takes several years of research and evaluation to reach the final stage (Zurita-Cruz et al. 2019).

Another example of a toxicity strategy that has been implemented can be seen in the history of the regulation of chemicals. Chemical substances are nowadays all around us, but their use, handling, bioaccumulation and physico-chemical properties have been studied for several years beforehand. In addition, toxicity studies have been carried out to determine the safe use of these chemicals. Information on these substances has been collected through the Chemical Abstracts Service (CAS), where a unique number is assigned to each chemical. CAS is a worldwide authority on chemical information, where data for each chemical is collected, organized and published (Chemical Abstracts 2024). Despite this, in some other countries chemicals emerged that did not have CAS number information. In those years a concern and uncertainty arose about not having the information for those chemicals. Therefore, international organizations since 1992 have started a series of initiatives as health and environmental care measures. Thus, several conventions and efforts emerged, such as the Stockholm Convention on Persistent Organic Pollutants, the Montreal Protocol on Substances that Deplete the Ozone Layer, among others, and the participation of organizations such as the Food and Agriculture Organization of the United Nations (FAO), among others. In this way, substances have been regulated and substances that are highly harmful to health and the environment have been identified. This led to the prohibition of certain chemical substances, such as pesticides. Following international treaties, conventions are made and signed to ban highly harmful substances (WHO 2014; Mendoza Cantú and Ize Lema 2017; Wang et al. 2021).

The constant advancement, search for new substances or materials, is to find innovative technologies, which can change various aspects of science. However, as has been addressed, their effect or change on what they may be exposed to must be considered and investigated from a multidisciplinary approach. The search for strategies on nanotoxicity is still in progress, as there are several reports that some nanomaterials may be toxic. But, as well as materials or substances that are involved in human health and the environment, nanomaterials should be further studied by researchers before they are used in environments with many variables, having contact with organisms and the environment (Medina-Pérez and Fernández-Luqueño 2018).

In the case of nanomaterials, there are areas or regions with technical and some legal regulations; in some regions they are necessary and in other regions they are not obligatory (Saldívar Tanaka 2020). In general, however, and taking the examples of the regularization of other substances, the technical strategies to be followed are summarized below:

- To know the physicochemical properties of the nanomaterial.
- To study the pharmacognosy of the nanomaterial.
- To carry out in vitro toxicity tests, different models.
- Determine the dose of the effect of the nanomaterial.
- Establish exposure limits.
- Assess risks and potential adverse effects.
- Conduct in silico testing.

- Conduct in vivo tests in controlled environments.
- Bioaccumulation testing.
- According to the material carry out further research.
- Scale-up to the final product.

In the case of international, national or regional organizations, this could be developed in legal guidelines:

- Generate more monitoring committees for toxic substances.
- Conduct and confirm assessments of proposed nanomaterials.
- Regulatory frameworks.
- Comprehensive policies.
- International conventions.

Nanotechnology has generated and can generate many contributions to different sciences, but it is necessary to be cautious about the technologies and their consequences when they are in contact with the population and the environment. Science is constantly changing, according to the continuous discoveries that generate knowledge that generates or answers more research questions. Thus, information must be updated in a multidisciplinary way and constantly followed up. Health and environmental care is a team effort where the whole population interacts in one way or another.

9 Conclusion

Nanotoxicology is an essential field for understanding the risks associated with the use of nanomaterials, particularly in sectors like agriculture. While nanomaterials offer significant benefits, such as improving crop yields and promoting sustainability, their widespread use raises concerns about safety and environmental impact. The small size and high reactivity of nanoparticles allow them to easily move through soil, water, and air, potentially affecting ecosystems and causing bioaccumulation in the food chain. Although the mechanisms of nanomaterial toxicity are still under investigation, evidence suggests they can cause cellular damage, oxidative stress, and disruptions in biological processes. It is crucial to continue research on the biological effects of these materials and their potential long-term toxicity. The use of omics approaches, such as genomics and proteomics, is key to identifying biomarkers and molecular responses to nanomaterial exposure. Moreover, developing specific regulatory frameworks for nanomaterials, inspired by international agreements like the Stockholm Convention, is necessary to ensure their safe use, minimizing risks to human health and the environment. In agriculture, continued research will be essential to apply sustainable and safe practices when using nanomaterials.

References

Abishagu A, Kannan P, Sivakumar U, Manikanda Boopathi N, Senthilkumar M (2024) Metal nanoparticles and their toxicity impacts on microorganisms. Biologia 79(9):2843–2862. https://doi.org/10.1007/s11756-024-01760-z

Abraar SAM, Rajhu NM, Vardhan TV, Agrawal A, Saxena KK, Savithiri V, Buddhi D, Senthilvel K, Ramesh B (2023) Experimental investigation on metallurgical and mechanical properties and Wear behavior of Al5032/SiC nanocomposites. Mater Res 26:e20230178

Adarsh S, Shilpa S, Ameena M, Susha VS, Sreelekshmi K, Renjan B, Umkhulzum F, Sethulakshmi VS, Shanavas S (2024) A review on nano herbicides: the future of weed management. J Adv Biol Biotechnol 27(7):1244–1253

Adhikari S, Adhikari A, Ghosh S, Roy D, Azahar I, Basuli D, Hossain Z (2020) Assessment of ZnO-NPs toxicity in maize: An integrative microRNAomic approach. Chemosphere 249:126197

Aktar SN, Shah MH, Mohanty B, Pramanik K, Hosen MS, Ddoza RMB, Kumawat SM, Sharma NK, Sayyed RZ (2025) Beneficial microbes in plant health, immunity and resistance. In: Mathur P, Roy S (eds) Plant microbiome and biological control, Sustainability in plant and crop protection, vol 20. Springer, Cham, pp 1–17. https://doi.org/10.1007/978-3-031-75845-4_1

Alhasany AR, Leiby HR, Noaema AH (2021) Effectiveness of spraying nano-fertilizers of iron and potassium on the growth and yield of faba bean crop (Vicia Faba L.). Int J Agric Stat Sci 17(1)

Al-Khayri JM, Rashmi R, Surya Ulhas R, Sudheer WN, Banadka A, Nagella P, Aldaej MI, Rezk AA-S, Shehata WF, Almaghasla MI (2023) The role of nanoparticles in response of plants to abiotic stress at physiological, biochemical, and molecular levels. Plants 12(2):292

Alvarado K, Bolaños M, Camacho C, Quesada E (2019) Nanobiotechnology in agricultural sector: overview and novel applications. J Biomater Nanobiotechnol 10(02):120–141

Asmatulu E, Andalib MN, Subeshan B, Abedin F (2022) Impact of nanomaterials on human health: a review. Environ Chem Lett 20(4):2509–2529

Behzadi S, Serpooshan V, Tao W, Hamaly MA, Alkawareek MY, Dreaden EC, Brown D, Alkilany AM, Farokhzad OC, Mahmoudi M (2017) Cellular uptake of nanoparticles: journey inside the cell. Chem Soc Rev 46(14):4218–4244

Bodkhe M, Chalke T, Kulkarni S, Goswami A (2024) A review on sustainable applications of nanobiosensors in various fields and future potential. BioNanoScience:1–21

Bright JP, Maheshwari HS, Thangappan S, Perveeen K, Bukhari NA, Mitra D, Sayyed RZ, Mantinu A (2025) Biofilmed-PGPR: a next-generation bioinoculant for plant growth promotion in rice (*Oryza sativa* L.) under changing climate. Rice Sci 32. https://doi.org/10.1016/j.rsci.2024.08.008

CAS (Chemical Abstracts) (2024) CAS history. [https://www.cas.org/about/cas-history]

CCSC, 2020. (Comité Científico de Seguridad de los Consumidores): dictamen sobre el cobre (nano) y el cobre coloidal (nano); versión preliminar de 27-28 de octubre de 2020; versión final de 5 de marzo de 2021, SCCS/1621/2020

CCSC, 2021. (Comité Científico de Seguridad de los Consumidores): dictamen científico sobre la seguridad de los nanomateriales en los productos cosméticos; versión preliminar de 6 de octubre de 2020; versión final de 8 de enero de 2021, SCCS/1618/20, Corrección de errores de 8 de marzo de 2021

CCSC (Comité Científico de Seguridad de los Consumidores): dictamen sobre la hidroxiapatita (nano); versión preliminar de 4 de enero de 2023; versión final de 21-22 de marzo de 2023, SCCS/1648/22

Cervantes-Avilés P, Huang Y, Keller AA (2019) Incidence and persistence of silver nanoparticles throughout the wastewater treatment process. Water Res 156:188–198

Chaithra M, Nishmitha K, Patel PS, Mitra D, Gangaraj R, Sayyed RZ (2025) Fungal bio-stimulants: cutting-edge bioinoculants for sustainable agriculture. In: Mathur P, Roy S (eds) Plant microbiome and biological control, Sustainability in plant and crop protection, vol 20. Springer, Cham, pp 289–307. https://doi.org/10.1007/978-3-031-75845-4_13

ɔorty R, Mukhopadhyay A, Paul S, Sarkar S, Mukhopadhyay R (2023) Nanocomposite-ɹed smart fertilizers: a boon to agricultural and environmental sustainability. Sci Total ɲviron 863:160859. https://doi.org/10.1016/j.scitotenv.2022.160859

ɩa PM, Fadeel B (2016) Emerging systems biology approaches in nanotoxicology: towards a mechanism-based understanding of nanomaterial hazard and risk. Toxicol Appl Pharmacol 299:101–111

Dang F, Wang Q, Cai W, Zhou D, Xing B (2020) Uptake kinetics of silver nanoparticles by plant: relative importance of particles and dissolved ions. Nanotoxicology 14(5):654–666

de Oliveira JL, Campos EVR, Goncalves da Silva CM, Pasquoto T, Lima R, Fraceto LF (2015) Solid lipid nanoparticles co-loaded with simazine and atrazine: preparation, characterization, and evaluation of herbicidal activity. J Agric Food Chem 63(2):422–432

Dethier J-J, Effenberger A (2012) Agriculture and development: a brief review of the literature. Econ Syst 36(2):175–205. https://doi.org/10.1016/j.ecosys.2011.09.003

Diario oficial de la unión europea, 2024. Reglamento (CE) n.o 1223/2009 del Parlamento Europeo y del Consejo en lo que respecta al uso de los nanomateriales copolímero de estireno/acrilato, copolímero de acrilato/estireno de sodio, cobre, cobre coloidal, hidroxiapatita, oro, oro coloidal, ácido tioetilaminohialurónico de oro, acetilheptapéptido-9 oro coloidal, platino, platino coloidal, acetiltetrapéptido-17 platino coloidal y plata coloidal en productos cosméticos. 5 p. http://data.europa.eu/eli/reg/2024/858/oj

Dimkpa CO, Bindraban PS (2017) Nanofertilizers: new products for the industry? J Agric Food Chem 66(26):6462–6473

ECHA (2023) How to prepare registration dossiers covering nanoforms. https://doi.org/10.2823/2782

Ekner-Grzyb A, Chmielowska-Bąk J, Szczeszak A (2021) Influence of GdVO4: Eu3+ nanocrystals on growth, germination, root cell viability and oxidative stress of wheat (*Triticum aestivum* L.) seedlings. Plants 10(6):1187

El-Naggar ME, Abdelsalam NR, Fouda MMG, Mackled MI, Al-Jaddadi MAM, Ali HM, Siddiqui MH, Kandil EE (2020) Soil application of nano silica on maize yield and its insecticidal activity against some stored insects after the post-harvest. Nano 10(4):739

European Chemicals Agency (2022) Appendix for nanoforms applicable to the Guidance on Registration and Substance Identification Reference: ECHA-21-G-06-EN Cat. Number: ED-08-21-370-EN-N ISBN: 978-92-9468-030-3. 10.2823/40

European Commission (2024) Commission Regulation (EU) 2024/858 of 14 March 2024 amending Regulation (EC) No 1223/2009 of the European Parliament and of the Council as regards the use of the nanomaterials Styrene/Acrylates copolymer, Sodium Styrene/Acrylates copolymer, Copper, Colloidal Copper, Hydroxyapatite, Gold, Colloidal Gold, Gold Thioethylamino Hyaluronic Acid, Acetyl heptapeptide-9 Colloidal gold, Platinum, Colloidal Platinum, Acetyl tetrapeptide-17 Colloidal Platinum and Colloidal Silver in cosmetic products. Official Journal of the European Union, L 2024/858, 15 March 2024. https://data.europa.eu/eli/reg/2024/858/oj

Ferioun M, Zouitane I, Bouhraoua S, Elouattassi Y, Belahcen D, Errabbani A, Louahlia S, Sayyed R, El-Ghachtouli N (2025) Applying microbial biostimulants and drought-tolerant genotypes to enhance barley growth and yield under drought stress. Front Plant Sci 15:1494987. https://doi.org/10.3389/fpls.2024.1494987

Foyer CH, Noctor G (2005) Redox homeostasis and antioxidant signaling: a metabolic interface between stress perception and physiological responses. Plant Cell 17(7):1866–1875

García-Sánchez S, Bernales I, Cristobal S (2015) Early response to nanoparticles in the Arabidopsis transcriptome compromises plant defence and root-hair development through salicylic acid signalling. BMC Genomics 16:1–17

Gharde Y, Singh PK, Dubey RP, Gupta PK (2018) Assessment of yield and economic losses in agriculture due to weeds in India. Crop Prot 107:12–18

Ghosh M, Bandyopadhyay M, Mukherjee A (2010) Genotoxicity of titanium dioxide (TiO2) nanoparticles at two trophic levels: plant and human lymphocytes. Chemosphere 81(10):1253–1262. https://doi.org/10.1016/j.chemosphere.2010.09.022

Ghosh M, Bhadra S, Adegoke A, Bandyopadhyay M, Mukherjee A (2015) MWCNT uptake in Allium cepa root cells induces cytotoxic and genotoxic responses and results in DNA hypermethylation. Mutat Res / Fundam Mol Mech Mutagen 774:49–58

Giese B, Klaessig F, Park B, Kaegi R, Steinfeldt M, Wigger H, von Gleich A, Gottschalk F (2018) Risks, release and concentrations of engineered nanomaterial in the environment. Sci Rep 8(1):1565

Grillo R, Rosa AH, Fraceto LF (2014) Poly (ε-caprolactone) nanocapsules carrying the herbicide atrazine: effect of chitosan-coating agent on physico-chemical stability and herbicide release profile. Int J Environ Sci Technol 11:1691–1700

He X, Deng H, Hwang H (2019) The current application of nanotechnology in food and agriculture. J Food Drug Anal 27(1):1–21

Hernández MT, del Carmen Berrocal M, Ly-Pen D, Aguilar MV (2021) Introduction to toxicology. In: Toxicology for the health and pharmaceutical sciences. CRC Press, pp 1–19

Hildebrandt R, Skubacz K, Chmielewska I, Dyduch Z, Zgórska A, Smoliński A (2022) Implementing silica nanoparticles in the study of the airborne transmission of SARS-CoV-2. Molecule 27(12):3896

Hochella MF Jr, Mogk DW, Ranville J, Allen IC, Luther GW, Marr LC, McGrail BP, Murayama M, Qafoku NP, Rosso KM (2019) Natural, incidental, and engineered nanomaterials and their impacts on the earth system. Science 363(6434):eaau8299

Huang X, Auffan M, Eckelman MJ, Elimelech M, Kim J-H, Rose J, Zuo K, Li Q, Alvarez PJJ (2024) Trends, risks and opportunities in environmental nanotechnology. Nat Rev Earth Environ 5(8):572–587

Hung NTQ, Ky NM, Nhut HT, Van Tan L (2021) Investigating community health effects of bioaerosol pollution in Ho Chi Minh megacity, Vietnam. Environ Eng Manag J 20(5)

ISO (2024) Nanotechnologies. https://www.iso.org/committee/381983/x/catalogue/p/1/u/0/w/0/d/0

Jameel M, Shoeb M, Khan MT, Ullah R, Mobin M, Farooqi MK, Adnan SM (2020) Enhanced insecticidal activity of thiamethoxam by zinc oxide nanoparticles: a novel nanotechnology approach for pest control. ACS Omega 5(3):1607–1615

Jane PJ, Saigeetha S, Samrot AV, Ponniah P, Chakravarthi S (2021) Overview on toxicity of nanoparticles, it's mechanism, models used in toxicity studies and disposal methods–a review. Biocatal Agric Biotechnol 36:102117

Ji F, Shi C, Shu Z, Li Z (2024) Nanomaterials enhance pyroptosis-based tumor immunotherapy. Int J Nanomedicine 19:5545–5579

Jia H, Chen S, Wang X, Shi C, Liu K, Zhang S, Li J (2020) Copper oxide nanoparticles alter cellular morphology via disturbing the actin cytoskeleton dynamics in Arabidopsis roots. Nanotoxicology 14(1):127–144

Jordan JT, Oates RP, Subbiah S, Payton PR, Singh KP, Shah SA, Green MJ, Klein DM, Cañas-Carrell JE (2020) Carbon nanotubes affect early growth, flowering time and phytohormones in tomato. Chemosphere 256:127042

Kabir E, Kumar V, Kim KH, Yip AC, Sohn JR (2018) Environmental impacts of nanomaterials. J Environ Manag 225:261–271

Keller AA, Adeleye AS, Conway JR, Garner KL, Zhao L, Cherr GN, Hong J, Gardea-Torresdey JL, Godwin HA, Hanna S, Ji Z, Kaweeteerawat C, Lin S, Lenihan HS, Miller RJ, Nel AE, Peralta-Videa JR, Walker SL, Taylor AA et al (2017) Comparative environmental fate and toxicity of copper nanomaterials. NanoImpact 7:28–40. https://doi.org/10.1016/j.impact.2017.05.003

Khalid P, Suman V, Hussain M, Arun A (2015) Toxicology of carbon nanotubes-a review. J Environ Nanotechnol 4(4):62–75

Khan I, Saeed K, Khan I (2019) Nanoparticles: properties, applications and toxicities. Arab J Chem 12(7):908–931

Khiyami MA, Almoammar H, Awad YM, Alghuthaymi MA, Abd-Elsalam KA (2014) Plant pathogen nanodiagnostic techniques: forthcoming changes? Biotechnol Biotechnol Equip 28(5):775–785

Kica E, Bowman DM (2012) Regulation by means of standardization: key legitimacy issues of health and safety nanotechnology standards. Jurimetrics 53(1):11–56

Kumah EA, Fopa RD, Harati S, Boadu P, Zohoori FV, Pak T (2023) Human and environmental impacts of nanoparticles: a scoping review of the current literature. BMC Public Health 23(1):1059

Kumar N, Kumar R, Shakil NA, Das TK (2016) Nanoformulations of pretilachlor herbicide: preparation, characterization and activity. J Sci Indust Res 75(11):676–680

Kumari M, Pandey S, Mishra SK, Giri VP, Agarwal L, Dwivedi S, Pandey AK, Nautiyal CS, Mishra A (2020) Omics-based mechanistic insight into the role of bioengineered nanoparticles for biotic stress amelioration by modulating plant metabolic pathways. Front Bioeng Biotechnol 8:242

Kumari A, Chokheli VA, Lysenko VS, Mandzhieva SS, Minkina TM, Mazarji M, Rajput VD, Shuvaeva VA, Sushkova SS, Barakhov A (2023) Genotoxic and morpho-physiological responses of ZnO macro-and nano-forms in plants. Environ Geochem Health 45(12):9345–9357

Lee Y-L, Shih Y-S, Chen Z-Y, Cheng F-Y, Lu J-Y, Wu Y-H, Wang Y-J (2022) Toxic effects and mechanisms of silver and zinc oxide nanoparticles on zebrafish embryos in aquatic ecosystems. Nano 12(4):717

Lin D, Xing B (2007) Phytotoxicity of nanoparticles: inhibition of seed germination and root growth. Environ Pollut 150(2):243–250. https://doi.org/10.1016/j.envpol.2007.01.016

Liu Y, Hardie J, Zhang X, Rotello VM (2017) Effects of engineered nanoparticles on the innate immune system. Semin Immunol 34:25–32

Lu S, Chen J, Wang J, Wu D, Bian H, Jiang H, Sheng L, He C (2023) Toxicological effects and transcriptome mechanisms of rice (Oryza sativa L.) under stress of quinclorac and polystyrene nanoplastics. Ecotoxicol Environ Saf 249:114380

Manke A, Wang L, Rojanasakul Y (2013) Mechanisms of nanoparticle induced oxidative stress and toxicity. Biomed Res Int 2013(1):942916

Manna S, Roy S, Dolai A, Ravula AR, Perumal V, Das A (2023) Current and future prospects of "all-organic" nanoinsecticides for agricultural insect pest management. Front Nanotechnol 4:1082128

Mays V, Smith N, Pham C, White M, Wu Q, Berry J, Linan A, Wait DA, Kovacs L (2024) Attenuation of photosynthesis in nanosilver-treated Arabidopsis thaliana is inherently linked to the particulate nature of silver. Heliyon 10(6):e27583

Medina-Pérez G, Fernández-Luqueño F (2018) Nanotoxicity: Challenges and opportunities. Mundo Nano. Revista Interdisciplinaria En Nanociencias y Nanotecnología 11(20):7–16

Mehrian SK, De Lima R (2016) Nanoparticles cyto and genotoxicity in plants: mechanisms and abnormalities. Environ Nanotechnol Monit Manag 6:184–193

Meldrum K, Guo C, Marczylo EL, Gant TW, Smith R, Leonard MO (2017) Mechanistic insight into the impact of nanomaterials on asthma and allergic airway disease. Part Fibre Toxicol 14:1–35

Méndez-López A, García-Santiago JC, González-Méndez LM, Martínez-Amador SY, Leal-Robles AI, Sánchez-Vega M (2024) Impact of nanofertilizers on cucumber yield and quality under greenhouse conditions. Rev Mex Cienc Agríc 15(4)

Mendoza Cantú A, Ize Lema IAR (2017) Chemical substances in Mexico. Perspectives for their sound management. Revista Internacional de Contaminación Ambiental 33(4):719–745

Merghany MM, Shahein MM, Sliem MA, Abdelgawad KF, Radwan AF (2019) Effect of nanofertilizers on cucumber plant growth, fruit yield and it's quality. Plant Arch 19(2):165–172

Mishra S, Sundaram B (2023a) Efficacy and challenges of carbon nanotube in wastewater and water treatment. Environ Nanotechnol Monit Manag 19:100764

Mishra S, Sundaram B (2023b) Fate, transport, and toxicity of nanoparticles: an emerging pollutant on biotic factors. Process Saf Environ Prot 174:595–607

Mu Q, Jiang G, Chen L, Zhou H, Fourches D, Tropsha A, Yan B (2014) Chemical basis of interactions between engineered nanoparticles and biological systems. Chem Rev 114(15):7740–7781

Muzammil S, Ashraf A, Siddique MH, Aslam B, Rasul I, Abbas R, Afzal M, Faisal M, Hayat S (2023) A review on toxicity of nanomaterials in agriculture: current scenario and future prospects. Sci Prog 106(4):00368504231221672

Namasivayam SKR, Priyanka S, Lavanya M, Shree SK, Francis AL, Avinash GP, Bharani RSA, Kavisri M, Moovendhan M (2024) A review on vulnerable atmospheric aerosol nanoparticles: sources, impact on the health, ecosystem and management strategies. J Environ Manag 365:121644

Oancea F (2022) Development of safe nanoagrochemicals—the nanoporous route. Chem Proc 7(1):68

Oberdörster G (2010) Safety assessment for nanotechnology and nanomedicine: concepts of nanotoxicology. J Intern Med 267(1):89–105. https://doi.org/10.1111/j.1365-2796.2009.02187.x

Oberdörster G, Oberdörster E, Oberdörster J (2005) Nanotoxicology: an emerging discipline evolving from studies of ultrafine particles. Environ Health Perspect 113(7):823–839

Oberdörster G, Stone V, Donaldson K (2007) Toxicology of nanoparticles: a historical perspective. Nanotoxicology 1(1):2–25. https://doi.org/10.1080/17435390701314761

Peijnenburg WJGM, Baalousha M, Chen J, Chaudry Q, Von der Kammer F, Kuhlbusch TAJ, Lead J, Nickel C, Quik JTK, Renker M (2015) A review of the properties and processes determining the fate of engineered nanomaterials in the aquatic environment. Crit Rev Environ Sci Technol 45(19):2084–2134

Pettersen T, Aamlid T (2019) Testing of Nano-Gro on established turf at fairway mowing height (15 mm). NIBIO Rapport

Pradhan DD, Rawal S, Asif M, Behera SR, Panda U (2024) Effect of zero tillage and different weed management practices in direct seeded rice (Oryza sativa L.) in Indo gangetic plains: a review. J Sci Res Rep 30(6):543–557

Qiao H, Liu W, Gu H, Wang D, Wang Y (2015) The transport and deposition of nanoparticles in respiratory system by inhalation. J Nanomater 2015(1):394507

Quan C, Sun Q, Lippmann M, Chen L-C (2010) Comparative effects of inhaled diesel exhaust and ambient fine particles on inflammation, atherosclerosis, and vascular dysfunction. Inhal Toxicol 22(9):738–753

Rajput V, Minkina T, Mazarji M, Shende S, Sushkova S, Mandzhieva S, Burachevskaya M, Chaplygin V, Singh A, Jatav H (2020) Accumulation of nanoparticles in the soil-plant systems and their effects on human health. Ann Agric Sci 65(2):137–143

Rastogi S, Sharma G, Kandasubramanian B (2020) Nanomaterials and the environment. In: The ELSI handbook of nanotechnology: risk, safety, ELSI and commercialization, pp 1–23

Ray PC, Yu H, Fu PP (2009) Toxicity and environmental risks of nanomaterials: challenges and future needs. J Environ Sci Health C 27(1):1–35. https://doi.org/10.1080/10590500802708267

Ray PD, Huang BW, Tsuji Y (2012) Reactive oxygen species (ROS) homeostasis and redox regulation in cellular signaling. Cell Signal 24(5):981–990

Remédios C, Rosário F, Bastos V (2012) Environmental nanoparticles interactions with plants: morphological, physiological, and genotoxic aspects. J Bot 2012(1):751686. https://doi.org/10.1155/2012/751686

Risom L, Møller P, Loft S (2005) Oxidative stress-induced DNA damage by particulate air pollution. Mutat Res Fundam Mol Mech Mutagen 592(1–2):119–137

Rodriguez-Hernandez AG, Aguilar Guzmán JC, Vázquez-Duhalt R (2018) Membrana celular y la inespecificidad de las nanopartículas.¿ Hasta dónde puede llegar un nanomaterial dentro de la célula? Mundo Nano Revista Interdisciplinaria En Nanociencias y Nanotecnología 11(20):43–52

ROYAL-Royal Society and Royal Academy of Engineering (2004) Nanoscience and nanotechnologies: opportunities and uncertainties. https://royalsociety.org/news-resources/publications/2004/nanoscience-nanotechnologies/

Saarimäki LA, Federico A, Lynch I, Papadiamantis AG, Tsoumanis A, Melagraki G, Afantitis A, Serra A, Greco D (2021) Manually curated transcriptomics data collection for toxicogenomic assessment of engineered nanomaterials. Sci Data 8(1):49. https://doi.org/10.1038/s41597-021-00808-y

Saldívar Tanaka L (2020) Soft regulation, technical standards and international regulatory harmonization, for nanotechnology. Mundo Nano. Revista Interdisciplinaria en Nanociencias y Nanotecnología 13(24)

Sethi G, Behera KK, Sayyed R, Adarsh V, Sipra BS, Singh L, Alamro AA, Behera M (2025) Enhancing soil health and crop productivity: the role of zinc-solubilizing bacteria in sustainable agriculture. Plant Growth Regul. https://doi.org/10.1007/s10725-025-01294-7

Shang Y, Hasan MK, Ahammed GJ, Li M, Yin H, Zhou J (2019) Applications of nanotechnology in plant growth and crop protection: a review. Molecule 24(14):2558

Sharma S, Shree B, Sharma A, Irfan M, Kumar P (2022) Nanoparticle-based toxicity in perishable vegetable crops: molecular insights, impact on human health and mitigation strategies for sustainable cultivation. Environ Res 212:113168

Shawon ZBZ, Hoque ME, Chowdhury SR (2020) Chapter 6: Nanosensors and nanobiosensors: agricultural and food technology aspects. In: Pal K, Gomes F (eds) Micro and nano technologies. Elsevier, pp 135–161. https://doi.org/10.1016/B978-0-12-820702-4.00006-4

Shukla RK, Kumar A, Vallabani NVS, Pandey AK, Dhawan A (2014) Titanium dioxide nanoparticle-induced oxidative stress triggers DNA damage and hepatic injury in mice. Nanomedicine 9(9):1423–1434. https://doi.org/10.2217/nnm.13.100

Singh N, Bhuker A, Jeevanadam J (2021) Effects of metal nanoparticle-mediated treatment on seed quality parameters of different crops. Naunyn Schmiedeberg's Arch Pharmacol 394(6):1067–1089

Sly PD, Schüepp K (2012) Nanoparticles and children's lungs: is there a need for caution? Paediatr Respir Rev 2(13):71–72

Souza I d C, Morozesk M, Mansano AS, Mendes VAS, Azevedo VC, Matsumoto ST, Elliott M, Monferrán MV, Wunderlin DA, Fernandes MN (2021) Atmospheric particulate matter from an industrial area as a source of metal nanoparticle contamination in aquatic ecosystems. Sci Total Environ 753:141976

Sun D, Hussain HI, Yi Z, Siegele R, Cresswell T, Kong L, Cahill DM (2014) Uptake and cellular distribution, in four plant species, of fluorescently labeled mesoporous silica nanoparticles. Plant Cell Rep 33:1389–1402

Sun TY, Bornhöft NA, Hungerbühler K, Nowack B (2016) Dynamic probabilistic modeling of environmental emissions of engineered nanomaterials. Environ Sci Technol 50(9):4701–4711. https://doi.org/10.1021/acs.est.5b05828

Thomas SP, Al-Mutairi EM, De SK (2013) Impact of nanomaterials on health and environment. Arab J Sci Eng 38:457–477

Tran T-K, Nguyen M-K, Lin C, Hoang T-D, Nguyen T-C, Lone AM, Khedulkar AP, Gaballah MS, Singh J, Chung WJ (2023) Review on fate, transport, toxicity and health risk of nanoparticles in natural ecosystems: emerging challenges in the modern age and solutions toward a sustainable environment. Sci Total Environ 912:169331

Tripathi DK, Singh S, Singh S, Srivastava PK, Singh VP, Singh S, Prasad SM, Singh PK, Dubey NK, Pandey AC (2017) Nitric oxide alleviates silver nanoparticles (AgNps)-induced phytotoxicity in *Pisum sativum* seedlings. Plant Physiol Biochem 110:167–177

Vimalrajiv B, Chinnamuthu CR, Subramanian E, Senthil K (2018) Effect of nanoparticles in combination with pendimethalin and hydrogen peroxide on growth parameters and nodulation of blackgram (Vigna mungo L.). Int J Commun Soc 6(3):2816–2819

Vochita G, Oprica L, Gherghel D, Mihai C-T, Boukherroub R, Lobiuc A (2019) Graphene oxide effects in early ontogenetic stages of Triticum aestivum L. seedlings. Ecotoxicol Environ Saf 181:345–352

Walker GW, Kookana RS, Smith NE, Kah M, Doolette CL, Reeves PT, Lovell W, Anderson DJ, Turney TW, Navarro DA (2017) Ecological risk assessment of nano-enabled pesticides: a perspective on problem formulation. J Agric Food Chem 66(26):6480–6486

Wang Z, Altenburger R, Backhaus T, Covaci A, Diamond ML, Grimalt JO, Lohmann R, Schäffer A, Scheringer M, Selin H (2021) We need a global science-policy body on chemicals and waste. Science 371(6531):774–776

Wei L, Liu J, Jiang G (2024) Nanoparticle-specific transformations dictate nanoparticle effects associated with plants and implications for nanotechnology use in agriculture. Nat Commun 15(1):7389. https://doi.org/10.1038/s41467-024-51741-8

WHO/FAO (2014) Report on the 8TH FAO/WHO joint meeting on pesticide management and 10TH Session of the FAO panel of experts on pesticide management. World Health Organization and Food and Agriculture Organization [in línea]. http://www.fao.org/agriculture/crops/thematic-sitemap/theme/pests/code/hhp/en/

Wu H, Li Z (2022) Nano-enabled agriculture: how do nanoparticles cross barriers in plants? Plant Commun 3(6)

Yadav A, Yadav K, Abd-Elsalam KA (2023) Exploring the potential of nanofertilizers for a sustainable agriculture. Plant Nano Biol 5:100044

Yang Y, Zhang N, You Q, Chen X, Zhang Y, Zhu L (2023) Novel insights into the multistep chlorination of silver nanoparticles in aquatic environments. Water Res 240:120111

Yuan X, Wang Z, Peijnenburg WJGM (2024) Molecular characterization and transcriptional response of Lactuca sativa seedlings to co-exposure to graphene nanoplatelets and titanium dioxide nanoparticles. Ecotoxicol Environ Saf 282:116761

Zhao J, Castranova V (2011) Toxicology of nanomaterials used in nanomedicine. J Toxic Environ Health, Part B 14(8):593–632. https://doi.org/10.1080/10937404.2011.615113

Zhao G, Guo Y, Sun X, Wang X (2015) A system for pesticide residues detection and agricultural products traceability based on acetylcholinesterase biosensor and internet of things. Int J Electrochem Sci 10(4):3387–3399

Zhang WY, Wang Q, Li M, Dang F, Zhou DM (2019) Nonselective uptake of silver and gold nanoparticles by wheat. Nanotoxicology 13(8):1073–1086

Zhou L, Forman HJ, Ge Y, Lunec J (2017) Multi-walled carbon nanotubes: a cytotoxicity study in relation to functionalization, dose and dispersion. Toxicol In Vitro 42:292–298

Zurita-Cruz JN, Barbosa-Cortés L, Villasís-Keever MÁ (2019) From research to practice: clinical phases for drug development. Rev Alerg Mex 66(2):246–253

MIX
Papier aus verantwortungsvollen Quellen
Paper from responsible sources
FSC® C105338

If you have any concerns about our products,
you can contact us on
ProductSafety@springernature.com

In case Publisher is established outside the EU,
the EU authorized representative is:
Springer Nature Customer Service Center GmbH
Europaplatz 3, 69115 Heidelberg, Germany

Printed by Libri Plureos GmbH
in Hamburg, Germany